"The earth's resources are finite, but we consume and denigrate them as if they are unlimited. This is a recipe for a catastrophic future. In *'One Planet'* *Cities*, David Thorpe not only sounds the alarm but provides us with richly detailed, real-world examples of how to live within our means."

F. Kaid Benfield, Senior Counsel for Environmental Strategies, PlaceMakers LLC, USA

"Wales may have its own legislation to create a harmonious future for people and planet, but we still need help to navigate the global evidence and help us on that path. With this book, David Thorpe has demonstrated that he can be that guide, helping us reach our own pathways at country, city or individual level on the basis of global science and evidence. Wherever you live, this book must be read and most importantly, acted upon."

Jane Davidson, Pro Vice-Chancellor for Engagement & Sustainability, University of Wales Trinity Saint David, UK

"Ultimately, humanity needs to fit within what our planet can renew. Cities play a critical role. David Thorpe gifts us with countless examples of what cities already do and what is possible to support this needed transition."

Mathis Wackernagel, co-creator of the Ecological Footprint and President of Global Footprint Network, California

"David Thorpe is a fabulous writer who brings home on a very human level what the challenges of a warming climate really mean. His technical and evidence-based understanding, combined with his gift for simple language, makes this work a compelling read… It [contains] a cornucopia of solutions and proven alternatives that can inspire action. This book is a must read for anyone who wants to understand the mechanisms in play with our changing planet and how we can hope to influence its current trajectory."

Tina Perinotto, Managing Editor and founder, The Fifth Estate, Australia

"Impressive work. An indispensable guide to what people are already doing around the world to ensure a sustainable future for us all."

Ian Crawley, CLT Technical Advisor and a Trustee of the National CLT Network, UK

"An excellent integration of social, economic and environmental regeneration."

Storm Cunningham, Publisher of REVITALIZATION, USA

"I am sure the book is going to be a success."

> Pedro B. Ortiz, ex-Deputy Mayor of Madrid, Senior Consultant on Metropolitan Management and Planning for IGOs, USA

"This collection is a mosaic of challenges and strategies, examples and ideas that point the way to further research."

> Margaret Robertson, Lane Community College, USA

"David Thorpe's fascinating book is packed with ideas and inspiration; a gifted storyteller drawing together the strands of a new and hopeful collective story."

> George Marshall, founder of Climate Outreach, UK

"A must-read for 2019. David continues his campaign to a saner way of life with this latest book."

> Mal Williams, Executive Director at Zero Waste International Trust, UK

'ONE PLANET' CITIES

This book addresses the crucial question of how the essential needs of the growing human population can be met without breaking the Earth's already-stretched life-support system.

With four out of five people predicted to be urban dwellers by 2080, *'One Planet' Cities* proposes a pathway to genuine sustainability for cities and neighbourhoods, using an approach based on contraction and convergence. Utilising interviews with key players, including the Global Footprint Network, World Future Council, WWF, mayors and government officials, and case studies from across the globe, including Europe, North and South America, Australia, South Africa, China and India, David Thorpe examines all aspects of modern society from food provision to neighbourhood design, via industry, the circular economy, energy and transport through the critical lens of the ecological footprint and relevant supporting international standards and indicators. Recommendations on managing supply chains and impacts, how the transition to a world within limits might be financed, and a deep examination of the Welsh Government's pioneering efforts follow. It concludes with an imagined vision of what a genuinely sustainable future might be like, and an appeal for 'one planeteers' everywhere to step up to the challenge.

This book will be of great interest to practitioners and policymakers involved in governance, administration, urban environments and sustainability, alongside students of the built environment, urban planning, environmental policy and energy.

David Thorpe is a lecturer in 'one planet' development and governance at the University of Wales Trinity Saint David, UK, a consultant on renewable energy and sustainable building, and founder/patron of One Planet Council. He is the author of numerous articles and books, including *Passive Solar Architecture Pocket Reference* (Routledge, 2017), *Solar Energy Pocket Reference* (Routledge, 2017) and *The One Planet Life* (Routledge, 2014).

'ONE PLANET' CITIES

Sustaining Humanity within Planetary Limits

David Thorpe

First published 2019
by Routledge
2 Park Square, Milton Park, Abingdon, Oxon OX14 4RN

and by Routledge
52 Vanderbilt Avenue, New York, NY 10017

Routledge is an imprint of the Taylor & Francis Group, an informa business

© 2019 David Thorpe

The right of David Thorpe to be identified as author of this work has been asserted by him in accordance with sections 77 and 78 of the Copyright, Designs and Patents Act 1988.

All rights reserved. No part of this book may be reprinted or reproduced or utilised in any form or by any electronic, mechanical, or other means, now known or hereafter invented, including photocopying and recording, or in any information storage or retrieval system, without permission in writing from the publishers.

Trademark notice: Product or corporate names may be trademarks or registered trademarks, and are used only for identification and explanation without intent to infringe.

British Library Cataloguing in Publication Data
A catalogue record for this book is available from the British Library

Library of Congress Cataloging-in-Publication Data
Names: Thorpe, Dave, 1954- author.
Title: "One planet" cities : sustaining humanity within planetary limits / David Thorpe.
Description: Abingdon, Oxon ; New York, NY : Routledge, 2019. | Includes bibliographical references and index.
Identifiers: LCCN 2018059887 (print) | LCCN 2019007673 (ebook) | ISBN 9780429463402 (Master) | ISBN 9781138615090 (hardback) | ISBN 9781138615106 (pbk.) | ISBN 9780429463402 (ebook)
Subjects: LCSH: Sustainable urban development. | City planning--Environmental aspects. | Sustainable development. | Sustainable living.
Classification: LCC HT241 (ebook) | LCC HT241 .T56 2019 (print) | DDC 307.1/16--dc23
LC record available at https://lccn.loc.gov/2018059887

ISBN: 978-1-138-61509-0 (hbk)
ISBN: 978-1-138-61510-6 (pbk)
ISBN: 978-0-429-46340-2 (ebk)

Typeset in Bembo
by Taylor & Francis Books

Printed and bound by CPI Group (UK) Ltd, Croydon, CR0 4YY

To 'one planeteers' everywhere:
all people working for a fair and just future for humanity
that repays civilisation's debt to nature.

&

To the future generation: Dion, Nemos, Sam and Bea.

CONTENTS

List of illustrations   xi
Abbreviations   xii
About the author   xvi
Acknowledgements   xvii
Foreword by Herbert Girardet   xxi

Introduction: It's time to end humanity's war on nature   1

1. Do the stories we tell influence the future we will live in?   6
2. The ultimate problem: Humanity's limits to growth   10
3. Ecological footprinting and other standards   24
4. Choosing which standards to use   49
5. Feeding cities while saving the planet   69
6. Regenerative cities   95
7. Zero carbon cities   104
8. Transforming industry   134
9. Managing water in the age of change   150
10. 'One planet' neighbourhoods   169
11. Buildings in a 'one planet' city   184
12. Mobility in a 'one planet' city   199

13	How smart is a 'smart city'?	215
14	Financing the way to a 'one planet' future	224
15	A menu of case studies	240
16	Wales and 'one planet' development	262
17	Six steps to a 'one planet' city	285
18	One day in a 'one planet' city: A short story	292

Index of people   *296*
Index of places   *298*
Index of topics   *302*

ILLUSTRATIONS

Tables

4.1	Ecological footprint and greenhouse gas emissions of 12 materials	63
5.1	2017 Worldwide average CO_2 emissions per food type, per person, per year	74
7.1	Cities which have set targets for up to 100 per cent renewable energy or electricity	115
7.2	Cities that have achieved 100 per cent renewable energy	116
7.3	Some targets for renewable share of total energy, all consumers	117
7.4	Some targets for renewable share of electricity, all consumers	117
7.5	Some heat-related mandates and targets	117
7.6	Targets for government self-generation/own-use purchases of renewable energy	118
7.7	Some targets for renewable electric capacity or generation	118

Boxes

7.1	List of the main currently commercial renewable energy technologies	105
7.2	A summary of energy storage options	128
16.1	The national indicators for Wales' Well-being of Future Generations Act	269

ABBREVIATIONS

AD	anaerobic digestion
AHS	Active Heat Storage
ARCO	Action Research for CO-development
ASEAN	Association of Southeast Asian Nations
bioSNG	bio-synthetic gas (or bio-syngas)
BRT	bus rapid transit
BVT	Bournville Village Trust
$CaCO_3$	calcium carbonate
CCC	Committee on Climate Change
CCHP	combined cooling, heating and power plant
cCR	carbonn® Climate Registry
CCS	carbon capture and storage
CCTV	closed-circuit television
CCU	carbon capture and utilisation
CFCs	chlorofluorocarbons
CH_4	methane
CHP	combined heat and power
CIC	Catalan Integral Co-operative
CLEANKER	CLEAN clinKER
CLT	community land trust
CO_2	carbon dioxide
CO_2e	carbon dioxide equivalent
COP	(United Nations International Panel on Climate Change) Conference of the Parties
EDGAR	Emission Database for Global Atmospheric Research
EeMAP	Energy Efficient Mortgages Action Plan
EEMRIO	environmentally extended multi-regional input-output

EF	ecological footprint
EJ	exajoule
EU	European Union
FAO	Food and Agriculture Organization
ft	feet
g	gram
G20	Group of 20
GDP	gross domestic product
GFN	Global Footprint Network
gha	global hectare
GHG	greenhouse gas
GIIN	Global Impact Investing Network
GJ	gigajoule
GM	genetically modified
GNI	gross national income
GPC	Global Protocol for Community-Scale Greenhouse Gas Emission Inventories
GPI	Genuine Progress Indicator
GPS	Global Positioning System
Gt	gigatonne
GVA	gross value added
GW	gigawatt
ha	hectare
HCA	Homes and Communities Agency
HDI	Human Development Index
HVO	hydrotreated vegetable oil
ICT	information communications technology
IEA	International Energy Agency
IO	input-output
IoT	Internet of Things
IPBES	Intergovernmental Science-Policy Platform on Biodiversity and Ecosystem Services
IPCC	International Panel on Climate Change
ISO	International Organization for Standardization
km	kilometre
kWh	kilowatt-hour
LCA	life-cycle assessment
LCI	life-cycle inventory
LCIA	life-cycle inventory impact assessment
LED	light-emitting diode
LEED-ND	Leadership in Energy and Environmental Design – Neighborhood Development
LEILAC	Low Emissions Intensity Lime And Cement
LEZ	low-emission zone

LPG	liquid petroleum gas
LRS	land rent scheme
m	metre
MDGs	Millennium Development Goals
MJ	megajoule
MPa	megapascal
MRIO	multi-regional input-output
MSWIA	municipal solid waste incinerator ash
Mt	metric ton
MWh	megawatt-hour
NASA	National Aeronautics and Space Administration
NETL	National Energy Technology Laboratory
NGO	non-governmental organisation
NPK	nitrogen, phosphorus and potassium
NPV	net present value
NPV+	net present value plus
NYC	New York City
NZB	net zero building
OLED	organic light-emitting diode
PAR	Participatory Action Research
PCB	polychlorinated biphenyl
PM	particulate matter
POTEs	Ordinary Energy Transition Pioneers
PRINCE	Policy-Relevant Indicators for National Consumption and Environment
PV	photovoltaic
PVC	polyvinyl chloride
RCP	Representative Concentration Pathway
REAP	Resources and Energy Analysis Programme
ScoT	Le Schéma de Cohérence Territoriale, or Montpellier Territorial Coherence Scheme
SDC	Sustainable Development Commission
SDGs	Sustainable Development Goals
SEEA	System of Environmental Economic Accounting
SEI	Stockholm Environment Institute
SIC	Standard Industrial Classification
SLNC	Saffron Lane Neighbourhood Council
SNAP	Supplemental Nutrition Assistance Program
SUDS	sustainable urban drainage systems
tCO_2e	tonne of carbon dioxide equivalent
TCPA	Town and Country Planning Association
toe	tonne of oil equivalent
TW	terawatt
UN	United Nations

UNEP	United Nations Environment Programme
UNESCAP	United Nations Economic and Social Commission for Asia and the Pacific
UNFCCC	United Nations Framework Convention on Climate Change
UN-Habitat	United Nations Human Settlements Programme
UNIDO	United Nations Industrial Development Organization
UV	ultraviolet
WCCD	World Council on City Data
WEF	World Economic Forum
W/mK	watts per metre-Kelvin
World GBC	World Green Building Council
WRI	World Resources Institute
WWAP	United Nations World Water Assessment Programme

ABOUT THE AUTHOR

David lives in Wales, UK, where he is a founder/patron of the One Planet Council and lectures in 'One Planet' development and governance at the University of Wales Trinity Saint David. He is also an author, journalist, consultant, leader and inspirational speaker in the fields of environmental matters, rural and urban sustainability, energy management and carbon-free energy. His guiding principles include taking a holistic perspective, adherence to proven techniques, and eco-minimalism.

He also writes fiction and film scripts. After working for Marvel Comics, he won the UK HarperCollins-Saga Magazine 2006 Children's Novelist competition with *Hybrids*. When not writing and teaching he enjoys growing vegetables and cycling.

ACKNOWLEDGEMENTS

Huge thanks to the many people who have assisted with and inspired this book:

- Professor Herbert Girardet, co-founder, World Futures Council, for inspiration and the foreword
- Mathis Wackernagel, co-founder, President of the Global Footprint Network, for inspiration and valuable feedback
- Helen Adam, for endless support, proofreading and copyediting
- For supporting the writing of the book by publishing edited sections of it as journalistic articles: Tina Perinotto, Managing Director, The Fifth Estate
- For giving me the opportunity to interview many of the participants and for publishing earlier versions of some of the material when I was responsible for coordinating the website: the late, lamented Robin Carey, Managing Director, Sustainable Cities Collective
- Jane Davidson, Director, INSPIRE Wales at the University of Wales Trinity Saint David, the best Environment Minister the UK has so far had, and whom I am proud to call a mentor
- All members of the One Planet Council, courageous, pioneering, inspiring, one planeteers
- For reading and commenting on the book: Frank Jackson
- For networking and ideas: Andy Middleton, Chief Exploration Officer, TYF Group; Jessica McQuade, Head of Policy and Advocacy, WWF Cymru
- For stimulating the climate fiction chapters: The Free Word Centre, London; Professor Esther Eidinow, Chair in Ancient History, Bristol University; Ty Newydd writers' retreat in North Wales; Bristol Climate Writers; and Mark Goldthorpe of the website Climate Cultures
- For the teaching opportunities that have refined the ideas: Paul Harries, Head of School of Architecture, University of Wales Trinity Saint David; Dr Louise

Emanuel, Senior Lecturer, School of Business, Finance and Management, University of Wales Trinity Saint David; Jessica Wood, MRes Ecosystem & Environmental Change, Imperial College London, organiser of the Future Cities Symposium; The Centre for Agroecology, Water and Resilience at Coventry University
- My publisher, including Hannah Ferguson, the Senior Editor of Environment and Sustainability, Routledge, Taylor & Francis Group, and my editor, Matthew Shobbrook, for commissioning this and all my previous titles with this publisher

Thanks to the following, who are interviewed and quoted in the book:

- Matthew Lynch, World Council on City Data
- Pedro B. Ortiz, ex-Deputy Mayor of Madrid
- Margaret Robertson, academic and author
- Eray Ragip, for telling me about his time growing up in Cyprus
- Dan Bloom, the creator of the clifi label, in Taiwan
- George Marshall, author, campaigner, founder, Climate Outreach Network
- Marco Matteini, project leader, United Nations Industrial Development Organization
- Lucy Young, Science and Policy Adviser, WWF-UK, editor of the *Living Planet Report*
- Carina Borgstrom-Hansson, Senior Adviser at WWF on One Planet Cities
- John Devaney, Standards Publishing Manager, Sustainability, British Standards Institute
- Bernard Gindroz, Chair of ISO/Technical Committee 268, International Organization for Standardization
- Dan O'Neill, Associate Professor in Ecological Economics, Leeds University
- Kate Raworth, Environmental Change Institute, Oxford University
- Kirstin Miller, Executive Director of EcoCity Builders, Vancouver, Canada
- Gunhild Stordalen, EAT Forum, Sweden
- Henk de Zeeuw, RUAF Foundation
- Dickson Despommier, Emeritus Professor of Microbiology and Public Health at Columbia University
- Milan Kluko, Director, Green Spirit Farms
- Oscar Rodriguez, Architecture & Food
- Henry Gordon-Smith, Association for Vertical Farming
- Professor Dr Ranka Junge, ZHAW School of Life Sciences and Facility Management, Institute of Natural Resource Sciences, Switzerland
- Jack Ng, CEO, Sky Urban Solutions, Singapore
- Dr Toyoki Kozai, Professor Emeritus at Chiba University in Chiba, Japan
- Rebecca Ruth Seidel, Director, Wholesome Dairy Farms
- Illtud Dunsford, Director, Cellular Agriculture

- Dr Peter Holzer, Senior Researcher, Institute of Building Research & Innovation
- Dan Sadler, H21 Project Manager, Northern Gas Network
- Kit Oung, Energy and Resource Productivity Consultant
- Steven Fawkes, Director, EnergyPro Ltd, London Energy Efficiency Fund
- Theresa Williamson, Executive Director of NGO Catalytic Communities
- Ian Crawley, Training Officer, British National Community Land Trust Network
- Professor Vanessa Watson, Professor of City Planning in the School of Architecture, Planning and Geomatics, Cape Town University, South Africa
- Professor Per Heiselberg, The Faculty of Engineering and Science, Aalborg University, Denmark
- Mike Rainbow, Associate Principal, Ark Resources
- Margaret Robertson, author of *Sustainability Principles and Practice*
- Florian Lorenz, Director of think tank Smarter Than Car, co-founder of the Low Carbon City Foundation
- Amanda Ngabirano, urban planner at Makerere University, Uganda
- Stefanie Holzwarth, mobility expert at UN-Habitat
- Claudia Adriazola-Steil, Health & Road Safety Programme Director, World Resources Institute
- Claes Tingvall, Vision Zero
- Dario Hidalgo, Director for Integrated Transport at EMBARQ
- Storm Cunningham, Publisher, Revitalization News
- Melanie Nutter, former Director of the San Francisco Department of Environment
- Ciaran Mundy, Manager of The Bristol Pound
- Sophie Howe, Future Generations Commissioner for Wales
- Glyn Jones, Chief Statistician, Welsh Government
- Gretel Leeb, former senior civil servant, Welsh Government
- Alison Smith, civil servant, Governance and Performance, Permanent Secretary's office, Welsh Government
- Anne Meikle, head of WWF Cymru

Amongst others interviewed but not quoted directly were:

- Chuck Wolfe, author, *Urbanism Without Effort*
- Rob Hopkins on the Spread of the Transition Towns Movement
- Tim Campbell on Beyond Smart Cities
- Patricia Lee on Greening Cities, Air Quality and Building Efficiency
- Pooran Desai, BioRegional, on creating one planet towns
- Maria Gallucci on Bloomberg's climate change legacy
- Jeremiah Owyang on Sustainability and the Collaborative Economy
- Neal Gorenflo on the New Paradigm of Sharing

- Stéphane Dupas (Research Scientist, Laboratoire Evolution, Génomes Comportement, Ecologie, on the ecological and evolutionary trajectories of biodiversity) on the IMAGINE European Energy Cities Project
- Tony White, author, on fighting climate fear with fiction

If I have overlooked anyone, I hope they will forgive me. In reality this book is the product of a large community of like-minded people with refreshing faith in the future and the commitment to work towards it.

FOREWORD BY HERBERT GIRARDET

To sustain or to regenerate?

The challenge to create an urbanising world that is compatible with the living systems of our home planet is greater than ever. Cities, as the heartland of national economies and of consumerism, have vast impacts across the world. Can the environmental performance of cities be massively improved, whilst creating the basis for a new, green urban economy? This is the primary focus of David's book.

I have written extensively about the concept of sustainable cities. But recently I have come to the conclusion that 'sustain' is a very passive term. We live on a profoundly damaged planet. Sustaining it in this depleted condition is no longer good enough. The relationship between the way we live in our human settlements and the world beyond must become much more pro-active.

'Sustainable' is also a term that is both corrupted and grossly overused. There are many other terms that can be used to define the way we should manage our cities – such as liveable, resilient, smart, intelligent and responsible, for instance. I now prefer to use the term 'regenerative'. *Regenerative cities are environmentally enhancing, restorative urban systems that maintain a responsible relationship with the natural resources on which they depend.* They are not only energy efficient and low carbon, but also designed to have positive impacts on their local and regional ecosystems.

Cities tend to have an egocentric tendency, believing that they can declare *independence from nature*, which is a rather catastrophic way of thinking. Cities are, in fact, dependent systems, relying upon a vast array of ecosystem services somewhere out there. It's not good enough to simply look at a city in terms of its morphology as planners tend to do; we need to take a closer look at the linkages – pipes and cables connecting the city to the wider hinterland, the power stations, and food and water sources.

This fact highlights the importance of appropriate national policies, which I believe are key to creating a world of regenerative cities. The market itself is lazy! It doesn't want to innovate unless it's for profit and, by itself, is not fit to create truly liveable and environmentally compatible, regenerative cities.

At the other end of the scale are citizens. Environmentally conscious behaviour patterns are still difficult to instil in people, but we're slowly getting there. We are now 'amplified' by a vast array of technologies and gadgets – I use the term 'amplified man'. Although many devices have become ever smaller and more energy efficient, we use more of them than ever before – cars, washing machines, mobile phones, industrial robots, aeroplanes, etc. Greater awareness of our impacts is badly needed.

A crucial driver for making a regenerative development turnaround happen is to realise that there is huge economic potential in doing so. A lot of cities have suffered from job losses due to automation, industrial relocation, etc. But renewable energy can be supplied more locally, and recycling can become an integral part of the functioning of urban systems, and this means that jobs can come back to the urban arena. But it won't happen automatically – it has to be driven by both national- and community-level enabling policies.

David's book is an impressive survey of the options available to us. It also highlights the great urgency of the task at hand. With a world caught up in a wide range of existential crises, cities need to be a central focus of action. There is no time to lose!

Herbert Girardet

INTRODUCTION

It's time to end humanity's war on nature

We have two years to prevent an ecological meltdown that threatens the existence of the human species. That is according to someone who should know: Cristiana Paşca Palmer, the chair of the world's Secretariat of the Convention on Biological Diversity.

Cristiana Paşca Palmer told a story in September 2018[1] about how on the beautiful Micronesian nation of Palau, tourists have polluted and damaged the coral reefs, poached the island's tropical wildlife and caused much other environmental damage. She asked the question: "Can a nation continue to welcome tourists while simultaneously protecting itself from them?"

Her answer was the "Palau Pledge". This, she said, "requires all the island nation's visitors to sign – right in their passports – an agreement not to damage or exploit the natural resources. A short video on the pledge is being run on all flights to Palau. The effort has been paired with law enforcement and reporting of destructive behaviour, and received global acclaim from celebrities and activists, winning multiple international awards".

Ms Palmer went on to say: "How we convey this understanding of our connection to nature and nature's connection to us will be critical for the post-2020 biodiversity framework." Even more recently, she went further, issuing the unprecedented warning that the world has two years to secure this deal – to halt a "silent killer" as dangerous as climate change.[2]

> People in all countries need to put pressure on their governments to draw up ambitious global targets by 2020 to protect the insects, birds, plants and mammals that are vital for global food production, clean water and carbon sequestration. The numbers are staggering. I hope we aren't the first species to document our own extinction.

Global wildlife populations have fallen by 60 per cent in just over four decades, documents WWF's *Living Planet Report 2018*.[3]

This matters to people living in cities because cities are currently behaving like giant vacuum cleaners, sucking up resources from the rest of the planet. If they don't stop and reverse this process, they will die, like the Ouroboros, the mythical snake that ate its own tail. But unlike the Ouroboros they won't be reborn. Civilisation will end.

I hope that the 196 member states of the Convention on Biological Diversity can decide on a new framework for managing the world's ecosystems and wildlife in Beijing in 2020, and that it sticks, unlike previous impressive-sounding global agreements to reduce biodiversity loss which have largely been ineffectual.

But you must not think that biodiversity only exists in some faraway beautiful spot. You have to think about it in your own backyard, your own town, neighbourhood or city. You have to begin not just saving it, but regenerating it now, in order to end the war on nature.

Let me illustrate how the war on nature happens by relaying the testimony, mixed with my own experience, of a friend from the island of Cyprus, in the Eastern Mediterranean.

It happens gradually, little by little, as individuals destroy pockets of nature that they think are in their way. It seems like not so much at the time: a tree here, a woodland there, to build this house, factory, supermarket or road. Perhaps it seems like progress. However, with everybody else doing the same around them, soon very little nature is left, and a more sustainable way of life has vanished forever.

My friend Eray was a small child in the 1960s. At that time hardly anyone lived on the Cypriot coast; nowadays, of course, almost all the island's inhabitants do. His mother owned quite a lot of land, with plantations of vines, and trees bearing figs, almonds, olives, walnuts, carob, and more. They would eat and sell the harvests.

He remembers as a child visiting the north-west coast, near Polis, where the ancient forest stretched for miles. "It came right down to the sea," he said. As the "trend" of tourism spread from the south of France and Italy to Cyprus, he recalls locals chopping down these trees so that a beach could be exposed. "As a result, there was nothing to stop the sea at high tide coming further inland where the salty water soon also killed the eucalyptus trees."

This stretch of coast was where loggerhead sea turtles would come ashore and lay their eggs, burying them in the sand. After they emerged from the eggs the hatchlings would crawl back to the sea to begin the cycle again.

"But local people loved smashing the eggs," he said. "Eventually some of us had to mount a round-the-clock guard for three months on the beach to stop them, or stealing the eggs, as these turtles had become an endangered species."

In the early 1990s, Eray made an attempt to start a branch of Greenpeace on the island, to stop the continuing environmental destruction. "But nobody was interested," he said.

For centuries, both Greek and Turkish Cypriots had lived a sustainable life as peasants. They carved terraces along the hillsides to manage the water flows and stop soil being washed away, where they planted their olive trees. Their goats grazed on the wild sage and rosemary, their milk becoming haloumi cheese and anari, a sweetened cottage cheese.

But after Cyprus was divided by the 1974 Turkish invasion young Greeks from the villages in the hillsides moved to the coast to build hotels and restaurants to take advantage of the emerging British trend for tourism, furnished by the increasing availability of cheap air travel. They abandoned the peasant ways of their parents and grandparents, which they considered old-fashioned and hard work. They would get rich quick and buy big cars, televisions and computers.

To the north of the Green Line, the population increased almost tenfold as Turks from the mainland seized farmland that had belonged to the Greek Cypriots, and began building over that, too.

In the early 1990s my first wife, a Greek Cypriot, took me and our first son to her mother's ancestral village, Kilani, in the Troidos mountains. I recall the first time I met her uncle and aunt. It was about ten o'clock in the morning and he was coming down from the hills with his donkey, whose yoke carried two pails of milk freshly obtained from their goats.

We followed him into their courtyard where his wife had already prepared a fire in a stone fireplace. He poured the milk into an aluminium pot suspended over the fire and we watched while she made haloumi, followed by anari. She gave us each a delicious bowlful and I can still taste it today. They were in their eighties, their faces as brown and lined as the cracked mud of the land. Another aunt living nearby made wine from her grapes and there was a grappa still in her backyard.

They were the last generation to live like this. My then wife's sister and her husband had bought some land next to hers on the edge of the village and were restoring the disused terraces to farm the land organically. Everybody thought they were mad to eschew pesticides, herbicides and artificial fertiliser, because this was the modern way. But most of the other terraces across the Troidos mountains were to fall into disuse and start crumbling away.

Centuries of work gone in a generation.

Bulldozers were removing plantations in Eray's mother's orchards at this time to make way for development. Eray attempted to rally support to save the trees, but when the day came he was the only one to turn up. "I climbed to the top of the tree," he told me, "but this did not stop the chainsaws. They cut down the tree with me in it and I fell onto a nearby roof."

Walnut trees are an investment in the future: it takes at least a decade before they begin to produce nuts. All his mother's other plantations were soon gone too. The ancient forest in the north-west has mostly vanished also. Cyprus has a water shortage problem now, and its soils are washed away when it does rain because the terraces no longer provide the same defence.

The now middle-aged generation who abandoned the terraces of their parents, these days have their second homes in the old villages in the hills, where they come at weekends with their kids and complain about how stressed they are. Many do not live as long as their parents' generation, who had a healthier diet and plenty of exercise.

I tell this short story to illustrate the connection between the loss of nature, and its consequences. But it is not too late. Trees can be replanted. Wildlife can return. Birds that used to be common like the Cyprus warbler and the Cyprus wheatear, both now endangered, could increase their populations were their habitats to be restored and protected, as could the Cyprus mouse, the only endemic species of rodent in the Mediterranean left.

The island of Cyprus is a microcosm of the globe. This process of abandoning traditional ways for an urban, consumerist lifestyle has been happening for decades around the world and continues relentlessly. It is called 'development' or 'civilisation', and has always been assumed to be a good thing.

Through Cristiana Paşca Palmer's lens we can reframe it as a war on nature, an attack on traditional ways of life in which people live in a kind of balance with their natural environments, that were more or less sustainable. As people became – and become – richer, they pour more concrete, cover more land, destroy more habitats.

I don't see this process stopping deliberately any time soon. So for the readers of this book I say this: use your power to regenerate nature wherever you live and work, even if it is in the city.

It's time to start paying back what we have taken, if for no other reason than because our own survival depends upon it. We are choosing to flock to cities. Nature can thrive anywhere. It just needs to be given space. How to do this is what this book is about. We must make all cities regenerative, 'one planet' cities, so we live within our means, and become 'one planeteers'.

The solutions already exist. People are making them happen. Policies to support them must be based on evidence measured against standards – indicators of success. They must not be based upon ideology, belief systems or loyalties to one group or another, because we are all in this together.

We must favour local, low-impact, grass-roots projects instead of large, high-impact, expensive ones. We must curb excessive consumption, directing wealth to where it can give real social and environmental benefit. That's not to say we can't live the good life, but we should have enough for that and no more ... leaving sufficient for everyone else to have the same.

And we must employ our imaginations to visualise and communicate what these more successful futures could be like – including the potential pitfalls and compromises.

These are the themes of this book. It sets out the nature of the problems facing not just us but many other species and ecosystems with which we share this lovely planet; species which have as much right to exist as we do.

It attempts to explore, through a discussion of the types of stories we tell ourselves, how we reinforce and can reshape our apprehension of the potential futures ahead. And it attempts to evaluate, against explicit criteria, some of the many brave efforts around the world to overcome our problems.

It is a constant source of surprise to me (and surprise that I remain surprised) that more is not said about these huge, survival-threatening challenges in the popular media, nor about the heroic efforts to meet them being undertaken by courageous pioneers everywhere.

Slow, unfolding disasters do not make headlines. Only when a major report is issued do they reach the screens, and shortly afterwards they are again forgotten. Meanwhile, the progressive forces act mainly on the sidelines, and at grass-roots level, where news editors do not look, and business as usual is largely permitted to continue.

When I was aged 11 I won a national essay-writing competition for a piece about suburban sprawl covering fields near my house. Since then I have spent much of my professional life, as a specialist news editor, journalist, publisher and author of non-fiction titles, publicising positive examples of sustainable actions and technologies in the hope that others would replicate or take them up. In my private life I do what I can, and I write fiction, some of which is climate fiction.

Unfortunately, I believe now that adults are not inclined to change their behaviour unless drastic events force them to. Drastic events are happening now to millions of people around the world, but because they are often the poor, it does not motivate the wealthier, who have more impact. This is why legislation and enforcement are vital. We have curbed many other behaviours now seen as disgraceful, like human genocide; can we do the same with over-consumption and animal genocide?

Notes

1 Palmer, C.P., Opening remarks at the Pre-COP 14 Meeting for the Pacific Countries, Convention on Biological Diversity, 24 September 2018, Apia, Samoa. See: www.cbd.int/doc/speech/2018/sp-2018-09-24-samoa-en.pdf
2 Watts, J., "Stop biodiversity loss or we could face our own extinction, warns UN", *The Guardian* Online, 3 November 2018. See: www.theguardian.com/environment/2018/nov/03/stop-biodiversity-loss-or-we-could-face-our-own-extinction-warns-un
3 Grooten, M. and Almond, R.E.A. (Eds), *Living Planet Report – 2018: Aiming Higher*, WWF, Gland, Switzerland. See: www.wwf.org.uk/updates/living-planet-report-2018

1

DO THE STORIES WE TELL INFLUENCE THE FUTURE WE WILL LIVE IN?

Imagining the future is the first stage in the task of creating it. Unless we can picture in realistic detail the kind of future that we want, we don't have a chance of working towards it. This scene-setting chapter wonders about the effect of negative visions of the future, why we are so poor at imagining positive futures, and what can be done about this.

I know from my own introspection that fear is a massive motivator for behaviour ... but what kind of behaviour? Fear of being a victim of crime, for example, is a prominent reason given in Michael Moore's 2002 documentary *Bowling for Columbine* for the huge disparity between homicide rates in Canada and the USA, many other factors being equal. What fuels this fear? According to Moore it is the daily dosage of crime reportage meted out to the American public in the media. This drives a perception that crime rates are much worse than they really are, and a consequent perceived need to arm oneself and shoot first. Gun ownership thrives, along with an obsession with security.

In other words, he says, the moral, social and political fabric of American society is being skewed by the distorted picture of the world being drip fed into the American psyche. In this negative feedback loop, each random mass shooting and each deliberate homicide reinforces a feeling of threat and the conviction that possession of loaded firearms is the best form of personal security, a feeling that is precisely opposite to the reality. For, as Moore's documentary portrays, in Canada, where levels of gun ownership are approximately equal and the population is also racially mixed, many people do not even bother to lock their doors and murder rates are extremely low. News media and politicians there do not fuel the inevitability of violence as a means of solving problems, instead focusing on the need for mediation, negotiation and compromise.

Similarly, how else can we explain the fact that it's only really in the USA that climate scepticism reaches epic, violent proportions – where political polarity fuelled by literally fake news, paid for by fossil fuel companies,[1] convinces scientifically illiterate people that they know better than 97 per cent of the world's top climate scientists?

The conclusion I draw from this is that the stories we are told about the world out there define the way we prepare ourselves to face it. And, as Dan Bloom[2] has it, fiction has the power to reach parts of the human psyche inaccessible to politicians and scientists. We writers like to believe we can change minds.

But are we just speaking to the converted? Let's look at it from the writer's point of view. Many of us are thinking: what kind of world do we want to live in? What kind of future will our children inhabit? What is the best future we can imagine?

But others aren't. From Fritz Lang's 1927 film *Metropolis* and Charlie Chaplin's 1936 film *Modern Times*, through George Orwell's 1949 book *Nineteen Eighty-Four*, George Lucas's 1971 film *THX 1138*, Mega-City One from *Judge Dredd*, conceived in 1977, to Ridley Scott's 1982 film *Blade Runner*, they have all set the template for many other stories and films such that in the popular imagination the sprawling mega-cities of the future will largely be over-populated, polluted, broken places, featuring dark towers, high levels of surveillance and crime, their citizens treated little better than battery-reared animals, with no room for nature. The Netflix series *Altered Carbon* is the latest to succumb to this cliché.

If that's the popular image, does this make a dystopic metropolis into a self-fulfilling prophecy, subconsciously, if not consciously, reinforcing the mindsets of planners and architects? Does it soften up the public, preparing them to acquiesce in the face of grim and unimaginative design, polluted air, poor policing and service levels, corrupt or inefficient governance, long commute times, constant noise, high levels of personal danger?

Where would you rather live: utopia or dystopia?

William Gibson, in his 1984 cyberpunk thriller *Neuromancer*, describes Night City, a fictional city located between Los Angeles and San Francisco on the west coast of the USA, as being "like a deranged experiment in social Darwinism, designed by a bored researcher who kept one thumb permanently on the fast-forward button". Dystopian par excellence, it has inspired a roleplay game, *Cyberpunk 2020*, and a detailed guide book – not bad for a fictional city. Night City is an arcology – a portmanteau of 'architecture' and 'ecology' – a design concept for very densely populated habitats, coined and popularised by architect Paolo Soleri. But it turns out that he and other architects have conceived highly sustainable and desirable arcologies. Soleri's concept appears as early as 1969 in his *Arcology: City in the Image of Man*.[3] Attempts have even been made to build them.

Soleri intended his Babel IIB arcology as "an anti-consumptive force and a city form that is the only choice compared to pathological sprawl and environmental destruction". It was designed for a population of 520,000, at a height of 1,050 metres. Besides residential spaces it includes gardens and waste-processing plants, everything you need: parks, food factories, etc.

How funny that Gibson took the idea and then reverted it to pathological sprawl and environmental destruction. It just goes to show that the devil gets the best tunes. Which, I submit, is part of our problem, as we collectively, culturally, try to imagine the future.

It's revealing that the world's most sustainable cities often also feature in lists of the most popular ones, according to polls. That's because differing architectural designs and urban place-making can generate quite contrasting lifestyles and therefore social experiences, not to mention differing scenarios for greenery, wildlife, and greenhouse gas emissions.

Psychologically, there are many aspects to people's reluctance to engage with the profound implications of climate change and sustainability in a way that's appropriate and proportionate. George Marshall's brilliant research in *Don't Even Think About It: Why Our Brains Are Wired to Ignore Climate Change* [4] documents some of these. It's not just the jargon, it's peer pressure, culture, near-sightedness, fear, lack of awareness, vested interests, to name a few. It entails huge leaps of imagination to envision what life will be like for different people in the scenarios, and the responses that will be required or demanded.

Yet stories are fundamentally how humans understand and spread wisdom, as well as entertain themselves. Because of this I think there is some responsibility not to paint self-fulfilling, disempowering dystopic futures or preach about environmentalism to the converted, but to dare to imagine inspiring and realistic future visions, as settings for potentially popular fictional narratives, that demonstrate how humanity might successfully meet climate change's challenges and make a better world/solve multiple challenges. Not a tall order, is it? Creators have the power to empower and give hope. And that includes architects and designers.

Lateral thinking, through transposing stories from one's personal experience, or from other sources, into one's professional life, can help. That's something many professionals working in city design or administration too often fail to do. It's box mentality. To create a better world we need means that we must really evoke them to make it easy for people to imagine living there – a better way of life in many ways. We also need to imagine how to negotiate the pitfalls that potentially may divert or frustrate us along the way …

… and move beyond dystopias and utopias. Reality is more complex; all possible pasts, presents and futures have positive and problematic aspects, due to trade-offs and unintended consequences. At the end of this book you will find a story that explores this, set towards the end of this century, in a world that successfully sustains 11 billion people, most living in mega-city conurbation strings, based on some of the examples explored in these pages. Perhaps by then you will be able to think up your own scenarios, especially involving your own neighbourhood. To navigate the future successfully, we need to craft the right stories.

This book sets out a menu, exploring aspects of climate-friendly and equitable, regenerative cities already happening somewhere on Earth. If all of these were all happening together, in the same city, it would be the ultimate city, "the best city and a good enough city for any world", to paraphrase Raymond Chandler. I sometimes call this future hyperworld '2084', to invert the pessimism of Orwell's *Nineteen Eighty-Four*. For a while, it was a working title for this book. Herbert Girardet calls it Ecopolis.

This menu, like pieces of a jigsaw puzzle, will challenge existing practice, especially in policy, planning and land use, amongst city professionals everywhere. Many of these examples are rarely reported in mass media. I am aware of the limits of the conditions under which these solutions can thrive. And I know that 'what gets measured gets saved', so I'm big on using metrics to determine what is *truly* sustainable, which is why Chapters 3 and 4 cover standards, measurement and verification.

For a long time, my motto has been: with imagination we can change the world. Let's ask what's rarely asked: what is it that prevents rational solutions from being (properly) implemented and what drives the adoption of what *is* implemented? How much can we change the future? After reading this book, I hope you'll have developed your own ideas.

Notes

1 Mulvey, K., Shulman, S., et al., *The Climate Deception Dossiers*, Union of Concerned Scientists, 2015. See: https://bit.ly/2qMUkNc
2 Journalist who coined the term 'cli-fi', meaning fiction about climate change.
3 Soleri, P., *Arcology: City in the Image of Man*, MIT Press, 1969.
4 Marshall, G., *Don't Even Think About It: Why Our Brains Are Wired to Ignore Climate Change*, Bloomsbury, 2014.

2

THE ULTIMATE PROBLEM

Humanity's limits to growth

Here is something that not many people really consider. It is, perhaps, the biggest elephant in the global room. The world's population is now 7.5 billion and is predicted to peak at 11.2 billion by 2100.[1] But the ability of our lovely planet Earth to support life depends on us staying within a number of 'planetary boundaries' that are essentially limits to growth. At the same time more and more people are aspiring to a more comfortable lifestyle, and the United Nations (UN) is campaigning for its 2030 Sustainable Development Goals to be met to ensure that every person alive has their human rights protected and a decent standard of living. These goals are not, on the whole, linked to planetary boundaries, adding more pressure on the world's environment.

How to resolve this complex problem is the most important question facing humanity.

Planetary boundaries

What are these planetary boundaries? The environmental non-governmental organisation WWF, assisted by the Stockholm Resilience Centre, identifies nine of them in its biennial comprehensive survey, the *Living Planet Report*:[2] extinction rate and loss of biosystem services; biogeochemical flows – e.g. pollution from nitrogen and phosphorous fertilisers; stratospheric ozone depletion; ocean acidification; atmospheric aerosol loading; and 'novel entities' – polluting substances such as plastics and toxins that have potentially irreversible effects on living organisms and the physical environment.

The last edition, published in 2018, stated:

> Current analysis suggests that people have already pushed at least four of these systems beyond the limit of a safe operating space. Attributable global impacts

and associated risks to humans are already evident for climate change, biosphere integrity, biogeochemical flows and land-system change. Other assessments indicate that freshwater use has also passed beyond a safe threshold.

I'll go into more detail about these boundaries below, but whichever way you consider it, humanity as a whole is presently not doing very well in terms of taking care of its planet and its future. As measured by a related metric, ecological footprinting (which the report goes on to discuss), humanity now needs the regenerative capacity of 1.7 Earths to provide the goods and services it collectively uses. The next chapter talks more about this topic.

The point I want to emphasise here is that the ecological footprint of high-income nations dwarfs low- and middle-income countries. In other words, the more people spend, the greater their environmental impacts. This is a real, objective correlation, because supplying all of those goods and services that people buy has an effect on the Earth's resources and ecosystems.

Many people aspire to the standard of living led by citizens of North America, Scandinavia and Australia. But WWF's reports show that it's simply not possible for 11.2 billion people to achieve it without destroying the living systems that support life on Earth.

One thing is abundantly clear: the greater part of anybody's ecological footprint is due to consumption; both the level of consumption and the type of consumption. A model where biocapacity is explicitly addressed and future scenarios can be enumerated, as with the International Panel on Climate Change's (IPCC) models for climate change, would be useful. According to David Rapport, a Canadian long-time developer of ecological footprinting methodology:

> The survival of humankind [...] depends upon far more than only the resource demands placed on the earth ... The challenge to humankind [...] lies not only in reducing the size of the ecological footprint, but also in increasing the prevalence of healthy ecosystems. [...] [M]aintaining ecosystem health is essential to sustain the ecological goods and services upon which EF [ecological footprint] accounting is based.[3]

This is why, for example, One Planet Development in Wales (see Chapter 16), besides requiring residents to reduce their ecological and carbon footprints, requires improvement to bioproductivity and biodiversity.

Doughnut economics

This topic is addressed by Kate Raworth, of Oxford University's Environmental Change Institute, in her book *Doughnut Economics* [4] in which she proposes the idea of a 'safe and just space' framework. This combines the concept of planetary boundaries with that of social boundaries. With the ring doughnut metaphor, the inner edge of the doughnut represents the sufficient level of resource use to meet people's needs, and the outer edge of the doughnut represents the planetary limits. For the sake of fairness and sustainability, people should live within these two limits, i.e. within the doughnut itself; this is the 'safe and just space'. Anything less and the quality of life is reduced; anything more and we are living beyond our means.

This assumes that any allocation of resources to people should be sustainable, fair and efficient. Are there any communities in the world at the moment that occupy this space? To answer this question, four researchers writing in the journal *Nature* looked at what quality of life could be achieved universally if global resources were equally distributed, building on Raworth's work.[5] They – Dan O'Neill and his fellow researchers at Leeds University – believe that this approach is genuinely sustainable because it requires sufficient natural capital to be maintained at the same time as meeting essential human needs.

The first important conclusion they came to is that you can't satisfy people's social needs without healthy, functioning ecosystems and the resources that they provide. While this might seem obvious, it is not always reflected in the behaviour of nations or companies. Another topic they considered is the way in which resources are used to achieve the same social outcomes. Travelling by bike, car or train will all get you from A to B, but each will use different quantities of resources per person.

Their main conclusion is not terribly optimistic. "Of the 150 or so countries that we looked at we found no country that was achieving meeting the social needs of its people in a way that was sustainable."

No country is sustainable.

"We found countries that do well on the social indicators like the UK and Nordic countries don't do particularly well on the planetary boundaries," continued O'Neill.

Which country does best? The answer may be surprising: Vietnam. "It does well on six of the eleven social indicators but it still transgresses one of the planetary boundaries – its CO_2 emissions are too high." If it could reduce or eliminate these, its people would live within the doughnut.

The researchers created a website [https://goodlife.leeds.ac.uk] which allows people to play with adjusting different levels of satisfaction on 11 social variables and see what kind of difference it makes to the levels of resource use. "If you try to reduce the level of resource use to one that is sustainable, then the level of social satisfaction is not that high. It's a pretty dire picture," admits O'Neill.

What the website model doesn't do is allow you to change the way those resources are used, i.e. the model expresses the way governance and economics are now. It clearly demonstrates that they must transform in the future if we are to meet human needs sustainably, especially if there are more people on the planet. "We hope that it is possible that this is achievable," O'Neill told me. "Otherwise that doughnut of Kate Raworth's is vanishingly thin. It's a bit of a wake-up call for us. If we are serious about achieving sustainable development goals then we need to shift our economies away from the idea of pure growth."

O'Neill is being disingenuous. The doughnut is not just thin, it does not exist anywhere right now. No country is satisfying all of the needs of its citizens within planetary boundaries.

His team's future work is looking at ways in which resource use can be much less wasteful than it is now, where especially the rich, who are the most wasteful, do without what should be regarded as unnecessary luxuries. "Countries in the north should actually 'de-grow'," O'Neill says.

> We find that the economy is developed to the point where most social needs are met, then beyond that emissions continue to increase because wealth is being pursued for its own sake without any particular increase in social satisfaction. Countries like the USA and the UK can move a long way back on that curve and not lose a lot of well-being. This leaves some space for the poor countries to catch up with the little ones. But even if we did that it is still not enough with the present way we are using resources. We need to be provisioning systems that are much better at translating resources into meeting human needs. That is the point of our research.

High consumption equates to a high ecological footprint and is popularly associated with a better quality of life. Yet rich people are not necessarily happy. Can a 'one planet city' make people happy without excessive consumption? I believe so.

There are three top-level requirements to get us out of this hole we have dug ourselves into:

1. It is necessary to reduce humanity's ecological footprint (and other footprints according to location, such as the water footprint) in order to meet the outer edge of the doughnut and position our ways of living within the biocapacity limitations.
2. It is necessary to pursue the Sustainable Development Goals in order to define the inner edge of the doughnut so everyone has a decent quality of life.
3. It is necessary to increase the rate of regeneration in order to pay back our debt to nature and increase biocapacity.

Only then can all people thrive within the means of the planet, i.e., within the doughnut.

The book argues that attaining these top-level requirements should be the aim of all government strategies, whether at local or national levels, otherwise civilisation cannot continue indefinitely. This is the absolute meaning of the word 'sustainability'.

Contraction and convergence

This idea is summed up in one phrase: contraction and convergence. Contraction requires gradually reducing the overall level of consumption both within nations and over the whole world, while convergence requires reducing the difference between the consumption levels of the poorest and the richest (the wealth gap), to spread the benefits of technological progress to everyone. It is a simple idea, but politically very difficult to implement.

Convergence is necessary for all societies to have a hope of reaching together the inner circle of the doughnut. Contraction is necessary for all societies to have a hope of reaching the outer circle. It is just the same with greenhouse gas emissions.

It becomes therefore necessary to define goals, metrics to measure whether they are being achieved, to identify and implement systems that can keep track of these, and strategies to ensure that they keep within that level. So Chapters 3 and 4 examine the existing metrics, standards or indicators that might help achieve these objectives.

How we talk about the future

This problem is something that most futurologists don't seem to consider. Books about the future seem largely to assume that technological progress will continue to solve human problems and make things better for everyone. Perhaps that's because most of these books are written by highly educated people living very comfortable lives in developed countries, who haven't explored what life is like for the very large majority of people on the planet. I'm thinking about books like Peter Diamandis and Steven Kotler's bestseller, *Abundance: The Future Is Better Than You Think.* [6]

We certainly can't trust economists. In a collection of essays called *In 100 Years: Leading Economists Predict the Future,* [7] none of these so-called experts consider inequality and these limits to growth. Of popular science books that do consider ecological aspects of the future, climate change is by far the most common topic, and there is a handful of books devoted to the sixth great wave of species extinction for which humanity is currently responsible. In popular non-fiction, the remaining boundaries are hardly worthy of a footnote. Yet, we need all of them to survive. Amongst other threats to our future, the rise of the machines, terrorism and climate change feature large in books and journalistic articles, but to a much lesser extent the fate of the biosphere upon which we depend for life. Even if the machines betray us and there is more terrorism, then human beings and much of civilisation would still survive if the biosphere were stable and thriving.

So we are curiously blind to the more fundamental existential threats to our future. Perhaps the reason is that it is just too huge a subject for us to conceptualise. Most futurologists take past trends and continue them into the future. For all the current talk about technological disruption, the fatal disruption of our life support system is something that does not enter popular discourse. So let us consider it.

Our life support system is failing: planetary boundaries in detail

Biodiversity, and nature's capacity to sustain humanity are being degraded, reduced and lost in every corner of the world, with the exception of a number of positive examples where lessons can be learned. This was the depressing conclusion of a major report published in March 2018 by the Intergovernmental Science-Policy Platform on Biodiversity and Ecosystem Services (IPBES – often described as the 'IPCC for biodiversity', an independent body comprising 129 member governments).[8]

Human pressures on habitats, our overexploitation and unsustainable use of natural resources, our pollution of the air, land and water, climate change, our increasing numbers, and the impact of invasive alien species are all forcing this process.

Less than one-quarter of the Earth's land surface remains free from substantial human impacts. Projections into the future show that these remaining relatively untouched areas are vulnerable, and that by 2050 only 10 per cent of land surface will be left substantially free from direct human impact. These areas will be in deserts, mountainous areas, tundra and polar locations – places unsuitable for human use or settlement. For example, all the plausible future scenarios explored in IPBES' assessment of Africa highlight that the drivers of biodiversity loss will increase.

1 Land degradation

Land degradation and the loss of biodiversity and ecosystem services is the most pervasive phenomenon affecting the planet now. It is already negatively impacting the well-being of at least 3.2 billion people according to the IPBES. It is pushing the planet towards a sixth mass species extinction, and costing more than 10 per cent of the annual global gross product.

At the current rate of soil and water loss, 4 billion people will be living in drylands in 2050. Wetlands are particularly degraded, with 87 per cent lost globally in the last 300 years, and 54 per cent since 1900.

Decreasing land productivity, among other factors, makes societies, particularly in drylands, vulnerable to instability. Every 5 per cent loss of productivity caused by degradation is associated with a 12 per cent increase in the chances of violent conflict, according to IPBES.

Land degradation is also a major contributor to climate change. There are vicious feedback effects: more extreme weather events will accelerate soil erosion on degraded lands, increase the risk of forest fires and change the distribution of invasive species, pests and pathogens. But there is widespread lack of awareness of this process and that in itself is a major barrier to action.

As stated above, the main factors driving this process are high-consumption lifestyles in more developed economies, plus rising consumption in emerging economies. Feeding this consumption is causing the rapid expansion and unsustainable management of croplands and grazing land. The consumers who benefit from the overexploitation of natural resources are among those least affected by the direct negative impacts – and therefore the least aware of it.

This over-consumption leads to more extractive industry, infrastructure, industry, urbanisation, vulnerability to fire and erosion, and invasive species.

It is true that intensified land-management systems have greatly increased crop and livestock yields in many areas of the world. But they can result in high levels of land degradation, including soil erosion, fertility loss, excessive ground and surface water extraction, salinisation and oxygen-starvation of aquatic systems.

2 The sixth great extinction event

The second planetary boundary is to do with the number of species we share the planet with. Having a diverse and large number of species of all kinds gives the planet's life support systems greater resilience – which means an ability to withstand shocks and changes. But instead of looking after these species, human beings are responsible for the sixth great extinction event on the planet. This level of extinction is on a par with the extinction of dinosaurs, and the rate of extinction is unprecedented.

Since humanity crossed the 'one planet' ecological footprint boundary in the early 1970s, the numbers of wild terrestrial vertebrate species have declined by 38 per cent and that of freshwater vertebrate species by 81 per cent. Overall, biodiversity has declined by a staggering 58 per cent (give or take a few percentage points) between 1970 and 2012, according to the Global Living Planet Index.

The result is that the net primary productivity of ecosystem biomass and of agriculture is lower than it would have been under its natural state on 23 per cent of the global land area. By 2050, land degradation and climate change together are predicted to reduce crop yields by an average of 10 per cent globally and up to 50 per cent in certain parts of the world.

Coupled with projected population increases this can only point towards mass starvation. Already, nearly 800 million people worldwide lack access to adequate nutrition.

3 Climate change

Climate change is the planetary boundary that most people are aware of. But even if countries do what they have already promised to meet targets enshrined in the Paris Agreement,[9] then greenhouse gas emissions are still in line to cause a rise of between 3°C and 4°C in temperature by the end of the century. This would be catastrophic.

A 3°C rise would cause serious drought in Southern Europe every ten years, water shortages for 1 to 4 billion more people, increased floods for 5 billion people and strong risk of hunger for 150 to 500 million people, of whom 1 to 3 million more would die from malnutrition. There is also a higher risk that the West Antarctic ice sheet, and the Atlantic thermohaline circulation, would collapse, reversing the Gulf Stream.

At a 4°C rise Southern Africa and the Mediterranean could lose 30–50 per cent of their water. In the tropics, strong heatwaves would cause the loss of forests, cropland, and extreme drought. Mass migrations would result.

Sea-level rise threatens coastal areas. The most recent evidence from NASA (the National Aeronautics and Space Administration) shows that ice loss from the West Antarctic and Greenland ice sheets is happening faster than expected, leading to acceleration in sea-level rise sooner than was projected by earlier IPPC reports.

The implication is grave for hundreds of coastal cities and communities. If, as now predicted, the global sea level rises by an average of 3 feet this century, 17 per cent of Bangladesh will be flooded, displacing over 18 million of its citizens. Many more will be threatened by salinisation of the soil which they use to grow their crops, rendering it useless. In the USA alone, up to 670 coastal communities are under threat.

Climate change threatens food supplies in many ways. Firstly, many regions where most of the world's crops are grown are also areas where the climate could become too hot or dry or wet for them to thrive. It is predicted to cause a greater variety of crop pests. Droughts will cause livestock to starve. Acidification and temperature change of the oceans will have a severe impact on fish populations.

For example, the crops wheat, rice, maize, and soybean provide two-thirds of human caloric intake. They are mostly provided by the USA, China, Brazil, Russia and India. By the end of the century it is predicted that for each 1°C rise in temperature, yields could fall by up to 22 per cent for wheat, 10 per cent for rice, 27 per cent for maze, and 11 per cent for soya beans.[10]

4 Water

The fourth planetary boundary is the availability of fresh, clean water. Four out of five people in the world already live in areas where there is a threat to water security; 40 per cent are already affected by water scarcity that threatens crop growth and livestock.

By 2025, seven years away from the day I write this, half of the world's population are predicted to be living in water-stressed areas. Some 80 per cent of countries say they don't have enough money to meet national drinking-water targets.

Much water is lost from agricultural land by not returning organic matter to the soil, which can then retain water like a sponge, instead relying on artificial fertilisers. Much is lost by deforestation. Industry uses a great deal, and pollution renders it undrinkable. A higher consumption of meat means more animals, and this, too, increases water demand. Cities are poorly designed to manage rainfall. This all adds up.

5 Nitrogen and phosphorus pollution

In the recent past, the Green Revolution gave us synthetic fertilisers that have been sustaining about half of the world's population. They have therefore contributed to a growth in population from 1.6 billion in 1900 to today's 7.5 billion. But the production of these fertilisers is dependent upon fossil fuels, resulting in rising atmospheric CO_2 concentrations.

The use of the nitrogen fertilisers has also caused excess reactive nitrogen to pollute soils and watercourses, affect drinking-water quality and air quality (smog, particulate matter, ground-level ozone) and caused eutrophication of freshwater and coastal ecosystems (dead zones where nothing can live). Stratospheric ozone depletion is another consequence of the human-caused excessive nitrogen cycle.

Much of the nitrogen and phosphorus applied to soils finds its way to the sea, where it pushes marine systems into higher-risk conditions. Take the Gulf of Mexico's "dead zone"; this is caused by large quantities of nutrient run-off into the Mississippi River and other catchment areas.

6 Stratospheric ozone depletion

The ozone layer filters out harmful ultraviolet (UV) radiation from the sun. The thinning of this layer gives rise to a higher incidence of skin cancer, cataracts and immune system disorders in humans, and harm to terrestrial and marine biological systems as well.

7 Novel entities

This refers to toxic and long-lived chemicals and plastics. They include organic pollutants, heavy metal compounds and radioactive materials. Persistent organic compounds have caused dramatic reductions in bird populations. PCBs (polychlorinated biphenyls) concentrate in fatty acids and therefore work their way into milk given by all female mammals including humans to their young. Micro-plastics are clogging our seas.

8 The deaths of the oceans

Around a quarter of the CO_2 that humans release into the atmosphere ends up dissolved into the oceans where it forms carbonic acid. The proportion of this acid in the seas has increased by 30 per cent since pre-industrial times. Beyond a certain concentration, it becomes harder for organisms such as corals, some shellfish and plankton species to grow and survive.

The amount of fish in the oceans has halved since 1970, due to overfishing and pollution. Most marine life carries plastic that they either ingested directly or by eating smaller marine creatures. Did you know that if you consume an "average amount" of seafood you will ingest around 11,000 plastic particles per year? Most of us have micro-beads of plastic in our bodies from the food we eat – with unknown health consequences. Over-fishing, acidification and other types of toxic pollution threaten all life in the ocean and any bird, animal or person who eats it.

9 Chemical pollution

Chemical pollution affects the atmosphere as well as the planet's waters. CFCs (chlorofluorocarbons) threaten the protective ozone layer high in the atmosphere. Atmospheric aerosols such as haze, particulate air pollutants and smoke have many effects on ecosystems, including weather and climate change, and on human health.

These interconnected problems

Many of these problems are related and so the solutions to them are also related. When the chair of IPBES, Sir Robert Watson, announced the publication of the IPBES report, he said:

> Although there are no 'silver bullets' or 'one-size-fits all' answers, the best options are found in better governance, integrating biodiversity concerns into sectoral policies and practices (e.g. agriculture and energy), the application of scientific knowledge and technology, increased awareness and behavioural changes.
>
> It is also clear that indigenous and local knowledge can be an invaluable asset, and biodiversity issues need to receive much higher priority in policy making and development planning at every level. Cross-border collaboration is also essential, given that biodiversity challenges recognize no national boundaries.

The fact is that richer, more diverse ecosystems are better able to cope with disturbances such as extreme events and the emergence of diseases. They are our 'insurance policy' against unforeseen disasters and, used sustainably, they also offer many of the best solutions to our most pressing challenges.

Population: the other question we tend to avoid

The question of the impact of population increase on social and environmental pressures is another challenge we try to avoid. As I said at the beginning of this chapter, there are now 7.5 billion souls upon the planet and half as many again are predicted to be around by 2100. The topic is understandably a sensitive one. For example, we cannot blame the inhabitants of countries with fast-growing populations since there are structural reasons why the birth rate is so high. It is better to understand and address these reasons, which are linked to poor education, extreme poverty and inequality. Addressing these is proven to result in lower birth rates. They are the subject of several Sustainable Development Goals.

Guess which is the world's most crowded city? According to UN-Habitat (the UN Human Settlements Programme),[11] it's Dhaka in Bangladesh. Some 44,500 people are crammed into each square kilometre of space, with more migrating in every day. The sewers frequently block and overflow. Electricity supply is patchy. Rubbish piles are everywhere. Hours are lost every day in traffic jams. It is simply dysfunctional. Dhaka is also one of the poorest cities, with a gross national income (GNI) per person of just US$1,330.

If anywhere on the planet could be considered dystopia, this is it. Most people in the world would definitely not want to live there, and yet more are migrating in every day because they imagine life would be better than in the countryside.

Professor Nurun Nabi is project director in the Department of Population Sciences at the University of Dhaka. He says, "There are cities bigger in size than Dhaka in the world. But if you talk in terms of the characteristics and nature of the city, Dhaka is the fastest growing megacity in the world, in terms of population size."[12]

How is the world going to support so many people within planetary boundaries? Many experts, such as James Lovelock and Mayer Hillman, are unafraid to say that it cannot and that we are doomed. At some point the environmental shit is going to hit the human fan. Some days I wake up and think they are right, and some days I come down on the side of hope. That is why I collect examples of proven success. That is why much of this book contains these stories, of people trying to make their corner of the world a better place, sustainably, and I feel that if everybody follows suit in one way or another then there will be hope, if not certainty.

Cities can be densely populated without being overpopulated. Take the small island city of Singapore. With 5.7 million people, it has high population density – 7,908 per square kilometre – but it is not overpopulated because its residents are well catered for by a well-functioning administration. Its residents live in high-rises, often with rooftop gardens. The average GNI per capita is $51,880 (2017). Its wealth and good governance are illustrated by a simple fact: all births in the city are attended by a qualified medical person (in Bangladesh it is 42 per cent – 2016 figures). The annual growth rate is just 1.3 per cent.

Unfortunately, though, it has a high ecological footprint. If everybody in the world were to live that way we would need 3.5 planets. But it turns out that the largest component of Singapore's ecological footprint is carbon emissions. If Singapore went 100 per cent renewable energy it would have an ecological footprint of 0.86 Earths.[13] That is well within the limits. It would be even better if everybody went vegetarian, or even vegan, because there would be less impact on the oceans and grazing lands.

It has no plans at the moment for doing so, and its efforts currently to become more energy-efficient will not save enough energy to compensate for the growth in demand.[14] Its aspirations to cut emissions are currently rated 'highly insufficient'. Its main challenge is lack of space; as a small city state it cannot install hydropower or wind power. But it can install solar, for example offshore, and has plans to generate a quarter of its energy demand this way by 2025. Sensible use of heat pumps and hydrogen-powered vehicles could factor this up to 50 per cent. Beyond that, it may just have to power down.

But we were talking about population. Back in Dhaka, sprawling shanty towns are home to 40 per cent of its people, where communicable diseases spread easily and fires often break out. This capital city regularly tops lists of the world's least liveable cities. The cause is unplanned urbanisation that is out of control, coupled with weak governance. The deeper cause is poverty. Alleviating that is a huge challenge and the subject of *A World of Three Zeroes* [15] and other books by Nobel

Prize winner Muhammad Yunus, who details plenty of proven solutions to poverty that he has personally been involved in.

China is well known for its population control policy. Its annual population growth rate has dropped from almost 2 per cent in 1955 to 0.39 per cent today. That's the lowest on record. Meanwhile the country has experienced "the greatest mass migration in human history", from the countryside to the cities. In the 25 years from 1990 to 2015 its urban population grew by 463 million (over one-third of its present population), about half of whom came from villages.

The fullest evidence of the effect upon a person of moving to the city from the countryside comes from China and compares the happiness of such migrants with those who remain in the countryside, and with urban dwellers who have always lived there. It finds that migrants have roughly doubled their work income by moving from the countryside but are less happy than those they left behind.[16] On the whole, however, Chinese city dwellers are happier than their rural counterparts.[17] But over that period China's ecological footprint has grown from 0.75 Earths to 2.21 Earths due to that increase in income. But, again, the vast majority of that is due to carbon emissions (1.53 Earths). Being 100 per cent renewable would bring it back down to 0.68.

Of course it is not as simple as that, because there are different types of impact and, although they are interrelated, all need to be proportionately reduced. But we will come to that later.

The implications for cities

Cities suck up resources from outside their boundaries, spitting out waste, in the current way they operate, and the rate at which they do this is increasing.

In 2008, the global urban population overtook the rural population. Human beings are now increasingly likely to live in cities than not, giving rise to the 21st century being christened the 'urban millennium'. By 2030, the world is projected to have 43 mega-cities containing over 10 million inhabitants, most in developing regions. With 73 million residents being added to cities every year it is expected that by 2050 two-thirds will be living in urban areas, and by 2100 up to 85 per cent. (Of course, anything could happen between now and then, forcing a population drop or a rush to re-colonise the rural areas.) "As the world continues to urbanise, sustainable development depends increasingly on the successful management of urban growth, especially in low-income and lower-middle-income countries where the pace of urbanisation is projected to be the fastest," says the United Nations' *2018 Revision of World Urbanization Prospects*.[18]

Harvard University economist Edward Glaeser has made a special study of population growth in relation to mega-cities (usually defined as having a population of at least 10 million), in particular low-income mega-cities. According to him, most countries with a per capita income of less than $1,000 had urbanisation rates of under 10 per cent in 1960, but by 2011 it had risen to 47 per cent. Glaeser concludes that, "The challenge of developing world mega-cities is that poverty and weak governance reduce the ability to address the negative externalities that come with density".[19]

Key to this is lack of data. Administrations do not know who has arrived or where they live. Addresses are often non-existent. Therefore taxes cannot be collected and projects funded. The poor are discounted because their economic value is seen as negligible. This is a catch-22 situation.

"The new, poor mega-cities still grapple with the same adverse urban externalities that have troubled western cities for centuries, such as contagious disease, traffic congestion and crime. They also exhibit the high costs of housing that come from density. Yet they must face these problems with neither economic wealth nor capable government," Glaeser says. He concludes: "The mega-cities of the developing world have significant problems that are impossible to eradicate given the current combination of weak institutions and poverty. Yet, it seems likely that the process of urbanisation itself is the most probable path towards the prosperity and institutional strength that will eventually lead to more liveable cities."

High incomes lead to unsustainable over-consumption. But equally, low incomes are unsustainable.

These are related problems, giving rise to the notion of 'contraction and convergence' of footprints, central to attaining 'one planet' status, and which we'll explore throughout this book. Addressing poverty and income inequality is vital to solving these problems, and education is the main tool to achieving this. Glaeser observes that in countries with average per capita income levels under $5,000 there is now no correlation between governmental effectiveness and urbanisation, unlike in the past in now-developed countries.

The poorest countries in the world, where per capita incomes are under $1,550, populations are over 10 million, and urbanisation is over one-third, are the Democratic Republic of the Congo, Zimbabwe, Mali, Haiti, Pakistan, Senegal and Côte d'Ivoire. Of these, the Democratic Republic of the Congo contains the poorest megacity in the world, Kinshasa, where around 12 million inhabitants have an average GNI each of $430 (2016 World Bank figures). The population density is 20,000/km^2 (51,000/mile^2). UN-Habitat estimates that 390,000 people immigrate to Kinshasa each year, but no one really knows because no one can count them.

Cities come in all types. Whatever happens to the resources, the pollution and waste caused by their consumption is extremely important. For cities now and in the future to be truly sustainable, let alone regenerative, the supply chains which feed them, and the consequences of their consumption, must enhance, not deplete the world's ecosystems.

But how might we measure this? That is the subject of the next two chapters.

Notes

1 *World Population Prospects: The 2017 Revision*, UN Department of Economic and Social Affairs, 2017. See: https://population.un.org/wpp/
2 Grooten, M. and Almond, R.E.A. (Eds), *Living Planet Report – 2018: Aiming Higher*, WWF, Gland, Switzerland. See: www.wwf.org.uk/updates/living-planet-report-2018

3 Rapport, D.J., "Ecological footprints and ecosystem health: Complementary approaches to a sustainable future", *Ecol. Econ.* 2000, 32, 367–383.
4 Raworth, K., *Doughnut Economics: Seven Ways to Think Like a 21st-Century Economist*, Chelsea Green, 2017.
5 Daniel W. O'Neill, Andrew L. Fanning, William F. Lamb and Julia K. Steinberger, "A good life for all within planetary boundaries", *Nature Sustainability*, February 2018, 1, 88–95. See: www.nature.com/natsustain and https://goodlife.leeds.ac.uk/about/
6 Diamandis, P., and Kotler, S., *Abundance: The Future Is Better Than You Think*, Free Press, 2012.
7 Palacios-Huerta, I. (Ed.), *In 100 Years: Leading Economists Predict the Future*, MIT, 2013.
8 The assessment reports are available at: www.ipbes.net
9 See www.climateactiontracker.org
10 Zhao, C., et al., "Temperature increase reduces global yields of major crops in four independent estimates", *PNAS*, 29 August 2017, 114 (35), 9326–9331. See: https://doi.org/10.1073/pnas.1701762114
11 See: www.weforum.org/agenda/2017/05/these-are-the-world-s-most-crowded-cities/
12 See: www.theguardian.com/cities/2018/mar/21/people-pouring-dhaka-bursting-sewers-overpopulation-bangladesh
13 Global Footprint Network (using UN data): globalfootprintnetwork.org
14 Climate Action Tracker: climateactiontracker.org
15 Yunus, M., *A World of Three Zeroes: The new economics of zero poverty, zero unemployment, and zero net carbon emissions*, Scribe, 2017.
16 Helliwell, J., et al., *World Happiness Index 2018*, United Nations.
17 Ibid.
18 See: population.un.org/wup
19 Glaeser, E.L., *A World Of Cities: The Causes and Consequences of Urbanization in Poorer Countries*, Working Paper 19745, National Bureau of Economic Research, Cambridge, MA, USA, 2013. Available at: www.nber.org/papers/w19745.pdf

3

ECOLOGICAL FOOTPRINTING AND OTHER STANDARDS

To build a 'one planet' future we must understand what our goals are, and choose metrics to measure whether they are being achieved. It's vital to identify and implement systems that can keep track of progress, together with strategies to ensure that we stay on track. (Plenty of 'sustainable' projects have been based on optimism at most, and green-washing at the least, with the results not being quite what were hoped for!) So, before delving into examining efforts to work towards 'one planet' living at scale in the subsequent chapters, this and the next chapter attempt to answer the fundamental question:

How do we measure genuine sustainability?

The UN Conference on the Human Environment in Stockholm on 5 June 1972, at which 114 governments were represented, was the first time the concept of planetary limitations was explicitly raised at a global level. The conference's motto was "Only one Earth".[1] At the conference, Swedish Prime Minister Olof Palme criticised industrial countries for causing ecological and economic damage at the expense of developing countries. The conference began a process of environmental policy becoming a universal thread within international diplomacy and the founding of the United Nations Environment Programme (UNEP).

It was another 15 years before the formulation of the definition of sustainability most commonly adopted by governments around the world now. This is derived from the document *Our Common Future*, also known as the Brundtland Report:

Sustainable development is development that meets the needs of the present without compromising the ability of future generations to meet their own needs.

This definition was only formally adopted a further five years later, by the Earth Summit held in Rio de Janeiro in 1992. This momentous event resulted in several agreements.[2] But this form of words just pushes the definition back a stage: how might we *know* whether development is meeting the needs of the present without compromising the ability of future generations to meet their own needs? To answer this question, UNEP was charged in 1992 with finding ways and means to do so.

The Sustainable Development Goals

The Sustainable Development Goals (SDGs) are the latest iteration of that quest. They are a set of 17 ambitious goals to be achieved by 2030. They replace the successful Millennium Development Goals (MDGs) that ended in 2015.

The 17 goals are:

1. No poverty
2. Zero hunger
3. Good health and well-being for people
4. Quality education
5. Gender equality
6. Clean water and sanitation
7. Affordable and clean energy
8. Decent work and economic growth
9. Industry, innovation and infrastructure
10. Reduced inequalities
11. Sustainable cities and communities
12. Responsible consumption and production
13. Climate action
14. Life below water
15. Life on land
16. Peace, justice and strong institutions
17. Partnerships for the goals.

The total number of targets under these headings is 169. Their status is supported by standards and indicators – 232 of them.[3] The useful thing about the indicators is that many of them are quantitative – meaning they rely on objective evidence. Progress to attaining them is coordinated under the Cape Town Global Action Plan for Sustainable Development Data and the UN Data Forum.

Useful and necessary as they are, however, along the road to the agreement upon their definition many of them have lost the link to planetary boundaries. They no longer make explicit reference to the carrying capacity of the Earth.

To understand why, let's examine the goal which attempts to tackle sustainable development from an urban perspective, Sustainable Development Goal 11. This aims to "make cities and human settlements inclusive, safe, resilient and sustainable" by 2030.[4]

26 Ecological footprinting and other standards

Goal 11 has ten aims, with 15 indicators. Examples include:

- The proportion of urban population living in slums, informal settlements or inadequate housing (down from 28 per cent in 2000 to 23 per cent in 2014).
- The proportion of population that has convenient access to public transport, by sex, age and persons with disabilities.
- The ratio of land consumption rate to population growth rate (up from 1.22 between 1990 and 2000 to 1.28 between 2000 and 2015 – i.e. cities are becoming less dense as they grow).
- The proportion of cities with a direct participation structure of civil society in urban planning and management that operate regularly and democratically.
- The proportion of persons victim of physical or sexual harassment, by sex, age, disability status and place of occurrence, in the previous 12 months.

The related Goal 12, to "ensure sustainable consumption and production patterns", contains 11 aims with 13 indicators, such as:

- Material footprint, material footprint per person, and material footprint per gross domestic product (GDP) (up globally from 8 metric tons per person in 2000 to 10.1 metric tons per person in 2010).
- A global food loss index.
- The national recycling rate, tons of material recycled.
- The number of countries implementing sustainable public procurement policies and action plans.

You can see that some of these are about human rights. Others reference environmental impacts. Material footprint and material footprint per person, derived from indicator 12.2.1 by the United Nations Economic and Social Council, is designed to assist in measuring the target to achieve the sustainable management and efficient use of natural resources by accounting for a country's per capita share of global materials extraction (biomass, fossil fuels, and ores) in relation to its demand for them. It is meant to stand as an indicator for the material standard of living/level of capitalisation of an economy. It is not meant to stand as a measure of absolute impact in relation to the planet's carrying capacity, although it can be used to calculate this.

As the UN's rationale says, it needs to be seen alongside the country's domestic material consumption, since together they cover the two aspects of the economy, production and consumption. As we will see in the next chapter, countries can outsource their impacts by importing goods – and therefore the industrial processes and impacts – instead of producing them themselves, so it is possible for them to have a very high material consumption level because they have a large primary production sector for export, or to have a very low level because they have outsourced primary production. The material footprint corrects for both cases by accounting for both imports and domestic production. For the UN's SDGs, the global material flows database is based on country material flow accounts from the European Union and Japan and estimated data for the rest of the world.

In any case, these UN goals are voluntary and non-enforceable. Presently there is much talk about them, but less practical action. This is gradually changing, faster with some goals than others.[5]

As an example of how a goal might relate to planetary boundaries or to remaining within local levels of resource supply, but not yet fully address it, consider SDG 6: "Ensure availability and sustainable management of water and sanitation for all." Its key sub-goal in this regard is: "by 2030 [to] ensure sustainable withdrawals and supply of freshwater." Indicator 6.4.2 is the only one for checking this: "freshwater withdrawal as a proportion of available freshwater resources." This is vague, as these "resources" can change in relation to underground aquifer levels, watercourse supplies and rainfall levels. Many cities are depleting the water levels within their nearby underground natural aquifers. Climate change will see dramatic variations in rainwater and river levels. Many cities and countries have data gaps that are yet to be filled.[6]

The Global Pact for the Environment

So the goals are all very well but there is no necessary requirement for countries to strive to achieve them. Therefore how to make the goals enforceable and legally binding has been the subject of much discussion. This has resulted in a Global Pact for the Environment that is scheduled to be approved by the end of 2020, under a commitment made by France's President Emmanuel Macron in 2017. The current draft contains 26 articles.[7] The first of these states that "Every person has the right to live in an ecologically sound environment adequate for their health, well-being, dignity, culture and fulfilment".

The other articles expand on this and describe monitoring and compliance procedures. The Pact's defining characteristic is that its legal approach is based on human rights law. It recognises the inter-related nature of all the challenges associated with reaching the Sustainable Development Goals. This is because "too often, international environmental law has been unable to evolve because it is too granular, with statutes written with a mind only to specific problems," according to Manuel Pulgar-Vidal, WWF head of climate and energy practice.[8]

The final draft will contain definitions backed up by standards and evidence.

Such a pact would – in principle – be legally binding on nations, the first time such a thing has happened. If it does happen it will be a wonderful thing. It would be a means to ensure the sustainability of local, national and global governance. It would be potentially enforceable, although how is not yet clear.

Speaking at a high-level UN conference on the topic, Professor Jeffrey Sachs, the special adviser to the UN secretary-general on the Sustainable Development Goals, said such a pact "is absolutely reasonable and necessary by the standards of the crises that we face", amongst which he listed the present rate of species extinction – 1,000 times faster than the natural rate. Translation: an emergency.

So, besides referencing planetary limits, goals should ideally be legally binding. The goals and how to achieve them would be defined as a type of standard, with indicators. Before evaluating the ecological footprint indicator, let's discuss what we mean by standards.

The relevance of standards

Indicators define measurements that are used to check progress against standards. Standards are very useful, and they need precise and universally understood definitions in order to be effective.

Internationally agreed standards are contained in documents that describe requirements, specifications, guidelines or characteristics that can be used consistently to ensure that materials, products, processes and services are fit for their purpose.

Frequently, they describe methodologies that can be applied in particular processes that lead to agreed outcomes (goals), and how progress towards those outcomes can be measured. They can and are used as a basis for the development of national and international regulations, and are typically developed with the involvement of governments worldwide, to take into account the needs of policymakers.

To be confident about the success of any project it is ideal to measure progress against pre-set goals and targets from an already established baseline. This process of checking and correction is an example of the use of continuous feedback loops – as frequently found in our daily life, such as in the use of a thermostat to control indoor temperatures.

Standards are amazing things once you focus on them. Every piece of equipment that you use will contain components designed so that, when connected up, they will be compatible, comply with necessary regulations, and be safe and reliable: anything from a battery to a light fitting, and hundreds and thousands of components besides – all because of standards.

For one to be defined and agreed, a technical committee made up of specialists from many different countries often spends years solving the problems they set themselves collectively – an incredible example of international cooperation that happens because it's in everybody's interest to do so.

Sustainability standards

In the sustainability world there is a variety of standards to choose from. While not often pertaining to absolute measures of biocapacity, they can be deployed for particular purposes within the struggle to attain genuine sustainability. The International Organization for Standardization (ISO) offers over 18,600 standards – practical tools for all aspects of sustainable development. Some popular and relevant ones to be considered 'one planet' cities would be:

- Energy Efficiency – 50001
- Environmental Management – 14001

- Life Cycle Analysis – 14040
- Sustainable Cities and Sustainable Communities – 37100.

Most of these and others are discussed below. These standards are thrashed out by technical committees, made up of experts from different countries, who constantly check with their national governments and other academics, as well as representatives of industry, that they are on the right track. The standards are then tested extensively before being adopted. Every few years they are updated. If this process did not happen there would be no buy-in from anyone to use the standards.

ISO 50001

To see how standards can work and how successful they can be, take ISO 50001 *Energy management systems – Requirements with guidance for use*. ISO 50001 constitutes a series of standards and procedures that are used to conduct an energy assessment in a given building, company or industrial facility, and to train personnel in implementing an Energy Management System. It is now widely established, and its use is mandated in the energy policies of many nations around the world by organisations to reduce energy use and carbon emissions.

ISO 50001 was designed "to enable an organisation to establish the systems and processes necessary to improve energy performance, including energy efficiency, use and consumption", and is applicable to all types and sizes of organisations irrespective of other conditions and in all sectors. It will help an organisation reduce the energy component of the ecological footprint.

It dovetails with other management standards such as ISO 9001 (quality management), which many organisations already use, and ISO 14001 (environmental management), which covers pollution and resource efficiency.

My own experience from working with organisations implementing this standard is that success is dependent upon their management giving their support and encouraging buy-in from all employees. All employees are encouraged to come up with ideas to improve efficiency. The standard allows an organisation to be free to set its own objectives, targets and action plans, which is important, as every organisation is different. Most – substantial and initial – savings are made with little or no cost.

Progress to reducing consumption is monitored by metrics that include kilowatt-hours, litres of fuel per week, and output per unit of energy. These can be converted into carbon emission savings, derived from the fuel and electricity source.

Its implementation is based on a Plan-Do-Check-Act continual improvement feedback cycle:

- Plan: conduct an energy review, establish a baseline, policy, targets and action plan;
- Do: implement the energy management action plans in day-to-day operations to reduce energy consumption;

- Check: monitor energy performance and check it against the objectives; report the results;
- Act: use this information iteratively in a process of continual improvement.

Energy savings of 4 per cent to 40 per cent are typical in the first one or two years of adopting an Energy Management System, with little or no capital investment. Having such a system, like having an environmental management scheme, brings many co-benefits, making the organisation more robust.

The process is not dependant on particular skilled employees who might leave the organisation, but works by embedding a set of business processes to ensure the optimum use of available resources, and that all personnel involved have clear roles and responsibilities. This approach also helps companies to identify and create fundable higher capital cost projects to save more energy, and to manage risk.

According to Marco Matteini, a senior officer of the United Nations Industrial Development Organization which runs an Industrial Energy Efficiency Programme in over 18 countries, "an Energy Management System produces significant no and low cost energy savings in its own right, along with the associated multiple non-energy benefits at enterprise, social and national levels, and is also the foundation for unlocking, and optimising, the investment opportunities in industrial energy efficiency".

The advantages of standards

It can be seen how such standards are not prescriptive but supportive. They save an organisation from figuring out everything from scratch, and enable its performance to be compared to that of others. Comparisons must be made against baselines: the starting point, established for each project.

The results can be both absolute and relative. For example, it is now relatively easy to both state the annual carbon emissions of a country or a city (absolute) *and* the percentage improvement on previous years (relative).

Using absolute measurements makes it possible to compare one project with another. For example, the IPCC chose 1990 to be the baseline year from which countries should measure their carbon emission reductions. All claims by countries of emissions reductions are made with reference to the amount of emissions in that year compared to the current year in question, using absolute measurements in units of tonnes of carbon dioxide equivalent greenhouse gases per year.

A measurement of the overall sustainability of a town or city would incorporate this indicator amongst others. One can compare one city's performance against another, just as one can compare the energy performance of a building or the health of its occupants against that of another building.

Standards therefore make it possible to have measurement, goals and verification for particular indicators. It's easy to see how the a Plan-Do-Check-Act continual improvement feedback framework would be applicable for all the other factors of an ecological footprint: water, material goods, energy, food, pollution, bioresources, and waste.

The ecological footprint

The ecological footprint (EF for short) is a convenient overview *indicator*. It captures the impact of our activities on nature (as opposed to the social and economic impacts) in relation to what the planet can provide. This distinguishes it from other uses of the word 'footprint' (such as in 'carbon footprint'), which refer to the impact of an activity, but not in relation to the carrying capacity of the Earth. This concept is fundamental to the concerns of this book.

The following section outlines the history of this indicator, its limitations and usefulness, and the current state of thinking and data supporting or impeding its application.

The concept of an ecological footprint was originally proposed by William Rees, then a professor of planning and resource ecology at the University of British Columbia.[9] He brought the concepts of human carrying capacity and natural capital to develop a framework to evaluate each city's 'ecological footprint', arguing that cities compete with each other for resources all over the world, with wealthy nations grabbing more than their fair share. He passionately believed that conventional economics couldn't even detect, let alone offer, policy advice on this effect, because it was so abstracted from reality.

A not-for-profit consultancy, the Global Footprint Network (GFN), nowadays collects and processes data to make EF calculations.

According to Mathis Wackernagel, its president:

> Ecological Footprint accounting tracks how much nature we use compared to how much nature can regenerate. The demand side – the Ecological Footprint – adds up all the biologically productive land and water areas an individual, population or activity uses to produce all the resources it consumes, to house all its infrastructure, and to absorb its waste given prevailing technology and resource management practices. The availability side – the biocapacity – adds up the biologically productive areas that exist in a region or on the planet.

Ecological footprint is measured in 'global hectares'. A global hectare is a biologically productive hectare that is an average of the different productivities or regeneration rates of the world's productive land area for a given year; cropland is more productive than pasture land, for example. The ecological footprint divides the 'biocapacity' of land (supply) by human consumption levels (demand).

Calculating biocapacity

The component of the footprint that represents the biocapacity supply of land is calculated using a value called its Net Primary Productivity. Since not all land is equal there has to be a way of reconciling this in the accounting system. This is done by the GFN using a ratio system. This ranks "the maximum agricultural potential of land of a specific land-use type (like cropland) and the average productivity of all biologically productive lands on Earth".[10] The accounts obtain this information from the Global Agro-Ecological Zones assessment made by the Food and Agriculture Organization (FAO), a United Nations agency, combined with information about actual areas of cropland, forest and grazing area derived from FAOSTAT, a UN database provided by FAO. This model divides all land globally into five categories, each of which is assigned a suitability score:

- Very Suitable – 0.9
- Suitable – 0.7
- Moderately Suitable – 0.5
- Marginally Suitable – 0.3
- Not Suitable – 0.1

The calculation assumes that the most productive land is put to its economically most productive use. Of course this may not happen in practice. GFN's National Footprint Accounts also employ an 'intertemporal yield factor' that captures the change of productivity over time.

The biocapacity supply is a measure of the pollution land can absorb and the services and resources it can provide to the population under investigation. The demand is a factor of this given population level and its consumption level, which is determined by its use of cropland, forests, grazing land and fishing grounds for providing resources and absorbing carbon dioxide from burning fossil fuels. The result can be an average of hectares per person (or nation), as if it were distributed equally between everyone (in the world or a given country) and as if all land were equal in what it could provide. A hectare is 2.47 acres or 10,000 square metres (m^2) or 0.01 km^2.

The other data sources used by GFN for their accounting purposes are statistics submitted by countries' officials to the UN each year from 1961. GFN calculates the EF of countries annually and publishes them on an open data platform at data.footprintnetwork.org.

According to GFN estimates for 2018, the world average biocapacity was 1.6 global hectares (gha) per person, and the world average EF was 2.8 gha per person. This indicates that demands placed on the nature were 1.75 times higher than what can be renewed. As population and consumption levels climb, the demand for biocapacity increases. Globally, it has grown from 7 billion gha in 1961 to 22 billion gha in 2018.

In 2018, the level of the EF per capita of the world's lowest-consuming countries in Africa and the Indian sub-continent was 1.7 gha. Does this imply that the only way the planet can support humanity sustainably is for everyone to live as the population of these countries currently does? This is the crucial question that this book aims to answer.

The GFN also provides information for WWF's biennial *Living Planet Reports*. Using this process, Mathis Wackernagel has seen that in his lifetime (he is 56) the footprint of humanity's consumption has grown from 70 per cent of what Planet Earth can provide to 164 per cent.

How is it possible to go over 100 per cent? By depleting our natural resources. "Ecological deficits are possible because we can use the global commons, have net imports and over-use our own biocapacity," he says. "Some countries like Sweden have more biocapacity in their country than what it takes to support them. Switzerland uses four times more than what Swiss ecosystems can renew. Egypt uses three Egypts. Because of higher per person income, Switzerland feels the crunch less. If you have more income than the others you can buy from them more than the other way round."

The limit to sustainability

GFN claims the world total EF tells us that, given current population and available land area, any jurisdiction containing an EF below 1.7 gha per person "makes its resource demands globally replicable". Over that figure is unsustainable. It adds that: "The United Nations Human Development Index (HDI) tells us that an HDI higher than 0.7 is considered 'high human development'. Combining these two indicators gives clear minimum conditions for sustainable human development and shows how much more we need to 'think inside the box'."[11]

The HDI combines indices of life expectancy, education and per capita income indicators in order to rank countries in terms of development. It provides a simple indication of progress with respect to the social and well-being aspects of the Sustainable Development Goals.

Some 71 per cent of the world's population currently lives in countries with both an ecological deficit and below average income. Wackernagel says:

> Most people want to increase income. Low-income people typically need more basic material things that make their lives better such as more food, shelter, energy. But if they do not have the resources nor the money to buy them from elsewhere, such conventional development cannot last. Because of math, not everyone can beat the average. Striving for an average that beats what's available on average is just self-defeating. Given this mathematical truth, we need to reduce the average demand – and this implies contraction and convergence. Given our massive overshoot, resource security is becoming a key parameter for lasting economic success. Therefore we need robust resource accounting to help us track what kind of development truly works, and enables scalable solutions.

GFN uses these data to calculate 'Earth Overshoot Day' each year. It defines this as the day on which humanity passes the limits of biocapacity by over-consumption, assuming that at the beginning of each year the slate was wiped clean. This point is reached earlier every year; in 2018 it fell on 1 August, one day earlier than the previous year. Individual countries have different dates: Germany's 2018 Earth Overshoot Day fell on 2 May, and the full range begins with the most unsustainable country – Qatar, whose day fell on 9 February – to the least unsustainable – Vietnam, whose day fell on 21 December.[12]

No country in the world operates completely within the Earth's ability to sustain the same level of lifestyle for everyone; a sobering thought. How can this be reversed? Only by conscious effort, education, political will. This book is motivated by the subject of how to make this effort. I believe and hope that if each community were to set their own targets and design their own pathways, then it could be done. How fast? As an example, Wackernagel suggests that if the world's population were together able to postpone Earth Overshoot Day by 4.5 days each year, then by 2050 we would all live again within the ecological capacities of our planet.

Critiques of ecological footprinting

But is EF a good metric to use? After Rees, the ecological footprint concept was eagerly taken up by a number of countries and campaign groups. WWF adopted it for a while, as did Wales as a country-level indicator of sustainability (see Chapter 16).

At first glance, ecological footprint analysis seems to offer much of what we need, but there are several definitions, and it has come under scrutiny.

In a 2010 review of 50 international users and over 150 original papers on its applications over the previous decade, Thomas Wiedmann and John Barrett of the Centre for Sustainability Accounting in York[13] found that EF has great advantages as a strong communication tool because it is easily grasped and understood.

> The main strengths of the Ecological Footprint – as confirmed by the expert survey – are (a) its ability to condense the size of human pressure on different types of bioproductivity into one single number, (b) the possibility to provide some sense of over-consumption, and (c) the opportunity to communicate results to a wide audience.

Wiedmann and Barrett advise that advances in linking bioproductivity with ecosystem services and biodiversity, and the method of environmentally extended input-output analysis provide a number of advantages for improving EF calculation. The latter method is now being used in Sweden and we shall look at it below. They believe it would be advantageous "to align the Ecological Footprint with the UN System of Environmental and Economic Accounting" (see below), and that it should be "part of a basket of indicators". Although as a single aggregated indicator it is very useful for communication, it does not "provide enough detail to undertake a meaningful assessment of regenerative capacity compared with demand".

Carina Borgstrom-Hansson has worked in the Swedish WWF office for 12 years, during which time she has worked on EF, also with the Stockholm Environment Institute (SEI). She developed a measurement tool and a Swedish version of the EF in 2010, which was used for a while with cities.

She told me that EF is less popular now with WWF

> because of difficulties with keeping the data behind the tool up to date. It is time-consuming and expensive. There are also challenges with application of the model due to its incorporation of different components; besides carbon emissions there are air, water and soil quality and biodiversity. For this reason Sweden narrowed its focus to simply measuring carbon emission impacts as these are updated annually anyway, for reporting purposes under the international climate change agreements. It is now not applying any tools developed by the SEI.

(However, this has severe limitations: see the section on Sweden's PRINCE project in Chapter 4.)

Lucy Young, who advises on the *Living Planet Reports* for WWF, told me:

> I think GFN are doing good work and the ecological footprint makes a useful contribution to the discourse on environmental limits. Being a science-based organisation we consider the robustness of a number of tools and what they bring to the picture. No one metric can tell us all we need to know. Hence the Living Planet Report 2018 used a number of different measures and models to show how humans are impacting biodiversity. And our focus is increasingly on understanding solutions.

She admits the EF is a good communication tool: "We have to get our heads around what we consume because it can have a huge impact in other countries. Anything one can do to encourage that message to be understood is great but we also need to make our tools better to help us find policy-relevant solutions." Carina agreed. "The data they refer to is often limited, depends on what the nations are reporting." But she adds that they are complementary: "We have different roles to play."

A recent WWF discussion paper concluded:

> The EF has cemented its place as a pioneering and important step towards providing a framework and metric for measuring environmental limits. We can expect to see it continue to be used. However, it will be increasingly important to understand what it can and can't do, and how to make the most of it alongside the significant and growing generation of new tools now emerging.[14]

I put these criticisms to Mathis Wackernagel and he responded:

> We do not say the Ecological Footprint measures everything. It just compares human demand against regeneration. It measures a necessary condition for sustainability. Sustainable development is a multifaceted process. It is not possible to represent its multiple features with one indicator alone. The footprint cannot, and never will, indicate sustainability as a whole. It just tracks one key

aspect: how much demand does this development put on ecosystems, and how much can ecosystems regenerate. Globally, human demand has to be less than what the planet's ecosystems can renew.

He agreed that, "Ecological Footprint accounting can be improved, and needs to be complemented with other relevant measures to monitor sustainability. An assessment of the linkages between Ecological Footprint accounting and other measures such as wealth accounting, longevity, or happiness."

He adds that the Global Footprint Network and others have worked on making the footprint more relevant to cities. He also notes that the method and calculations behind the footprint accounting are publicly available.[15] "We make the national templates available for free to academics (we only charge for commercial use). The underlying concept is quite simple: add up all demands on nature that compete for space. Footprint is an aggregator, an interpretation lens." He also emphasises that to calculate the ecological footprint of a product, you need a life-cycle assessment (LCA) first. "With those LCA data points then you can calculate its Footprint."

Wackernagel further defends his version of the EF by arguing that it focuses on biocapacity, which he calls "the mother of all limitations". He says the concept's key selling point is that it forces us to ask: how much nature do we have and how much do we use? "The Swiss government just recalculated independently the GFN's Ecological Footprint numbers.[16] It had been done previously over a dozen times (such reviews are far more rigorous than most 'peer reviews' in science). It concluded that for the time series they examined, they could reproduce the results within 1–3 per cent accuracy."

WWF also states that "accounting for our actions in terms of carbon and footprint reduction, however statistically difficult, should be a pre-requisite of a nation aspiring to one planet living".

It is clear that the EF is a valuable tool, alongside further indicators that are sufficiently mature, objective, transparent, open and verifiable, to match the importance and effectiveness of carbon and energy accounting methodology.

The need for verifiability

This demonstrates how it is vital that these standards and their indicators are both verifiable and transparent. It is also an urgent task to create easily updatable, openly available data streams, and present the data in a way that people can use at all levels, from government downwards. A chosen methodology must be as robust as we have seen ISO 50001 to be.

Let us look at a few candidates.

Carbon footprinting

Carbon footprinting is an important part of the solution. A carbon footprint is the total greenhouse gas (GHG) emissions produced by an individual, event, organisation or product, expressed as carbon dioxide equivalent. It is typically defined as:

> A measure of the total amount of carbon dioxide (CO_2) and methane (CH_4) emissions of a defined population, system or activity, considering all relevant sources, sinks and storage within the spatial and temporal boundary of the population, system or activity of interest. Calculated as carbon dioxide equivalent using the relevant 100-year global warming potential (GWP100).[17]

Greenhouse gases can be emitted through land clearance and the production and consumption of food, fuels, manufactured goods, materials, wood, roads, buildings, transportation and other services. Data collection methodologies for countries are defined under the United Nations' 1996/2006 IPCC Guidelines for National Greenhouse Gas Inventories.

Since nations are required by the United Nations' IPCC to collect and supply this information on an annual basis, it is relatively easy to use this indicator as a proxy for general ecological impact. Some organisations do so, for convenience. But it is a mistake to assume it covers all impacts of human activities. To find out why, read the section in Chapter 4 on Sweden's PRINCE project.

Nevertheless cities and countries should all strive to reduce their carbon footprint to zero or below zero, i.e. to store away atmospheric carbon dioxide in order to reduce the amount in the atmosphere that is causing global warming. (This topic is covered in Chapter 7.)

There are two different ways of measuring the greenhouse gas emissions of a city: you can look at the quantity of the emissions it produces, or the quantity of emissions that results from everything that's consumed there. The Global Protocol for Community-Scale Greenhouse Gas Emission Inventories: An Accounting and Reporting Standard for Cities,[18] or GPC for short, uses the former method.

PAS 2070 is an international standard[19] that contains both methodologies in order to measure a city's carbon footprint. The direct plus supply chain methodology captures territorial GHG emissions and those associated with the largest supply chains serving cities; it is consistent with the GPC. The consumption-based methodology uses input-output modelling to estimate direct and life-cycle GHG emissions for all goods and services consumed by residents of a city. It was sponsored by, amongst others, C40 Cities Climate Leadership Group, New York City, the Greater London Authority and ICLEI – Local Governments for Sustainability. It captures production emissions associated with the largest supply chains serving cities, such as for city infrastructures and from activities within the city boundary and indirect emissions from the consumption of grid-supplied electricity, district heating and/or cooling, transboundary travel, and the supply chains from consumption of key goods and services produced outside the city boundary (e.g. water supply, food, building materials).

In other words emissions are allocated to the final consumers of goods and services, rather than the original producers of those emissions. But it does not assess the impacts of production of goods and services within a city that are exported for consumption outside the city boundary, visitor activities, or services provided to visitors, because this would lead to double accounting (on a global scale, of the emissions both where a product was manufactured and where it was consumed).

Energy intensity

A related concept is energy intensity. This measures productivity in terms of the quantity or value (of anything) produced, divided by the amount of energy needed to produce it. It is typically measured in the context of fossil fuel-sourced energy rather than renewable energy since the purpose of doing so is to reduce carbon emissions.

Countries and industries measure the rate of improvement in their energy efficiency by quantifying their energy intensity. Energy intensity is reduced if more goods and services are produced with less energy. Producing more with less of anything is a way of reducing an ecological footprint. See Chapter 8 on industry for more detail.

LEED-ND

The Leadership in Energy and Environmental Design – Neighborhood Development (LEED-ND) certification programme, currently in version four, has been applied to district-scale projects since 2009. It is the leading certification programme of its type in the USA. It is a standard whose indicators are both qualitative and quantitative: smart location, good design, the reduction in vehicle miles travelled by making jobs and services accessible by foot or public transit, green building and green infrastructure, efficient energy and water use, the protection and conservation of habitat, wetlands, water bodies, and prime agricultural lands. Its checklist criteria are acknowledged to be less rigorous than carbon neutrality, but there is a requirement for district-scale projects to reduce energy and material flows. It does not relate directly to absolute biocapacity.

Water footprint

The water footprint forms part of the ecological footprint, that is the total volume of freshwater consumed by an individual, community or business to produce goods and services, or produced by a business. It is measured in volume consumed (evaporated) and/or polluted over a given period of time. Methodologies for measuring the water footprint are covered by the standards ISO/TR 14073:2017 and BS EN ISO 14046:2016. It includes a water scarcity index that maps demand onto local supply limits.

Material flow accounting

This provides economy-wide data on material use, i.e., the use of resources and raw materials. In effect it is a resource footprint, but does not relate this to absolute biocapacity; additional calculations are needed to achieve this and translate it into an ecological footprint. In combination with the economic activities of a nation, it can be used to determine the material intensity of an economy or production process. The lower its material intensity, the more efficient it is.

For a given city or country, data are typically divided into those materials produced within its boundaries and those which are imported. They include not just the physical materials such as the products of mining operations, but hidden ones including the impact on land, soil water and air quality.

They are used in environmentally extended input-output analysis (see Chapter 4).

The UN System of Environmental Economic Accounting (SEEA)

The SEEA framework is a standard rule book used by all countries, giving rise to standardised data.[20] It provides the data used by the Global Footprint Network to calculate countries' ecological footprints. It follows a similar accounting structure to the well-established System of National Accounts, in order to be compatible with each country's already-collected conventional economic statistics associated with production. To them are added environmental data areas of agriculture, forestry and fisheries, air pollution, energy, environmental activities, ecosystem land accounts, material flows and water. It is flexible enough to be both adaptable to different countries' priorities and policy needs, and to provide a common framework, concepts, terms and definitions.

The SEEA does not itself contain indicators, but provides policymakers with all kinds of information, using which they can choose indicators for their own purposes.

Environmental data come from the SEEA Central Framework. This is the first international standard for environmental-economic accounting, which was developed in 2012 and adopted by the UN Statistical Commission. To this is added an experimental ecosystem accounting system which is evolving in line with developing knowledge in ecosystem accounting; and a set of applications and extensions for decision-makers and policymakers, analysts and researchers. Users are able to drill down to specific, quite detailed topics, and to use the information gained for specific subject or policy areas, and for reporting purposes.

This makes it ideal for our purposes, although to it would need to be added data on consumption and flows. We discuss this more, and how one nation is circumventing this problem, in the next chapter.

Data limitations

Because of the enormous amount of information to be collected from all over the world, the most recent collated information is always about four years old. This means that there is a significant time lag between implementation of a potential policy and beginning to receive feedback on its effectiveness.

Data are also limited to those countries which collect them – typically the developed countries – and not all countries are yet able to collect all of the required information. At the time of writing, the most recent statistics are from 2014, at which time data existed for 54 countries, representing 64 per cent of those countries that responded to the UN's call for information.

In other words, the world has begun a journey to collect and systematise information on how sustainable it is, but the results are still patchy and there is a long way to go.

ISO 14040: Life Cycle Analysis

A life-cycle assessment quantifies and assesses the emissions, resources consumed, and pressures on health and the environment attributed to different products over their entire life-cycle. This should be the big target of measuring the sustainability of any activity or development. It quantifies all physical exchanges with the environment, whether these are inputs (resources, materials, land use and energy), or outputs (emissions to air, water and soil).

Life-cycle assessment is standardised through a range of ISO documents, including ISO 14040:2006 and ISO 14044:2006, which cover principles, framework requirements and guidelines, and ISO/TR 14047:2012 and 14049:2012, which help with applying the earlier standards, the impact assessment and inventory analysis. They describe the definition of the goal and scope of the LCA, the life-cycle inventory analysis (LCI) phase, the impact assessment (LCIA) phase, the interpretation phase, reporting and critical review stage, limitations, and conditions for use of value choices.

The LCA process may be divided into four key steps:

- Identify goal and scope by defining boundaries and the functional unit
- Model the processes and resources involved in the system, collate the life-cycle inventories of these processes and resources, and generate any new inventory required
- Adjust life-cycle impacts in terms of mid points and endpoints
- Evaluate and interpret results and generate the report for decision-making.

Life-cycle assessment is complicated enough for a single product. A building is an assembly of many different products, and a town or city may contain millions. Clearly this approach by itself from the bottom up will be tough, but data behind LCAs are constantly improving.

Genuine Progress Indicator

The Genuine Progress Indicator (GPI) is applicable to existing settlements. In one form (there are several) it uses 26 indicators: seven economic, nine environmental and ten social. It represents an attempt to find an alternative to GDP as a measure of progress. This is because GDP, as an indicator, does not take into account the state of the natural environment or social welfare, from which much economic value is taken. Therefore a nation can both deplete its natural capital such as soil and biodiversity, and have poor welfare and health support systems, but still report a good GDP and therefore seem to be performing well.

It is calculated by adding together the amount of money spent on consumption (GDP), plus assigning values to housework, volunteer work, higher education and infrastructure, and from this deducting the cost of the loss of natural resources such as pollution, soil quality and biodiversity, and the costs of crime, automobile accidents, underemployment, and so on.

Amongst the administrations who have made use of it include the province of Alberta in Canada, the government of Canada, Maryland, Baltimore, Finland and Vauban, Germany, which was a case study in my book *The One Planet Life*.[21]

There have been great hopes for it to become a 'beyond-GDP' indicator to be adopted by nations as giving a more accurate picture of their state of health, but its methodology and data sources are still evolving. There is not yet much evidence of administrations using it to formulate and implement policy. Because GDP still forms a significant component of the calculation, it has been called "an imperfect measure to support a post-growth, post-consumerist narrative".[22] Critics have questioned the practice of putting monetary values on environmental and social factors. Is this really necessary for policy design? It does complement the UN SEEA described above, however.

BS 8904:2011 Guidance for community sustainable development

This provides guidance for community sustainable development and it is intended to offer recommendations and guidance to help communities to improve their sustainability. It is accompanied by a free micro-guide for developing sustainable communities[23] which begins by advising communities to jointly create a shared vision, identifying needs and solutions, following this up with a strategy and plan to implement them. It suffers from vagueness and is principally a community engagement tool.

A companion standard is ISO 37101, 'Sustainable development in communities – Management system for sustainable development – Requirements with guidance for use'. According to John Devaney, who works for the British Standards Institute on sustainability issues, it has not won many supporters. "The idea of applying a management system like this to a city has not appealed to city leaders," he says. "The way they see it, such a process is useful for corporations which are managed in [a] top-down way, but cities are the opposite: cities only control a small amount, and by consensus." Devaney says that "a small club of cities from France, China, Germany, and I think Russia has formed to support use of the standard, but its popularity compared to ISO 37120 is low".

Such a view may seem at odds with the way local government is seen by civilians. City administrations are responsible for much procurement and regulation. Progressive cities like Bristol in England have adopted a way of coordinating the plans and sustainability aspects of different departments and are able to use modern technology to consult with their populations.

The European Union's sustainable towns and cities programme

The 'Aalborg Charter' is possibly the world's oldest urban sustainability initiative, dating from 1994. Over 3,000 local authorities from more than 40 countries signed the Charter, resulting in the largest European movement of its type and starting the European Sustainable Cities and Towns Campaign,[24] formed in 2016. It works with a group of mayors called the Covenant of Mayors, which is allied with the Global Compact of Mayors.

Cities sign up to a list of ten transformational aims, from decarbonising energy systems and reducing overall energy consumption, to action on transport, biodiversity, natural resources, public space, affordable housing, social inclusion, and jobs and the economy. There is a commitment to monitor and document this process and there is an award scheme to recognise good achievements.

The initiative is accompanied by a data portal containing plenty of information about a great many cities in Europe, but the types of data are not yet aligned terribly accurately with the sustainability aims; for example, there are no data on access to public transport, cycling or health facilities, and assessments of flood risk are highly inaccurate.

Associated with this is the Reference Framework for Sustainable Cities,[25] a still-active online toolkit for European local authorities working towards an integrated management approach and another tool to help city leaders develop and implement plans and strategies for attractive and sustainable cities. It is an online European framework of 30 sustainable objectives that supports the delivery of the Leipzig Charter (a successor to the Aalborg Charter), and of the European common vision for sustainable cities. Importantly, it contains a monitoring tool that allows cities to choose which indicators to adopt, and report and share the data at will.

The scheme is managed by the French Ministry of Housing and Sustainable Homes, the Council of European Municipalities and Regions, CEREMA (a public body in support of national and local authorities in the field of sustainable development), and the French network of planning agencies (FNAU). Most of the 50 or so projects which use it are in France, Germany and Spain.

ISO 37120: sustainable development of communities

Objective indicators are also the intention behind ISO 37120 Sustainable Development of Communities: Indicators for City Services and Quality of Life. It is a globally applicable standard created by cities, for cities, building on the work of the Global City Indicators Facility, which worked with over 250 cities worldwide. Published in 2014, it was updated in 2018 to include new indicators, including for culture, urban agriculture and food. However, it does not relate these to biocapacity.

"Currently, nearly a hundred cities have implemented, or are in the process of implementing, the standard, and they have been very vocal in terms of what new indicators they want and need, which we have incorporated into the latest version," according to Bernard Gindroz, chair of ISO/TC 268, the technical committee that developed the standard. The updated version includes more comprehensive indicators on housing, such as vacancy rates and living space sizes, all essential if future cities are to manage growing populations effectively.

ISO 37120 outlines 19 sectors and services provided by a city: economy, education, energy, environment and climate change, finance, governance, health, housing, population and social conditions, recreation, safety, solid waste, sport and culture, telecommunication, transportation, urban/local agriculture and food security, urban planning, wastewater, and water. For each of these, it gives a key indicator to be reported and a profile indicator, plus an assortment of supporting indicators. Of the indicators, these are explicitly about environmental matters:

1. Total residential electrical use per capita (kilowatt hours per year: kWh/year)
2. Energy consumption of public buildings per year (kWh/m^2)
3. Percentage of total energy derived from renewable sources, as a share of the city's total energy consumption
4. Fine particulate matter ($PM_{2.5}$) concentration
5. Particulate matter (PM_{10}) concentration
6. Greenhouse gas emissions measured in tonnes per capita
7. Percentage of city population with regular solid waste collection (residential)
8. Total collected municipal solid waste per capita
9. Percentage of city's solid waste that is recycled
10. Percentage of city population served by wastewater collection
11. Percentage of the city's wastewater that has received no treatment
12. Percentage of the city's wastewater receiving primary treatment
13. Percentage of the city's wastewater receiving secondary treatment
14. Percentage of the city's wastewater receiving tertiary treatment
15. Percentage of city population with potable water supply service
16. Percentage of city population with sustainable access to an improved water source
17. Percentage of population with access to improved sanitation
18. Total domestic water consumption per capita (litres/day).

However, of the ISO 37120 indicators, very few of these are absolute measures that relate to planetary limits; only greenhouse gas emissions measured in tonnes per capita. Thus, it is possible for cities to score 100 per cent on an indicator, such as 'wastewater receiving tertiary treatment', or 'percentage of municipal waste collected', or 'percentage of homes serviced with water', but for this to be done in an unsustainable way: for example, the waste nutrients not recycled in a circular fashion, and the amount of water being extracted exceeding the capacity of sustainable supply. Such cities may therefore be deemed to be sustainable while not being genuinely so.

It is claimed that ISO 37120:2018 can be used by any city, municipality or local government wishing to measure its performance in a comparable and verifiable manner, irrespective of size and location or level of development. Its development period saw input from over 30 countries in the ISO standards development process.

Matthew Lynch, a sustainability professional working with cities and business for ten years, told me that:

> ISO 37120 is a unique city-led standard. Hundreds of cities pooled what they believed were the most-important, common performance metrics that cities have to get right and decided that it makes sense to measure these in a globally comparable way to enable learning and solution sharing. The standard has a sustainability focus but does not address planetary boundaries explicitly. However, this could become part of future development of the standard in future. And it is important to note that ISO 37120 itself says that conformance with this standard does not confer a status with respect to sustainability – only that the city has reported its data in accordance with the definitions and calculation methodologies.

At the time of writing it is about to be joined by two companion standards: ISO 37122: Indicators for Smart Cities, and ISO 37123: Indicators for Resilient Cities. ISO 37106, Sustainable cities and communities – Guidance on establishing smart city operating models for sustainable communities, provides a toolkit of 'smart' practices for managing governance, services, data and systems across the city in a collaborative and digitally enabled way. The standards are part of the ISO 37100 series, which includes ISO 37101, the overarching management systems standard for sustainable development in communities, and ISO 37106:2018, which gives guidance on establishing smart city operating models for sustainable communities.

Resource consumption and the circular economy is becoming an increasingly important focus of cities and their sustainability efforts. A publication by Brussels Capital Region, UNEP, the World Council on City Data and Ecocity Builders explores how ISO 37120 indicators can play a role in providing key performance metrics to drive resource efficiency and the transition towards circularity. In this area, Matthew Lynch noted: "The transition to a circular economy presents complex measurement challenges for cities, but these challenges are important and incredibly urgent to address. ISO 37120 can build a foundation to bring cities along the road to producing high quality data that is hopefully a stepping stone to putting good tools in place to make the requisite deep reductions in resource consumption at the local level".

The European Union's Seventh Environment Action Programme

The European Commission monitors progress on the continent in support of its environmental actions using the European Environment Agency's indicators on the state of the environment as well as indicators used to monitor progress in achieving existing environment and climate-related legislation and targets such as the climate and energy targets, biodiversity targets and resource efficiency milestones.

These correspond to aspects of the programme's three thematic priority objectives:

1. To protect, conserve and enhance the Union's natural capital (nine indicators to do with protecting biodiversity)
2. To turn the Union into a resource-efficient, green and competitive low-carbon economy (13 indicators to do with waste minimisation and greenhouse gas emissions)
3. To safeguard the Union's citizens from environment-related pressures and risks to health and well-being (seven indicators to do with pollution).

Each year progress against these indicators is registered on a scorecard. It shows whether the trends are improving, deteriorating or stable, and whether or not the set targets for each indicator are to be met.

The Welsh Government

In a similar way, as part of its work towards its Well-Being of Future Generations Act, the Welsh Government has placed ecological footprinting as one of five overarching indicators for sustainable development, under which more specific indicators can sit:

1. Economic output – gross value added
2. Social justice – percentage of the population in relative low-income households
3. Biodiversity conservation – status of priority species and habitats
4. Ecological footprint – national EF against the UK and global average
5. Well-being – a standard set of 36 health questions which ask respondents about their own perception of their physical and mental health.

The aim of the process is to establish a quantifiable link between human consumption, ecosystem functioning and bioproductivity, that is as detailed as possible. See Chapter 16 for more detail.

How popular are standards?

Since standards are usually voluntary, though in some countries encouraged by tax incentives, they are not as widely adopted as they should be. The degree of adoption tends to correspond to the length of time they have been in use.

By the end of 2017 the number of certifications to ISO 50001 (energy efficiency) had reached nearly 23,000, a 13 per cent increase from 2016, six years after its introduction, but the rate of increase had slowed down in that year, attributed to the fact that it was being reviewed. An updated version (ISO 50001:2018) has since been published. The single biggest adopter country by far was Germany, with the rest of Europe comprising around 80 per cent of the remainder. In China in 2017, the number of ISO 50001 certifications grew by over 40 per cent, making China the fourth highest globally and the largest outside Europe. But a Chinese national energy management standard (GB/T 23331) also exists, to which there were in total 2,552 certifications by the end of 2017.[26]

For comparison, ten years after introduction, 60,000 ISO 14001 (environmental management) certifications and over 120,000 ISO 9001 (quality management) certifications had been made. China is responsible for the largest percentage of these.[27]

Obviously, in comparison with the number of businesses and organisations, globally, these standards have been adopted by the tiniest fraction, yet those organisations which have done so, have seen their efficiency greatly increase.

ISO 37120 is showing rapid uptake by cities, with more than 60 cities globally having been awarded the World Council on City Data (WCCD) ISO 37120 Certification by 2018 with a further 40 in progress. According to Matthew Lynch:

> WCCD ISO 37120 Certification is incredibly popular with cities because it provides independently-verified, globally comparable data for cities to promote their successes and learn from others. It is also crucial for economic development and investment attractiveness because it measures key attributes that sophisticated investors use for making their investment location decisions – such as a skilled workforce, good public transportation, clean air and good public safety. This was evident in the recent competition for hosting the second Amazon Headquarters in North America.

Choosing which standards to use

We can draw several conclusions from our overview of these standards: they vary according to the problem being addressed; problems with data collection and methodology exist which are still being dealt with; and many of them use different indicator sets, and definitions.

The British Standard Institute's John Devaney is part of the team responsible for sustainability standards. He says:

> What option is chosen depends on what suits a city and/or what they are most comfortable with. Process standards can be more powerful, and help develop strategy, vision, objectives and targets, but they take commitment and resources. Reporting standards give a quick indication of how your city is doing against a raft of issues that are commonly agreed to be important, and they allow ranking of city performance.

But there is a bottom line, represented by what is meant by 'one planet'. It means keeping within the capacity of what the planet's ecosystems can renew. Genuine sustainability requires humanity's demand on nature to stay within this. The main criteria for the 'one planet' city are to explicitly refer to local and planetary boundaries so as not to make undue demands upon nature, and, if possible, to pay back more than they take – replenishing resources in a fair and safe manner. The next chapter looks in more detail at how some cities and communities are taking this forward, and measuring their progress.

Notes

1 *Report of the World Commission on Environment and Development: Our Common Future*, Gro Harlem Brundtland, Oslo 1987. See: www.un-documents.net/our-common-future.pdf
2 Agenda 21, the Rio Declaration on Environment and Development, the Statement of Forest Principles, the United Nations Framework Convention on Climate Change, and the United Nations Convention on Biological Diversity.
3 See: https://unstats.un.org/sdgs/
4 The cross-cutting nature of urban issues impacts other Sustainable Development Goals too, including 1, 6, 7, 8, 9, 12, 15 and 17. UN-Habitat's New Urban Agenda, adopted as the outcome document from the Habitat III Conference in 2016, seeks to offer national and local guidelines on the growth and development of cities through to 2036.
5 Relevant Sustainable Development Goals: in particular SDG 11 for Sustainable Cities and Communities has the following targets set for 2030: ensure access for all to adequate, safe and affordable housing and transport systems; safeguard natural heritage; reduce the adverse per capita environmental impact of air quality and municipal and other waste management; provide universal access to safe, inclusive and accessible, green and public spaces; support positive economic, social and environmental links between urban, peri-urban and rural areas by strengthening national and regional development planning; by 2020, substantially increase the number of human settlements adopting and implementing integrated policies and plans towards inclusion, resource efficiency, mitigation and adaptation to climate change. SDG 12 on resource consumption has the targets for 2030 to: achieve the sustainable management and efficient use of natural resources; halve per capita food waste at the retail and consumer levels and reduce food losses along production and supply chains, including post-harvest; substantially reduce waste generation through prevention, reduction, recycling and reuse; ensure that people everywhere have the relevant information and awareness for sustainable development and lifestyles in harmony with nature. SDG 13 on climate change has a target to "integrate climate change measures into national policies, strategies and planning". SDG 7 on energy has the targets for 2030 to: increase substantially the share of renewable energy; double the global rate of improvement in energy efficiency. SDG 15 on land has the targets for 2020 to: ensure the conservation, restoration and sustainable use of ecosystems and their services, in particular forests, wetlands, mountains and drylands; sustainable management of all types of forests, halt deforestation, restore degraded forests and substantially increase afforestation and reforestation; significantly reduce the impact of invasive alien species; integrate ecosystem and biodiversity values into national and local planning, development processes; take urgent and significant action to reduce the degradation of natural habitats, halt the loss of biodiversity; significantly increase financial resources to conserve and sustainably use biodiversity and ecosystems. By 2030: restore degraded land and soil.
6 Chen, R., et al., *Tackling the Challenges of SDG Monitoring: A Roadmap Outlining the Costs and Value of a Water Sector Monitoring System*, WHO/UNICEF, July 2015. See: https://sustainabledevelopment.un.org/content/documents/2086Tackling%20the%20challenges%20of%20SDG%20Monitoring.pdf
7 See: https://unstats.un.org/sdgs/
8 A summary of the conference is at http://ccsi.columbia.edu/2017/09/26/world-leaders-discuss-a-global-pact-for-the-environment/
9 Rees, W., "Ecological footprints and appropriated carrying capacity: What urban economics leaves out", *Environ. Urban*, 1 October 1992, 4 (2), 121–130. See: https://doi.org/10.1177/095624789200400212

10 Wackernagel, M., Galli, A., Hanscom, L., Lin, D., Mailhes, L., and Drummond, T., "Ecological Footprint Accounts: Criticisms and Applications", pp. 521–539 in *Routledge Handbook of Sustainability Indicators*, Simon Bell and Stephen Morse (Eds), Routledge, 2018.
11 Global Footprint Network, *Measuring Sustainable Development*. See: http://data.footprint network.org/#/sustainableDevelopment?cn=all&type=earth&yr=2014, where you can interact with a chart displaying the performance of all nations on the HDI with respect to the EF.
12 See: www.footprintnetwork.org
13 Wiedmann, T. and Barrett, J., "A Review of the Ecological Footprint Indicator – Perceptions and Methods", *Sustainability*, 2010, 2, 1645–1693; doi:10.3390/su2061645
14 Leyden, R., *WWF – Metrics for Environmental Limits*, internal discussion paper, 2014.
15 See: www.footprintnetwork.org. All the data are downloadable, plus the assumptions. There is also a personal footprint calculator.
16 Environmental Footprints of Switzerland, Federal Office for the Environment. See: www.bafu.admin.ch/bafu/en/home/topics/economy-consumption/economy-and-con sumption-publications/publications-economy-and-consumption/environmental-footp rints-of-switzerland.html
17 Wright, L., Kemp, S. and Williams, I., "'Carbon footprinting': Towards a universally accepted definition", *Carbon Management*, 2011, 2 (1), 61–72. doi:10.4155/CMT.10.39
18 Fong, W., Sotos, M., Doust, M., et al., *The Greenhouse Gas Protocol*, World Resources Institute, 2014. See: https://assets.locomotive.works/sites/5ab410c8a2f42204838f797e/content_entry5ab410fb74c4833febe6c81a/5ab4110fa2f42204838f79ba/files/GPC_Sta ndard.pdf?1526450598
19 See: http://shop.bsigroup.com/Browse-By-Subject/Environmental-Management-a nd-Sustainability/PAS-2070-2013/
20 Available at: https://seea.un.org/
21 Thorpe, D., *The One Planet Life*, Routledge, 2014.
22 Hayden, A. and Wilson, J., "Taking the First Steps beyond GDP: Maryland's Experience in Measuring 'Genuine Progress'", *Sustainability*, 2018, 10 (462). doi:10.3390/su10020
23 See: http://shop.bsigroup.com/upload/265921/BSI_BS8904_Sustainable_Communities _Guidance_A48pp_AW_LR.pdf
24 See: www.sustainablecities.eu
25 See: rfsc.eu
26 CNCA (Certification and Accreditation Administration of the People's Republic of China), "National Certification and Accreditation Administration announced the 2017 quality management system certification and energy management system: Announcement of certification supervision and inspection results", CNCA, Beijing, 2018. See: www.cnca. gov.cn/xxgk/ggxx/2018/201801/t20180118_56141.shtml (accessed 10 May 2018).
27 *Market Report Series: Energy Efficiency 2018, Analysis and Outlooks to 2040*, IEA, 2018.

4

CHOOSING WHICH STANDARDS TO USE

The last chapter gave an overview of some approaches to account for the impact of human activities on the world that is our home. Let's now look at which ones cities and some countries are actually using, whether they are effective, and which could be useful for governments in transitioning towards a 'one planet' world.

One size doesn't fit all

For city officials new to this field, the multitude of competing standards is confusing. For example, the Global Partnership for Sustainable Development Data collects data relevant to evaluating progress to the United Nations Sustainable Development Goals. Meanwhile UN-Habitat, a United Nations body specifically addressing issues faced by cities, collects another batch of statistics for the City Prosperity Initiative that is about making cities more prosperous; this is being used by about 300 cities around the world. Meanwhile the World Council on City Data and Global City Indicators Facility coordinates data for the smart city agenda – the technological drive to digitise city services.

Since cities come in all shapes and sizes, the biggest or most prosperous will have very little in common with smaller, or more impoverished ones. Each may have different priorities and different data requirements and collection abilities, not to mention access to expertise and finance. The many networks to which cities can belong each offer means to attract expertise and finance, but these may have strings attached. Ultimately, cities will choose whatever means are appropriate for the task at hand, but for the sake of future generations they must refer to planetary boundaries and assist administrations in charting a course to 'one planet' cities that meet citizens' essential needs within both local and global biocapacities.

Carbon accounting

To begin with, let's examine the greenhouse gas emissions of cities, a large part of their ecological footprint. The organisation C40 Cities has calculated emissions of its member cities[1] and found that on average they prove to be 60 per cent higher when calculated based on consumption of goods and services than on production of emissions. It used the standard PAS 2070. The vast majority of C40's relatively wealthy member cities – 80 per cent (63 out of a total of 79) – have larger consumption-based greenhouse gas emissions than production-based emissions. Only for 16 member cities – mostly in South and West Asia, South-East Asia and Africa – is the reverse true. This is what you would expect because they consume less per inhabitant (average wages are low) and are highly industrialised.

Another study comparing 13,000 cities, published in 2018, corroborates this.[2] Again using consumption-based analysis, it found that most of the members of C40's club have the highest emissions in the world ... as do China's large industrial cities which make the things that everybody else buys. This study uses income as a proxy for emissions.[3] It finds that carbon footprints are highly concentrated: just 100 cities drive 18 per cent of all global emissions.

For example, London, Manchester and Birmingham combined contribute more than 20 per cent of the entire UK's output; and Chicago, New York and Los Angeles combine to account for nearly 10 per cent of the USA's overall footprint. The website ranks all 13,000 cities.[4] The top five emitters are Guangzhou, New York, Hong Kong, Los Angeles and Shanghai.

What should city leaders do about this? The article's researchers recommend that state and local governments wanting to reduce emissions can benefit by better understanding the distribution and drivers of their carbon footprints to identify their hotspots of consumption and production emissions, to "isolate carbon intensive nodes".

The C40 report recommends that leaders in these cities use consumption-based greenhouse gas inventories *and* their sector-based greenhouse gas inventories to help them better consider how to reduce overall emissions. Its co-author, Michael Doust, who is the programme director for C40 Cities, says: "We're missing the other side of the coin if we only measure emissions involved in the production of food, energy, or other products and services. Knowing what the consumption emissions are and where allows cities and residents to make better decisions on how to reduce their carbon emissions."[5]

So, as Dan O'Neill discovered, what we're seeing is that wealthy cities like London, Paris, New York, Toronto and Sydney that no longer have large industrial sectors can claim to have significantly reduced their *local* emissions, but their *actual* emissions have risen substantially, with their citizens' personal carbon footprints larger than most people's in the world because of their high levels of consumption.

Quantifying the metabolism of cities using material flows

To tackle the impacts of consumption beyond just carbon, and to account for the entire ecological footprint, we need to identify these impacts and compare them to the available biocapacity. A new tactic aimed at doing this is being piloted around the world and could in the future help to control resource flows through cities to increase their regenerative capacity and reduce their negative impacts.

This tactic uses both citizen engagement and open source data to monitor resource flows, to assist in the design of urban policy, and to facilitate training, all to help citizens improve the efficiency of their use of resources. Areas such as Vauban in Freiberg, Germany, where these frameworks have been used as a part of the integrated design of sustainable neighbourhoods, have achieved substantial reductions in energy intensity compared to conventional designs.

Like the human body, cities are living, ever-evolving organisms. They have their own metabolisms, converting inputs of energy and material goods into work and, in doing so, creating waste and heat. They maintain their state of high complexity in the same way as living systems, by causing a larger increase in the entropy of their environments. In thermodynamic terms, metabolism maintains order by creating disorder. The amount of disorder is a measurement of their efficiency and sustainability. It follows that applying a framework to successfully model urban system flows will help with understanding the relationship between human activities and the natural environment.

Urban metabolism was first defined as "the sum total of the technical and socio-economic processes that occur in cities, resulting in growth, production of energy and elimination of waste".[6] Different ways of obtaining the information to perform the modelling have been tried out: getting households to measure their consumption; adopting a more top-down approach based on utility and waste information; and combinations of these. The resulting flows of energy and materials can then be turned into network diagrams, also known as Sankey diagrams.[7]

In the City of Vancouver, Canada, Sankey diagrams featuring energy sources, sinks, converters and demands, have been found useful for helping planners to identify the best strategy to move toward more renewable energy sources and energy efficiencies. They've also been applied to water and waste.

Most cities' diagrams would nowadays be linear, because resources are imported and then discarded when used – a wasteful and destructive practice. To make cities more sustainable, more loops are needed, involving the reuse and recycling of materials and energy, turning wastes into new inputs for new processes, and reducing cities' dependence on imports for resources.

Different methods have been offered to quantify resource flows: accounting approaches; input-output analysis; ecological footprint analysis; life-cycle analysis; and simulation methods. But as this is a young discipline there is as yet no consensus as to which is the best one. There presently is a lack of standardisation of methods and few guidelines on how to design a sustainable urban metabolism.

Last year UNEP published an assessment of the discipline[8] and offered its own advice, amongst which were that:

- All cities should assess their basic urban metabolism;
- They should use both top-down and bottom-up approaches to capture data;

- They should link spatial and temporal issues in urban metabolism assessments;
- Adopt multiple scales of analysis from household level upwards;
- And involve as many people as possible.

EcoCity Builders

This approach has been taken down to the neighbourhood level by a non-governmental organisation (NGO) called EcoCity Builders[9] who have successfully implemented it in Cusco, Peru, and Medellin, Colombia.

Medellin is known for being progressive; it was named the most innovative city in the world in 2013 by the Urban Land Institute due to its advances in politics, education, and social development. EcoCity Builders' activities involved citizens and city authorities charting their own material flows using techniques they had piloted with students in Cairo, Casablanca, and Lima, using an open source web app. Participants went on neighbourhood scoping trips before consulting with community leaders. They discovered that almost half of household waste in the city was organic. As a result they decided to research methods of constructing home composting modules, which they co-designed with community members and piloted in four communities.

Similar results were found in Bogotá, but also the discovery that the city's proposed formal recycling plan would produce more emissions than the current informal recycling system, which depends on the residents of informal settlements in the 'hidden economy'. As a result, city government, local universities, and community groups in Cusco and Medellin began working together to develop Neighbourhood Sustainability Plans.

Urban metabolism has also been considered in the United Kingdom. A 2015 review for the UK Government's Foresight Future of Cities Project recommended that "it is essential to assess these flows on a full life-cycle basis, in other words to take account of effects that ripple up and down supply chains whether they occur within or outside the city boundary". Material and energy flows depend on the structure and topography of a city; on the quality of its buildings and infrastructure; and on its social structure and the behaviour of its inhabitants. Input-output models have been used, for example, to assess urban-scale direct and indirect water consumption throughout the UK.

Around the world, generally, urban metabolism studies show a higher consumption of food and goods in the city centres and higher consumption of fossil fuels and construction materials in the sprawling suburbs. This should alert planning policy designers to curtail such activity and adopt a more sustainable strategic urban plan. On the energy front, heat dissipation – due to waste incineration as the predominant solid waste management strategy – is the largest contributor to energy loss. UNEP recommends capturing and reusing this heat for local electricity generation or heating.

Young cities tend to have larger inputs than outputs, as they are growing and need resources to build stocks (e.g. buildings and infrastructures); over time, this accumulation of built stocks contributes to waste flows, as it breaks down. As cities mature, the inputs and outputs become more similar in magnitude, and reuse and recycling initiatives have more impact on the city's metabolic intensity.

The greatest gap in mainstreaming urban metabolism, particularly for cities that are growing rapidly and are still building, or have yet to build, the infrastructures needed to support their populations and industries, is the need to link studies and action to public policies. This represents a significant opportunity for cities in the developing world to leapfrog the unsustainable infrastructure systems associated with cities of the global North, and instead embed sustainable infrastructures.

Kirstin Miller is an executive director of EcoCity Builders, which began in Berkeley, California in the early 1990s. She told me:

> We work in Morocco, Egypt, Abu Dhabi, Peru, Argentina and Canada. We have these tools we have been organising in a programme called Urbinsight which we run in partnership with universities and NGOs around the world. Fieldwork and training take participants through the process of data collection from providers and by audits and surveys, with the PAR framework, that can be customised according to users, the level of detail, software or social science aspects.

PAR, or Participatory Action Research, is a system of working around the structural barriers that are often found in trying to obtain information about material flows. By working at a grass-roots level it also provides opportunities for communities to directly lead the research process, thereby leading to community-generated solutions in urban planning and public policy.

Personally I've noticed that wherever grass-roots, community-level opinions and desires are listened to and are matched by the aspirations of top-level administrators and leaders, success is almost certainly guaranteed. One without the other will usually fail. If you look at all the successful projects from neighbourhood to city scale around the world, this is what you see.

Miller continues: "Following the processing of this data there is an implementation phase. So far we have conducted them in pilot cities in Latin America, working with UNEP, using research to implement replicable plans at neighbourhood level with a goal to keep improving it. An app is being developed to map this spatially using GIS [geographic information system]."

Using this method it is possible to find out how a typical household is using resources over a time period, including seasonal differences, to identify the pathways taken by the resources – metabolic flows – and the differences between districts. "Many cities don't have any idea at that level," Miller says. "It creates an iterative process. Others can watch … and say, 'let's try it in our neighbourhood'. In Cusco we started working in a historic neighbourhood downtown, then another area started working with it, then they decided to try it in a French area in a hybrid planned/unplanned rapidly urbanising area, and through another iteration repeated it in another area."

It sounds labour-intensive, but if done with volunteers and students this can cut costs. Besides, many neighbourhoods are similar. "There are neighbourhood typologies in a city, so if you can't audit everyone you can get a pretty finely detailed neighbourhood by analysing some people's flows," Miller says. "There will be some hybrid neighbourhoods with different typologies and also informal neighbourhoods. We work with eco-city footprint folks, and an expert called Jenny Moore, to bring specific data and convert it to a footprint. We can translate this data into global hectares or into metric tons of very specific amounts, that are local flows, tracking what is coming in to a city or household. This is not necessarily translated into a 'global fair share' but is one of our indicators in global carrying capacity."

Miller has conducted fieldwork in most of the cities named above plus others in South America. "We have worked with UNEP and WCCD to think about how they can turn around and regionalise economies in alignment with the anti-globalisation movement. The European Union didn't use all the Urbinsight tools but did use the metabolic tools, to work with water and building sectors."

When I spoke to her she was about to leave for Kathmandu.

> We are going to reintroduce materials we developed before, with a mandate that is written into Nepal's new constitution for eco-cities. In Nepal we have developed long-term partnerships with organisations; with one which oversees the work, and with the local governments from the three kingdoms. Their Institute of Technology has been bringing information and training that we've developed in the past into graduate streams. Now they have a working national government and they want further training.

The above cities are in developing countries, but the Urbinsight tools have been used in developed cities too, including Vancouver and Brussels.[10] The latter project focuses on the opportunities available to the Brussels Capital Region along its pathway to a circular economy, seeking to illuminate the question: "How can an approach that is both predictive and measurable be established and maintained?" It has two objectives: to connect Brussels' work on developing a local, circular economy to global environmental goals; and to provide recommendations (including a set of proposed indicators) to help Brussels monitor its progress along the way.

The project is a collaboration between the Brussels Capital Region and the UNEP-led 'Global Initiative for Resource Efficient Cities', in which EcoCity Builders and the WCCD are partners. The online portal includes an interactive map (see the above footnote), with layered open data, and an Urban Metabolism Information System that currently includes flow data on minerals, metals and timber.

Tackling globalisation

Miller mentioned globalisation. Globalisation has resulted in goods often being transported thousands of kilometres away to their customers, with consequences that would seem absurd to an observer from space. Many countries export and import the same commodities. For example, France in 2016 imported just over US $12 billion of medicaments and exported $15.7 billion of the same. It also exported just over half the value of petroleum oils that it imported.

Wales exports sheep, of which it hosts a greater number than its human population, yet at the same time imports them from New Zealand on the opposite side of the world. What is the benefit of exporting a product only to import it again from elsewhere? The transport component of moving such goods around the world is high, but at the same time the effect on the local economy – of cash leaving it for another country, thereby losing the 'multiplier effect' benefit (see below) – may not be balanced by profits from exports.

To find out how to minimise these types of emissions and other negative impacts, cities and countries would need first to obtain more information about the life-cycle impacts of the important products and services.

The PRINCE project – measuring impacts in Sweden

Several countries are attempting this task of tracking the total environmental pressures linked to consumption in a national economy as a first step in seeking to reduce them. One of these is Sweden, long a pioneer in environmental issues.

Its PRINCE project[11] is developing a system of macro-level consumption-based indicators for the country to monitor progress towards Sweden's Generational Goal. This is: to pass to the next generation a society in which the country's major environmental problems have been solved, without increasing environmental and health problems outside its borders.

PRINCE stands for 'Policy-Relevant Indicators for National Consumption and Environment'. These indicators include greenhouse gas emissions, hazardous chemicals, agriculture and other land-based production such as nitrogen and phosphorus pollution, emissions from land use change, water use and pollution, and fish.

The indicators were chosen to be compatible with Sweden's national statistics, and easily updatable on a regular basis without substantial new research each time, using the latest techniques. They reveal the estimated environmental pressures – along the whole supply chain of each product group – associated with the production of goods and services consumed in Sweden, in order to make it possible to identify, at a broad scale, the categories of products with the largest 'footprints' for the different environmental indicators ('product group hotspots'), allocated to different types of consumption – household consumption, government consumption or investments.

For example, phosphates mined in Morocco might be imported to Latvia for production of fertiliser, applied to vegetable crops in Germany, that are then consumed in Sweden. The PRINCE indicators would capture the related environmental pressures, and allocate them to each country or region.

What has PRINCE found? Although still at an early stage, it has so far discovered that:

- By far the largest share of the greenhouse gas footprint of Swedish consumption falls outside Sweden. Those emissions from Sweden are primarily linked to construction, food products, electricity and heat services, as well as direct emissions from households due to the use of vehicle fuel and other fossil fuels. Another large share is emitted in Russia – notably due to production of the fossil fuels Sweden imports.
- In 2010–2014 Swedish consumption was associated with the loss of around 7,300 hectares of tropical forest every year (averaged). Nearly half of this deforestation occurred in Latin America, the remainder in Asia and Africa. The resulting carbon emissions amounted to 3.9 metric tons (Mt) of CO_2/year, a substantial contribution to the overall carbon footprint of Swedish food consumption.
- 14 per cent (12 Mt CO_2) of Sweden's total CO_2 footprint came from transport activities in 2011, and most of these emissions – 7.8 Mt – were linked to transport activities embodied in goods consumed in Sweden, rather than the direct consumption of transport services in Sweden (4.2 Mt).
- For the water footprint, food and agricultural products have the largest water consumption footprints, followed by household use and textiles. The largest shares of the footprint lie in Sweden, followed by China and the rest of Asia and the Pacific, and then Spain and the rest of the Middle East.

How will this information be used? In a number of ways: national government, the European Union (EU), or an industry association or federation might use it to identify why the hotspot is appearing and what could be done to reduce the pressures or mitigate the impacts.

Policy options that could come out of this include campaigns targeting consumption patterns in Sweden, new or better-enforced environmental standards on products, duties and subsidies, clauses in trade agreements, development assistance, technology transfer, or regulation of production in Sweden. Such measures could be at national level, bilateral or through organisations like the EU, the Organisation for Economic Co-operation and Development or the United Nations.

Regular updating of the indicators could reveal emerging hotspots, as well as cases where action targeting a hotspot has led to positive results.

For me, the PRINCE hotspot results hold two other important lessons: firstly, it's not enough just to track carbon – it cannot be a proxy for the whole environmental footprint. Each environmental macro-indicator follows a quite different pattern. It may have origins in different parts of the world and be associated with different product groups, suggesting that more specific, targeted enquiries will be needed to obtain feedback and reduce these separate footprints of consumption.

Secondly, its approach represents a useful tool to help the drive to reduce the impacts of consumption to 'one planet' levels, as long as it draws upon accurate and timely specific ecosystem service data sources.

Sweden is one of the least sustainable societies in the world because of the impact of its high consumption levels, but at least it is under no illusions and is beginning to do something about it.

Extended input-output modelling

This is how the PRINCE system chose the ecosystem service data sources it is based upon. It begins with extended input-output (IO) modelling. This is judged to be a practical and cost-effective way to create consumption-based environmental accounts at a national (or regional) scale because much of the data is already collected in many countries at national level – at least aggregated to the level of industrial sectors.

The PRINCE system uses multi-regional input-output (MRIO) models to accommodate complex supply chains. These collect three classes of information: the quantity and type of goods and services consumed; their place of origin – including the supply chains right back to the production of raw materials and other inputs in many regions; and their environmental impacts, including transport. As one review of ecological footprint indicators concluded:

> The main advantage of input-output analysis lies in its unambiguous and consistent (SEEA-compatible) accounting of all upstream life-cycle impacts and the good availability of expenditure data that allow a fine spatial, temporal and socio-economic breakdown of consumption footprints.[12]

The SEEA is the UN System of Environmental Economic Accounting (see page 39).[13] Most nations collect data showing the environmental impacts of domestic industrial and commercial production. To these data can be added the environmental impacts of goods and services imported from other countries, as detailed below.

This produces what is called an 'environmentally extended MRIO' (EEMRIO) model. The model reveals comparisons of environmental impacts between different countries of origin for the same goods and services, reflecting the differences in national environmental policies, practices and technologies. The importing nation can then in principle choose the points of origin with the least environmental impact. This model does not show social impacts, but it could be extended to do so.

Because the most closely monitored impacts around the world are to do with climate change – emissions of greenhouse gases – these data are much easier, and therefore cheaper, to collate than other impacts. This has affected the development and choice of models that assess ecological footprints in most countries that have evaluated them.

The team evaluating the best model to use in Sweden, that incorporates that country's already very detailed national accounts for consumption-based monitoring, decided to compare the frequency with which the data sources are likely to be updated, their compatibility with the national accounts in terms of economic structure (industrial and service sector breakdown), and the degree of granularity of the country/regional-level data.[14]

The team judged that the easiest method would be to use the data available from the origin countries on the environmental impacts of imports and then adapt these data to the conventions of their own EEMRIO analysis, based on a model called

EXIOBASE v3.3. This is one of the most comprehensive such databases available.[15]

In many developed countries the production of complex commodities such as equipment and vehicles often involves the manufacture of components, which are exported, combined with other components, and re-imported. An input-output model needs to account for these. In Sweden's case it was decided to deal with this using data from the EEMRIO component, rather than Sweden's own national statistics, even though the latter may yield a more accurate result, because in an economy like Sweden's, the environmental impacts of such supply chains are small, and would not have a significant impact on the results. The final model chosen was, on balance, simpler and more cost-effective to maintain.

Different countries may have different solutions depending on their economies. Whichever is chosen, it permits the comparison of the global environmental impacts of countries and the chance to explore how to reduce them.

MRIO databases

Besides EXIOBASE there are other MRIO databases: Eora,[16] the Global Trade Analysis Project[17] and the World Input-Output Database.[18] But only EXIOBASE has a clear environmental and resource focus and much detail in primary production. Comprehensive surveys of impacts look at these other databases and compare them, although this may be time-consuming. Taking all these together allows users to account for emissions, employment, energy, resource and water use.

EXIOBASE 3 provides the level of detail necessary for a macro-level policy screening of policies designed to control the environmental impact of consumption. It holds data for 200 products and 163 industries from 44 countries (28 EU members plus 16 major economies). Future work will increase the number of countries covered. It uses the accounting principles proposed in the SEEA manual.

This itself follows a similar accounting structure to the System of National Accounts,[19] an internationally agreed standard set of recommendations on how to compile measures of economic activity. They are also related to indicators used to monitor progress under the Sustainable Development Goals. These accounting principles therefore permit the integration of environmental and economic statistics. They can be adapted to countries' priorities and policy needs while at the same time providing a common framework, concepts, terms and definitions.[20] EXIOBASE also allows for building scenarios with a detailed modelling of consumer behaviour.

An earlier version of EXIOBASE, v2, baselining from 2007, is currently available for free under a public licence. The data informing the model, and its methods of harmonisation, are constantly being updated by the international team of consultants behind EXIOBASE. In a 2015 article about EXIOBASE in the journal *Sustainability*, the team conclude that:

> The large inequalities that exist around the world are being exacerbated by trade as cheap labour and lenient emissions regulations in poorer countries are

exploited by the wealthiest countries. In trying to derive policy and drive consumer action on these issues, we need to raise consumer awareness of the indirect impacts of their consumption habits. Only by doing so can we ensure that informed choices are made and that the problems we strive to solve are not simply shifted from one place to another.[21]

Different classification frameworks, country resolution, and sets of included satellite data make it difficult to compare available EEMRIO databases against one another. One way to overcome this problem is to aggregate the available EEMRIO tables into a common classification scheme. A webpage has been set up by the Industrial Ecology Programme at the Norwegian University of Science and Technology to do just this.[22]

It allows one to compare and visualise EXIOBASE 3 accounts against the other available EEMRIO databases. It collects data from multiple international databases and enables users to extract environmental information of countries and regions. Spatially disaggregated footprint data can also be provided to sustainability consultants and professionals, for use in projects or tools. All data and results available on this webpage are open source.[23]

It's important to understand that there is a conceptual difference between this approach and the ecological footprint approach of the Global Footprint Network. It is this:

> Ecological Footprint accounting addresses the key question: How much of the biosphere's regenerative capacity do human activities demand? This measure can then be compared to the biosphere's available regenerative capacity. By doing so, the Ecological Footprint framework accounts for (1) the magnitude of humanity's physical metabolism and (2) the demand such metabolism places on the Earth's ecosystems. Thus, it captures a necessary, but not sufficient, condition for sustainability. The Ecological Footprint framework is not a measure of total human impact but a proxy for human pressure on ecosystems.[24]

PRINCE and the EXIOBASE method may reveal information about the human impact, but not on its own relate this in an absolute sense to planetary limits, the biosphere's available regenerative capacity.

REAP

The Stockholm Environment Institute's (SEI) Resources and Energy Analysis Programme[25] (REAP) is an early example (2010) of EEMRIO. The SEI has used it to calculate Wales' ecological footprint, which the Welsh Government employed as a basis for its One Wales: One Planet policy document,[26] using a set of spreadsheets. It can be applied at the local, regional and national levels, and generates indicators on carbon dioxide and greenhouse gas emissions measured in tonnes per person; the ecological footprint required to sustain an area in global hectares per person; and the material flows of products and services through an area measured in thousands of tonnes. It includes a scenario tool that can model the impacts of policy and create plausible scenarios of the future.

Like the above examples, it combines tables from national economic accounts with data from environmental accounts. REAP is a two-region input-output model that distinguishes between products produced in the UK and products imported from the 'rest of the world'. Its economic input-output tables describe the flow of goods and services between the UK and 'rest of the world' for 178 individual sectors over a year. The sectors cover a range from agricultural and manufacturing industries to transport, recreational, health and financial services. They are classified using the Standard Industrial Classification (SIC) System.

Environmental data from the supply chain of a product can be assigned to the point of consumption using the economic relationships depicted in the input-output table. REAP uses UK Environmental Accounts together with International Energy Agency and global databases (GTAP – Global Trade Analysis Project, and EDGAR – Emission Database for Global Atmospheric Research),[27] to distinguish between the environmental impact of industrial sectors in the UK and the rest of the world. By modelling the combination of industries needed to produce different products, the impact per pound sterling spent of 356 product groups and over 50 distinct consumption categories can be calculated.

The SEI then came up with Reap-Petite,[28] which applies this on a smaller scale – individuals and groups can use it – with the output being in carbon emissions. The Welsh Government commissioned a team to use SEI's data to build a separate spreadsheet[29] that also helps to determine per capita ecological footprint at the smallest possible level, that of a household, for use in its One Planet Development planning process. The Welsh Government has a stated policy aim (in One Wales: One Planet[30] and its Well-being of Future Generations Act – see Chapter 16) of aiming to only use the resources commensurate with there being one planet within one generation, and is currently working out how to get there.

This approach involves different calculations and assumptions to the macro level deployed by the Global Footprint Network for its ecological footprint calculations. At the national level, reporting is often done on a production basis, whereas on an individual or household level it is done on a consumption basis.

Spearheading countries

Developed countries with smaller populations, particularly in Northern Europe, seem to be spearheading attempts to understand how to reduce the ecological footprint of consumption. Smaller populations evidently make management of policies and data easier. Besides Finland, Sweden and Wales, a further example is Switzerland.

The Swiss Government has recently completed its third review of Ecological Footprint calculations. The report, conducted by the Swiss Federal Statistical

Office, found that Switzerland consumes 2.9 times the amount of natural resources that are available per capita worldwide. It concluded: "The consumption of fossil fuels accounts for almost three quarters of the Swiss ecological footprint. It has also grown substantially more than any other factor of the ecological footprint in the last decades. Another major factor is our use of arable land, forests and natural meadows, which accounts for 25.3 per cent of the total ecological footprint."[31]

The above Swiss EF review also concluded: "The ecological footprint does not take into account the destruction of ecosystems, the loss of biodiversity or renewable or non-renewable natural resources. Neither does it take account of fresh water consumption, pollution by heavy metals, or emissions from pollutants of low degradability. The ecological footprint is therefore not an accurate indicator for sustainability."

In Norway, a study linked an environmentally extended multi-regional input-output database with a Norwegian consumer expenditure survey to analyse the behavioural changes needed by households to reach different emission reduction targets.[32]

The team examined the effects of 34 different behavioural changes. If households implemented all of them it would result in a decrease of 58 per cent in their carbon footprint. This would save them some money, which they would probably spend on something else, reducing this saving to between 23 and 34 per cent.

This effect of efficiency is called the rebound effect, and is also a consequence of energy-efficiency actions that save money. If households were really careful about how they spent this money (by avoiding purchases associated with exploitation of fossil reserves, for example), the team calculated that the decrease would rise back up to 45 per cent, meaning that curbing the rebound effect is crucial to achieve real reductions in household carbon footprints without compromising the level of economic activity.

I hope that in the future, as the world decarbonises and more genuinely sustainable policies are put in place by governments and city authorities to transition to a 'one planet' world, households will find more options for choosing the most sustainable products and services.

Other ways to measure and reduce the urban ecological footprint

For cities, another way to figure out the material consumption part of the ecological footprint is to look at what gets thrown away. This gets round the challenge of sourcing data on consumption because many cities regularly collect information about the composition of municipal solid waste. Four researchers from Israel and British Columbia have looked at ways of working out the impact of what cities consume and opted to base it on the 'solid waste life-cycle assessment', using it as a proxy for material consumption.[33]

"High-income cities typically have eco-footprints several hundred times larger than their political or geographic footprints. As major consumers of resources and generators of waste, urban populations in these high-income cities are mainly responsible for humanity's current state of ecological overshoot," Kissinger and his colleagues concluded in their paper.

There have been three methods used to calculate ecological footprints at the sub-national scale: the compound, the component, and direct data.

The compound approach uses national per capita ecological footprint data scaled down on a per capita basis to the city and accounting for local energy sources. This is relatively straightforward because reliable data are more often available at the national than city level. But this won't reflect the impacts of local policy and actions, so an alternative is needed. This is the component method, which is to use economy-wide input-output sub-national analysis. This draws on data on local expenditures for some consumption items like food and materials, and relates them to carbon emissions in an extension of conventional monetary input-output analysis. These data are then combined with national ecological footprint assessment data.

A third alternative approach is to collect data directly from local government sources without relying on currency units as a proxy for consumption. It relies on data on locally generated energy and materials flow, thereby reflecting local resource consumption and waste generation patterns. This is helpful to local policymakers and planners who want to reduce environmental damage from activities within their jurisdiction, and monitor and assess the effectiveness of their policies and actions.

To simplify their task Kissinger and his colleagues narrowed their work to 12 materials commonly consumed in cities for which data on the life-cycle assessment existed in terms of energy use and/or carbon dioxide emissions, plus agricultural and forest land data, which they used to convert into each material's ecological footprint.

Which materials have the highest ecological footprint per unit of material? It turns out to be textiles (see Table 4.1), because of the large area of agricultural land and energy inputs required to grow cotton, as well as high amounts of energy used in production. Paper, cardboard and newsprint also have high ecological footprints due to the land area required to grow trees. Aluminium comes second, then nappies (diapers), polystyrene and steel. But textiles and aluminium make up relatively smaller proportions of urban waste streams compared to products like paper and diapers.

Cities will also want to measure the greenhouse gas emissions associated with consumption and these are found in Table 4.1. Cotton again tops the list, followed by aluminium, but paper, cardboard and newsprint are much further down.

Kissinger and colleagues said their results show that "it is important for local governments to identify both the quantities consumed and per unit production impacts of a material". "This direct connection can be used to communicate to citizens about stewardship, recycling and ecologically responsible consumption choices that contribute to urban sustainability," they advise.

Well, it can if the local government has the resources to do it. But for most of them, especially in the developing world, this is presently beyond their capability, so they will need assistance from some of the organisations supporting cities to become more sustainable.

TABLE 4.1 Ecological footprint and greenhouse gas emissions of 12 materials

Average GHG emissions		*Average ecological footprint*	
Material	CO_2e/tonne	Material	gha/tonne
Cardboard	890	Glass	0.24
Glass	990	HDPE plastic	0.20
HDPE plastic	1,015	PVC plastic	0.41
Newsprint	1,120	PET plastic	0.48
Printing paper	1,290	Steel	0.61
PVC plastic	1,920	Polystyrene	0.66
PET plastic	2,240	Diapers	1.14
Steel	2,530	Newsprint	2.23
Polystyrene	2,970	Cardboard	2.76
Diapers	3,580	Printing paper	2.82
Aluminium	10,840	Aluminium	2.42
Cotton fabric	21,500	Cotton fabric	10.20

The challenge of food's footprint

However easy it may be for individual administrations to quantify material flows, when it comes to the food component of the ecological footprint it will be harder.

A study by the Global Footprint Network of the ecological footprint of Mediterranean cities[34] using MRIO analysis showed that regardless of the overall size of the footprint, the size of the food component varied by a factor of two. Intuitively this might seem to be because, well, we all eat more or less the same amount. But if you then consider how much of the modern diet is produced and processed, compared to how it used to be before the introduction of artificial fertilisers and other inputs, and before the advent of the global-scale processing industry, you can begin to see how it may be possible to reduce it.

Mayukh Hajra is a team leader for the Indian Government of Himachal Pradesh, where he has developed methodologies for community-based carbon-environment assessment and used them at various locations in central India. He believes that if we are to sustainably feed the world, we need "a plurality of indicators that consider not just the rather simplistic measure of productivity that has been driving much of the agricultural sector, but also the wellbeing of the farmer and the health of the environment. Even more importantly, the set of indicators used for measuring success must be outcome oriented, so that we are able to track if our actions are leading to the desired impacts."[35]

Examples of indicators he gives include reduction in the use of chemical fertilisers, increase in the use of organic manure, and the amount of organic carbon in the soil. "Tracking systems must already be calibrated for the purpose," Hajra adds, which "will require an overhaul of our measurement systems". Implementing such

The local multiplier effect

Consuming goods produced locally has the additional benefit that the money spent doesn't leave the area but remains in the local economy to generate more wealth and jobs. If the same unit of currency is continually spent in local shops and businesses, then it creates more local jobs. As soon as the cash is spent in chain stores, or online, or on imported goods and services, it leaves the area and its benefit is lost. A higher proportion of money re-spent in the local economy therefore yields a higher multiplier effect because more income is retained locally, or within national borders. This supports more jobs, higher pay and more tax revenue for government, all of which may lead to better living standards and a reduced ecological footprint – at least from the transport component.

To measure the multiplier effect one would start with a source of income and track how it is spent and re-spent within a defined geographic area. A special tool to do this has been created by the New Economics Foundation. Called Local Multiplier 3, it helps an organisation improve its local economic impact, and can be used to assist the public sector (national or local) in considering the impact of its procurement decisions.[36]

The impacts of the consumption footprint of imported goods are often unknown to consumers because they occur so far away. As we have seen, many developed countries are managing to decouple environmental impacts from production and their economy as measured by GDP, but at the cost of increasing imports from developing countries, which negates this improvement. Of the various components (indicators) of the footprint – materials use, energy, land use, water and greenhouse gas emissions – materials use stands out as the only one growing in both absolute and relative terms relative to population and GDP. Conversely, land use is the only one showing absolute decoupling from both references. Energy, greenhouse gas emissions and water use show relative decoupling.

Using EXIOBASE data (see above), the net total environmental impact of consumption in developed countries that was displaced to other countries by importing goods between 1995 and 2011 (the last comprehensive set of global figures available at the time of writing) can be found. The impact rose during that period for all indicators: from 23 to 32 per cent for material use, 23 to 26 per cent for water use, 20 to 29 per cent for energy use, 20 to 26 per cent for land use, and 19 to 24 per cent for greenhouse gas emissions.

It again turns out that clothing and footwear constitute the commodity area of imports with the largest and the most rapid growth in environmental footprints.[37]

This reveals the advantage for developed countries of buying products made in their own country – replacing imported goods with locally produced goods; and that clothing and footwear should be prioritised, followed by food.

Once a country has minimised and rationalised the level of imported goods, its next step would be to minimise the impact of the goods and services produced within its borders. Ultimately, we have to dissociate environmental pressure from economic growth by rapid and radical improvement in resource efficiency, moving towards a closed-loop economy.

What standards should the aspiring 'one planet' city governors deploy?

Let's be absolutely clear: no country or city on the planet is yet completely sustainable, in social and environmental terms, whatever its stage of development. Administrators are not yet used to thinking long term or holistically. Policymaking has not caught up with the fact that humanity crossed the threshold of 'one planet' living and began living in deficit way back at the beginning of the 1970s. Almost 50 years have passed, and the longer we leave it, the more damage is done, the more species become extinct, the more habitats are destroyed, the more pollution is caused, and the more climate change becomes potentially irreversible. There is so much work to do, to reduce the upward curve of the graph of ecological impact plotted against the passage of the years.

This is why we need data, markers, indicators and standards to relate our activities to what the biosphere of our planet can tolerate. Contraction and convergence is the appropriate global, national and local strategy.

To this end, policymakers should first adopt the Sustainable Development Goals if they have not done so already, bearing in mind, however, that these do not relate consistently to planetary boundaries. Any of the above tactics then become useful as a starting point, depending on the problem to be addressed.

At the city level, it will be necessary to determine the metabolic and material flows, using tactics described above. On the supply side, a city, state or nation's biocapacity represents the productivity of its ecological assets (including forest lands, grazing lands, cropland, fishing grounds and built-up land).

On the demand side, the Ecological Footprint measures the ecological assets that a given population requires to produce the natural resources and services it consumes (including plant-based food and fibre products, livestock and fish products, timber and other forest products, space for urban infrastructure, and forest to absorb its carbon dioxide emissions from fossil fuels).

Both measures are expressed in global hectares – globally comparable, standardised hectares with world average productivity. Each city's, state's or nation's ecological footprint can be compared to its biocapacity. If a population's ecological footprint exceeds the region's biocapacity, that region runs an ecological deficit. A region in ecological deficit meets demand by importing, liquidating its own ecological assets (such as overfishing), and/or emitting carbon dioxide into the atmosphere.

This will identify the origins, destinations and impacts of consumption. It would then be possible to model the effects of changes of policy and practice towards a circular economy – perhaps drawing on the ideas in the following chapters – upon the related biocapacity.

Tracking the Human Development Index against the ecological footprint over a time period can indicate the direction of progress. (For examples of such charts, see the Global Footprint Network's analysis of Mediterranean cities.[38]) Government agencies at all levels can manage their capital investments in a fiscally responsible and environmentally sustainable way by using ecological footprint accounting and GFN's Net Present Value Plus (NPV+) tool.

The traditional net present value (NPV) formula adds up revenue and expenditures over a period of time and discounts those cash flows by the cost of money (an interest rate), revealing the lifetime value of an investment in present terms. GFN's NPV+ tool adds to this calculation currently unpriced factors, such as the cost of environmental degradation, and benefits like ecological resiliency. All costs and benefits – even those where no monetary exchange occurs – thereby can be seen as "cash flows", and can be evaluated using different future scenarios. This will provide a more accurate and useful guidance on the long-term value of the investment, because it makes reference to the ecological footprint of the project in question.

The ecological footprint can therefore help to identify which issues need to be addressed most urgently to generate political will and guide policy action. It can improve understanding of the problems, enable comparisons across regions and raise stakeholder awareness. By identifying footprint "hot-spots", policymakers can prioritise policies and actions, often in the context of a broader sustainability policy. Footprint time trends and projections can be used to monitor the short- and long-term effectiveness of policies.

NPV+ helps governments and public agencies more accurately measure the long-term value of their investments in infrastructure and natural capital. It uses multiple scenarios that can provide a basis for capital decisions by allowing more informed assessment of risks and opportunities. Therefore, by understanding where the best long-term value is, policies can be oriented toward better outcomes, building wealth, avoiding stranded assets and leaving a better legacy for future generations.

In addition, complementary to this, PAS 2070 can assist with monitoring cities' carbon footprints of consumption and production. The ISO standards covering environmental management, energy management and life-cycle analysis will help put in place procedures for reducing the impacts of specific projects and processes. Chapter 17 presents a summary of the steps towards declaring as a 'one planet' city.

At the same time, policymakers, citizens and politicians need to engage in the urgent task of awareness-raising about the issues.

We've established some principles for evaluating the level and type of collective activity that can sustain our needs with negative impacts within the regenerative capacity of nature. So now let's get on with the job of examining and evaluating some of the many claimed examples of sustainability in urban environments. The following chapters describe incredible, brave work being done around the world to strive towards the holy grail of 'doughnut economics', wherein true sustainability and meeting human needs combine.

I find them truly inspiring – experimental and sophisticated beacons feeling the way towards a shared 'one planet' future. We all need good news and positive stories to counteract the insistent messages of doom, don't we?

Notes

1. C40 Cities Climate Leadership Group in partnership with the University of Leeds (United Kingdom), the University of New South Wales (Australia), and Arup, co-authored by the IPCC: *Consumption-Based GHG Emissions of C40 Cities*, 2018. See: www.c40.org/researches/consumption-based-emissions
2. Moran, D., et al., "Carbon Footprints of 13,000 Cities", *Environmental Research Letters*, June 2018, 13 (6). See: iopscience.iop.org/article/10.1088/1748-9326/aac72a
3. As does the Welsh government's ecological footprint calculator, an Excel spreadsheet downloadable from their website.
4. See: citycarbonfootprints.info
5. See: www.c40.org/researches/consumption-based-emissions
6. Kennedy, C., et al., "The Changing Metabolism of Cities", *The Journal of Industrial Ecology*, 2007, 11 (2), University of Toronto. See: https://doi.org/10.1162/jie.2007.1107
7. Sankey diagrams are a specific type of flow diagram. They summarise all the resource flows taking place in a process from start to finish. The thicker the line or arrow, the greater the amount of resource involved. The routes of the arrows show what processes the resources travel through.
8. Musango, J.K., Currie, P. and Robinson, B., *Urban metabolism for resource efficient cities: From theory to implementation*, UN Environment, Paris, 2017.
9. See: ecocitybuilders.org
10. See: gis.ecocitybuilders.org/portal/apps/MapSeries/index.html?appid=6719873278b6478295ea3015c086c3f2
11. See: www.prince-project.se
12. Wiedmann, T. and Barrett, J., "A Review of the Ecological Footprint Indicator – Perceptions and Methods", *Sustainability*, 2010, 2, 1645–1693. doi:10.3390/su2061645
13. Available at: https://seea.un.org/
14. More on the thinking behind this decision-making process can be found in these two sources: Hoekstra, R., Edens, B., Zult, D., Wu, R. and Wilting, H., *Environmental Footprints: An Methodological and Empirical Overview from the Perspective of Official Statistics*, 2013. See: www.eframeproject.eu/fileadmin/Deliverables/Deliverable6.2.pdf; and Elena Dawkins, E. and Trimmer, T., *Environmental Footprinting With Multiregional Input-output Models*, Swedish Environment Institute, 2016. See: www.prince-project.se
15. See: www.exiobase.eu
16. See: worldmrio.com
17. See: www.gtap.agecon.purdue.edu
18. See: www.wiod.org
19. See: unstats.un.org/unsd/nationalaccount/sna.asp
20. Stadler, K., Wood, R., Bulavskaya, T., Södersten, C.-J., Simas, M., Schmidt, S., Usubiaga, A., Acosta-Fernández, J., Kuenen, J., Bruckner, M., Giljum, S., Lutter, S., Merciai, S., Schmidt, J. H., Theurl, M. C., Plutzar, C., Kastner, T., Eisenmenger, N., Erb, K.-H., de Koning, A. and Tukker, A., "EXIOBASE 3: Developing a Time Series of Detailed Environmentally Extended Multi-Regional Input-Output Tables", *Journal of Industrial Ecology*, 2018. doi:10.1111/jiec.12715
21. Wood, R., et al., "Global sustainability accounting-developing EXIOBASE for multi-regional footprint analysis", *Sustainability* (Switzerland), 2015, 7 (1), 138–163.
22. See: www.environmentalfootprints.org
23. Stadler, K., Wood, R., et al., 2018, op. cit.

24 Galli, A., et al., "Questioning the Ecological Footprint", *Ecological Indicators*, 2016, 69, 224–232, Elsevier.
25 See: www.sei-international.org/reap
26 See: http://wales.gov.uk/docs/desh/publications/090521susdev1wales1planeten.pdf
27 See: http://themasites.pbl.nl/tridion/en/themasites/edgar/index.html
28 See: www.reap-petite.com
29 See: http://wales.gov.uk/docs/desh/publications/121115calculatoren.xls
30 The research behind it is here: http://wales.gov.uk/about/foi/responses/dl2013/aprjun/susdev1/dlsus5/?lang=en
31 Swiss Federal Statistical Office, Switzerland's ecological footprint. See: www.bfs.admin.ch/bfs/en/home/statistics/sustainable-development/ecological-footprint.html
32 Bjelle, L.E., Steen-Olsen, K. and Wood, R., "Climate change mitigation potential of Norwegian households and the rebound effect", *Journal of Cleaner Production*, January 2018, 172, 208–217. https://doi.org/10.1016/j.jclepro.2017.10.089
33 Kissinger, Meidad, et. al., "Accounting for the Ecological Footprint of Materials in Consumer Goods at the Urban Scale", *Sustainability*, May 2013, 5, 1960–1973.
34 Baabou, W., et al., "The Ecological Footprint of Mediterranean cities: Awareness creation and policy implications", *Environmental Science & Policy*, March 2017, 69, 94–104. https://doi.org/10.1016/j.envsci.2016.12.013
35 Hajra, M., *Sustainable Agriculture – Calling for a Plurality of Indicators*, in Development Alternatives, December 2017. See: www.perspectives.devalt.org/?p=2521
36 See: www.nefconsulting.com/our-services/evaluation-impact-assessment/local-multiplier-3/
37 Rapport, D.J., "Ecological footprints and ecosystem health: Complementary approaches to a sustainable future", *Ecol. Econ.*, 2000, 32, 367–383.
38 *How can Mediterranean societies thrive in an era of decreasing resources?* Global Footprint Network, 2015. See: www.footprintnetwork.org/2015/10/26/med/

5

FEEDING CITIES WHILE SAVING THE PLANET

The 'one planet' city would go a long way towards not only feeding its citizens healthy food from within, and from its hinterlands, it would do so in a way that replenishes natural resources.

According to the influential Gunhild Stordalen,[1] of the EAT Forum in Sweden, food is the main issue around which coalesce many others: climate change, poor health, social inequality, soil loss, biodiversity loss. Addressing the issue of food will help to address all the other issues. "Food is the biggest driver of climate change. As 4 billion more people will be added to the planet this century and more people become affluent, more will eat meat and there is no scientific consensus on solving these interconnected problems," she says.[2] "We need action to change this and to end the disconnect between consumption and production." She thinks this can be done by collaboration across sectors. "We need new business models as much as new practices." This chapter explores these new models and practices.

Researchers at the Global Footprint Network looked at 17 Mediterranean cities and found huge variations in their ecological footprints. Valletta, Genoa and Athens had the highest per capita EF in the region (4.8 to 5.3 global hectares (gha) per person); while Tirana, Antalya and Cairo had the lowest (2.1 to 2.9 gha per person). The main Footprint drivers were food consumption (accounting for ≈30 per cent of resource requirements in all the cities), transportation (≈20 per cent) and consumption of goods (≈15 per cent). Food is both the largest and the hardest component to reduce.

The researchers concluded: "Differences are primarily driven by different purchasing power, different levels of technology – which allow for a lower resource intensive consumption – and differences in lifestyles."

In Europe, food consumption constitutes 27 per cent of the total EF; it's about 30 per cent to one-third in most parts of the world. How do we reduce the overall level of impact? Here is a five-step plan:

1. Reduce food waste
2. Change our diets
3. Reduce the processing impacts
4. Improve the way food is grown and take care of our soils
5. Make it more local.

Reducing food waste

Every year, people harvest roughly 13 billion tonnes of biomass globally to use as food, energy and materials. Dame Ellen MacArthur is a truly inspirational woman who sailed single-handed around the world before forming an organisation to rid the world of waste, because she saw how it contaminates even the most remote seas. She calls cities "great aggregators" of resources and materials much of which ends up as 'waste' – especially nutrients from food. The opportunities to collect and reuse these is described in her foundation's publication, *Urban Biocycles*.[3]

The Ellen MacArthur Foundation report calls a collection of industries which generate organic waste the 'biocycle economy'. It includes agriculture, forestry and fishing at the primary stage; food processing, textile manufacturing and biotechnology in the processing stage; and retail and resource management in the consumption stage. Globally it is worth about US$12.5 trillion, or 17 per cent of global GDP, a sum which is predicted to rise by 55 per cent by 2050.

There is a great deal of waste in the biocycle economy – about a third of all food and plant matter produced, not to mention losses to nature and the damage to the environment. The greenhouse gas emissions produced by the world's wasted food are only just less than those of China and the USA. Nutrient run-off caused by fertiliser use has resulted in aquatic dead zones all around the world.

But the value to be found in rectifying the situation is huge, and estimated by the World Economic Forum to be potentially as high as $295 billion by 2020. Cities produce about 1.3 billion tonnes of solid waste per year, roughly half of which is organic, a figure set to almost double by 2025. They therefore have a major role to play in harvesting the benefits from the biocycle economy.

Nowadays, virtually none of these materials are looped back into the biosphere, resulting in the degradation of rural soils and increasing reliance on destructive synthetic fertilisers. The report claims that in theory, nitrogen, phosphorus and potassium (NPK) nutrients recovered from food, animal and human waste streams on a global scale could contribute nearly 2.7 times the nutrients contained within the volumes of chemical fertiliser currently used.

Amsterdam has calculated that high-value processing could lead to added value of €150 million, as well as 900,000 tonnes of material savings and a reduction of 600,000 tonnes in CO_2 emissions annually. Jobs would be created by waste separation and return logistics, nutrient recovery, and the construction of biorefineries – which use techniques such as thermal treatment, biological processes and enzymatic conversion to produce fuels, power, heat and value-added chemicals from biomass.

Some cities like Milan in Italy have already implemented programmes to recover and obtain value from food waste and wastewater. Anaerobic digestion is the most common type of biorefinery, producing biogas and compost. This is part of what is termed the circular economy. Governments and city halls can encourage it along by changes in regulation, including redefinitions of what constitutes waste, and by bringing in taxes on the environmental damage caused by the existing system.

Avoidance of waste in the first place is better than reclamation, however. Tackling food waste should follow the same waste reduction hierarchy as general waste. At the top of this pyramid is prevention. In the 'developed' world, around half of food waste takes place in homes and so public education is required. Retailers are responsible for less food waste, but there is much more waste in the supply chains.

One area where it is easier to tackle food waste is in businesses where food is prepared in advance, such as hotels, educational institutions, hospitals and companies with staff canteens. Here, over 70 per cent of food waste by value is lost before it reaches the customer's plate.

Solutions are being developed which apply the principles of energy management to food management in such a situation. In energy management, what is measured gets saved. So in one proprietary system, a weighing scale is installed below the waste bin which automatically records the weight of food thrown away. A member of staff records the type and origin of the waste. A computerised system assigns costs to the food discarded on a daily and weekly basis. A baseline is established at first (experience shows that the amount of waste recorded is typically double what was previously assumed). Using the information gathered, organisations then institute a reduction strategy, training employees, and commonly reduce waste by up to 70 per cent with a huge improvement to the bottom line.

IKEA, which serves 650 million meals each year, has just begun using a quantitative system of this type and aims to reduce food waste by half by the end of 2020. Staff awareness is increased. A bonus is that their workers take the information gained home, where they often apply it to reduce waste too.

'Think•Eat•Save'

On a larger scale there exist two standards for measuring and reducing food waste. The 'Think•Eat•Save' tool helps governments, local authorities and businesses in developed countries with the prevention and reduction of food and drink waste. It was produced by UNEP with help from the UK's Waste and Resources Action Programme, FAO and the SAVE FOOD Initiative.[4]

It begins by mapping and measuring food waste and from the information gained supports the development of national or regional policies and measures. In-depth modules then focus on programmes for food waste prevention in retail, households and in the food supply chains. This approach is delivering significant reductions in food waste in the UK, Norway and Japan, saving many billions of dollars per year.

Food Loss and Waste Protocol

A more advanced, second-generation standard is the 'Food Loss and Waste Protocol'.[5] This was developed out of experience with the one above by the World Resource Institute. This voluntary standard enables countries, cities, companies, anywhere in the world to produce inventories of the extent and origins of food waste and where it goes to in order to inform strategies for minimising it. It offers robust measurement techniques, consistent accounting and reporting requirements, and universally applicable definitions.

Governments and international organisations in developing countries may not have the data-gathering capacity needed for this standard. Instead they are advised to invest in food storage infrastructure and to assist with the coordination of action to improve food production, storage and distribution with the aim of reducing food waste. A quarter of food rots or passes its best on its route to market, and so investment in sustainable cold and frozen supply chains is strongly recommended.

Cities with fast-growing middle-class populations can reduce waste and help residents save money by setting up consumer food waste prevention campaigns, such as the UK's Love Food Hate Waste campaign, which helped London to save up to eight times the campaign costs in waste disposal savings.

As for companies, they should support and participate in sector agreements to develop savings in their supply chains.

Changing diets

Why are some diets better than others for the planet? Mostly, it's to do with making the most efficient use of land space, preserving and enhancing soil quality, and protecting or improving the environment, for example in respect of water pollution, habitat protection, or climate change. For example, palm oil is widely used in processed foods, but often associated with the destruction of rainforests, so avoiding it will help preserve rare species such as orangutans. A diet which avoids products of the meat industries will contribute to achieving all of the above goals.

But not all meat is 'bad'. That reared on natural grassland (23 per cent of the Earth's surface) is fairly benign, since this system of farming supports the soil and sequesters carbon. By contrast, indoor, intensive rearing of animals, and the clearing of forests to make grassland or grow animal feed, have the opposite effect. Intensively reared animals need feed (soya and grain) and a staggering third of agricultural cropland is used to grow this feed; the world's future, higher population can only be supported by feeding more people directly from this produce. The ecological footprint of animal feed is huge, especially when rainforests are cleared to provide this food.

Eating less of this type of meat would have multiple benefits: on animal welfare, the land, biodiversity, transport impacts, greenhouse gas emissions and human health. Systemic patterns in the food system – agricultural subsidies, trade agreements, commodity markets – need to alter to effect this change. So do mental habits – such as the belief that higher economic status means higher levels of consumption, especially of meat.

The latest study of the ecological footprint of foodstuffs confirms previous research, that a vegan diet based on vegetables, fruits and cereals is environmentally convenient and more sustainable.[6] Seventy foodstuffs were evaluated on their use of water and their effect on global warming. Vegetables and cereals were discovered to consume more water, on the whole, than animals, pulses and oilseeds.

Animal-based foods cause more greenhouse gas emissions, require more energy, and demand a much more elaborate food supply chain than vegetarian foods. Funnily enough, pumpkins seem to have the best performance for both water footprint and ecological footprint, while beef meat has the lowest position (local conditions notwithstanding).

Looking at carbon emissions of food alone, a simple average is shown in Table 5.1 for different sample foodstuffs.

It's important to take into account nutrition, and a food with a higher environmental impact does not always have better nutritional quality. The sweet spot for any location in the world will combine optimum nutrition and taste, and minimum environmental impact within what a given culture will accept as palatable.

In practice, what type of food people consume is affected by their income and cultural norms; convenient access to cheap, healthy, tasty and culturally acceptable food, alongside efforts to improve awareness are all vital to shift habits to ones that support more sustainable diets. 'Food deserts' is the name given to urban neighbourhoods and rural towns without ready access to fresh, healthy and affordable food. These must be identified and targeted. Messages about the healthiness and environmental impacts of food products need to be well publicised and consistent. Taxes on sugar and salt levels help but are just a start.

New York City

New York City (NYC) is tackling all of these problems at once with a system called Health Bucks. It was started by Lilliam Barrios-Paoli, the city's deputy for health and human services. She explains that, "While we're a thriving metropolis that is proud of its rich culinary depth, New York has too many residents who are unable to even eat."[7] New York City also has Green Carts, Healthy Bodegas, many farmers' markets in low-income neighbourhoods, urban farms, mobile markets, nutrition guidelines for schools and childcare centres, and an incredible array of organisations committed to making changes.

SNAP (Supplemental Nutrition Assistance Program) is the largest programme of the US Government to tackle food poverty, participated in by many farmers' markets. It has multiple benefits: recipients of the aid get increased access to healthier and fresher foods, farmers increase their customer base and sales, and more locally grown food is eaten, reducing the carbon cost of food miles.

TABLE 5.1 2017 Worldwide average CO_2 emissions per food type, per person, per year

Food type	Emissions
Lamb and goat	35.02 kg
Beef	30.86 kg
Pork	3.54 kg
Nuts, including peanut butter	1.77 kg
Fish	1.60 kg
Milk, including cheese	1.42 kg
Rice	1.28 kg
Poultry	1.07 kg
Eggs	0.92 kg
Soybeans	0.45 kg
Wheat and wheat products	0.19 kg

Source: Food and Agriculture Organization of the United Nations, Emissions Intensities Items: Data collected on 2018-03-09. Last update of FAO: 12 December 2017. Year of data points: 2014. More data: www.nu3.com/c/food-carbon-footprint-index-2018.

Note: Emissions intensities are from FAO (when available).

There are 148 farmers' markets throughout New York City. If buyers present their Benefits Transfer card they receive free fruits and vegetables; for every $5 spent they can obtain an additional $2. The scheme also supports pre-purchased produce boxes – a form of Community Supported Agriculture – and Nutrition Education Workshops. This has the additional benefit of keeping money in neighbourhood economies, since the US Department of Agriculture reckons that every $1 in food stamps spent generates $1.79 in local economic activity.

City Hall works with many community groups, some part of a programme by the Laurie M. Tisch Illumination Foundation called Healthy Food & Community Change that also fosters affordable housing, jobs, developing neighbourhood economies, and expanding educational opportunities. Healthy food is seen as an integral aspect of what makes a healthy, vital community.

There are also public-private partnerships, which grew out of the successes of the NYC Green Cart Initiative. The Green Cart permit system allows individuals to operate a food cart selling whole, raw fruits and vegetables in designated areas of the city. Frozen or processed produce is not allowed, and for food safety reasons, vendors cannot cut, slice, peel or process fruits or vegetables.

A Fruit and Vegetable Prescription Program also operates in New York, issuing prescriptions for healthy food to overweight and obese children who are at risk of developing diet-related diseases, such as type 2 diabetes and heart disease. Some 58 per cent of adults in NYC are overweight or obese, and 40 per cent of elementary school students. It's the poorest communities that remain disproportionately burdened by diet-related diseases. Health care providers and farmers' market partners work together to identify and enrol overweight and obese children as participants. They include the NYC Health and Hospitals Corporation.

The city could go even further, using its power as one of the largest purchasers of food in the USA to force suppliers to make available good-quality and affordable food. Lilliam Barrios-Paoli says that public institutions can really make a difference this way through their buying power.

> The sheer volume that public institutions buy is enormous – public hospitals, public schools and other entities serve a huge number of New Yorkers, and disproportionately the people served by those institutions are the poor, the unemployed, the underemployed, low-wage workers, and families who are scraping by and need extra support or a boost. To make change the agencies and stakeholders have to sit together to identify opportunities and impediments.[8]

Education and procurement power to reduce impacts

Peter Seggers runs Blaencamel Farm[9] in Wales. He practises agro-ecology with market gardening using a Community Supported Agriculture model, and feeds his 300-strong local community year-round with organic produce. He agrees that the trick to ending the disconnect between consumers and producers lies in educating his consumers about the add-on benefits of sustainably produced food – fighting climate change, feeding the soil, improving biodiversity. Year-round productivity without using artificial fertilisers is obtained by feeding the soil and using polytunnels. He employs thermophilic composting and he composts absolutely everything he can in order to feed microbes to the soil. This increases the nutrients in the food, making it healthier, and gives greater protection to the crops from disease. Food produced this way might be more expensive, but through passionate communication, Seggers persuades his customers it is worth it to avoid the damage done by intensive farming.

At a larger scale, the food and catering industries themselves can also force their suppliers to improve their ecological footprints as a condition of renewing their contracts. This gives them a vital role in broadcasting these messages too. The trend to transform procurement, processing and supply chains is already underway throughout the developed world, with established and new catering and retail chains offering more healthy and sustainable vegetarian and vegan choices, and sourcing from organic and more local suppliers.

The food store chain Iceland is striving to completely phase out palm oil from its supply chains, for example, saying "there is no such thing as sustainable palm oil", because of its role in deforestation of old tropical rainforests.[10] It has also pledged to remove all plastic packaging from its products by 2023.

The chain Marks & Spencer, as part of its 'Plan A', has a Farming for the Future[11] programme to encourage farmers and growers to minimise their impact on the environment in terms of soil, water, pesticides and energy, and to enhance biodiversity. It says this benefits everyone, long term: "This will make their businesses more resilient and profitable, ensuring that they can continue to deliver quality and innovation for the long term whilst reducing their impact on the world around them. In turn, this delivers security of supply for us."

Making it more local

Feeding our cities in a way that regenerates the planet's ecology is one of the biggest challenges facing humanity. In the old days cities were smaller and were fed from food produced in their hinterlands. With the advent of globalisation, food is now shipped in to cities from all over the world and people have lost connection with the way their food is produced and processed and supplied.

The ecological and health impacts of intensive animal rearing and agriculture, production which strips out much of the nutritious value of food, and the carbon emissions of transportation and processing add to the unsustainability of this business model. While consumers sometimes benefit from cheap food, their health frequently suffers. This is apparent in increased rates of obesity and related health problems in countries which have shifted from a Mediterranean-style diet of fresh vegetables and a small amount of fresh meat and fish to one based on processed foods.

"You can't separate the sustainability of the city from its hinterlands," says Henk de Zeeuw. He works for the RUAF Foundation, a global network in the field of (intra- and peri-) Urban Agriculture and City Region Food Strategies.

> Cities must think more about their surroundings and how they relate to them, particularly their food and water supply lines. In the food crisis of 2007–10 it was the poor who were the most affected. Climate change and concerns about the resilience of the food system are driving this, and the multiple roles agriculture and forestry can play in sustainable cities.

He adds:

> It is important to support the urban poor. This stimulates the regional economy and reduces dependence on the global food markets. Growing food locally also reduces the ecological footprint of the city, helps it to adapt to climate change, enables resource recovery, and the productive reuse of urban organic waste and wastewater in the region. Furthermore it can enhance urban biodiversity and enable outdoor recreation and eco-education in the city. It should also support urban green infrastructure.

The successful cities such as Bulawayo (which started in 2007) formulate strategies which contain a wide variety of aims, and institutional arrangements with different actors, from the bottom up and top down. Many cities with policies include urban agriculture and forestry in urban land-use planning. They provide technical and financial facilities for processing and marketing farmers' produce and stimulate innovation in intra- and peri-urban farming, for example in water saving and safety. They also support the public to buy local food.

Some cities require smart food labels that explain to potential buyers the low food miles. They also enable the productive use of urban waste with composting facilities and wastewater use, perhaps for irrigation. They help to reduce food waste by creating food donation programmes or food waste recycling, and they include urban agriculture and forestry in city climate change plans, for example to reduce flooding by planting trees.

They also support multifunctional agriculture and promote the conversion of farms to organic cultivation methods. They support production projects by the urban poor and disadvantaged and promote the consumption of nutritious food with kitchens, training and the provision of food vouchers to the most vulnerable households, perhaps by using menu labelling with choices. A city action plan would begin with a stakeholder inventory and analysis and continue to awareness raising. It would then establish a working group, mapping and analysis of the local food system, and joint visioning and scenario building.

It would be inclusive of all sectors of the population and proceed to define strategies and create the legal operations and financial frameworks. "Otherwise it will definitely fail," said de Zeeuw, speaking from a lifetime's experience. "Many have failed because they are not coordinated correctly at these later stages."

De Zeeuw also says that it is vital to generate media attention and replicate and upscale the successful initiatives. "But projects must not be encouraged to become dependent on central funding. That is the way to failure. They must have a position independent of the local or national government while having good links with them. Larger-scale, long-term programmes can provide a framework for small initiatives on the ground."

So, we can solve several problems at once by growing much more food within cities and near to them.

Recolonising the countryside

Encouraging 'one planet' food production on a wider scale would be a way to provide employment and livelihood because replacing fertilisers and pesticides with labour means more people living and working in the agriculture, horticulture and agroforestry industries in the countryside. Many remote areas would benefit from higher levels of population density, provided that it was introduced in a sensitive and regenerative manner. Land workers are land managers, stewards of biodiversity and soil quality; what could be a greater responsibility? Why should we accept the idea that it is inevitable that cities will keep expanding at the expense of the countryside?

Wherever food comes from, to make a success of local food provision in cities, a lot of work is needed by many diverse stakeholders: from business, academia, civil society and public authorities. According to the World Bank this means: "a network of organisations, enterprises and individuals focused on bringing new products, new processes and new forms of organisation into economic use, together with the institutions and policies that affect their behaviour and performance."[12]

Recognition of the importance of agricultural land and the encouragement of local sustainable food provision within legal frameworks is vital.

Some regional authorities are leading the way. In France, the Montpellier Territorial Coherence Scheme (ScoT) defines agricultural areas to be protected for a period of 15–20 years, stipulating the preservation of agricultural and natural spaces as a condition for new projects. In Spain, the Catalan Council of Ecological Agriculture Production has, since 2001, set a regulatory mandate on technical norms for agro-ecological production and labelling.

In Italy, the Bologna Metropolitan Agriculture Project, part of the Metropolitan Strategic Plan, strengthens the urban-rural relationship by identifying the peri-urban area as a buffer zone between urban and rural needs. Quality produce is identified by a 'km 0', label and promoted with direct sales, a sustainable distribution system and educational farms.

Back in France, Pays d'Arles, the hinterland area of Marseilles, is the main agricultural supplier of the metropolis, of a large and diverse set of crops, due to the history of the region and to a favourable environment. But production has been declining for many years because of urbanisation. Legislative planning tools developed to halt this include Agricultural Charters and a Territorial Coherence Scheme.

In Albania, the national Rural Development Strategy specifically values urban agriculture and allows for projects to create small parcels for individual urban agriculture and green and healthy districts. And in northern Greece, the Thessaloniki Metropolitan Area, the authorities have started supporting the spontaneous use by citizens impoverished by the prolonged recession of open spaces in and around cities to grow crops and keep animals. Although there is not yet a legal framework for metropolitan, peri-urban and urban agriculture, the National Strategic Reference Framework legitimises and fosters the development of community gardens in Greek cities.

Much work has to be done to develop relevant policies and the new types of relationships that are required to heal the vast and historical disconnect between purchasers and producers of food. Sources of investment for suppliers, logistics and distribution are needed. In most cases much time, some voluntary, is spent, for example arranging the appropriate procedures for a school or hospital to obtain locally produced organic food and be confident that it is a safe and reliable supply.

Most existing policies have targeted actions with specific goals – such as addressing health or food waste. These can later be incorporated into more general, integrated food policies, as understanding of the issues increases amongst participants. The policies form just one part of the broader-scale food systems change that is ongoing. This transformative process consists of overlapping policies at local, national, regional and global levels.

Hundreds of cities around the world have food policies or governance structures, often focused on specific food-related issues, which then have spin-off benefits in related areas.

Tackling obesity

Lack of access to nutritious food, for example, is a problem tackled by the Public Policy on Food Security, Food Sovereignty and Nutrition in Medellin, Colombia,[13] which includes improving agricultural production in the city districts.

Tackling obesity is a frequent headline stimulus for a food policy, as in the Healthy Diné Nation Act,[14] a North American Navajo Nation tax on junk food. Amsterdam has a plan to eradicate overweightness and obesity by 2033. Unlike obesity programmes in other cities, its policy contains integrated actions across the departments of public health, health care, education, sports, youth, poverty, community work, economic affairs, public spaces and physical planning, and organisations from outside local government. It seeks to address the structural causes, including the living and working conditions that make it difficult for people to ensure children eat healthily, sleep enough and exercise adequately. It aims to make the healthy choice the easy choice, and create a healthier urban environment. The policy was driven by one man: Eric van der Burg of the liberal-conservative People's Party for Freedom and Democracy. The policy was given no initial budget, in order to show what could be achieved through cooperation and taking joint responsibility, and has succeeded handsomely.

Burkina Faso

The need to tackle climate change is behind the protection of the 'greenways' in Bobo-Dioulasso, Burkina Faso.[15] The city has set up agroforestry projects on urban greenways and the sustainable management of peri-urban forest border zones to provide windbreaks, shade, run-off retention and a source of fuel and fodder, market gardening, and recreational and environmental education areas.

San Francisco

Waste reduction is behind the efforts of San Francisco, USA.[16] In 2002 the city, which produces about 5,000 tons of waste a day, set a goal of 75 per cent diversion by 2010 and Zero Waste to landfill, incineration or high temperature technologies by 2020. It passed a Mandatory Recycling and Composting Ordinance, requiring everyone to separate garbage into recyclables, compostables and landfill bins.

The city implemented the first and largest urban food scraps composting collection programme in the USA, covering both commercial and residential sectors. The compost is looped back to be used by local farmers and wineries in Napa and Sonoma counties. In 2017 it was deemed not due to meet its 2020 target, and so revised its systems and re-educated residents on an "often confusing and potentially anxiety-causing sorting process". Even if it misses 2020, it will achieve its goal soon.

Cape Town

Reviving the local economy and providing jobs, especially for women, is the prime motivator for other cities such as Cape Town, South Africa's urban agriculture policy.[17] This gives support, land access and training to women in Cape Flats, who grow food partly for social causes such as crèches and schools.

It's also behind the Central Market programme[18] in Valsui, Romania, where, because traditional systems have remained unchanged before being reinvented for the 21st century, 4,000 small agricultural farms covering just over a hectare each (70 per cent of the municipality total area) satisfy the entire food demand of the 73,000 locals.

Malmö

Projects which use existing policy responsibilities and public food procurement contracts to achieve new ends and rethink urban planning systems to achieve multiple win-wins, are found in the Policy for Sustainable Development and Food in Malmö, Sweden,[19] and the wide range of different urban challenges addressed by the Toronto Food Strategy, Canada (see below for more on this).

Malmö has a target of making all of its food purchasing sustainable, using the Brundtland Report definition and deploying its own 'eat S.M.A.R.T. model'. This combines health and the environment without increasing costs. S.M.A.R.T. stands for: Smaller amount of meat; Minimise intake of junk food/empty calories; An increase in organic; Right sort of meat and vegetables; Transport efficient.

Milan Urban Food Policy Pact

As of the end 2017, 160 cities have signed up to the Milan Urban Food Policy Pact,[20] which commits them to working towards "sustainable food systems that are inclusive, resilient, safe and diverse" – and to encouraging others to do the same by sharing experience of suitable governance practices. Appropriately, they are developing first a monitoring framework for their Urban Food Action Platform.

This is one of several networks of cities or for cities to support them in transitioning to more sustainable food policies. At the global level, the UN Committee on World Food Security has compiled policy recommendations on urbanisation and rural transformation: implications for food security and nutrition,[21] which are particularly useful for cities in developing countries.

The C40 Food Systems Network[22] is another, and this works in cooperation with the EAT Forum to support the efforts of 80 cities around the world with procurement guidelines, advice on fostering urban agriculture, sustainable food transportation and supply chains, and on reducing food loss and waste. The ambition of the EAT Foundation[23] itself is "to reform the global food system and enable us to feed a growing global population with healthy food from a healthy planet". This trailblazing organisation is science-led and hosts an annual Forum where bright minds from science, politics, business and civil society exchange ideas on how to shift food systems towards sustainability, health, security, and equity within the boundaries of our planet.

Nations and regions are setting up their own more local networks. EUROCITIES' food working group, consisting of members of elected local governments in 130 European cities, has published its own collection of ideas and best practice on urban food.[24] These attempt to tackle the main five governance barriers that prevent action: poor integration between city departments; unclear division of competences between local authorities, regions and national level; a lack of multi-level governance and policy coherence; little joining-up of research, practice and policy; and the need for more inclusion of citizens in decision-making. The 130 cities are developing more holistic and participatory structures which are having the effect of shortening some supply chains, but there is a long way to go.

The concept of City Region Food Systems has taken off in Europe, where it involves a "network of actors, processes and relationships to do with food production, processing, marketing, and consumption" across a regional landscape comprising urban, peri-urban and rural areas.[25] This is particularly exciting because it aims to maximise ecological and socio-economic links and co-governance by both urban and regional players.

For example, the 2017 People's Food Policy, produced and supported by over 80 food and farming organisations, includes a call for "the establishment of statutory Food Partnerships in each regional, metropolitan and local authority in England built on broad civil society and cross-sector participation".[26] Amongst its supporters is the UK Sustainable Food Cities Network, which has 54 members that are developing cross-sector partnerships to promote healthy and sustainable food. It runs award schemes to recognise the high achievers and campaigns.

The Association des Régions de France signed the Rennes Declaration for Territorial Food Systems[27] in July 2014, through which they committed to promoting agriculture and food policies for territorial development, economic development, and sustainable use of natural resources. And in the Netherlands, 12 cities, one province and three ministries signed the CityDeal 'Food on the Urban Agenda'[28] in early 2017. Not only will the cities and province include food in their own plans and strategies, but they are also collaborating to build an integrated food strategy for the whole country.

These are all very exciting and represent just the tip of an iceberg, but we are only at the beginning. The potential is huge for involving cities' populations in providing their own food, and linking them back to the origin of their food, and so to nature – a connection all too easily lost in urban life.

Can we produce enough food?

If the amount of food produced per hectare at present can be increased then it will be possible to feed more people with less land as the population of the world increases. Big agriculture uses the argument that it alone can do this to justify its continuing grip on food policy and research, and to push for more genetically modified (GM and GE) crops and chemical inputs.

But are they right? Firstly, reducing chemical inputs (nitrate and phosphate fertilisers and pesticides) should be an important additional policy aim. This is because they are produced using fossil fuels, reduce biodiversity, pollute our watercourses, harm health and, year-on-year, deplete soil fertility and soil carbon.

In the 19th century Paris fed itself by obtaining six to seven harvests a year from its surrounding hinterlands. It did this by using thermophilic (heat-generating, from manure) composting in conjunction with a high labour input and what we'd call today agro-ecological methods.[29]

Of course, Paris would need a lot more land now as it's far bigger, but six to seven harvests a year? That's more than big agriculture can manage. It turns out there are four strategies in addition to dietary change to get more from less land and feed more people in cities. Firstly, GM and GE crops are acceptable and useful for resisting pests and drought as long as the seeds are 'open source' and do not lock farmers in to using specific products. They are more resilient and do improve productivity. Then there are vertical farms, hinterland sourcing, and the use of agro-ecological practices. To satisfy the taste for meat, research is also developing on artificial meat and insect protein.

Urban farms are already located in many American cities, including Chicago, Baltimore, New York City, San Francisco, Los Angeles, Sacramento and West Sacramento. They are typically found on properties ranging in size from 1 acre to 3 acres.

Vertical farming

Vertical farming involves growing plants in layers, usually indoors, in a controlled environment. In an online debate on the subject which I chaired, the panellists agreed that it is capable of producing more food per acre, perhaps as much as 70 to 90 times as much, with vastly reduced energy losses and better use of space and resources. Vertical farms growing leafy vegetables are able to produce around 12 harvests per year. They're being developed in many places, such as the CityFARM project of Massachusetts Institute of Technology Media Lab's City Science Initiative.

Dickson Despommier is an emeritus professor of Columbia University and has been amongst the first evangelists for vertical farms, writing an early key text.[30] He sees the growth of urban farming as exponential. "The future of agriculture is growing soil-less. Hydroponics, aeroponics and drip irrigation. Everybody is on board for this. We do it by applying the indoor growing strategies we already have to growing vertically in cities. It uses 70 per cent less water, allows restoration of damaged ecosytems, remediates greywater. We can use abandoned buildings."

Many companies are members of the Association for Vertical Farming, which has 100 members, of which 20 are corporate. Among them is Green Spirit Farms, founded by Milan Kluko in New Buffalo, Michigan. He also runs vertical farms in Detroit. "In exploiting the vertical plane, plant density is critical," he says. "The aim is to improve the traditional agricultural supply chain, which is too long and involves very few players, and produces much ecosystem degradation."

The figures speak for themselves.

> Our farms use 98 per cent less water than going in the open air, on average. Growing Romaine lettuce in California would use about 75,000 gallons per acre and in Arizona 1.5 million gallons. Green Spirit Farms uses 800 gallons per acre that is 0.3 gallons per head of lettuce. As it is sold locally there are also significant carbon savings from not transporting it in refrigerated containers across the country. We can grow lettuces in mixed varieties and produce 17 harvests per year.

For dealing with pests, which do not infest plants very often, ladybirds are ordered which arrive in a box "about the same size as an ice cream container". When released, they get to work eating aphids and die out when all the aphids have been consumed. "So there is significant reduction in the use of pesticides and herbicides, basically to virtually zero." Some farms install highly reflective groundcover around the greenhouse, which disorients pests and prevents them from entering the greenhouses.

In New Jersey, New York State, Aero Farms[31] has built the largest indoor farm so far. It is 130 times more productive per square foot annually than a field farm, uses 95 per cent less water, 40 per cent less fertiliser, and no pesticides because it is in an enclosed environment and water can be reused. Plants can be organically grown nearby to the market. Crops that usually take 30 to 45 days to grow, like leafy gourmet greens, take as few as 12. With aquaponics and hydroponics, potatoes and other root vegetables, and fish such as tilapia can be provided. A billion-dollar industry, its popularity is expected to soar over the coming years as costs come down.

"We need to replace the old forms of agriculture and encourage the eating of more vegetables because they are a really efficient conversion of energy from the sun into something we can eat, that powers us. Vegetables need more respect," says Oscar Rodriguez, a registered architect and Building Integrated Agriculture specialist. His company, Architecture & Food, pursues how Building Integrated Agriculture can inform the architecture of resilient and equitable cities in an economy free of fossil fuels. "Our type of agriculture is not subsidised. Its advantage is that it is close to the people it serves."

According to his desktop study, in the short term, London could in theory meet 6 per cent of its fruit and vegetable needs by committing all of its available roof area to indoor production. With a suitable level of commitment, in the mid term, widespread construction of greenhoused hydroponic production systems on available flat roof areas in commercial zones added to containerised production in residential zones could almost satisfy the food requirements of the whole of London. If there were to be an extensive programme of residential roof reinforcement to support a mix of containerised and intensive production on the roofs of residential areas, alongside greenhoused hydroponic production in commercial zones, London could even beat this and raise a surplus.

Empty warehouses can easily be converted to vertical farming. Milan Kluko uses a former injection moulding factory in New Buffalo: "We cleaned and renovated it in 90 days and had a harvest 90 days later. The systems we use mean that the equipment can be moved from one place to another relatively easily."

Some foods cannot be adapted to indoor growing, specifically grains and sugar. Indoor farming is appropriate for fresh foods. The panel also agreed that for animal welfare reasons, they would not support intensive indoor animal rearing. "We needed to eat less meat if we're going to make it in the future anyway," said Kluko. "I'm opposed to high-density meat farms in cities."

Vertical farming uses specialised LED (light-emitting diodes) lighting of specific frequencies that the plants particularly like. This maximises the growing potential in an energy-efficient way. Where current vertical farms use traditional LEDs, OLEDs, which produce less heat and have a higher energy efficiency, are being phased in. OLEDs (organic light-emitting diodes) are a flat light-emitting technology, made by placing a series of organic thin films between two conductors. Frequencies can be adjusted to growing plants' needs from a broad sun-like spectrum range.

Vertical farms exist in an increasing number of cities around the world such as Anchorage, Berlin, Singapore and Tokyo, providing fresh vegetables, mushrooms, herbs and salads, fish, crabs and other foods. China is leading the world in vertical farming in cities. It's been pioneering the research for many years. Atop the Chinese Academy of Agricultural Sciences in Beijing, indoor patches of tomatoes, lettuce, celery and bok choy yield between 40 and 100 times more produce than a typical open field of the same size.

Professor Qichang Yang (Chinese Academy of Agricultural Sciences, China), director of the Research Centre for Protected Agriculture & Environmental Engineering, observes that since China's population is projected to reach 1.5 billion in 2030, and agricultural land is diminishing due to natural disasters and poor land management by 300 km^2/year while the demand for food is increasing, cultivated land is down to only 0.08 ha/person, a figure which is just 40 per cent of the world average.

So partly automated indoor farming is an essential feature of the future, he believes, and not just in China, because the same pattern is being repeated throughout the world. Because of the climate variabilities in China their vertical farms need to be capable of producing food in winter in temperatures as low as -55°C. Climate control technologies being developed to achieve this include hydroponics and LED lighting, heat pumps and Active Heat Storage (AHS) such that when the temperature outside is -12°C, inside it is 17°C.

"Compared to conventional farming, a plant factory with natural solar light can increase productivity per unit area by between two and ten times. With artificial light this rises to 40 times and by adding vertical farming this goes to 1000 times," Yang says, quoting research from the Ministry of Agriculture.[32] Plant factories in

China are mushrooming and catching up with the number in Japan. "Stereoscopic culture and nutrient solution management with control systems are deployed to supply nutrients. Plants are bathed in light continuously in the days before harvesting to reduce the nitrate content in the leaves."

He demonstrates at conferences a hydroponic system that grows tuberous root vegetables, the tubers hanging down from the ceiling so that they may be harvested without destroying the plant, unlike conventional farming. He is fond of showing a slide of a sweet potato plant that is ten years old and has produced 500 potatoes this way.

China is planning a 100-ha urban farming district in Shanghai, which has a population of 24 million. The Sunqiao Urban Agricultural District,[33] designed by US-based firm Sasaki Associates, between Shanghai's main international airport and the city centre, will use urban farming as a living laboratory for innovation, interaction and education. There will be a range of techniques, including algae farms, floating greenhouses, green walls, and vertical seed libraries. An interactive greenhouse, science museum, aquaponics showcase, and market will help to educate children about where their food comes from.

While some people have tried to synergise the growth of fish and salads with aquaponics, the panel on my webinar thought that on the whole this did not work economically. "You either need to focus on the needs of the vegetables or the fish. They are not the same. So if you focus on one, the other becomes less efficient. Certain species are not right, but home systems do work well." There are many of these on the market. "There is also a case for using aquaponics in warm countries where it can be done outdoors very efficiently."

Henry Gordon-Smith is a sustainability strategist focused on urban agriculture, water issues and emerging technologies, and a board member of the Association for Vertical Farming. He believes that the increased use of robotics will grow the industry, but it is not the only way. Particularly where there are plants being grown with different varieties together and human judgement needs to be deployed, only people will be able to harvest. "We are creating lots of green collar jobs. We do use some automation but not a lot. There is a place for automation that we need to discern, between efficient versus effective production, where effective means socially effective and creates employment in the neighbourhood. We like the human input. It's not a green industry if humans are cut out."

Aquaponics

Not everyone agrees with the above view on aquaponics. Aquaponics is a combination of aquaculture (fish farming) and hydroponics (growing plants without soil). From a small unit it's possible to harvest both fish (usually tilapia) and green vegetables, while using the waste from the fish to feed the plants and the plants to clean the water for the fish. Basel, Switzerland hosts the first commercial example in Europe, situated on a rooftop of LokDepot, the brainchild of the Institute of Natural Resources Sciences.

According to Ranka Junge of Zhaw Zurich University, the benefits are: nutrient utilisation; low water consumption; edible plant production as well as fish. But the disadvantages are that you need to know about both fish and plant production; the complex system involves a lot of expertise. "For this reason, doing it on a commercial scale is hard," she said. "But the advantages are no use of pesticides or antibiotics, making food local, adding greenery in the city, nearly closed nutrient cycles, reduced energy input, perfect food safety control, around a 90 per cent reduction in water use, and that vertical farming increases efficiency."

Roman Gaus and Andreas Graber, who run the farm, say it occupies just 26 m^2 and has been operating since winter 2012. The main input is the fish feed: 1 kg produces a fish harvest of 700 g and 5–10 kg of tomatoes. The water is evapo-transpired, condensed and returned (cleaned thereby) to the fish. Fish produce ammonia and the plants clean the water. This amount of fish produces 2 litres of sludge, which is vermicomposted (composted with worms), giving nutrients to the plants. No artificial lighting is used. In a year, 20.9 megawatt hours (MWh) of electricity, 32.2 MWh of heat plus 763 m^3 of water produce 3,401 kg of salad and 706 kg of fish. The top line is it produces 2.7 kg fish and 13.1 kg vegetables per acre.

There is no environmental pollution and the food is organic and healthy, produced with respect to animal welfare, fresh and sustainable. Ranka Junge has calculated that on this basis 3 m^2 of rooftop space could feed one person 12 per cent of their diet. The first commercial farm was started in London in 2018, in an abandoned warehouse.

Now let's look at a few places in the world where vertical farming is being used.

Singapore

Singapore has one of the highest population densities on the planet. Some 90 per cent of food is imported and just 8 per cent of vegetables are grown locally. In the last ten years there's been a 16 per cent reduction in land used for farming to just 1 per cent of land area. Food resiliency is therefore at crisis point.

To tackle this, Jack Ng, founder of Sky Urban Solutions, built one of the world's first commercially viable vertical farms in 2012. His four-storey farm grows tropical leafy vegetables, and root crops such as yams and carrots. The design principles maximise the use of real light (artificial lighting would be used in higher latitudes). The rotating towers provide for light to reach all of the plants equally. The design also alleviates the impact to the environment; has high productivity; consumes low levels of resources; and improves workers' quality of life – "The work should not be back-breaking!" says Jack.

Rainwater is collected from the roof. Nutrients and water are recycled, and it is mostly powered by a gravity-fed waterwheel. The design also includes cold chain management and organic waste recycling. As a result it uses very little energy, with a cost in 2015 of just $3/month. Three-quarters of the greenhouse is netted so natural ventilation may be used instead of fans. The only major input is seaweed fertiliser from China.

"We have to take action now for our future in the area we believe in," says Jack. "Don't think about it, go out and do it. I am working not just for Singapore but for the whole world."

Japan and Taiwan

Chiba University academic Toyoki Kozai told me in 2014 about the astonishing growth of plant factories with artificial lighting in Japan; there were 168 in March 2014. Of these, a quarter were making a profit, and half were breaking even. The ones that were not profitable were run by people who were more skilled in growing than business, and would be receiving management training. "Annual productivity per unit land area is roughly 100-times that of open fields," he says, "which is increasing with scale-up, partly due to shortening the culture period and other environmental control procedures. But the biggest factor is the use of tiers. Quality is improved and there is less loss due to pests and diseases."

The largest produces 10,000 heads of lettuce per day or over 7,000 heads/m^2/year. They are sold at €1.4/bag (70–80 g). The latest farm produces 1,000 heads per day; it is in Fukushima and run by the Agricultural Co-operative. Besides greens, they grow medicinal mushrooms, which thrive on CO_2.

The concept has been exported to Mongolia. There, it is -40°C in winter, but the lamps used generate heat and no further heating is required. Suitable plants are short in height and grow fast (10–30 days), grow well at high density under artificial light and have a high value as fresh, tasty and pesticide-free. They include salads, herbs, turnips, radishes, angelica, ginseng, carrots and wasabi roots. Products made from them in Taiwan include paste for baby food, juice, cosmetics, sauces, powders, herbal teas and supplements.

Planning and regulation

Negotiating planning and regulations for new urban farming initiatives currently varies according to locality because of the inexperience of officials in dealing with this type of planning application. To support urban farming in the future, planning codes must change. Some cities are leading the way. "Boston has an app that helps you navigate the planning codes. New York City should have one and all cities should run workshops and say these are the structure we would like to see if you want to apply to grow food in the city," says Gordon-Smith. "There is a greenhouse amendment in New York City – you can put one on top of a building – but in the UK you need a permitted development order."

"Cities need to develop typologies on the type of building that can be used with codes," says Rodriguez. "Building integration of agriculture is about to what degree operational and utility costs get reduced to make the buildings and the entire city more efficient." Kluko agrees: "It's about nomenclature. We need definitions and to approve the components of each project. Planners don't yet understand, it needs to change soon."

Detroit

Detroit is a post-industrial ghost town. It is reinventing itself using urban farming. The authorities were forced to face the fact that locals had begun using empty lots and warehouses to grow their own food – not permitted at the time. So in 2012 the Detroit City Plan was updated to feature urban agriculture as a desirable activity, acknowledging the environmental, economic and social benefits. This was followed by 2013's Detroit Future City Strategic Framework, which made food production a priority for all city stakeholders to accelerate economic revival, address land-use issues, improve city services, and foster civic engagement. It gave urban agriculture a zoning ordinance, thereby formally permitting, promoting and regulating certain types of food production as a viable land use.

This planning barrier is often a vital issue facing those who want to practise urban farming, especially indoors with vertical farms. Entrenched attitudes and vested interests are still being dealt with. This is in the nature of all attempts at "system change". But Mayor Mike Duggan has stated a preference for vacant land to be put in the hands of local residents for growing food. Detroit contains several vertical farms, often built in disused warehouses. These grow food organically in controlled conditions, recycling water and nutrients, and using LED lighting while training unemployed young people in valuable skills and giving them employment.

One interesting side effect has been this: much salad used to be imported from California, over 2,000 miles away. Not only was the transport impact huge, but that state continues to experience chronic drought. 95 per cent of salad is water. So by eating locally grown salad, Detroit (which receives over twice the annual rainfall) is helping to conserve the water present in California.

Artificial meat

As developing countries grow wealthier and adopt a Western diet their citizens tend to eat more meat. This is incompatible with feeding the growing world population within planetary limits. Even if the meat industries were to increase yield by half and reduce waste by half there will not be enough cropland available in 2050 to feed everyone without replacing mammal products with poultry, aquaculture.[34] The alternative is a mass conversion to veganism, which liberates land presently given over to animals themselves and growing animal feed to be converted to growing food for humans. But a third way is being suggested: to continue to satisfy the taste for meat, why not grow artificial meat in factories or laboratories? These (currently hypothetical) products would not require so much land.

Artificial meat is made by growing meat cells on a growth medium and is also known as cellular agriculture. Investors are pouring millions into partnerships with biologists and chemists in the hope of developing products that will prove popular and economically viable. This is far from farming, but fourth-generation dairy manager and cheesemaker Rebecca Ruth Seidel of Wholesome Dairy Farms in Pennsylvania believes that, "Cellular agriculture should not be viewed as a threat by the agricultural community. Rather, it should be viewed as yet another tool to feed a growing population on a planet with limited resources."

This is a view that is gaining ground. I talked to a multi-award-winning charcutier, and organic farmer, Illtud Dunsford, who supplies gourmet restaurants. He has formed a new company in south Wales called Cellular Agriculture. He said, "There is an ethical problem with artificial meat in that with most business models it requires stem cells from the embryos of unborn calves taken from slaughtered heifers." His view is that it is possible to produce artificial meat from a different, genetically engineered source. "After going to a conference on this subject with a friend to discover that we were the only farmers there I've teamed up with financiers and biologists, and that is what my company is developing – something more sustainable and ethical than what others are doing. It would completely remove the need for dependence on the dairy industry."

It may take a few years, but the feeling is that there is a potentially huge market out there, and it is the only way for meat to stay on the menu for most current meat eaters. Real meat may well become an exotic, expensive menu item. With their intention to let people 'have their cake and eat it', fake steaks, bacon and burgers have the same relationship to their real counterparts as vaping technology does to cigarettes for smokers: both would allow old habits to continue but with reduced damage.

Some studies estimate that artificial meat will probably create fewer greenhouse gas emissions and less pollution than livestock farming, especially when the livestock is factory-farmed. It is also more ethical. However, others are less sure: although they could require smaller quantities of agricultural inputs and land than livestock, this could come at the expense of more intensive energy use as biological functions such as digestion and nutrient circulation are replaced by industrial equivalents.[35]

Insects are also likely to be increasingly factory-farmed and processed into palatable products, though the ecological footprint of this process, given that they also need feedstock, is currently unclear.

Further cities to learn from

Belo Horizonte; Nairobi's Urban Agriculture Promotion and Regulation Act; and Greater Toronto's Golden Horseshoe Food and Farming Action Plan. Shanghai provides an additional example as it is forging ahead with vertical farming. The Marta farm, 20 km outside Havana, Cuba, is another world-leading example of regenerative urban farming.[36]

Belo Horizonte

In south-eastern Brazil, Belo Horizonte's government-led alternative food system is a successful experiment in providing sustainable nutrition to people on low incomes while creating jobs. It runs in parallel to the conventional, market-led system. Programmes are delivered in partnership with civil society, private companies and municipal departments, and reach around 300,000 citizens – 12 per cent of the population – every day.

In 2015 the School Meals programme served 155,000 children in the public school system, while the Popular Restaurants served over 11,000 meals per day. Food is sourced locally from 133 school vegetable gardens and 50 community gardens, educating children about the link between food and the land. The Straight from the Country programme supports 20 family farmers, and 21 grocery stores are in the ABaste-Cer programme.[37]

Nairobi

The Nairobi Urban Agriculture Promotion and Regulation Act 2015 was passed after sustained campaigning by urban farmers to build supportive relationships with national civil servants. Previously, 20 per cent of Nairobi's population were either growing food crops or rearing animals for food – illegally. Now, the Nairobi City County Government is responsible for training farmers, ensuring their access to organic waste, and for developing marketing infrastructure as well as monitoring and regulating quality and hygiene standards, animal welfare and traceability. This promotes job creation, value addition and value chain development.

One of the many barriers to overcome, common to many cities, was the perception that the nutrition benefits resulting from rearing animals for food in the city outweigh the public health risks.

Toronto

The hinterlands-sourcing approach is being taken in the Golden Horseshoe region that runs around the western shores of Canada's Lake Ontario, including the Greater Toronto area and neighbouring communities. It is densely populated with rapidly expanding cities of educated, affluent professionals. In 2011–2012 seven municipalities adopted a common ten-year plan to help the food and farming sector remain viable by guaranteeing markets for their sustainable produce and thereby giving incentives to invest. This Golden Horseshoe Food and Farming Plan 2021 has five objectives:

1. to grow the food and farming cluster
2. to link food, farming and health through consumer education
3. to foster innovation to enhance competitiveness and sustainability
4. to enable the cluster to be competitive and profitable by aligning policy tools, and
5. to cultivate new approaches to supporting food and farming.

Its broad aims and membership have achieved much since then but also seen conflict arise between advocates of small-scale, ecological agriculture and so-called "big agriculture". Face-to-face meetings are sorting out differences but so far no major food company is represented.[38]

A community planned around an urban farm

This is a truly 'one planet' idea: California's first farm-to-table new home community is a mixed-use community with 583 residences on the site of a former tomato cannery in Davis, on the outskirts of Sacramento near San Francisco. The development, known as The Cannery, covers 7.4 acres, of which the farm is 5.5 acres, including 4 acres of farmland with organic vegetables, poultry and orchard fruit, retail shops, a recreation centre, outdoor amphitheatres and miles of trails. There is more detail in a case study in Chapter 15.

The project includes a teaching academy for sustainable farming. The centre launched the California Farm Academy in 2009 to help those wanting to break into a career in agriculture. Crops from the urban farm are sold through a farm shop. Mary Kimball, the executive director of the Center for Land-Based Learning, the organisation managing the urban farm, says, "We have three beginning farmers, all graduates of the California Farm Academy, who have started new farming businesses and are now providing the residents of the town very local produce."

"To see The Cannery today becoming a viable farm community is not only personally exciting for me, but also one of the most fulfilling accomplishments in my career," said Craig McNamara, founder of the Center for Land-Based Learning.[39] "The Cannery Urban Farm honours what I believe in most: connecting eaters directly to food." There are several "agrihoods" in North America, such as Agritopia in Phoenix[40] and Serenbe in Atlanta,[41] she says, but they tend to have different arrangements with the farmers. "As far as we can tell, The Cannery is the nation's first farm-to-table housing development focused on beginning farmers."

The Academy offers a full range of training and internship opportunities, focusing not just on growing but on conservation and access to land and equipment, the most expensive barriers to starting a farming business. "The cost of land is really expensive," says Hope Sippola, one of the farmers. "The only way to make it affordable is to lease land through the centre." This is one example of how the centre partners with public and private landowners to provide low-cost lease opportunities. Weekly vegetable subscriptions via veggie boxes – Community Supported Agriculture – are also offered.

The Cannery is managed by New Home Company. Kevin Carson, northern California president, says: "We have worked extremely hard over the past several years to get to this moment. The Cannery is unlike any other community in the western United States and it has truly been a rewarding experience to contribute to such an innovative concept."

Sacramento Region has had a push in recent years to establish itself as America's Farm-to-Fork Capital. In 2015, Sacramento's Elk Grove City Council expressed unanimous support for a plan to introduce urban, commercial farming within non-agriculturally zoned areas in the city. If the idea becomes a reality in Elk Grove, similar farms could be operated on currently vacant, unimproved or otherwise underdeveloped parcels in the city. They can be for-profit, non-profit and/or social enterprises. Their products can be sold at such places as on-site stands, farmers' markets, grocery stores and restaurants. These farms can also contribute to food banks.

The future of food

Planning new communities around urban farms reconnects city dwellers to nature, and the process of growing food, from which they are all too often distant and alienated.

In a 'one planet' future, sugar, coffee, tea, cocoa, grains and large-scale pulses like soya will have to continue to be grown at scale, outside cities. They will ideally be grown organically using natural pest-control techniques and perhaps be more resistant, GM strains. Most other vegetables can be produced in cities in controlled, organic, pesticide-free conditions that conserve water and provide meaningful jobs.

Both vertical farms and agro-ecological smallholdings are more productive than traditional farms. Data from a conversion of a Welsh sheep farm to this type of horticulture[42] have shown a 30-fold increase in productivity, without subsidy. In addition they improve biodiversity and soil fertility year-on-year through composting and employ more people to keep a closer eye on each square metre of land. Animals (dairy, fowl, pigs) are often used productively as part of the growing cycle.

In cities, growing areas using both methods should be encouraged through education.[43] Mini-farms can be extremely small and fit well within and around urban areas. Ribbon developments can draw their sustenance from the land either side. Growing may occur on rooftops and backyards (also tackling urban air quality and over-heating). 'Patchwork farm' arrangements such as Farmdrop[44] in London link multiple producers of organic produce via mobile technology to coordinate direct sales to customers.

Cities can use their procurement and legislative powers to support local agri-food businesses, shortening supply chains, favouring environmental, health and social benefits. Overall this will reduce the cost burden, increase resilience and harness the multiplier effect. The future of food can be nutritious, more local, regenerative, healthier and available to all.

Notes

1 See: www.eatforum.org/person/dr-gunhild-anker-stordalen/
2 At the Harmony in Food and Farming Conference, Llandovery College, June 2017.
3 Walker, D., et al., *Urban Biocycles*, Ellen MacArthur Foundation, 2017. See: www.ellenmacarthurfoundation.org/publications/urban-biocyles
4 See: www.wrap.org.uk/sites/files/wrap/Prevention per cent20and per cent20reduction per cent20of per cent20food per cent20and per cent20drink per cent20waste per cent20in per cent20business per cent20and per cent20households.pdf
5 See: http://flwprotocol.org/

6 Borsato, E., et al., "Sustainable patterns of main agricultural products combining different footprint parameters", *Journal of Cleaner Production*, 2018, 179, 357e–367. DOI 10.1016/j.jclepro.2018.01.044.
7 *NYC Daily News*, 17 March 2014.
8 *NYC Daily News*, 17 March 2014.
9 See: http://blaencamelbox.com/
10 See: https://eia-international.org/supermarket-iceland-removes-palm-oil-from-its-products-over-sustainability-concerns/
11 See: https://corporate.marksandspencer.com/plan-a/food-and-household/capacity-building-initiatives/farming-for-the-future
12 World Bank, *Agricultural innovation systems: An investment source book*, The World Bank, Washington, DC, 2012.
13 See: www.fao.org/in-action/food-for-cities-programme/news/detail/en/c/447315/
14 See: www.npr.org/sections/thesalt/2015/04/01/396607690/navajos-fight-their-food-desert-with-junk-food-and-soda-taxes
15 *Urban Agriculture* magazine, March 2014. See: www.ruaf.org/sites/default/files/Multiple per cent20use per cent20of per cent20green per cent20spaces per cent20in per cent20Bobo per cent20Dioulasso.pdf
16 See: www.epa.gov/transforming-waste-tool/zero-waste-case-study-san-francisco
17 See: www.iol.co.za/capetimes/news/urban-agriculture-means-food-plus-social-cohesion-2096857
18 Jégou, F., *Sustainable Food in Urban Communities – Developing low-carbon and resource-efficient urban food systems*, URBACT, 2013. See: www.strategicdesignscenarios.net/downloads/Publications/130201_ePublication_v7-light.pdf
19 See: http://malmo.se/download/18.d8bc6b31373089f7d9800018573/Foodpolicy_Malmo.pdf
20 See: www.milanurbanfoodpolicypact.org
21 *A Vision for City Region Food Systems*, FAO's Food for the Cities Programme/RUAF Foundation, 2017. See: www.fao.org/cfs/
22 See: www.c40.org/networks/food_systems
23 See: eatforum.org
24 De Cunto, A., et al., *Food in Cities*, European Commission Directorate-General for Research and Innovation and Directorate Policy Development and Coordination Unit, 2017. See: ec.europa.eu/research/openvision/pdf/rise/food_in_cities.pdf
25 IPES-Food, *What makes urban food policy happen? Insights from five case studies*, International Panel of Experts on Sustainable Food Systems, 2017. See: www.ipes-food.org
26 Butterly, D., Fitzpatrick, Ian, et al., *A People's Food Policy*, June 2017. See: www.peoplesfoodpolicy.org
27 See: regions-france.org/wp-content/uploads/2016/10/Declaration-of-Rennes-For-Territorial-Food-Systems-June-2015-English1.pdf
28 See: www.ruaf.org/projects/dutch-city-deal-food-urban-agenda
29 Girardet, H., *The Gaia Atlas of Cities: New Directions for Sustainable Urban Living*, Gaia Books Ltd, 1996, ISBN-10: 1856750973.
30 Despommier, Dickson, *The Vertical Farm: Feeding the World in the 21st Century*, St Martin's Press, 2010, ISBN 978-0-312-61139-2.
31 See: www.inc.com/kevin-j-ryan/aerofarms-disruptive-25-2017.html
32 Presentation at a conference at Nottingham University's Centre for Urban Agriculture on Vertical Farming and Urban Agriculture, September 2014.
33 See: www.sasaki.com/project/417/sunqiao-urban-agricultural-district/
34 Roos, E., et al., "Greedy or needy? Land use and climate impacts of food in 2050 under different livestock futures", *Global Environmental Change*, November 2017, 47. doi.org/10.1016/j.gloenvcha.2017.09.001
35 Mattick, C., et al., "Anticipatory Life Cycle Analysis of In Vitro Biomass Cultivation for Cultured Meat Production in the United States", *Environ. Sci. Technol.*, 2015, 49 (19), 11941–11949.

36 See: www.havanatimes.org/?p=116725
37 Hawkes, C. and Halliday, J., "What Makes Urban Food Policy Happen?" *International Panel of Experts on Sustainable Food Systems*, 2017. See: www.ipes-food.org
38 See: www1.toronto.ca/wps/portal/contentonly?vgnextoid=75ab044e17e32410VgnVCM10000071d60f89RCRD for more information.
39 See: http://landbasedlearning.org/
40 See: http://agritopia.com/
41 See: http://serenbe.com/
42 See: www.oneplanetcouncil.org.uk/wp-content/uploads/2017/01/OPD-briefing.pdf
43 See: http://urban.agroeco.org/ for one example.
44 See: www.farmdrop.com

6

REGENERATIVE CITIES

Soil and the natural world are frequently overlooked in discussions about cities but healthy soil and biodiversity are just as much the foundation of 'one planet' cities as they are of the countryside.

The concentration of power and lock-ins to trade

Sustainable Development Goal 15 sets a target for zero land degradation and measures to protect and prevent the extinction of threatened species by 2020. We need these targets because of the way that cities devour their vast quantities of food.

At present just a handful of mega-companies dominate global food production.[1] In food trading there are only four agricultural firms: ADM, Bunge, Cargill, and Louis-Dreyfus, who exert huge influence.

Because of their drive for a purely economic version of 'efficiency', three-quarters of the world's food is generated from only 12 plants and five animal species, according to the UN FAO.[2]

Large-scale monoculture operations mean high volumes of chemical inputs equating to pollution, carbon emissions and soil degradation. The annual cost of soil degradation is put at US$300 billion.[3] As soil quality reduces, so does productivity, leading into a vicious cycle of dependency upon chemical inputs. Its ability to retain atmospheric carbon also decreases, thereby worsening climate change. In Africa, two-thirds of agricultural land is degraded.

Massive land-use intensification and conversion from the natural state accelerates the current extinction crisis through habitat loss and, with just a few crops, a dramatic loss of genetic diversity. So now, although the world's 7.6 billion people only make up 0.01 per cent of the Earth's biomass, humanity has caused the loss of over 80 per cent of all wild mammals and half of plants, while livestock has increased dramatically.

To remedy this we need "policies and programmes to diversify diets and improve micronutrient intake; and developing and deploying existing and new technologies for the production, processing, preservation, and distribution of food", says the International Assessment of Agricultural Knowledge, Science and Technology for Development.[4]

WWF's *Living Planet Report* [5] acknowledges this and adds that to achieve the necessary transformation we need a change of mindset – to adjust our 'systems thinking': "individual consumers can change their purchasing behaviour, [and] people with greater political or economic influence can formulate strategies for policy change."

One solution beyond those covered in the previous chapter is regeneration. Writer, campaigner and academic Herbert Girardet has long been an advocate of sustainable cities but in the last 15 years he has been championing the concept of regenerative cities. Cities 'regenerate' by encouraging natural ecosystems to thrive within their jurisdictions. Underpinning their success is what he and others call a 'regenerative economy' – a form of capitalism that repairs the damage done by current and past economic systems.

"Our own bodies replace their 100 trillion cells every seven years or so. Life is continuously renewed through the action of great variety of interconnected forces. Could modern economies operate in a similarly regenerative manner?" he asks. In a recent paper[6] he adds, "It is astonishing that current accounting practices still do not factor environmental externalities into national balance sheets. We have to urgently address this fundamental systems problem. How can a regenerative approach prevail?"

He argues that accounting for the damage we do to the natural world implies assuring a restorative relationship between the realms of economy and ecology; the efficient, circular processing of materials in the technical system; mainstreaming efficient, renewable energy systems across the world; and the creation of appropriate policy frameworks for thriving green economies.

The challenge of externalities

The crucial issue of the fact that economics does not account for the damage to the environment caused by business has been discussed for decades, with little to show for it. It is becoming apparent that if externalities are directly experienced at the *local* level, there's a greater chance that they will eventually be addressed, particularly where people's health is at stake. For example, effective measures have been taken to deal with effects of smoking, asbestos pollution and contaminated drinking water. This is where cities can intervene.

Three areas need urgently addressing. First is large-scale regeneration of forest ecosystems and farmland soils. This can be achieved through procurement and supply chain contracts. Second: river water contamination caused by cities not returning the nutrients contained in urban sewage back to farmland, forcing farmers to replace these lost nutrients with damaging artificial fertilisers. This can be remedied by creating a market for these reclaimed nutrients. Third is curbing plastic pollution, which must be combatted by substitution of plastics by more sustainable materials.

All of these are aspects of a circular economy. Girardet supports a shift from the present linear systems of mine, use, throw away to circular systems which involve reclaiming and reusing materials endlessly, as advocated by the Ellen MacArthur Foundation and by Anders Wijkman, co-president of the Club of Rome, chairman of the Swedish Association of Recycling Industries and a member of the World Futures Council. In 2012 Anders published the influential report to the Club of Rome, *Bankrupting Nature*,[7] and in 2017 co-authored – together with Ernst von Weizsaecker – *Come On! Capitalism, Short-termism, Population and the Destruction of the Planet.*[8]

Pavan Sukhdev, president of WWF, argues that to encourage this move we need to overcome the problem of "the economic invisibility of nature". He says that to align economic and ecological systems it is crucially important to understand that the value of ecosystems services is vast, and probably more than double the value of the global economy.[9]

Under the right conditions, biological resources will always tend to renew themselves. Regenerative development (a step beyond sustainable development) requires understanding the integral and interdependent nature of living systems, with the sun powering all life processes on Earth, says Girardet. Then we must craft and implement policies, incentives and everything else possible to stimulate the regenerative capacity of nature.

City administrators often view nature – such as street trees – as a cost, not an opportunity to improve cities. They worry about litigation, operating costs and engineering problems. A study of Australian cities[10] recommended that this could be remedied by a better understanding of the many benefits and costs of green infrastructure, employing new valuation frameworks. For example, Blacktown City Council informed residents of one street that by adjusting the number and type of street trees, they could eventually reduce the average yearly household electricity bill by AU$249. The City of Brisbane estimated that street trees contribute $1.67 million to the economy by improving air quality, capturing rain and storing carbon. The City of Sydney uses a register of significant trees to protect trees that "contribute to the environmental, cultural, [and] social character of the city".

There are potential conflicts between roots and service cabling and pipes beneath streets. Some councils now provide lists of tree species they recommend for planting near power lines. The San Francisco Public Utilities Commission uses a tool to estimate the extent green infrastructure will affect the financial, social and environmental outcome of urban precinct projects. Applying it to the city's Green Square revealed that doubling the tree canopy can result in improved property value, biodiversity, health and well-being, suitability for walking, amenity, calming of traffic and reduced energy consumption.

With such a scientific approach we can encourage and support nature within cities. We can relocate human beings within nature – as our ancestors were for hundreds of thousands of years – but instead of being in forests or savannahs we will dwell within dense, artificial habitats; artificial as in 'designed', not unnatural.

Living within ecosystems is not just essential for our survival and well-being but is in our genes and heritage, and we ignore this at our peril. A designed regenerative city would combine the best knowledge of ecosystems with the best knowledge of city management. A marriage of the past and the future. An antithesis to the soulless, concrete and steel monstrosities featured in dystopian futures.

Regenerative cities pay back the debt they take from nature. The concept takes its model from nature, where there is no waste and everything is endlessly reused without an increase in entropy. Entropy is a form of disorder, defined in this case by resources emerging from their consumption within cities in a less useful, downcycled state which at their worst cause damaging pollution. Fossil fuels and their plastic products are a polluting, entropy-increasing and therefore ultimately destructive way to power industry.

By contrast, living things have the ability to up-cycle nutrients and biological waste back into materials useful to us. The use of thriving organisms and clean, renewable energy are regenerative. They will never run out and, as with nature, create no waste since everything is constantly re-cycled. "Nature does not set out to be sustainable but achieves it because it is regenerative. If the economic system is to prosper long term, it must operate like nature; regeneratively."[11]

When Herbert Girardet was working in Adelaide in 2003 he argued that a vigorous move towards regenerative development could greatly stimulate South Australia's economy. "A city region that takes active measures to improve the efficiency of its use of resource also reduces its reliance on imported resources – it re-localises parts of its energy and food economy and brings a substantial part of it back home," he wrote in 2018. "These concepts became the basis for a wide range of new policy initiatives by the government of South Australia. Today South Australia is a world leader in solar and wind energy, battery storage of electricity, solar-electric transport, organic waste recycling and composting, wastewater irrigation, urban-fringe organic farming, etc."[12]

He adds: "In the US the concept of a 'regenerative economy that supports the mutual thriving of people and planet'[13] has been getting much traction in a growing environmental community." He predicted that the Commonwealth Secretariat (representing one-third of humanity) would support that idea in its April 2018 proclamation, but sadly it did not.

Since cities are increasingly responsible for the majority of environmental damage caused by human beings, then Girardet argues that it is absolutely necessary for them to adopt new policies that foster the closed-loop approach, where either there are no negative outputs, or they become the inputs for processes which generate positive outputs.

Girardet proposes amongst other things that citizens of regenerative cities consume only about 2,000 watts (2 kW) on an average ongoing basis. This is because the human body uses only about 100 watts but nowadays people use between 6,000 watts in Europe, and 12,000 watts in America. This way of looking at energy flows and entropy both gives us a dramatic picture of what fossil fuels have enabled us to do but also a means of measuring how successfully we might be reducing our input. It balances the urban metabolism. There is work going on in Zurich, Basel and other Swiss cities to try to achieve this limit.

Ecopolis

In his book *Creating Regenerative Cities*,[14] Girardet explores the concept of Ecopolis, a fictional city embodying these ideals. In Ecopolis all energy is renewably generated and water is supplied in a sustainable manner. Food is mainly grown organically, or, before the water is discharged into the environment, potential pollutants such as phosphates and nitrates are removed. Much food is also grown near to the city if not within it. Integrated transport is the norm with zero air pollution, much cycling and walking. Above all, Ecopolis returns to nature at least as much as it takes, since this is the meaning of regenerative.

Herbert discussed with me whether Ecopolis is a feasible proposition. He described cities where he has already worked to try to realise his vision, including London, where he worked on quantifying the metabolism of London. "I concluded that its ecological footprint was equivalent to an area of land the size of England. More recent work has doubled this figure to reach an area of land that is 300 times that of Greater London itself."

In Adelaide his work led to policy changes both in the government of Southern Australia and of Adelaide the city to attempt to reduce its ecological footprint. He worked on developing ideas for Chinese eco-cities including Dongtan, east of Shanghai, "which though unfortunately never realised, was hugely influential in policy terms for designing other Chinese eco-developments".

At one time when we spoke (2014–2015), he was hopeful that Bristol would become as green a city as Adelaide because of a confluence of green groups, an independent visionary mayor, George Ferguson, and other factors contributing to many eco-policies being adopted. But more recently with the election of a new mayor, some of the momentum towards making the city more sustainable has been lost, with an emphasis instead upon growth to accommodate the fact that it is an increasingly popular place to live (ironically because of its green policies) and is growing fast.

Cities require stable and consistent governance as much as countries. Girardet cites the example of Copenhagen in Denmark. "Copenhagen is a foremost city in the world that has done all the right things in terms of renewable energy, pedestrianised its city centre, creating cycle lanes, implementing energy efficiency and combined heat and power and distributed heat systems. It's true that we are seeing more retrofitting of existing cities to make them more regenerative than we are the building of new eco-cities but I believe that as the world becomes more aware of impending ecological disasters and climate change we should see more and more city leaders taking action to create regenerative cities," he says.

But at the same time he laments those cities in developing countries which emulate the old model of 'Petropolis' while attempting to be green. He criticises Abu Dhabi's Masdar City for not going nearly far enough. "It's never enough. Cities frequently do not have the full capability required: this often falls in the hands of national governments or the capitalist system itself, for the real enemy of the ideal of regenerative cities is high consumption levels."

The regenerative city concept is at the heart of The City We Need, a campaign of the World Urban Campaign and UN-Habitat.[15] Its Principle 6 states that:

> The City We Need is regenerative, energy and resource efficient, low-carbon, and increasingly reliant on renewable energy sources. It replenishes the resources it consumes and recycles and reuses waste. It manages water, land, and energy in a coordinated manner and in harmony with its hinterlands. It supports ecosystem restoration and city-regional food systems, including urban and peri-urban food production and community-based agriculture. It is endowed with multifunctional, adaptable infrastructure that supports local biodiversity while providing public space that improves quality of life. It recognizes the carrying capacities and limitations of the natural systems which support it, and values ecosystem services for the roles they play in urban health, environmental protection, aesthetics and livability.

This is pretty much what the rest of this book explores. To create regenerative cities could mean adapting and extending the idea of garden cities, as supported by the British Town and Country Planning Association.[16] How to create garden cities is more completely explored in Chapter 15, while the importance of green space to mental and physical health and nature in cities is explored in a section in Chapter 10.

Making space for nature

The share of land allocated to open spaces in most cities is currently insufficient, according to a 2018 UN report on cities.[17] It describes how bringing nature into cities has many benefits. For example, Colombo, Sri Lanka, is enhancing climate resilience and reducing flood risk by restoring wetlands.

Singapore, with 8,155 people occupying every square kilometre, is consistently ranked as one of the world's most liveable cities. One of the reasons for this is the prevalence of high-quality urban greenery throughout the city. This has been achieved by mandatory roadside tree planting, to create a pleasant environment: the trees, parks and other greenery help to reduce temperatures, filter air pollution, and mute street noise, while permeable street surfaces help to manage storm water run-off and prevent overflow from sewers.

Parks in Singapore are deliberately well-distributed and connected by hundreds of kilometres of pathways. Between 1986 and 2007, green cover grew from 36 to 47 per cent, despite a 68 per cent increase in population. This reduced average temperatures by between 0.5 and 5°C. In that part of the world, a drop of 1°C in air temperature lowers peak electricity demand for air conditioning by as much as 4 per cent. The government now requires property developers to replace any greenery lost during construction and very wisely covers half the cost of installing green roofs and walls on existing buildings.

Countering the prevailing global mood of pessimism, the UN survey concludes its analysis of the trend towards regenerative cities: "We know that we are grossly under-estimating the benefits of this new growth story. Current economic models are deeply inadequate in capturing the opportunities of such a transformational shift. Bold action could yield a direct economic gain of US$26 trillion through to 2030 compared with business-as-usual. And this is likely to be a conservative estimate."

Biomimicry

Another word for the application of nature's principles in cities in popular currency is biomimicry. Not surprisingly, this is about techniques that imitate the processes found in nature.

Jamie Miller is the founder of Biomimicry Frontiers and the Biomimicry Commons. He cites the examples of Velcro and a whale-inspired turbine blade as examples of emulating nature's forms. He says that nature has devised clever systematic strategies for transforming complex systems such as forests, and cities have similar levels of complexity.

What can we learn from nature to apply to cities? One thing, he says, is an adaptive capacity. "For example, a tree that can adapt its roots, bark, branching structure, leaves, and connections to other species will allow multiple, small-scale [changes] to occur simultaneously, which will reduce the potential for large-scale transformation or collapse. E.g. leaves falling or trees dying allow the stored nutrients to go back into the soil for other, perhaps more appropriate, species to emerge."

Miller believes we should reframe our built environments through the lens of biomimicry: "walls that could breathe like skin, buildings that could self-heal, small-scale and local fabrication of complex materials, or creating communities that generate no wastes in a circular economy. With emerging technologies like 3D/4D printing, big data, nanotechnology and green chemistry, these ideas are increasingly becoming more realistic as we're beginning to see incredible advancements of what is possible."

Eastgate Building, an office complex in Harare, Zimbabwe, has an internal climate control system originally inspired by the structure of termite mounds; without air conditioning it uses only 10 per cent of the energy of a conventional building of the same size. Modelling the nose of the Shinkansen Bullet Train after the beak of kingfishers, which dive from the air into bodies of water with very little splash to catch fish, resulted in a quieter train that uses 15 per cent less energy yet travels even faster than a previous design.

Forests and other complex ecosystems are able to use the millions of hourly interactions of multiple species at many different scales in self-organising alignment to ensure that every component is used efficiently without waste and the whole is in balance with the wider geo-regional environment. There is no central organising force.

Whilst cities themselves could not be run this way, since the financing of infrastructure and planning procedures need an executive management, their administration should be flexible enough to allow individuals and community groups to operate relatively autonomously within a rules-based system. This aspect of biomimicry is covered in more detail in Chapter 13, on smart cities, while we also revisit the topic in the chapter on water to see how the principles of regeneration and closed loops apply to water management in cities.

Towards closed-loop societies

China and the EU are cooperating on building more closed-loop societies. Research conducted by the Ellen MacArthur Foundation has found that applying circular economy principles to China's cities could make goods and services more affordable for citizens, and reduce traffic congestion and air pollution. Brussels, Belgium, Phoenix, Arizona, USA, and London, England are amongst developed cities trying to move towards a circular economy.

In Brussels it is led by a public body called Bruxelles Environnement that was set up to advise the government on issues including waste management, sustainable construction and air quality. The region has adopted a programme called 'Be Circular', which is a global strategy that aims to take Brussels from a linear to a circular economy.

To measure the efficacy of a transition to a circular economy, whether the body is a city or a company, the Ellen MacArthur Foundation has provided a set of indicators[18] which complement the Global Reporting Initiative guidelines. So far they are set to be applied at product and company levels. A 'Material Circularity Indicator' (how much materials are reused) is complemented by indicators for toxicity, scarcity and energy. They answer questions such as: How much input is coming from virgin and recycled materials and reused components? How durable and repairable is the product compared to an industry average? How much material goes into landfill (or energy recovery), or is recycled or reused, and how efficient are the processes to produce recycled input and to recycle material after use?

We humans are part of nature and it is bad for us to be too far removed from it. This is only one of the reasons why the 'one planet' city must take steps to pay back to nature more than what human actions have taken away in the past and which we continue to extract now. Nature should be welcomed everywhere because it is good for us. Everything around us is a part of many natural cycles, and to embrace them is to understand the true meaning of sustainability. The 'closed-loop' economy is nothing more than adapting these principles to the artificial ones that we create.

Notes

1 Gura, S. and Meinberg, F., *Agropoly – A handful of corporations control world food production*, Berne Declaration & EcoNexus, 2013. See: www.econexus.info/sites/econexus/files/Agropoly_Econexus_BerneDeclaration.pdf
2 See: www.fao.org/docrep/007/y5609e/y5609e02.htm
3 *The state of the world's land and water resources for food and agriculture – Managing systems at risk*, Food and Agriculture Organization of the United Nations, Rome, Italy, and Earthscan, London, 2011. See: www.fao.org/docrep/017/i1688e/i1688e.pdf

4 *Agriculture at a Crossroads, Synthesis Report International Assessment of Agricultural Knowledge, Science and Technology for Development*, Island Press, 2009. See: www.globalagriculture.org
5 Grooten, M. and Almond, R.E.A. (Eds), *Living Planet Report – 2018: Aiming Higher*, WWF, Gland, Switzerland. See: www.wwf.org.uk/updates/living-planet-report-2018
6 Girardet, H., *Regenerative economies for a sustainable world: A paper for the Club of Rome's 50th anniversary gathering* (circulated privately), April 2018.
7 Wijlkman, A. and Rockström, J., *Bankrupting Nature*, Routledge, 2012.
8 Von Weizsaecker, E. and Wijkman, A., *Come On! Capitalism, Short-termism, Population and the Destruction of the Planet*, Springer, 2018.
9 See: https://mro.massey.ac.nz/handle/10179/9476
10 *Green Infrastructure: A vital step to brilliant Australian cities. A Brilliant Cities Report*, AECOM, 2017.
11 Lovins, H., Wijkman, A., Fullerton, J., Wallis, S., Maxton, G., *A finer future is possible*, Club of Rome, 2016.
12 See: www.thesolutionsjournal.com/article/regenerative-adelaide/
13 See: https://regenecon.net
14 Girardet, H., *Creating Regenerative Cities*, Routledge, 2015, ISBN: 9780415724463.
15 *The City We Need 2.0, World Urban Campaign*, UN Habitat, 2016. See: www.worldurbancampaign.org/resources
16 Ellis, H., Henderson, K. and Lock, K., *The Art of Building a Garden City: Designing new communities for the 21st Century*, RIBA Publishing, 2017, ISBN: 9781859466209. See: www.tcpa.org.uk
17 UN Habitat, *Tracking Progress Towards Inclusive, Safe, Resilient and Sustainable Cities and Human Settlements SDG 11 Synthesis Report*, 2018.
18 See: www.ellenmacarthurfoundation.org/programmes/insight/circularity-indicators

7
ZERO CARBON CITIES

The 'one planet' city will strive towards being independent of fossil fuels in every sector by switching to renewable energy, and to absorb carbon dioxide from the atmosphere by 'locking it up' in its fabric, soils and trees, in order to restore the planet's equilibrium. The city will often find it cheaper to remove by design the need to use energy for many tasks, and to reduce existing energy use through improving efficiency, than it would be to build new energy generators.

The tragedy of fossil fuels

It is a great tragedy that the Earth's crust contains so much fossil fuel; not just because of global warming, but because of the millions of lives that have been lost in the wars that have been fought over access to oil in the last 120 years. The competition between nation-states for access to these resources has time and again brought turbulent geo-politics, conflict, suffering, widespread destruction and loss of life. The latter – not just for human lives – continues to be caused by air pollution, acid rain, oil spills and mining disasters. The presence of oil, coal and gas in a territory has been a curse as much as a blessing. So ending our reliance upon them will have multiple benefits beyond tackling climate change.

Before fossil fuels began to be exploited at scale, renewable energy technologies were in widespread use: hydropower was long established. Building design was more climate-appropriate, reducing the need for air-conditioning. Wood was the main heating fuel. Wind pumps irrigated the prairies of North America. Concentrated solar power was being exploited in Egypt, and in 1884 it had even been used to power a printing press in Paris. Battery-powered trams ran in the streets of New York. Solar water-heating was in widespread use in Florida. The train network was spreading in North America. All of these innovations were stunted and stalled by cheap oil and coal. If this had not happened who knows how much more advanced renewable technologies would be now?[1]

Instead, the world's economy is largely predicated nowadays upon the use of fossil fuels. The companies and economies which rely on them are as enthusiastic as ever to exploit them. Yet the scientific case for the likelihood with business as usual of a runaway greenhouse effect has been conclusively established. Humanity – or its leaders – are faced with a clear choice: to stick with the vested interests that promote fossil fuels; or to accelerate development of all renewable, sustainable technologies and practices.

Today, the words 'energy security' and 'resilience' are being used by those who want to see local, sustainable sources of clean energy replace dirty fossil fuels. This is because the sun, wind, marine, anaerobic digestion, geothermal and other renewable sources of energy are available abundantly, everywhere on the planet, with no need for conflict over their use. Deaths due to their use are rare. Nuclear power is an option too, but is expensive, linked to military use, has high risks associated with it, and leaves a dangerous legacy of radioactive waste for millennia.

Meanwhile renewable and energy storage technologies are evolving and becoming cheaper. New business opportunities such as demand management are accelerating the deployment of renewables. (See Box 7.1.) But renewable electricity is the easy bit. Providing heating and cooling for buildings and cooking, and power for aircraft, shipping and freight are more challenging. Doing more with less energy is part of the secret of going zero carbon: this is why energy efficiency is often called 'the invisible fuel'. The cheapest and cleanest energy choice of all is to remove the need for it. This is a design challenge.

In this chapter we look at cities which are pledging to reach or have already reached 100 per cent renewable energy, and what will happen to the gas networks.

BOX 7.1 LIST OF THE MAIN CURRENTLY COMMERCIAL RENEWABLE ENERGY TECHNOLOGIES

1. Municipal and industrial waste burnt in incinerators
2. Solid biofuels: crops (often timber); short-rotation coppice
3. Biogases: biomethane, e.g. from anaerobic digestion of biomass (AD); landfill gas (from rotting rubbish tips)
4. Liquid biofuels: biodiesel, bioethanol, biokerosene (from plants and algae)
5. Geothermal energy (deep boreholes)
6. Hydroelectricity (from dams and water currents)
7. Solar photovoltaic electricity (solar panels)
8. Solar thermal: panels, evacuated tubes, parabolic troughs and dishes, solar towers
9. Marine energy: tidal, wave, marine current turbines
10. Wind: offshore, onshore wind farms
11. Heat pumps: ground, water, air source for heating or cooling.

The energy trilemma

"If we change the way we power our cities we will change the way we power the world and in the process we will save it," said the former US Secretary of State John Kerry at the Second China-US Climate-Smart Low-Carbon Cities Summit in Beijing. Decisions upon energy policy are now increasingly determined with reference to the need to balance the challenges of energy security, affordability and sustainability in a given context. This is called the Energy Trilemma. "The trilemma means that policy should never focus on one single aspect. You have to look at the three challenges in a balanced way," says Christoph Frei, secretary-general of the World Energy Council.

The strategy for decarbonising urban areas will depend upon whether the area already exists or is yet to be built, which has a radically different impact upon the future of gas use in that context. New urban developments can be designed to be low or zero carbon by 2050. Existing urban areas require adaptation. Each country and region will have its own strategies, starting from a different place.

The power sector should be almost completely decarbonised by 2050, with no coal- or oil-fired generation and only some natural gas generation. Around 5,100 gigawatts (GW) of new capacity is needed between now and then around the world. Much of it will be renewable and locally supplied, within or close to cities. Electricity storage will be widespread to meet demand at all times. The technologies used for storage will vary but include hydrogen as power-to-gas. Disruptive technologies may well have appeared in regard to storage and renewable energy generation. One such on the horizon is perovskite photovoltaic solar cells.

There will be some gas-fired combined heat and power (CHP) networks, and non-CHP gas-powered generation may well have to have carbon capture and storage (CCS) attached, or the capture of CO_2 for use as a feedstock for fuels, chemistry and polymers (known as carbon capture and utilisation – CCU). Without this, gas consumption will fall by 2050 to only about 12 per cent of the 2010 level, to balance intermittent renewables and for use in industry. Under current assumptions, most of this fall will happen after 2030,[2] indicating the limit of gas as a transitional fuel, but these assumptions are subject to rapid change.

The role of cities

Urban areas account for up to 76 per cent of energy-related CO_2 emissions, while the world's cities produce almost half of all greenhouse gas emissions. Cities are therefore central to tackling climate change. Action taken by cities is considered to be one of the most cost-effective ways of attaining the goals of the Paris Agreement by the International Energy Agency.[3] "Cities and mayors are now a central part of the solution to climate change. City governments have demonstrated an ability to get to grips with climate change where others have failed," says Eduardo Paes, former mayor of Rio de Janeiro and a former chair of C40 Cities.[4]

Patricia Espinosa, executive secretary of the United Nations Framework Convention on Climate Change (UNFCCC), is on record as saying, "We cannot address climate change in any real and meaningful way unless cities are a part of the process."[5] And the mayor of Paris and chair of C40, Anne Hidalgo, has praised Chicago's Mayor Rahm Emanuel as "one of the strongest climate advocates in the U.S." At the Paris-hosted COP21 climate talks Hidalgo said: "Do not listen to the advocates of yesterday's world, they are accountable for today's destruction, and do not care about tomorrow."

Emanuel himself said, "Chicago has met 40 per cent of its Paris Agreement goals![6] City carbon emissions are down 11 per cent while jobs expand. Additionally, Chicago focuses on inclusive growth in our climate actions increasing quality of life for those normally left outside."[7]

Mayors like Emanuel are showing the way forward. Mayor Stephen Adler from Austin, Texas is another: "Nothing happening at national level can stop a city that is already in transition for climate change. Almost half of the goals of the Paris Accord have to be achieved at the city level – our path is very clearly set. Cities who are transitioning to a clean economy are seeing their economies improve. And this is not just because they are attracting dynamic people."[8]

What are these leading cities doing? Cities can use many strategies. Here are a few:

- Use their purchasing power to purchase only renewable energy
- Make their own buildings energy efficient
- Decarbonise their fleets and transit services
- Install solar electricity and thermal panels on the rooftops of buildings they own
- Use ordnances to deliver policies favouring energy efficiency and renewable energy
- Change planning regimes to support compact cities
- Install district heat mains.

In New York City, Mayor Bill de Blasio has led an acceleration of such measures put in place by his predecessor Michael Bloomberg. His administration's plan to make New York City's buildings efficient "will have the equivalent impact of taking 900,000 cars off the road".[9] NYC is aiming at the goal of limiting global temperature rise to 1.5°C. "We want to go farther, go deeper, go faster, because we have to. Climate change isn't a theoretical problem for New York City. It's a threat we're boldly taking on. This is the effort that never ends – we should not be afraid of the work that lies ahead," said de Blasio.

There is still a gap between national energy policy and climate change at a local level, in the USA and elsewhere; there are still plans to build fossil fuel infrastructure. But the mayor of Vancouver, Gregor Robertson, is of the opinion that, "it's still up to cities to hold federal government accountable, supporting our rural sisters and brothers in going renewable and to a clean economy. We have to take every step we can to support 100 per cent renewable energy."

Michael Bloomberg has continued to use his influence and wealth to support action against climate change since leaving public office. He supports the C40 Cities Bloomberg Philanthropies Awards[10] which celebrate the best projects that fight climate change in cities. Some recent winners include:

- Austin Texas' solar programme, which has expanded solar power at scale while limiting increases in electricity bills to no more than 2 per cent per year.
- Retrofit Chicago, which partners with local organisations to improve energy efficiency in the city's buildings.
- Qingdao's clean energy plan to eliminate 215 million tonnes of coal consumed, by retrofitting existing boilers and subsidising new clean energy infrastructure.
- Copenhagen's energy surveillance programme, which makes the city the first in the world to have a completely centralised building monitoring system.
- Bangalore, which has developed India's first Intelligent Transport System, for its 5.2 million daily riders.
- Buenos Aires' innovative Recycling Centre, which provides five different facilities to process different kinds of waste and repurpose surplus materials for soil stabiliser or compost.
- Oslo's Climate Action Plan, aiming to reduce emissions by 95 per cent by 2030. It has developed a climate budget which accounts for every single unit of CO_2 emitted by the city.
- Mexico City's Climate Action Programme (see below).

This is the tip of the iceberg. Cities all over the world are, bit by bit, taking action to make their futures more secure.

Coastal cities

Nowhere is this action more pertinent than in coastal cities. Limiting a global temperature rise to 2°C could still lock in up to six metres of global average sea-level rise as soon as 2200, leaving over a quarter of the populations of Shanghai, Hong Kong and Mumbai below the water line.[11]

Lord Krebs, president of the British Science Association and chair of the Adaptation Sub-Committee of the UK Climate Change Committee, agrees with the *Nature* analysis that the Paris Agreement will be lucky to limit the global rise to 3° C. Krebs says we should be preparing around the world for a 4°C rise.[12] This will lead to a global mean sea-level rise of 0.45 to 0.82 metres by 2080, rising much more in the subsequent century. As this is a mean figure, individual sites could experience three to four times higher.[13] Land would submerge that is now inhabited by half or more of today's population in Shanghai and Shantou, China; Haora, Kolkata and Mumbai, India; Hanoi, Vietnam; and Khulna, Bangladesh.

Thousands of other coastal settlements will be affected: global mega-cities with the top ten largest threatened populations include Hong Kong, Kolkata, Mumbai,

Dhaka and Jakarta. Three non-Asian cities with a high proportion of population at risk are Rio de Janeiro, New York and Buenos Aires. Limiting warming to 2°C would cut the threat by more than half in 13 mega-cities.

Coastal cities are therefore liable to lead the way in adapting to climate change and already have more stringent climate policies that favour renewables over fossil fuels. Docks and ports are vulnerable, and this has implications for the shipping of liquid natural gas and a reluctance among markets to be dependent upon this supply mode.

If docks are flooded, fuel (not to mention food) cannot be unloaded. Any city or region relying on this source of fuel for energy would find itself without power. Coastal gas-fired electricity power stations and substations at risk will be knocked out or relocated. Underground pipe networks in submerged areas will be less accessible for maintenance. City authorities planning for climate resilience and adaptation are anticipating these scenarios.

The design of new urban developments

New urban developments will require new infrastructure, and as time goes on, authorities are more likely to favour renewable sources of electricity and energy, and not rely on new gas pipeline networks or generation. Heat pumps and renewable energy district heat and power networks may well become the norm. A regulatory investment framework for district heating is desirable in most areas.

A 60 per cent reduction of energy use can be achieved in new urban developments by a different approach to planning – a compact city design that eliminates the need for long and frequent private car journeys.

This requires highly insulated buildings and neighbourhood plans that encourage the use of public transport, walking and cycling. Neighbourhoods or individual buildings can also generate their own renewable energy. If all cities adopt all of these tactics, an International Energy Agency (IEA) report says, they will help reduce global greenhouse gas emissions by 15 per cent – about 8 gigatonnes – by 2050.

The IEA believes that this approach has other benefits: it can increase access to modern energy services for inhabitants of developing economies, improve their living standards, and it avoids the 'locking-in' of carbon-intensive infrastructure by not copying the urban sprawl designs of developed nations.

Decarbonising existing urban areas

In existing cities, electricity supply, heating services, gas for cooking, and transport must be decarbonised. Existing urban developments will find new uses for existing gas infrastructure or replace it with electricity-powered alternatives. According to sustainable construction consultants Arup,[14] lower-carbon energy solutions such as wind, solar, geothermal, hydro, heat pumps, hydrogen, biogas, marine and nuclear power, zero net energy schemes, microgrids and district heating all have a part to play.

Arup does not explicitly consider natural gas to be part of the solution, but CHP is an affordable and more sustainable use of gas than for electricity alone, as capturing the heat as well as the electricity of a power station typically uses 75 per cent rather than 33 per cent of the energy content of the fuel.[15] Nevertheless all studies which see gas continuing to have a role, require there to be also some form of CCS or CCU.

The energy outlook until 2050 is addressed by the IEA's *Energy Technology Perspectives 2016 Towards Sustainable Urban Energy Systems*.[16] This uses three scenarios with a global temperature rises of 6°C, 4°C and 2°C. The latter is its main focus because it is the one that is in line with meeting the Paris Agreement goals, the Sustainable Development Goals, and the New Urban Agenda. It envisages a one-third rise in global energy demand. This will be met by wind, solar, hydro and nuclear power, with another 30 per cent from natural gas.

The impact of the Paris Agreement and air pollution control legislation, particularly in urban zones, will favour low-carbon fuel sources. Therefore by 2050, primary energy demand totals 663 exajoules (EJ), satisfied by renewables (44 per cent), fossil fuels (45 per cent, with gas at 17 per cent – less than now in relative and absolute terms), and nuclear (11 per cent). Electricity surpasses oil as the most important energy carrier, with a share of 28 per cent, increasing by 80 per cent from 2014. Oil (25 per cent), natural gas (16 per cent), biofuels and waste (15 per cent), and coal (11 per cent) make up the remainder of final energy demand. That's what the IEA thought in 2016, but as I read this again in the light of 2018's IPCC report on the need to curb emissions to zero by 2030 in order to limit global temperature rise to 1.5°C, I can see that there is too much fossil fuel in the mix. What we need more of is …

Energy efficiency

Energy efficiency is frequently described as the 'low-hanging fruit' because it is easier and often cheaper to save energy, or avoid its use, than to build equipment to generate and supply it. If you view it as a part of general resource efficiency – for example, reusing waste materials to save the purchase of new ones, and redesigning products, or a production line, so that fewer resources are needed – this will also result in the use of less energy. Similarly, water efficiency not only will result in lower water and sewerage bills, but require less energy in pumping or heating, producing double wins.

Cities and businesses that want to survive in the climate-changed and resource-scarce future will have to totally transform the way they work, with their customers, investors, employees and supply chains. Many that have already done so have discovered that managing their energy and resource use efficiently is transformative across the enterprise. We saw in Chapter 3 that there is an excellent procedure for implementing energy efficiency: ISO 50001, but other standards exist.[17]

We measure improvements in energy efficiency with a metric called primary energy intensity. When it falls, it tells us we are doing more with less energy. The latest figures show annual improvements in global energy intensity since 2011 have averaged 2.2 per cent, below par (=3 per cent) and not fast enough to meet the 2030 Sustainable Development Goal 7 target for energy efficiency.[18] The worst-performing parts of the world were the Western Asia and Northern Africa regions and the Latin America and Caribbean regions. Although industry still consumes the most energy worldwide, this sector was making the most progress toward improved efficiency, while in the high-income countries, transport is the highest energy consumer.

It is particularly important in low- and lower-middle-income countries for energy efficiency to be encouraged in new homes and buildings to avoid the 'lock-in' of inefficient buildings and appliances for decades to come.

Although the average thermal efficiency of fossil fuel-powered electricity generation continues to improve (mostly because of more efficient natural gas power plants), the average efficiency of coal- and oil-fired plants has not changed. The truth is that these need to be phased out urgently.

To improve the energy intensity of transport, it's necessary to reduce the need to travel, and to phase in electric and hydrogen-powered fuel cell vehicles, the latter especially for transporting freight. As a transition fuel towards the middle of the century, compressed and liquefied natural gas will replace diesel in road freight applications in certain markets, provided that certain issues (e.g. 'methane slip') are addressed.[19] I take a closer look at this in Chapter 12.

One of my main preoccupations is to find and implement ways of being confident that we are really making improvements when we take action. With energy efficiency, administrations can use particular metrics to measure progress. The World Bank and the IEA have come up with some suitable ones; examples include 'lighting energy use per unit of floor area in buildings', and 'energy use per unit of production in an industry subsector'.[20] Training courses are available for policy-makers and statisticians[21] in using these measurements.

For buildings, space heating, cooking and water heating are the largest buildings end-uses. But these are rising slower (25 per cent growth since 2000) than energy used for space cooling, lighting and appliances (over 45 per cent growth since 2000). Appliances are becoming more and more efficient, and lighting even more so, with the extremely low-energy and long-lasting LED lighting systems spreading rapidly. Yet this is being counteracted by the fact that people are owning more and more gadgets and lighting fixtures. All of this points to the need to curb consumption and decarbonise energy supplies.

Zero carbon buildings

Buildings are responsible for about 30 per cent of greenhouse gas emissions. Space heating (accounting now for over one-third of energy use in buildings around the world) will still be the largest consumer of energy in buildings around the middle of the century. Energy for space cooling (roughly 5 per cent of energy use in buildings today) is fast-growing (up to tenfold in some hot, rapidly emerging economies), and needs to be supplied sustainably – most likely by reversible heat pumps, which are a fantastic, simple technology I look at more closely below.

The climate-appropriate design of urban developments and the use of highly insulated, passive architectural design principles will reduce the need for energy for both heating and cooling. This is likely to occur sooner in commercial and other services buildings, where turnover and demolition rates are faster than for residential buildings.

Around one-fifth of cumulative energy savings up to 2030 can come from replacing equipment like boilers with more efficient versions, through minimum or compulsory energy performance standards, and from strong uptake of high-efficiency technologies like heat pumps. Some 15 per cent of these savings are expected to be in the developing world, switching from traditional biomass to solar technology and high-efficiency pellet stoves away from inefficient use of fossil fuels.

The World Green Building Council, which is a global network of over 70 Green Building Councils and their 32,000 member companies, has calculated how building energy use can contribute to the Paris Agreement goal to limit global temperature rise to below 1.5°C. It now advocates all new buildings operate at net zero carbon from 2030 and all buildings do so by 2050.

It has shown how the industry can achieve this in a report which outlines what businesses, governments and NGOs can do.[22] What is 'net zero carbon building'? The report defines it as follows: "a highly energy-efficient building with all remaining operational energy use from renewable energy, preferably on-site but also off-site production, to achieve net zero carbon emissions annually in operation."

By the middle or towards the end of the century most of the buildings in existence now will probably still be around, and they need to be retrofitted to become net zero carbon. As well as this, all new buildings will need to be built to a similar standard. One of the most popular standards by which this performance can be measured to make sure that we're actually achieving net zero carbon nowadays is called Passivhaus (although not everyone agrees with it).

The adoption of this standard is recommended by UNEP.[23] This type of building is often thought to be more expensive than conventional building, but that is no longer the case due, as with renewable technology, to increased innovation. UNEP says the price for new Passivhaus buildings in several countries is now comparable to standard construction costs, and if the costs saved are calculated over the lifetime of the building when compared to a conventional one, it is much cheaper because it is cheaper to run. This subject is examined more fully in Chapter 11.

Hot cities

Cities absorb heat. They become hotter in hot spells than the countryside around them, often to an unbearable extent. To reduce this 'heat island' effect, some cities are taking a regenerative approach: planting trees and other vegetation, installing ponds and lakes, replacing impervious surfaces with lawns and gardens, and increasing the albedo (reflectivity) of outside surfaces.[24]

But as the planet warms, and buildings in temperate climes are better designed to reduce the energy needed for winter heating, more and more energy is, by contrast, being used for cooling buildings in the summer.

This is a growing problem. This use of energy has tripled, globally, between 1990 and 2016, particularly in China and the USA, now consuming about 2,000 terawatt-hours (TWh) of electricity every year – two and a half times the total electricity use of Africa, and almost a fifth of all electricity used in buildings. Without intervention cooling-related energy demand will soar in the future, with dire implications. Chapter 11 relays present-day solutions for this issue.

The Global 100% RE city movement

Energy efficiency helps the world to reduce its demand for energy. At the same time we must increasingly supply the remaining energy requirements from sustainable sources. Already, a growing numbers of cities, communities and regions are proving that meeting all of their energy demand (except for transport) with renewable energy is viable.[25]

Furthermore, 608 cities, towns and regions from 62 countries, representing 553 million people – 8 per cent of the world population – at the time of writing have targets to reduce their greenhouse gas emissions. These areas are registered with the carbonn® Climate Registry (cCR),[26] in operation since 2010. Some examples of cities with these targets are as follows:

- Amsterdam, the Netherlands, committed to total decarbonisation of its district heating system in 2015 and set an immediate goal of increasing connections to a total of 230,000 houses by 2040 (a 70 per cent increase).[27]
- Tshwane, South Africa, is preparing the transition to renewable energy in the transport sector, including biogas recovery from waste to fuel the city-operated bus fleet running on compressed natural gas.[28] It has a target to reach 50 per cent renewable energy at community scale by 2030, and a clear political interest in aiming for 100 per cent renewable energy.
- A 'Carbon Free Island by 2030' declaration has been made by Jeju Province, South Korea, which has committed to reach 100 per cent renewable electricity and transport by 2030, and is connecting renewable energy to electric vehicles using smart grids to power both buildings and transport, using a battery-based energy storage system and fuel cell power plants.[29]
- Graz, in Austria, is to increase the share of solar thermal in its district heating network by 20 per cent through the installation of up to 500 MWh of new solar collectors.[30]
- Münster, in Germany, has become the first German city to divest from fossil fuels. It invested in hot water storage tanks in 2015 to utilise surplus grid electricity to generate heat for injection into the city's district heat network, as part of a plan to increase renewable energy.[31]

Numerous other cities around the world have committed to going 100 per cent renewable. Many already have achieved their goals, including the US cities of Burlington (Vermont), Aspen (Colorado) and Greensburg (Kansas), which all reached 100 per cent renewable electricity during 2015.

Further initiatives demonstrate the strength of this movement. The Climate Summit for Local Leaders, held in parallel with COP21, issued a declaration in support of a transition to 100 per cent renewable energy by 2050.[32] It was signed by nearly 1,000 mayors from around the world. This non-binding commitment builds on examples of leading cities such as Copenhagen, Frankfurt, San Francisco, Sydney and Stockholm.

The Global 100% RE [renewable energy] campaign published 12 criteria in 2015 to help define the concept of 100 per cent renewable energy for local governments[33] and to guide policymakers in initiating their transition to 100 per cent renewables. The 100 per cent RE Cities and Regions network[34] was also launched by the campaign in 2015, to enable peer learning and exchange among municipalities. The campaign works alongside Europe's 100 per cent RES Communities and RES Champions League.

In Germany, a pioneering country in this regard, 100 per cent renewable energy has been a movement for several years. A national framework empowers cities to transform their energy systems, and 140 regions, hosting one-quarter of the population, have set targets of one type or another. For example, the Fraunhofer Institute has modelled that it is feasible for Frankfurt to achieve 95 per cent renewable energy in power heating, cooling and local mobility from its hinterland, with the rest coming from outside the region.

Table 7.2 shows those locations that have already achieved 100 per cent.

Less ambitious programmes

There are many city and regional initiatives, such as the C40 Cities, Climate Group's States and Regions Alliance, and the Covenant of Mayors (a group of 6,317 city mayors committed to exceeding the European Union target of reducing CO_2 emissions by 20 per cent by 2020 and 4 per cent by 2030 compared to 1990 levels), working hard to achieve decarbonisation. Many initiatives exist, but they often lack specific objectives, and the means to deliver them, or scale, and they use differing criteria and baselines. A frequent additional challenge is lack of buy-in across all administrative departments.

New York City's 80x50 program

New York City's 80x50 program aims, like the UK's, for an 80 per cent reduction in emissions by 2050. In January 2016 the city passed a law on requiring all government buildings to switch to heat pumps in all new construction and retrofits when it is shown that doing so would be cost-effective. Hundreds of buildings in the city have already done so. The city goal is for over 900,000 of its 1.1 million buildings to make the switch by 2050. Heat pumps also eliminate the cooling

TABLE 7.1 Cities which have set targets for up to 100 per cent renewable energy or electricity

City	Target date for 100% total energy	Target date for 100% electricity
Aspen, Colorado, USA		2015
Burlington, Vermont, USA		Achieved in 2014
Byron Shire County, Australia	2025	
Coffs Harbour, Australia		2030
Copenhagen, Denmark	2050	
Frankfurt, Germany	2050	
Fukushima Prefecture, Japan	2040	
Greensburg, Kansas, USA		Achieved in 2015
Hamburg, Germany	2050	
Jeju Self-Governing Province, Republic of Korea		2030
Lancaster, California, USA		2020
Malmö, Sweden		2030
Munich, Germany		2025
Osnabrück, Germany		2030
Oxford County, Canada	2050	
Palo Alto, California, USA		(no date given)
Rochester, Minnesota, USA		2031
San Diego, California, USA		2035
San Francisco, California, USA		2020
San Jose, California, USA		2022
Seattle, Washington, USA		(no date given)
Skellefteå, Sweden		2020
Sønderborg, Denmark	2029	
Sydney, Australia	2030	
Ulm, Germany		2025
Uralla, Australia	(no date given)	
Vancouver, Canada	2050	
Växjö, Sweden	2030	

Taken from *Renewables 2016 Global Status Report*, REN21, Paris, France, available at: www.ren21.net.
This is not an exclusive list.

TABLE 7.2 Cities that have achieved 100 per cent renewable energy

City	Country	Continent	Administrative unit	Sectors (electricity, transport, heat)
Columbia, Maryland	USA	North America	city	Electricity, transport, heat
Dardesheim	Germany	Europe	town	Electricity, transport, heat
Dobbiaco	Italy	Europe	city	Electricity, heat
Feldheim	Germany	Europe	town	Electricity, heat
Flecken Steyerberg	Germany	Europe	village	Electricity, transport, heat
Kasese	Uganda	Africa	region	Electricity, heat
Knežice	Czech Republic	Europe	village	Electricity, heat
Linköping	Sweden	Europe	city	Transport
Mureck/Steiermark	Austria	Europe	region	Electricity, heat
Odenwald District	Germany	Europe	region	Electricity, heat
Perpignan Méditerranée	France	Europe	region	Electricity, heat
Saerbeck	Germany	Europe	village	Electricity
Saint-Julien-Montdenis	France	Europe	village	Electricity, heat
Schleswig-Holstein	Germany	Europe	region	Electricity
Thisted	Denmark	Europe	town	Electricity
Wallonie Picarde	Belgium	Europe	region	Electricity, transport, heat
Wildpoldsried	Germany	Europe	village	Electricity, transport, heat
Wilhelmsburg (Hamburg)	Germany	Europe	city district	Electricity, heat
Wolfhagen	Germany	Europe	city	Electricity

Source: Supplied by Anna Leidreiter of the World Futures Council.

towers needed by other cooling technologies, making buildings more storm and disease resistant.

The New York State Public Service Commission wants utilities to switch their pipelines from natural gas to heat pumps, along with other options, such as using the city's water mains for heat pumps. Other cities are expected to follow suit.[35]

Decarbonising Vienna

A plan for decarbonising Vienna[36] highlights issues that could apply to most developed world cities. For this city, the plan says that: traffic and heating/cooling

TABLE 7.3 Some targets for renewable share of total energy, all consumers

City	Target
Austin, Texas, USA	65% by 2025
Boulder, Colorado, USA	30% by 2020
Calgary, Alberta, Canada	30% by 2036
Cape Town, South Africa	10% by 2020
Howrah, India	10% by 2018
Nagano Prefecture, Japan	70% by 2050
Oaxaca, Mexico	5% by 2017
Paris, France	25% by 2020
Skellefteå, Sweden	Net exporter of biomass, hydro or wind energy by 2020

TABLE 7.4 Some targets for renewable share of electricity, all consumers

City	Target
Amsterdam, Netherlands	25% by 2025; 50% by 2040
Austin, Texas, USA	35% by 2020
Canberra, Australian Capital Territory, Australia	90% by 2020
Cape Town, South Africa	15% by 2020
Nagano Prefecture, Japan	10% by 2020; 20% by 2030; 30% by 2050
Taipei City, Taipei, China	12% by 2020
Tokyo, Japan	24% by 2024 [20% by 2024]
Wellington, New Zealand	78–90% by 2020

TABLE 7.5 Some heat-related mandates and targets

City	Target
Amsterdam, Netherlands	District heating for at least 200,000 houses by 2040 (using biogas, woody biomass and waste heat)
Chandigarh, India	Mandatory use of solar water heating in industries, hotels, hospitals, prisons, canteens, housing complexes, and government and residential buildings (as of 2013), and 100 MW of rooftop solar photovoltaics (PV) by 2022
Helsingborg, Sweden	100% renewable energy district heating (community-scale) by 2035
Loures, Portugal	Solar thermal systems mandated as of 2013 in all sports facilities and schools that have good sun exposure
Munich, Germany	80% reduction of heat demand by 2058 (base year 2009) through passive solar design (includes heat, process heat and water heating)
Nantes, France	Extend the district heating system to source heat from biomass boilers for half of city inhabitants by 2017
Osnabrück, Germany	100% renewable heat by 2050
Täby, Sweden	100% renewable heat in local government operations by 2020
Vienna, Austria	50% of total heat demand with solar thermal energy by 2050

TABLE 7.6 Targets for government self-generation/own-use purchases of renewable energy

City	Target
Belo Horizonte, Brazil	30% of electricity from solar PV by 2030
Cockburn, Australia	20% of final energy in city buildings by 2020
Ghent, Belgium	50% of final energy by 2020
Hepburn Shire, Australia	100% of final energy in public buildings; 8% of electricity for public lighting
Kristianstad, Sweden	100% of final energy by 2020
Malmö, Sweden	100% of final energy by 2020
Portland, Oregon, USA	100% of final energy by 2030
Sydney, Australia	100% of electricity in buildings; 20% for street lamps

TABLE 7.7 Some targets for renewable electric capacity or generation

City	Target
Adelaide, Australia	2 MW of solar PV on residential and commercial buildings by 2020
Esklistuna, Sweden	48 GWh from wind power, 9.5 GWh from solar PV by 2020
Gothenburg, Sweden	500 GWh by 2030
Los Angeles, California, USA	1.3 GW of solar PV by 2020
New York, New York, USA	350 MW of solar PV by 2024
San Francisco, California, USA	100% of peak demand (950 MW) by 2020

hold the greatest potential for decarbonisation; early installation of electric vehicle public charging infrastructure is essential; switching to green gas (biomethane and renewable hydrogen) is a good idea; district heating would provide the cheapest heating. The toughest nuts to crack are decarbonising heavy goods traffic (long distances, short stop times); and eco-retrofitting buildings (the most expensive decarbonisation measure to speed up, requiring support from the federal government).

Hydrogen production represents the second most expensive decarbonisation measure. Photovoltaics will be the most important renewable electricity source. Energy storage and demand-side management, highly efficient cogeneration plants and peak load power plants are also on the list.

The cost in the years up to 2050 was estimated at €28 billion. This sounds a lot but actually it is just under 1 per cent of Vienna's GDP each year. To achieve this would require a strong political will, increased funding from the federal budget and the close involvement of stakeholders, including citizens. But it is within reach.

100 per cent renewable energy in Italy

Over 40 Italian municipalities are already meeting 100 per cent of their energy demands with renewable energy. The change over the last ten years has been remarkable. Italy as a whole has brought its consumption of renewable electricity as a proportion of all electricity up from 1.5 to 13.5 per cent, largely due to a distributed production model with the addition of over 850,000 generation plants all over the country.

The number of municipalities with at least one plant supplying renewable energy has increased from 356 to 8,047. In 2,660 of these municipalities, the production of clean electricity often exceeds that consumed, allowing them to export to neighbouring areas for profit. The top municipalities have met all of both their thermal and electricity needs from renewables. They have achieved their results thanks to increasingly innovative and integrated solutions, smart grids, electric mobility, storage and a desire for reduced energy bills.[37] Italy was ahead of Germany and Greece and other front runners in the use of solar power in 2016, by providing 8.1 per cent of its population with electricity from this power source.

The top town for solar electricity is San Bellino, with 71.3 MW installed capacity, which by far exceed the electricity needs of resident households. The small Municipality of Seneghe heads the league table for solar thermal, with a proportion of solar panels in relation to the number of inhabitants of 1,955 m^2 per 1,000 inhabitants, distributed on public and private buildings. Some 850 municipalities use wind power, equivalent to the needs of 5.5 million households. In the many hilly areas of the country, 1,275 municipalities use mini-hydropower as a source of energy. Bioenergy is used by 3,137 municipalities and growing rapidly, while 535 municipalities make use of geothermal energy.[38]

The town of Pfitsch

The town of Pfitsch, in the province of Bolzano in northern Italy, with a population of 3,000, one of the 100 per cent renewable energy municipalities, deploys a mix of five technologies. Electricity needs are met by mini-hydroelectric and PV solar systems on the roofs of public and private buildings plus a hydroelectric plant built in 1927 and renovated between 1997 and 1998, yielding 21.7 MW.

Heating requirements are met through a district heating network, 52.9 km long, powered by a biomass plant that also contributes to the needs of the neighbouring municipalities of Sterzing and Racines. There are also two biomass plants, one using bioliquids, both connected to electricity and heat networks, and several solar thermal plants. A local energy supply company manages the networks.

The latest project is an anaerobic digestion biogas plant to process the waste of livestock farming in the valley into both electricity and thermal energy (fed into the grid and used in a yogurt-making enterprise). The waste material is sold as fertiliser, solving the problem of disposing of sewage waste, and returning the nutrients in the sewage, which would otherwise be lost, into the soil to grow the next generation of crops without the need for artificial fertilisers.

San Lorenzo Bellizzi

This small town's council decided in 2012 to use some property managed by local agricultural co-ops to install 15 MW of PV systems on their greenhouses. The revenue from feed-in tariffs, around €80,000 a year, has been redistributed to citizens through the exemption from another tax.

Developing country cities

Developing countries have different realities. Their cities must link their 100 per cent renewable energy strategy to their development strategy, i.e. making it a tool for development to enable access to international funding sources. This is India's approach. It's not just about technology transfer but a channel for new investment sources, and the additional advantage of securing energy security and developing new business models for local jobs.

Furthermore, the value of the energy produced stays within the city area. With the traditional energy model, this capital leaves the city to flow to the utilities, to pay for fossil fuels, which have contaminated the city and contributed to climate change. Instead, distributed energy production can be installed and reward citizens with a competitive price, clean air and jobs. There's an important opportunity to reduce energy poverty too, by implementing a plan that offers subsidised renewable energy to impoverished households.

The World Future Council and the ICLEI are two organisations which are active in this field, helping governments integrate these strategies into their national economic development plans to create jobs and train young people in the required skills.

India's 'Development of Solar Cities' programme

India's Ministry in the Urban Sector has a 'Development of Solar Cities' programme to support/encourage Urban Local Bodies to prepare a road map to guide their cities in becoming 'renewable energy cities' or 'solar cities'. The government has put aside around US$1 billion for the development of 60 of them, master plans for 49 of which have been prepared at the time of writing.

Initiatives include promoting solar water-heating systems in homes, hotels, hostels, hospitals and industry; deployment of pilot PV systems/devices; establishment of 'Akshya Urja Shops'; design of solar buildings; and promoting urban and industrial waste/biomass to energy projects. This is when rubbish or specially grown crops are burnt to supply heat and electricity in a combined heat and power plant.

(As an aside, I don't recommend the burning of waste for energy, unless it can be guaranteed to be pollution-free and absolute, non-recyclable waste. These plants, once built, create a market for waste, and can therefore inhibit other efforts to reduce waste and recycle more.)

The solar city programme aims to consolidate all the efforts and address the energy problem of the urban areas in a holistic manner. The initial aim was a minimum 10 per cent reduction in projected demand of conventional energy within five years. Past solar targets in India have been rapidly exceeded.[39] Already there is a new target for green townships of 25 per cent reduction within five years.[40]

Mexico City

Mexico City is a mega-city with a population of over 20 million, and a founder party to the new International Standard for sustainable cities, ISO 37120: Standard Sustainable Development of Communities: Indicators for City Services and Quality of Life.[41]

The city has adopted a goal to halve carbon dioxide emissions from 2000 levels by 2050. Mexico as a whole is aiming for 40 per cent of electricity to come from wind power alone by 2040. Policies for Mexico City[42] are being developed to encourage lower-emission vehicles (partly with a switch to biofuels) and public transportation, distributed renewable energy networks, to build more energy-efficient buildings, implement smart grid technologies, retrofit existing buildings and enforce building codes to support energy saving.

Heat pump-driven district heat networks

Heat pumps represent a significant competitor to natural gas for space heating and cooling. (They are sometimes known as geothermal energy, but strictly this applies to energy from deep boreholes that tap into hot rocks.) Heat pumps work like fridges; they connect a heating or cooling load (a building instead of a metal box) to either the earth, the air or a body of water.

Dr Peter Holzer, based in Vienna, Austria, is a specialist in eco-minimalism and building energy efficiency. He described to me how, once a heat pump is installed, it can be used to either remove heat from a space (cooling) and dump it in the ground, air or water, or to concentrate and therefore raise the temperature of the heat extracted from those places, in the target – for space or water heating. In other words, heat pumps can work forwards for heating space and backwards for cooling.

They're efficient: on average every unit of energy (preferably from renewable sources) used to drive the pump, results in three units of heat or cool. They can be installed in individual buildings or clusters of buildings, in which case they drive heat mains. The heating or cooling source is free. For heating, this renewable source may also derive from waste industrial heat.

Water-source heat pumps are particularly useful when developments are near bodies of water. A hotel complex in Greater London, for example, uses the River Thames as a heat source, while the town of Drammen in Norway heats most of its buildings from water with a source temperature of just 8°C.

The 45 MW Drammen district heating system serves over 200 large buildings. The heat was originally supplied from fossil fuel and biomass. Now a large heat pump is the primary source. This draws 75 per cent of the network heat from ammonia heat pumps with 15 per cent from biomass and 10 per cent from gas/oil. The fjord water is used to heat ammonia, at four times atmospheric pressure, until it evaporates (at 2°C).

The pressure is then increased to 50 times atmospheric pressure (50 bar), bringing the gas to 120°C. The heat is transferred via a heat exchanger to the water in the heating system, which (returning from the district heat main at 60°C) is heated back up to 90°C and sent round again. The ammonia returns to a liquid state and begins its own cycle over again. The pumps are driven by electricity, but the whole system produces three times more energy than is put in. The Dramman system has paid for itself very quickly and now saves the town around €2 million and 1.5 million tonnes of carbon emissions a year.

Elsewhere in the world, ground-source heat pumps are a most economic option. Switzerland, for example, already boasts over 25,000 such systems. Swiss public utilities use energy contracting to incentivise their adoption by paying for the installation and operation of systems then charging the property owner for the cooling or heating service.

Factors to be taken into account when comparing conventional heating and cooling technologies with heat pumps include: comparative greenhouse gas emissions, air pollution concentrations, annual electricity consumption and peak demand reduction, a possible revenue stream due to the peak demand reduction, fuel and power costs, net present value of all alternatives based on a 20-year life expectancy and capital costs, operations and maintenance, fuel costs, available federal, state and other non-city governmental funding assistance, and the social cost of carbon.

The role of industry

Decarbonising the industrial facilities we rely upon to make the products we consume is a huge challenge. Globally, the industrial sector is responsible for around one-quarter of total energy consumption and gas is widely used.

Over coming decades, as machinery is replaced and new plant installed, this will reduce energy demand and improve energy intensity (a measure of efficiency defined as GDP per unit of energy consumed). In combination with a switch to renewable energy, the industrial sector will be gradually decarbonised. Even so, gas use will still increase in the petrochemical and chemicals, iron and steel, and aluminium sectors, and grow slightly in the cement sector. Only in the pulp and paper sector is a decline currently foreseen. Energy efficiency is an important part of this process, because decarbonisation will be more gradual than in other sectors.

Often, industry is concentrated in a small number of urban areas. This makes it relatively easy to identify the big energy-consuming enterprises. Therefore, improving energy efficiency in the industrial sector is being prioritised in many countries.[43] This subject is covered more fully in Chapter 8.

Short-lived climate emissions

An important and often unreported task is limiting emissions of the so-called 'short-lived' climate pollutants. They have lifetimes roughly below 20 years, and include black carbon, ozone, methane, and many hydrofluorocarbons. Targeting these will limit the rate of short-term warming, and give countries more time to implement measures to limit long-term warming.

Where do these emissions come from? Fracking, and all exploration and distribution of fossil fuels, the waste sector (methane from landfill), rice paddy irrigation, livestock, diesel engines, shipping and air travel are examples of sources of these emissions. Improvements in the energy efficiency of refrigeration and air-conditioning appliances and equipment would also help.

Countries have to produce an action plan – called their Nationally Determined Contributions – to show how they are going to reduce emissions of greenhouse gases. But within their plans it is usually difficult or impossible to identify any actions to reduce emissions of these short-lived compounds, because targets are expressed in CO_2e (the equivalent of carbon dioxide, for comparison purposes). They often lack specific targets for methane, hydrofluorocarbons or black carbon. Another system is needed for tracking these.

The good news is that increasing energy efficiency can help reduce these short-lived emissions. For example, phasing out hydrofluorocarbon-based refrigerants can save between 340 GW and 790 GW of peak power load globally, avoiding about 98 gigatonnes (Gt) of CO_2 emissions by 2050. Efforts to reduce emissions of air pollutants such as sulphur dioxide, nitrogen oxides, and particulate matter, which will also reduce the burden on human health from outdoor air pollution, will have the same effect. This means increasing energy efficiency in the power sector, industry, transport, and using clean energy for cooking.

Substitutes for natural gas

What might be done with existing gas networks in a decarbonised future? Studies show that preserving and adapting these gas supply networks and infrastructure would be less socially and economically disruptive than abandoning them for an all-electric future.[44] But it requires using a different kind of gas: renewable gas.

There are a number of candidates for an alternative to natural gas that can ensure this continuity; each has pros and cons. Their sustainability, including their impact on the climate, needs careful scrutiny. In the past grave errors have been made, e.g. about the impacts of first- and second-generation biofuels, which turned out to be less sustainable than the fossil fuel they replaced.

Biogas

Biogas (biomethane, CH_4, usually with some CO_2 and hydrogen sulphide present) can be produced through AD of organic farm or catering/food waste, animal manure and sewage sludge, or dedicated crops such as maize, grass and crop wheat. The term also covers gas collected from emissions from landfill and sewage works. AD is proven technology. An advantage is that it also solves a waste problem. Other valuable outputs are: compost (liquid and solid), CO_2 (as feedstock). All of these provide additional income streams. Potential is high but limited locally by availability of year-round feedstock supply. The gas is used locally and, once treated, can be injected into the gas grid or used as fuel in natural gas vehicles, the numbers of which are growing rapidly. Using biogas to generate electricity is common but is a much less efficient use due to conversion losses.

Does it reduce carbon emissions? An analysis of an AD plant using energy crops and crop waste in Hungary found that it produced 33.39 g CO_2/kWh[45] – 93.7 per cent less than that from regular Hungarian energy production. But the result was heavily influenced by the distance the feedstock had to travel to the plant and the use of artificial fertilisers. For optimum results these should be minimised to under 10 km and zero, respectively.

Biogas production and use in electricity and heat production also leads to GHG reductions compared to composting of feedstock but reductions are not as high.[46] A further study[47] has concluded: "The reduction of direct methane emissions (from the decay of waste organic matter) together with the efficient utilization of the heat from cogeneration and the use of animal waste as input material can improve the GHG balance substantially." The use of biogas for transportation leads to reductions of GHG emissions from 49 per cent to 84 per cent compared to fossil fuels.

An alternative to liquid petroleum gas (LPG) is biopropane which is made from biomass, including used vegetable oil. Biopropane can be used to 'spike' biomethane for injection into the natural gas grid. The current cost is around three times that of LPG, but as a co-product of hydrotreated vegetable oil (HVO) biodiesel production in practice the cost is lower.[48] It can yield a carbon emission saving of 43–88 per cent on the benchmark fossil fuel it displaces.[49]

Bio-synthetic gas (bio-syngas or bioSNG) is processed from biomass or waste organic matter using gasification, which produces a mixture of gases. It is an intensive process. Syngas may also be converted into liquid advanced biodiesel, bioethanol or other fuels. Studies by the UK National Grid suggest between a third and half of UK domestic demand for gas, for example, can be met from bioSNG – around 100 TWh/annum by 2050. Growing crops for energy and not food, using marginal arable land or grassland, could provide a valuable source of feedstock for bioSNG plants. Potential is huge.

The UK Climate Change Committee and relevant government departments have agreed that it is wrong to use it to generate electricity since only 40 per cent to 65 per cent ends up as electrical energy. Biogas will be used directly, in the National Grid or locally, because it is two to three times more efficient to use organic waste for bioSNG than burn it for electricity. Biogas will play an important role in the UK's low-carbon energy future.

On the whole, it's advised that crops are not used as a feedstock for making biogas. A 2010 EU analysis found biodiesel made from North American soybeans to have a footprint four times higher than standard diesel.[50] A more recent comparative study[51] concluded that "the overall life cycle impacts for diesel, biodiesel, petrol, ethanol, natural gas, and biogas from grass are very close and roughly a factor 2 higher than for the other natural gas options assessed … Biogas based on waste is the best option".

There is as yet no standardised methodology for assessing the greenhouse gas emissions of all biofuels. For biogas, the choice of the time horizon in life-cycle carbon assessment is crucial and needs a strong justification. A team at Oak Ridge National Laboratory, USA, has developed a set of key indicators of bioenergy sustainability and proposed a method for their application to a bioenergy supply chain from feedstock production through to energy product consumption.[52] Thirty-five indicators under 12 categories are used to indicate the environmental and socioeconomic values that should be assessed across the entire supply chain. Important are: potential risks to biodiversity, soil fertility and productivity, and water quality and quantity.

In conclusion, bioSNG requires no changes to network infrastructure or appliances and ensures the long-term future of the gas grid – avoiding decommissioning costs. Waste from an average city could produce enough gas to meet one-quarter of domestic demand and using waste to make bioSNG is two or three times as efficient as waste to electricity. The technologies to support AD and bioSNG are tried and tested and low risk. BioSNG is a local solution, producing minimal waste emissions. A plant can be located close to the communities it serves.

Hydrogen

Another gas that could be used in gas grids is hydrogen. The first element in the periodic table and a constituent of water (H_2O), it is also found in methane (CH_4), which is 7.857 times more dense than hydrogen. Hydrogen gas carries 41.7 per cent less energy per unit of weight than natural gas, so just over three (3.277) times as much volume of uncompressed hydrogen is needed to obtain the same amount of energy. Therefore the gas needs to be delivered at a higher pressure to compensate. Customers' appliances, like cookers and boilers, would need adapting or upgrading to cope.

Over 90 per cent of today's hydrogen is mainly produced by steam reforming of fossil fuels – natural gas, oil or coal – which contain hydrogen as part of compounds. Carbon dioxide is removed from the fuel and usually just vented to the atmosphere; so to be sustainable it would have to be captured and stored. Hydrogen produced this way is two to three times the cost of the original fuel and a quarter of the original energy is lost in conversion.[53] The life-cycle carbon emissions of hydrogen production using steam reforming has been evaluated as between 269 g/kWh and 336.26 g/kWh. If 90 per cent of the carbon dioxide emitted by combustion is captured, it drops to 85.83 g/kWh – still hardly zero carbon. [Note: 1 kWh = 3.6 megajoules (MJ).] If the gas is produced from renewable electricity, then it is more likely to have a much lower carbon footprint. However, presently, this is around eight times more costly.

So there is a need to use CCS or CCU to produce the hydrogen from the steam reforming of methane. CCU holds more potential than CCS, since it allows local initiatives to proceed – with potential income streams for the CO_2 – without waiting for national governments to support an expensive CCS programme.

It's also possible to obtain hydrogen from water, splitting the hydrogen atoms from the oxygen atoms using electrolysis powered by renewable electricity, but it is expensive. As a form of energy storage for off-peak wind power it has possibilities as a niche market. It is currently much cheaper to store off-peak electricity in lithium batteries, in electric vehicles for example, than use it to make hydrogen, but the economics may change in the future.

Can we use the existing gas networks?

The existing gas networks can be adapted to carry biomethane or hydrogen.[54]

Biomethane cannot be produced in sufficient quantities to satisfy total demand but is the best option for satisfying demand local to a production facility (anaerobic digestion, landfill gas, etc.). It will most probably be used in combination with other sources such as SNG. "The most likely use for biomethane will be with hybrid heat pumps, using renewable electricity when available in the heat pump and using a mix of green gases, topped up at peak with natural gas when renewable generation is unavailable," says Chris Clarke, director of asset management, safety and environment, at Wales & West Utilities.

A hybrid heat pump can help reduce carbon emissions and may replace boilers using only mains gas and LPG as there's no need to replace existing radiators and pipework. It is most suited to properties with heat loads from 12 kW–20 kW, but can cover heat loads up to 27 kW. By monitoring external temperatures and internal heat and hot water demand, and the relative cost of gas and electricity, smart controls identify the most economical operating mode to maximise the use of an air-source heat pump when it is more cost-efficient than using the gas boiler.

However, hybrid heat pumps are not highly efficient (with an energy-efficiency rating label of only D or C) and can be expensive. The pumps will need to be powered by renewable electricity to be low-carbon.

Leeds' H21 gas to hydrogen conversion

A UK-wide conversion of the grid to hydrogen gas could, it's claimed, reduce greenhouse gas emissions associated with domestic heating and cooking – currently over 30 per cent of the UK's total emissions – by a minimum of 73 per cent, as well as supporting decarbonisation of transport and local electricity generation. The H21 Leeds City Gate project aims to test this and is currently undergoing £10 million of feasibility study work, driven by concerns of the North of England's gas distributor, Northern Gas Networks, about what could happen to its assets down the line, as the country reduces its GHG emissions to 80 per cent of 1990 values by 2050, which is the target.

The idea is that a hydrogen gas grid could use the existing underground natural gas pipe network, and that household appliances can be converted to run on hydrogen with far less disruption and expense than converting to renewable energy sources. Dan Sadler, H21 project manager at Northern Gas Networks, said: "Households won't be required to buy new appliances. The conversion process will be similar to that carried out in the 1960s and 70s when 40 million appliances across 14 million households were converted from town gas to natural gas. We'd have special teams, working street by street to make the conversion as smooth as possible for customers with minimal impact in the homes and the highways."

The project would be funded the same way as during that first conversion, allowing the costs to be paid back over time and, alongside energy-efficiency measures, would have a minimal impact on household energy bills. Household appliances will, however, need to be upgraded or modified.

Pure hydrogen embrittles many pipeline steels causing cracking and many pipes are made of iron, but they are slowly being changed to polypropylene at a cost of around £1 billion per year. This cost is spread across consumer bills. All being well, the conversion would take place from 2026–2029 and, if successful, there would be a rollout across the UK.

A commercial bioSNG demonstration plant, in Swindon, UK

The UK's National Grid has built the first automated plant in the world that produces bioSNG from residual household waste using a gasification process. It yields 22 GWh of grid-quality gas that is capable of heating 1,600 homes or of fuelling 75 heavy goods vehicles from 10,000 tonnes of refuse and waste wood per year. It is a step towards commercialisation of the technology required to produce significant quantities of renewable gas for injection into the gas grid.

Ontario's heat networks plan

Ontario's five-year Climate Change Action Plan, adopted in 2016, aims to make the region "one of the easiest and most affordable in North America for homeowners and businesses to install or retrofit clean-energy systems like solar, battery storage, advanced insulation and heat pumps".

Over four years the government will spend around C$7 billion, and it opted for heat pumps rather than natural gas because it was calculated that the $200 million subsidy demanded for installation of natural gas pipelines into 68 communities translated to an average of $25,600 for each connection, not including the cost of the furnace/boiler, hot water tank, or installation.

While setting up a heat pump network would cost about the same – for new equipment and putting the pipework in the ground – the ongoing running cost is 50 to 80 per cent less. Heat pumps were seen as cheaper and, if powered by renewable electricity, zero carbon. The government therefore chose to phase out natural gas. By 2030, the building code for residential and small buildings will eliminate combustion heating, and by 2050 combustion heating will be outlawed in all buildings.

Carbon capture and utilisation

To tackle climate change urgently we must take carbon dioxide from the air. CCS – burying emissions from coal and gas power plants underground – has been investigated but it is not cost-effective without a high global tax on carbon. CCU, however, for the production of fuels, chemicals and materials, is a possible better solution, but as yet it is little known. According to Dr Lothar Mennicken at the German Federal Ministry of Education and Research, "CCS is basically a non-profit technology, where every step is costly. CCU, however, has the potential to produce value-added products that have a market and can generate a profit."[55]

A report, *CCU in the Green Economy*, from The Centre for Low Carbon Futures,[56] shows that CCU can be profitable, with short payback times on investment. It says: "Although only a partial solution to the CO_2 problem, under some conditions using CO_2 for CCU rather than storing it underground can add value as well as offsetting some of the CCS costs."

The Carbon Storage Program of the National Energy Technology Laboratory (NETL) of the US Department of Energy supports four main CO_2 utilisation research areas: cement, polycarbonate plastic, mineralisation and enhanced hydrocarbon recovery. The first three represent a large potential market for CO_2 storage in materials, especially building materials.

According to Michael Carus, managing director of nova-Institute GmbH,[57] "If the European chemical industry were to meet its entire carbon needs from CO_2 rather than fossil sources such as oil, gas and coal, it would use or recycle 5.5 per cent of Europe's total CO_2 emissions, despite being responsible for just under 2 per cent of Europe's CO_2 emissions." Surely this is a great prize to aim for?

Energy storage

To satisfy the world's energy demands sustainably we will need a great deal of cheap ways to store energy. Unlike with fossil fuel power plants, we can't always generate power when we need it from renewable sources like wind, solar and tide. (See Box 7.2.)

BOX 7.2 A SUMMARY OF ENERGY STORAGE OPTIONS

How we store energy depends upon what we want to use it for. Here are the main types of energy storage solutions.

1. Chemical: for heat: biofuels (timber, biofuels and biogas)
2. Solid state batteries: for electricity: a range of electrochemical storage solutions, including advanced chemistry batteries
3. Electrochemical batteries: for electricity: flow battery (energy is stored in a liquid electrolyte for longer cycle life and quick response times)
4. Mechanical: Compressed Air Energy Storage (compress air to create a potential energy reserve for turbines); flywheels (harness rotational energy to deliver instantaneous electricity)
5. Thermal: capturing heat or cold in water, phase-changing materials, or high-density materials such as storage heaters

6. Pumped hydropower: creating large-scale reservoirs of energy by pumping water up an incline when energy is cheap, releasing it to drive a turbine for electricity and heat when needed
7. Hydrogen gas: by the electrolysis of water by renewable electricity: for fuel cells (electricity including electric vehicles) and for heat (in gas mains).

While batteries are viable for storing power produced during the day for use during night hours, and maybe up to a week later, they are not ideal over seasonal time frames. "You need other new technologies to convert cheap renewable energy into chemical fuel when the sun is shining or the wind is blowing," said Professor Steven Chu (a former energy secretary in President Barack Obama's administration) in *The Australian* newspaper.[58] "If you make really cheap hydrogen from renewables and store it underground, then you have something very different." While Chu's current research encompasses batteries, he believes that fuel cells hold more promise for urban power storage – particularly those based on liquid hydrocarbon. "We don't have that technically economical today, but that is also part of my research."

Electrochemical storage is also likely to mature in the mid-to-long term as a solution for energy storage over months. One promising technology uses an aqueous sulphur solution and is projected to reach an exceptionally low cost (about US$1/kWh) while providing moderately high energy density (29–121 watt-hours per litre). This is similar to the cost and performance of pumped hydro storage and compressed air energy storage, today's lowest-cost and most widely scaled energy storage technologies, but with a thousand times higher energy density at system level.[59]

Nordic settlements have for many years used solar-heated water stored underground to heat buildings in the winter. This requires a large area of solar collector and a suitable heat store, either a natural or excavated cavern or a large insulated tank.

Compressed air storage also enables a surplus of wind or solar power to be stored efficiently over long periods. A company called Storelectric's solution is to use surplus energy to compress and store air in salt caverns; underground reservoirs. When the energy is needed the air is released to drive turbines to generate usable energy. Together with NAM, the Netherlands' largest energy company, they are expected to build large-scale storage under the Netherlands and potentially the North Sea, to back up and store the massive amounts of intermittent energy that will be coming from large-scale offshore wind farms in the future.

Standards for cities

Measuring progress on all of the above is crucial for city authorities. Sustainable Development Goal 7 calls for access to affordable, reliable, sustainable and modern energy for all. Affordability is a challenge for countries that are still working to reach universal access to electricity. A World Bank report on progress to achieving SDG 7[60] says that this requires strong leadership commitment, backed up by sustained public financing. Most effort is required in heating/cooling and transport, where renewable energy use remains low. It would be a great help if fossil fuel subsidies were phased out.

Better and more available data are needed, especially on cooking fuels and particularly in developing countries. "There is still relatively little information on the energy efficiency of major consuming sectors, outside of the major economies, that is critical to inform policy interventions," says the UN.[61] Training is needed to support better gathering and handling of data in these countries.

A useful tool for measuring city and greenhouse gas emissions is PAS 2070, a standard sponsored by the Greater London Authority, which uses it to track its emissions (see page 37).[62] It specifies how to total the emissions of a city or an urban area, by following internationally recognised greenhouse gas accounting and reporting principles. PAS 2070 captures both direct GHG emissions – from sources within the city boundary – as well as indirect GHG emissions – from goods and services that are produced outside the city boundary for consumption and/or use within the city boundary. It uses either a direct plus supply chain methodology or a consumption-based methodology.

The former builds on the global protocol for community-scale greenhouse gas emissions developed by the World Resources Institute (WRI), C40 Cities Climate Leadership Group and ICLEI Local Governments for Sustainability. The latter captures direct and life-cycle emissions for all goods and services consumed by residents of a city, so that emissions are allocated to the final consumers of goods and services, rather than to the original producers of those emissions.

Conclusion

What we know now is that nations and cities must put more effort into tackling emissions and energy use because the world is failing to achieve all of the Sustainable Development Goal 7 targets under the current levels of ambition.

Cleaning up our cities from their reliance on fossil fuels will bring enormous benefits to air quality and health, making the 'one planet' city a much more pleasant place to be. Its citizens will suffer from fewer illnesses and live longer. They will not be reliant upon distant political events and wars for their energy and its price. And they will have much more control over their own destiny and the well-being of the planet.

Notes

1 See my book, Thorpe, D., *Solar Technology: The Earthscan Expert Guide to Using Solar Energy for Heating, Cooling and Electricity*, Routledge, 2011, ISBN: 1849711097; also Madrigal, A., *Powering the Dream: The History and Promise of Green Technology*, Da Capo Press, 2013, ISBN: 0306820994.
2 From McGlade, et al., *The future role of natural gas in the UK*, UK Energy Research Centre (UKERC), London, 2016. Available at: http://bit.ly/1p7hCLk
3 *Energy Technology Perspectives 2016* (ETP 2016), IEA. Available at: http://bit.ly/29mTbSf

4 From the introduction to *Climate Action in Megacities 3.0*, C40 Cities/Arup, December 2015.
5 On Twitter, 3:45 pm, 18 September 2017, https://twitter.com/c40cities/status/909790381902049280
6 See: www.cityofchicago.org/city/en/depts/mayor/press_room/press_releases/2017/september/40PercentParisAgreement.html
7 See: http://explorer.sustainia.me/cities/energy-saving-retrofits-for-aging-housing-stock
8 Speaking at a C40 conference on 18 September 2017 during Climate Week in NYC.
9 For more information see: www.nyc.gov/html/gbee/html/home/
10 See: www.c40.org/awards
11 Mengel, M., et al., "Future sea level rise constrained by observations and long-term commitment", *Proceedings of the National Academy of Sciences of the United States of America PNAS*, March 2016, 113 (10), 2597–2602. doi: 10.1073/pnas.1500515113. Available at: http://bit.ly/29rE6Ex
12 Lord Krebs, quoted in "The Floods Are Coming", a long essay in *The Guardian* newspaper, 7 July 2016, available at: http://bit.ly/29yujfK.
13 Current projections are for anthropogenic sea-level rise of 28–56 cm, 37–77 cm, and 57–131 cm in 2100 for the greenhouse gas concentration scenarios Representative Concentration Pathway (RCP) 26, RCP45, and RCP85, respectively. RCP85 corresponds to a radiative forcing value in the year 2100 relative to pre-industrial values of +8.5 W/m^2, leading to a mean and likely range of global temperature increase of 3.7°C (2.6 to 4.8°C) by 2081–2100. Cf. the IPCC's Fifth Assessment Report, part one: Physical Science Basis, available at: www.ipcc.ch/report/ar5/wg1/
14 *Low Carbon Energy, The Future Is Now*, Arup. Available at: http://bit.ly/29GbqXJ
15 According to the US Environmental Protection Agency: http://bit.ly/29DCx3r
16 *Energy Technology Perspectives 2016* (ETP 2016), IEA. Available at: http://bit.ly/29mTbSf
17 See: Thorpe, D., *Energy Management in Buildings: The Earthscan Expert Guide*, Routledge, 2013, ISBN: 9780415706469; and Thorpe, D., *Energy Management in Industry: The Earthscan Expert Guide*, Routledge, 2013, ISBN: 9780415706476.
18 *Energy Efficiency 2018, Analysis and outlooks to 2040*, OECD/IEA, Market Report Series. See: www.iea.org/efficiency2018/
19 Predicted in ETP 2016.
20 Efficiency indicators are published by the World Bank's Regulatory Indicators for Sustainable Energy (RISE) index at https://rise.worldbank.org, and by the International Energy Agency, www.iea.org/publications/freepublications/publication/energy-efficiency-indicators-essentials-for-policy-making.html
21 See: https://edx.iea.org/
22 Laski, J. and Burrows, V., *From Thousands to Billions – Coordinated Action Towards 100 per cent Net Zero Carbon Buildings By 2050*, World Green Building Council, 2017. See: www.worldgbc.org/news-media/thousands-billions-coordinated-action-towards-100-net-zero-carbon-buildings-2050
23 *The Emissions Gap Report*, United Nations Environment Programme, 2017, ISBN: 978-92-807-3673-1. See: www.unep.org/publications/ebooks/emissionsgapreport/
24 Gressot, M., et al., *City Footprint and Leverage Points for Resource Management*, Global Footprint Network, and Ecole de Technologie Supérieure, 2015.
25 *Renewables 2016 Global Status Report*, REN21, Paris, France. Available at: www.ren21.net
26 Chang, D.B. and van Staden, M., *carbonn® Climate Registry 5 Year Overview Report (2010–2015)*, Bonn Center for Local Climate Action and Reporting (carbonn® Center), ICLEI – Local Governments for Sustainability, 2015. Available at: http://bit.ly/1HK6D2K
27 Williams, D., "Amsterdam to totally decarbonize through district heating", *Cogeneration & On-Site Power Production*, 16 April 2015. See: http://bit.ly/29RnSjt
28 See Executive Mayor Kgosientso Ramokgopa talking about Tshwane's plans in a video at http://bit.ly/2ag21Ha
29 "Jeju aims to be carbon-free island", *The Korea Times*, 3 December 2015, at http://bit.ly/2aAdmOs. Detail and methodology is given in this presentation: Jeong, D.Y., *How*

To Formulate Carbon-Free Strategy, The Case of Gapado Island, Jeju, South Korea, Asia Climate Change Education Center, South Korea, 2015. Available at: http://bit.ly/2a1Ac09
30 Epp, B., "Austria: Up to 500 MWth for district heating in Graz", Solar Thermal World, 7 August 2015. See: http://bit.ly/2a0wfeQ
31 Morris, C., "Power to heat gets going in Germany", Renewables International, 23 June 2015. See: http://bit.ly/29SM83O
32 Dialogue Report: 100 per cent Renewable Energy In Cities, 7 December 2015, Cities and Regions Pavilion, TAP 2015, COP21, Le Bourget, Paris. Available at: http://bit.ly/1MYYlRz
33 Pia Buschmann (deENet Competence Network distributed Energy Technologies e.V.), Anna Leidreiter (World Future Council), Criteria for a Sustainable Transformation towards "100 per cent Renewable Energy", Global 100 per cent RE Campaign, October 2015. Available at: http://bit.ly/1OGWD9Z
34 100 per cent Renewable Energy Cities & Regions Network is at http://bit.ly/29ZzzDL
35 For example, the following topics are now included in the standard international professional reference for engineers and installers of heating, ventilation, air-conditioning and refrigeration – the ASHRAE (American Society of Heating, Refrigerating and Air-Conditioning Engineers) Handbook: thermal storage and district heating and cooling, unitary air conditioners and heat pumps, with sections on grid reliability, renewable power integration, heat storage, emergency cooling and water treatment.
36 Schimmel, M., Kube, M. and Petersdorff, C., "Strom. Wärme. Mobilität. Szenarien für die Dekarbonisierung im Großraum Wien bis 2050", Ecofys, May 2018.
37 Renewable Municipalities 2016 Legambiente, compiled with help from Enel Green Power. Legambiente is a long-established Italian environmental NGO. See www.legambiente.it/sites/default/files/images/rapporto_comuni_rinnovabili_2016.pdf
38 Many of the best developments are collected on a website: www.comunirinnovabili.it
39 Indian Government Solar/Green Cities programme, more details at: http://bit.ly/1RfuI19.
40 Development of Green Campus/townships/SEZs/ industrial towns, Institutional campus under the "Development of Solar Cities" programme, at http://bit.ly/29PRguj
41 Read more at http://bit.ly/2alZvxI
42 The efforts of Mexico City are cited in Energy Technology Perspectives 2016 (ETP 2016), IEA. Available at: http://bit.ly/29mTbSf. This has the theme of Towards Sustainable Urban Energy Systems.
43 Fawkes, S., Oung, K. and Thorpe, D., Best Practices and Case Studies for Industrial Energy Efficiency Improvement – An Introduction for Policy Makers, UNEP DTU Partnership, Copenhagen, 2016.
44 2050 Energy Scenarios: The UK Gas Networks role in a 2050 whole energy system, KPMG Consultants for Energy Networks Association (ENA), 2016. Available at: http://bit.ly/29P6CO2
45 Szabó, G., Fazekas, I., Szabó, S., Szabó, G., Buday, T., Paládi, M., Kisari, K. and Kerényi, A., "The Carbon Footprint of a Biogas Power Plant", Environmental Engineering and Management Journal, November 2014, 13 (11). Available at: http://bit.ly/29D2sWB
46 Uusitaloa, V., Havukainena, J., Manninenc, K., Höhnd, J., Lehtonend, E., Rasid, S., Soukkaa, R., Horttanainena, M., "Carbon footprint of selected biomass to biogas production chains and GHG reduction potential in transportation use", Renewable Energy, June 2014, 66.
47 Bachmaier, J., Effenberger, M. and Gronauer, A., "Greenhouse gas balance and resource demand of biogas plants in agriculture", Engineering in Life Sciences, Special Issue: Biogas, December 2010, 10 (6). Available at: http://bit.ly/29IeBee
48 RHI Evidence Report: Biopropane for Grid Injection, Prepared for DECC by Menecon Consulting in association with Atlantic Consulting, 2014.
49 RHI Evidence Report: Biopropane for Grid Injection, Prepared for DECC by Menecon Consulting in association with Atlantic Consulting, 2014.
50 339.9 kg of CO_2-equivalent per megajoule.
51 Pfister, S. and Scherer, L., "Uncertainty analysis of the environmental sustainability of biofuels", Energy, Sustainability and Society, 2015.

52 Dale, V. H., Langholtz, M. H., Wesh, B. M. and Eaton, L. M., "Environmental and Socioeconomic Indicators for Bioenergy Sustainability as Applied to Eucalyptus", Oak Ridge National Laboratory, Environmental Sciences Division, Center for BioEnergy Sustainability, TN 37831, USA, *International Journal of Forestry Research*, 2013. Available at: http://bit.ly/29BfOn1
53 Kukkonen, C. A., *SAE Paper 810349*, 1981.
54 MacLean, K., Sansom, R., Watson, T., Gross, R., *Managing Heat System Decarbonisation: Comparing the impacts transitions in heat infrastructure*, Final Report, Centre for Energy Policy and Technology, Imperial College, April 2016. Available at: http://bit.ly/2ahUnZw. *The Green Gas Book*, Parliamentary Labour Party Energy and Climate Change Committee, Summer 2016. Available at: http://bit.ly/29ZRE4s. *The "Future of gas" reports*, The National Grid, 2016. Available at: http://ngrid.com/2ahV2dH. *H21 Leeds City Gate Report*, Northern Gas Networks, West Wales Utilities, Amec Foster Wheeler and Kiwa, London, 2016. Available at: http://bit.ly/2a1Sn9q. *2050 Energy Scenarios: The UK Gas Networks role in a 2050 whole energy system*, KPMG Consultants for Energy Networks Association (ENA), 2016. Available at: http://bit.ly/29P6CO2
55 Dr Lothar Mennicken, German Federal Ministry of Education and Research, talks to SETIS, *Strategic Energy Technologies Information System magazine*, January 2016, at http://bit.ly/29SUn00
56 Professor Peter Styring, Chair, CO2Chem Network, The University of Sheffield; Heleen de Coninck and Daan Jansen, Energy Research Centre of the Netherlands; Jon Price, Director, Centre for Low Carbon Futures, 2012.
57 'CO2 is ready to go as a fuel and chemical feedstock', *Bio-Based News*, 21 November 2013. Available at: http://bit.ly/29YbzTl
58 Ross, J., "Obama energy secretary Steven Chu flat on battery plants", *The Australian*, 29 January 2018. See: www.theaustralian.com.au/news/health-science/obama-energy-secretary-steven-chu-flat-on-battery-plants/news-story/6fa3216e712fe7261a39a06f9f2056e9?nk=be52003a4c19f9804fa389f00a28828c-1542574753
59 Li, Zheng, et al., "Air-Breathing Aqueous Sulfur Flow Battery for Ultralow-Cost Long-Duration Electrical Storage", *Joule*, 2017, 1 (2). doi.org/10.1016/j.joule.2017.08.007
60 *Tracking SDG7: The Energy Progress Report*, International Bank for Reconstruction and Development/The World Bank, 2018.
61 *Tracking SDG7: The Energy Progress Report 2018 – A joint report of the custodian agencies*, United Nations Statistics Division, International Bank for Reconstruction and Development/The World Bank, 2018. See: http://documents.worldbank.org/curated/en/495461525783464109/text/126026-WP-PUBLIC-P167379-tracking-sdg7-the-energy-progress-report-full-report.txt
62 See: www.london.gov.uk/what-we-do/environment/leadership-and-policy/measurement-and-data

8
TRANSFORMING INDUSTRY

Besides food, industry is the other main component of humanity's ecological footprint that is very tough to deal with from the environmental angle. It is responsible for 31 per cent of global energy use and greenhouse gas emissions, and these emissions rose by almost a third from 2000 to 2010. Industry can also be polluting and wasteful, and benefits from being tightly regulated in order to become more sustainable – socially, economically and environmentally.

The task for 'one planet' cities is therefore a tough one: not only to follow the guidelines in the previous chapter for reducing energy demand and switching to renewables, but also transitioning to fully circular, 'closed-loop' production systems that do not waste resources or cause pollution, as described in Chapter 6.

Reducing consumption and its impact

Much of what we use – our buildings and the things in them, the clothes we wear, the gadgets we enjoy, and indeed much of what we eat, since it is often so highly processed – is a product of some kind of industrial process or series of processes. A process often begins with raw materials, whether they come from plants, from underground, are reclaimed from something else, or are a by-product of the petrochemical industry, as in the case of most plastics. They will be transported to a facility to be transformed and combined with other materials in highly complex processes involving energy, other materials, and cleaning agents, before being packaged and shipped to stores and homes. Every single step along this route has an impact; and every time we purchase a product we are in effect taking some level of responsibility for this process – even if we may be ignorant of it.

There are several ways of reducing these impacts. The first, for individuals and organisations, is to question whether one needs a given product. Only if humanity reduces its consumption levels will its collective ecological footprint reduce to a sustainable level. Any study of national ecological footprints reveals that the size of this footprint is directly proportional to average income and spending levels, and therefore consumption. Doing more with less has to be our way forward if we are to survive as a civilisation long term.

Assuming that a product has to be purchased, then the way money is spent is important, impacting on everything in the supply chain and life-cycle of the product's components. Procurement is at the heart of business, government and our personal lives, and ultimately connects them with many industries – including the waste and recycling industry.

For example, consider that a fleet of cars is being purchased. Without wishing to recommend this over any other vehicle, BMW's i series electric cars are made in a completely environmentally sustainable fashion; the factory is powered by renewable wind energy; all of the plastic in it is bioplastic – in other words, it is made from plants and has nothing to do with the petrochemical industry. It contains atmospheric carbon, rather than being responsible for the release of greenhouse gases in its manufacture. Were the electric car to spend its whole life being charged from renewable energy sources it would have almost no environmental impact in its use phase. When it eventually reaches the end of its life it is designed to be returned to the factory, where it can be taken apart and every bit used for something else.

Of course, one could decide to do without a car at all.

Support the circular economy

The life cycle of this car is an example of what's called the 'closed-loop' or circular economy, a way of imitating nature. Nature itself has no waste – for everything is endlessly reused. In nature, anything that dies, and any waste product from an organism, becomes food for something else in an endless recycling process which constantly renews its components. Similarly, in the closed-loop economy, the by-product or unwanted parts of an industrial process can always be used by another industrial process – even unwanted energy like heat.

The world is now firmly moving in this direction. For an organisation to be able to survive into the future, it has therefore to see all of its operations – its requirements in terms of materials, energy and water, and its fixed assets – as equal in importance to its core activity. Unilever is one company that has been doing this for a while. Its CEO, Paul Polman, was responsible for steering a Sustainable Living Plan, launched in 2010, that sought initially to double sales and halve the environmental impact of its products. It is working. The move was partly a reaction to the financial crisis, from a rules-based way of working back to one based on principles, but it has financial benefits.

The move requires a procurement strategy for equipment, supply chains and services that takes into account the overall lifetime impacts compared to compatible options. Lists of more sustainable equipment, together with standards, are becoming more and more widely available. For example, the EU Ecodesign Directive[1] covers all energy-related products sold in the domestic, commercial and industrial sectors, with the exception of transport.

According to the American Council for an Energy-Efficient Economy, supply chain optimisation can result in up to 60 per cent of energy intensity reductions.[2] For example, in food production and distribution, much perfectly good food is wasted due to spoilage, both in the supply chain and at the retail level. This means that all of the energy embedded in the food is wasted as well. By modelling the supply system throughout the chain, opportunities can be identified to significantly reduce waste by changing processing, handling, packaging and delivery systems. The result is frequently fresher food delivered faster and of a more consistent quality. There is less waste and greater saving.

We are all now aware that plastic waste is polluting the world's oceans, causing US$13 billion in damage each year to marine ecosystems, the United Nations estimates.[3] About 95 per cent of plastic packaging material value, or $80–$120 billion in economic value, is lost annually because of a short first use, according to a 2016 report by the World Economic Forum (WEF).[4] Some 72 per cent of plastic packaging is not recovered, according to the WEF report; 40 per cent ends up in landfill and 32 per cent leaks out of the collection system.

The circular economy has become a global imperative. China and the EU are cooperating and in the USA the American Chemistry Council's Plastics Division has committed to a goal of recycling or recovering all plastic packaging used in the USA by 2040, and along the way, to make all plastic packaging recyclable or recoverable by 2030. Steve Russell, its vice president of plastics, said when he announced the plan: "Together with our value chain partners we intend to transition to increasingly circular systems for designing, manufacturing, recycling, and recovering our plastic packaging resources."[5]

Waste makes no business sense. It is lost capital. The circular economy requires a significant shift in mindset, starting with the design process, which must ensure that goods not only have minimal environmental impact in terms of the use of water, energy and materials, but also that they can easily be repurposed or recycled through a 'cradle-to-cradle' approach. It also requires less packaging, higher rates of recycling and, ideally, for consumers to take sustainability into account when they buy. Products should be designed to be built to last, built to repair, and/or built to recycle, made from homogeneous materials to facilitate full recycling at end of life.

Progressive manufacturers can view this as a chance to cultivate new business opportunities. The business case for embracing the circular economy extends well beyond the creation of new, high-value products to give producers, for example, an opportunity to optimise their manufacturing processes, to increase vertical integration of the supply chain, reduce resource consumption, and gain greater control over the entire product life-cycle, including the ability to sustainably manage the end of life of a product.

Like BMW, many automakers are changing to producing no landfill waste. Lighting company Philips is offering its products as a service (lighting instead of light bulbs), which allows them to have greater control over their products' life-cycle. Luggage company Samsonite has introduced its ECO-Nu range made from recycled plastic bottles. Fairphone is a Dutch company that produces "the world's first modular, ethical smartphone", designed to be repaired and improved rather than disposed of at the end of its life.

Investors can support the development of a circular economy, according to Dr Emma Berntman of Hermes Equity Ownership Services, in three ways: by engaging with companies on improving resource efficiency, waste reduction and maximising recycling – within their own operations and throughout their supply chains; by engaging with policymakers and industry bodies to support regulations or incentives which encourage circular production; and by investing in businesses with good return prospects and which demonstrate a clear commitment to the principles underpinning a circular economy.[6]

The use of an open digital platform can transform a manufacturer into a network coordinator, working with stakeholders to track and trace production materials (including rare earth and noble metals, etc.) through the life-cycle, and to rapidly develop innovative products and services in collaboration.

Dame Ellen MacArthur, the inspiring yachtswoman-turned-prophet of the circular economy, says:

> We're trying to get governments to understand that the circular economy is about wealth and competitiveness for Europe. It's about looking at the entire system and at how products can be of service – like how we could share cars, how we can use the 'internet of things'. We did a big study looking at how the digital revolution will enable the European economy to save €900 billion. If we apply a circular lens, that could go up to €1.8 trillion. For us, it's always rooted in the economics.[7]

Industrial symbiosis

Many feedstocks and raw materials will become increasingly scarce and expensive in the future, and the cost of sending material to landfill prohibitive. Why not use waste products from other processes, industries or companies instead? This practice, part of the circular economy, is sometimes called 'industrial symbiosis'.

Symbiosis is another metaphor taken directly from nature, where two species may have a mutually beneficial – symbiotic – relationship with each other, such as an egret on the back of a buffalo that is performing a valuable service by ridding it of insect parasites. Often in an ecosystem one organism will thrive on the wastes of another organism. Industrial ecosystems are adopting this communal approach. If a single company or production facility cannot by itself reduce all of its waste to zero, it can partner with one or more other companies that can find a use for its waste products such as offcuts of various materials or chemicals.

The shift from using virgin feedstocks to recycling or reusing existing materials can already result in significant savings. Present-day examples include turning old tyres into rubberised asphalt, which lasts twice as long and requires half the volume of conventional asphalt. Plastics recycling is often cost-effective. In places, and for certain types of plastic, problems remain in finding customers willing to pay a price high enough to justify the collection and processing costs. In such cases, taxing the use of virgin products and enforcement of compliance with waste recycling streams do help.

In 2018 the CO_2 Value Europe consortium launched the H2020 BioRECO2VER project to pursue more efficient, sustainable processes for commercially producing platform chemicals like isobutene and lactate from carbon dioxide, although formidable technical and economic challenges remain.[8]

Industrial facilities that can take advantage of each other's waste materials may be encouraged by city authorities to situate next to each other in industrial estates. Even if facilities are not physically adjacent, in many areas, online networks exist which new enterprises can tap into. In these 'clubs', which usually operate online, members offer waste products that are otherwise costly to dispose of, selling them at profit to other members. If there is not one in a given area, they are relatively easy to set up.

Waste heat

Unwanted heat may also be reused. In some industrial processes, such as steel and cement industries, temperatures above 1,000°C are used in the production process and the waste heat from the process can be as high as 750°C. In CHP plants and boiler plants, the waste heat can be as high as 160°C to 180°C. Various options to recover and reuse the waste heat either as heat or power are available.

Depending on the temperature of the exhaust, various technologies can be used to convert waste heat to power – an approach sometimes called Waste Heat to Power. At high temperatures steam can be generated and used in steam turbines. This technology has been widely used in high-temperature processes in China where there have been over 700 installations in the cement industry alone.[9]

In parts of Sweden waste industrial heat is used to warm neighbouring houses. Elsewhere it is used to grow algae which sequester CO_2 and can produce ethanol, or it is used to heat greenhouses which may grow crops such as tomatoes.

Cities can make co-location of mutually beneficial companies a requirement of the granting of planning permission. Here again, standards can help. Covering almost every sector you can think of, from assistive products to zinc alloys, the International Organization for Standardization has over 200 technical committees that are developing standards. The ISO 14001 Environmental Management Systems standard helps any organisation find ways to improve the operational efficiency of its production processes. Like most sustainability standards, it is a framework for organisational change. It is open-ended, operational and pragmatic. It empowers all employees to think of creative ways to improve the organisation and take action for environmental protection, resource conservation and improved efficiency – doing more with less.

If a city or a nation were to make the adoption of standards such as this conditional upon the continued permission to operate within its boundaries, with a proper mentoring, inspection and enforcement regime, it could go a long way to reducing the overall impact upon the city itself, as well as making these industries more pleasant and creative places to work.

Data centres

Data centres, upon which we all depend for cloud computing, probably have the most potential for reducing energy use of any industrial or commercial sector. Computer giants like Apple run them on renewable electricity; for example, it has installed 10 MW of fuel cells running on biogas from a landfill site at its data centre in North Carolina. These places are famous for generating much waste heat; the average data centre has 50 times the power density per square metre of the modern office and many are increasingly feeding heat to nearby homes (in Finland), offices (IBM at Syracuse University, New York and Québecor, Canada), and even swimming pools (in a Swiss IBM data centre).

And what about a greenhouse? The idea of creating tropical conditions in a greenhouse to grow plants could be applied to commercial vegetable growing but at the Condorcet facility in Paris, France, the greenhouse is used to simulate a climate-changed, warmer Paris to find out what plants could grow well there. Built on a former industrial site entirely of natural materials, with no PVC (polyvinyl chloride) for ducting, its advanced filtration technologies remove the need for chemicals in air conditioning. The climate control system uses rainwater collected from the roof and stored in buffer tanks. Designed in a modular fashion to maximise the efficiency, it naturally draws in external air to provide free cooling. The roof is also painted white to reduce the amount of cooling needed. Excess heat is channelled into a semi-tropical arboretum in a greenhouse that is integrated within the building, which the French National Institute for Agricultural Research uses to research species most adaptable to changes in the prevailing climatic conditions in France.

Energy efficiency

Heat reclamation and reuse is one type of energy efficiency. All technologies for improving energy efficiency are mature and well proven. It mostly does not require developing new technology – simply adopting the best available ones now would save significant amounts of energy in every country in the world. The International Energy Agency has developed a set of indicators to measure energy use based on physical production in industry.[10] Based on these it predicts that manufacturing industry can improve its energy efficiency by 18 to 26 per cent, while reducing the sector's CO_2 emissions by 19 to 32 per cent, based on proven technology. The greatest opportunities lie in motor systems, followed by chemicals/petrochemicals. The highest range of potential sectoral savings for CO_2 emissions is in cement manufacturing, which is a case study at the end of this chapter.

According to the IEA, total investment in energy efficiency across all sectors was $237 billion in 2016. This sounds like a lot, but to achieve a '2°C scenario' – the higher target under the Paris Agreement – the cumulative investment between now and 2050 needs to reach about $39 trillion.

Economic development tends to follow a similar pattern from country to country. At very low incomes agriculture often accounts for a relatively high share of GDP. As economies grow, manufacturing's share of GDP grows and reaches a peak of about 20 per cent, at roughly $14,000 of GDP per capita. As incomes rise above that level the share of service industries in total GDP continues to rise and agriculture's share continues to decline. In addition, we find that some cities are built around light industry (such as clothing or electrical goods) and others around heavy industry (such as steel, concrete, chemicals, and fossil fuels).

How is energy efficiency measured? Overall energy efficiency is measured by energy intensity, which is defined as the amount of energy needed to produce a unit of GDP (usually expressed as tonnes of oil equivalent (toe) per US$1,000 of GDP in constant terms). The lower it is, the better, because it means more goods are being produced with less energy. Energy intensity globally has been falling steadily over the last decade at an average 1.5 per cent a year; although on a national scale it's also affected by the structure of a country's economy and fuel switching.

Sustainable Development Goal 7 has an aim to double the global rate of improvement in energy efficiency as measured by energy intensity by 2030. SDG 9 includes an aim to upgrade infrastructure and retrofit industries to make them sustainable, with increased resource-use efficiency and greater adoption of clean and environmentally sound technologies and industrial processes by 2030. The United Nations Industrial Development Organization (UNIDO) is coordinating the measurement of these goals.

How countries and industries can adopt energy efficiency

UNIDO has a huge amount of experience in running industrial energy-efficiency programmes in widely differing countries and sectors, as I know through my limited involvement with it, producing reports and policy recommendations.[11] My co-authors and I found that the potential to save energy – and water too – is huge, and most of it is either low-cost or pays for itself within a short time frame. It does not matter the size of the enterprise.

For example, at the large end of the spectrum Novacero, an Ecuadorian iron and steel manufacturer with sales in Central America, Bolivia, Peru and Chile, has an industrial facility in Cotopaxi that was consuming about 10,000 GWh per month at a cost of over $900,000 until it completed an energy management system programme (using the international standard ISO 50001, which we met in Chapter 3) at just the cost of one month's energy. This led it to save over 300,000 kWh per month. By continuing to implement further measures, it is continuing to save money and energy.

At the micro-level, energy-intensive brick manufacturers in Iran deployed a specially adapted energy management programme of changes which also led to significant energy savings.

Every country where UNIDO has helped governments implement an energy-efficiency policy has seen enterprises easily beat their own energy-saving targets. In the Philippines, the government realised it was alone in the region in having no energy-efficiency policy and accepted UNIDO's offer to develop bilateral programmes. It became an exemplary player in this field. Firstly it created an Energy Efficiency and Conservation Roadmap, with Minimum Energy Efficiency Performance Standards for industry and buildings, and energy labelling and efficiency standards. Then it went on to accredit energy service companies and develop roadmaps for each industrial sector up to 2030.

Its objective was to "make energy efficiency and conservation a way of life" and its goal was "to achieve average annual energy savings of 23 million barrels of oil-equivalent and five gigatonnes CO_2e emissions avoidance". Note how these two aims are about engineering a shift in culture, and attaining specific and measurable outcomes.

UNIDO also was responsible for influencing local governments in changing their planning policies to force any company to have an energy management system in place before they could apply for planning permission for the construction of a new facility.

To take another example, in Malaysia, five years ago there were subsidies for fossil fuel prices, which normally inhibits support for energy efficiency, because if there is cheap energy why should people bother to save it? But despite this and the absence of any energy-efficiency policy and having limited knowledge and experience of energy management, UNIDO influenced the government to pass a National Energy Efficiency Action Plan. By working with an accreditation body, SIRIM QAS, and industrial equipment suppliers, it began to train people in energy management systems. By 2017 149 factories were in various stages of implementing an energy management system. One-third of them had saved 4,865 GWh of electricity and 949,701 gigajoules (GJ) of thermal energy, and almost 3.4 million tonnes of CO_2 emissions. For an investment of just RM 27.55 million, factories reported saving RM 51.6 million a year (almost twice as much).

This just goes to show what is possible in countries that have had little experience of energy management in industry, because at the beginning of their industrialisation process they just wanted to industrialise with no thought of the consequences.

Here's how it is done – something any city or nation can emulate. Expert trainers are vital to kick-start a skilled energy management training sector in a given country. Firstly they're brought in from another country to train local people. They also train energy auditors to provide independent and objective support for efforts to improve energy management.

At the same time, exemplar companies are found and encouraged to pioneer the approaches to demonstrate the benefits. The choice of such a company is ideally based on the respect that company commands within its region, and the perception that it is locally owned. For example, in Vietnam, the well-known local Colusa Miliket Foodstuff Company became the first business in Ho Chi Minh City to become ISO 50001 certified, demonstrating a 24 per cent improvement in energy performance in 15 months. Its leading market position sent a strong message to other potential adopters, and so the programme gained momentum.

The majority of a company's energy savings at first come from operational changes alone rather than from capital-intensive equipment-related projects. Measurement and verification of results by metering is vital to a successful energy management system. The presence of an energy management system also inspires confidence in investors. However, at the present time, most investors themselves also need training in how to understand proposals for financing energy efficiency.

A further advantage of having national energy-efficiency policies is that they improve cooperation between industry and government. In Russia this gave rise to a Regional Centre of Competence, due to UNIDO's work, and this won recognition from the Ministry for the Economy. As its previous efforts to collect data from industry had failed, when UNIDO developed a web-based benchmarking platform that did just this, it was transferred to Russia's Energy Agency, which used it to rate companies for the allocation of subsidies, which resulted in a saving of $40 million.

Industry is helped by a plethora of standards for the most efficient equipment, from motors and drives, to steam systems, process heat, heat exchangers, cooling and refrigeration and so on. Part of the challenge for energy managers in industry is to see each step in an industrial process as one that can be improved, while at the same time looking at the whole system and what it is trying to achieve.

Energy manager Kit Oung says that while support from managers is essential, "Managers tend to focus purely on energy reduction and ignore the other benefits associated with the energy reduction efforts, e.g., a reduction in quality defects, improvement in working conditions, reduction in maintenance requirements, release in capacity for additional production, etc.". All of these things can drive sustainability.

As with most things in life, successfully transforming an organisation to become more efficient is not just about technology but working with people to get the best out of them. Oung, along with all of the energy managers I have interviewed, believes it is necessary to involve the whole organisation and give everybody responsibilities and appropriate resources; to challenge all established assumptions and integrate new ways of working into daily operations with performance measurements. "We should consider energy, carbon and water in the whole life-cycle of the business and its value chain," he says. "I'm happy when my work generates an interest in the people I meet and in sowing the seeds for continued energy reduction efforts when I leave."

Industrial cities

Cities are often constructed around particular industries. This is particularly true in China. Those which are centred on highly energy-intensive industries such as the power and the iron and steel sectors are usually the most polluting and energy intensive. More affluent cities may have lower emissions per GDP or per capita, but only because they rely on imports from nearby less affluent, industrial cities.

It is hard to reduce emissions from industrial cities in the absence of a carbon tax, as the polluting factories cannot just be shut down without huge and unfortunate consequences for employment and the economy. But researchers from the University of East Anglia[12] found that among the industrial sectors in each Chinese city they looked at, just a few factories or plants were responsible for a disproportionately large fraction of emissions: the top 2.5 per cent produced 70 per cent of total carbon dioxide.

By improving these 'super emitters' with changes to operational procedures, an energy management system and probably new technologies such as better boilers or low-carbon energy sources, it should be possible to drastically cut emissions without affecting economic growth. With aggressive technological changes, the researchers calculated that China could reduce its overall emissions by 2.4 billion tonnes, or 31 per cent of its 2010 emissions for the cities evaluated.

Of course, this approach could also work successfully for other countries.

Future industrial efficiency

Could we 'run out' of energy efficiency? In other words, is there a limit to how much energy industry can save? All the evidence is that the energy-efficiency resource is extremely large and is unlikely to run out in the foreseeable future. Many industrial enterprises have achieved high levels of energy savings over a long period of time. Work by the Next Manufacturing Revolution has shown that there are many enterprises that have achieved consistent and high levels of energy savings year-on-year for more than ten years.[13] Numerous examples from various industrial sectors demonstrate that companies in all sectors can continue to achieve energy savings over a long time.

For example, Owens Corning, a highly intensive manufacturer of insulation and related building products, embarked on an energy-efficiency programme in 1999 by declaring a target of reducing energy costs by 20 per cent. It formed an internal energy services company and introduced simple measurements of performance, encouraging ideas for improvements through competitions and awards. By 2003, annual energy costs had gone down by $40 million at a cost of $20 million, despite production increasing by 18 per cent and energy prices by 10 per cent. Three-quarters of the gains were down to ideas from employees. The 20 per cent target was achieved in ten years and it has since reduced energy intensity by a further 20 per cent. At the same time, globally it is now sourcing almost half of its energy renewably.

In the future, digitisation will become more and more commonplace, reducing the cost of metering and sensing, which is vital to knowing the levels of use of energy, water and resources. Combined with the Internet of Things, cloud computing and big data analytics, this is already having an impact on energy and resource management systems and markets. While there is presently much interest in digitising buildings with Building Energy Management Systems, within industry a shortage of personnel qualified to evaluate sector requirements and implement solutions is hampering progress, but the drive to be competitive should broaden this change.

The accelerating adoption of renewable energy by industry – including 'green' gas – demand management and distributed generation is driving a trend amongst manufacturers and industrial facility owners for greater flexibility in electricity supply as a way of reducing greenhouse gas emissions, reducing costs and providing extra revenue streams. In many countries a blurring of the lines is occurring between electricity suppliers and consumers. Improved energy management technology – with cloud computing, demand response and falling costs of storage, renewable and on-site generation – are all pointing to a future where industrial companies can become zero carbon – and create jobs and improve air and water quality at the same time. This will happen faster the more markets and governments consider energy efficiency as an energy resource in the same way as the supply-side resources – something that can be traded.

However …

While some industries are positively taking advantage of the above opportunities and protecting themselves from future shocks, others are actively lobbying against it, particularly oil companies, and especially through their trade associations. By 2018, nearly all of the world's largest 200 industrial companies[14] had directly or indirectly opposed climate policy since the Paris Agreement was signed, according to analysis by InfluenceMap, a UK-based think tank, of the lobbying activities of 200 of the world's biggest companies and 75 of the most powerful trade groups.[15] Among the 25 most influential companies on the list, IKEA, Apple, Unilever and SSE ranked the highest on positive engagement with climate policy, while the worst-performing companies include a host of fossil fuel firms such as Koch Industries, Chevron, ConocoPhillips, ExxonMobil, Shell and BP.

InfluenceMap found that 30 per cent of all companies analysed had directly lobbied against climate policy in the previous three years and that 90 per cent of them retain membership of trade associations which have actively opposed climate policy around the world. Edward Collins, project manager at InfluenceMap, concluded that trade associations lobby for much more polluting positions than their members support in public.

For example, in 2015 BP CEO Bob Dudley publicly supported a carbon tax and energy efficiency regulation, and Chairman Carl-Henric Svanberg used his opening speech to investors at the company's annual general meeting to talk about energy transition and BP's support for renewable energy subsidies.[16] That same year, BP lobbied to undermine EU renewable energy targets and subsidies in favour of gas.[17] BP is also a member of the Australian Petroleum Production & Exploration Association, FuelsEurope and the National Association of Manufacturing, trade groups which have consistently opposed ambitious climate change policy in Australia, Europe and the USA.

Sabido, a spokesperson from Corporate Europe Observatory, described the work of these trade groups as "lobbying vehicles" engaged in a "race to the bottom". He said their members sometimes had to accept other members' strong lobbying demands if they wanted the trade group to address their own concerns and issues. "We have to be clear that trade associations are a convenient front for companies, which can pursue their own branding strategies while the trading associations do the dirty work," he said. Sabido added that the "symbiotic relationship" between fossil fuel companies and governments across Europe had to end in order for policymakers to decide on ambitious climate policy without interference from fossil fuel lobbyists.

There is a point in time yet to come when shareholders and investors in these companies reach a consensus on the need to abandon these old, destructive practices and embrace the new sustainable economic model.

Case study: cement production

With global urbanisation adding a city the size of Shanghai to the planet every four months, and much of this involving cement, it's of crucial importance to tackle CO_2 emissions from the cement and lime industries. Cement is vital to modern construction; the massive ready-mix concrete industry, the largest segment of the cement market (a staggering 4.3 billion tonnes a year produced), is worth over $100 billion a year. The industry contributes 7 per cent (or 1.9 Gt annually) of total global anthropogenic CO_2 emissions and is the third most polluting industry in terms of greenhouse gas emissions, behind chemicals/petrochemicals and iron and steel (which can mainly be tackled by a switch to renewable forms of energy supply and materials reuse and recycling).

The best available technologies for the production of cement and lime presently have no carbon-capture capability, and reducing climate emissions in the cement and lime industries is mostly confined to improving kiln efficiencies and utilising non-fossil fuels. But efforts are underway to cut global warming impact of cement and concrete use around the world and by mid-century it is expected that all cement (or substitutes) will be carbon neutral.

Concrete is made from varying proportions of coarse aggregate bonded with cement that hardens over time. Most concretes used are lime-based concretes made from calcium silicate, such as Portland cement. The main ingredient is limestone or calcium carbonate ($CaCO_3$). Portland cement is made by heating the raw materials including the limestone firstly to above 600°C and then to around 1,450°C, to sinter the materials. This emits carbon dioxide and produces calcium silicate (($CaO)_3 \cdot SiO_2$). When it is turned into liquid cement with the addition of water and exposed to the air it absorbs carbon dioxide again, to reform into $CaCO_3$, and hardens. An average of 42 per cent of greenhouse gas emissions associated with the creation of cement are actually recouped from the atmosphere once the concrete is *in situ*.

Most of the greenhouse gas emissions associated with cement production arise because its production requires very high temperatures, but there are also significant emissions associated with mining and transportation. The IEA estimates that the global cement industry could save between 28 per cent and 33 per cent of total energy use by the adoption of best practice commercial technologies involving energy-efficiency savings in the production and supply chain such as heat recovery and reuse, relatively unexploited at present.[18]

As with any material, the supply chain impacts of each process during the production of concrete and its materials can be evaluated, and there are tools for this purpose. Substituting other materials such as wastes, geo-polymers, or granulated blast furnace slag in clinker can reduce both energy consumption and CO_2 emissions during cement production by about 40 per cent. Savings can also be made at the end of life of a concrete structure. Currently only 50 per cent of concrete is recycled for use in new building projects (compared to up to 99 per cent for structural steel). Down-cycling does help to reduce the use of aggregates, but does not help to reduce the supply of materials needed for new concrete.

Carbon capture and concrete

The European Commission is currently supporting five carbon capture and storage projects, of which three are in the cement industry, and two carbon capture and utilisation projects, which involve using carbon dioxide to produce methanol or ethanol.

One of these, the CLEANKER (CLEAN clinKER) project, is testing carbon capture by calcium looping at industrial scale in cement production. Here, CO_2 is separated from the furnace exhaust gas by the carbonation of calcium oxide, and an oxyfuel-operated calciner, in which a pure stream of carbon dioxide is generated. All but 10 per cent of the total carbon dioxide is hoped to be captured. A demonstration plant is being connected to an existing kiln line of the Buzzi Unicem plant in Vernasca, Italy. Adding this technology to an existing cement production plant will mean an increase in the cost of the final product, but the CLEANKER project anticipates this would only be €25 per tonne of cement.

Concrete absorbs carbon dioxide from the atmosphere as part of the normal weathering process. But if CO_2 is introduced in cement's manufacturing stage, under controlled conditions, it absorbs more of the gas, and better quality concrete is produced. CarbonCure is a CCU technology being pioneered in the USA, which does just this. The process makes it harder and stronger than conventional cement so that less is required to produce concrete. It has been proven in tests to reduce the initial setting time by 40 per cent and increase compressive strength by 10 to 14 per cent. A skyscraper built in 2018 in Atlanta, Georgia, incorporates 48,000 m^3 of concrete made with this technology, which it is therefore claimed diverted 680 metric tons of CO_2 from being released into the atmosphere.

Another new type of carbon-capture technology for cement production called Direct Separation is being trialled by the LEILAC (Low Emissions Intensity Lime And Cement) project. This aims to capture about 60 per cent of total CO_2 emissions without significant energy or capital penalty, and demonstrate that the technology works sufficiently robustly to scale up.

The technology is already proven in Australia at a commercial scale, for processing magnesite ($MgCO_3$), a similar ore to limestone. The company operating this facility, Calix, is the lead partner in the LEILAC project. Its pilot plant, at the HeidelbergCement plant in Lixhe, Belgium, aims to capture 95 per cent of carbon dioxide emissions and is currently on budget and schedule to be completed and verified by the end of 2020.

The LEILAC project anticipates negligible extra cost. The downstream processing costs of liquefying, transport and safe storage are not part of LEILAC as these steps are not cement or lime specific and are being developed in cross-sectorial approaches for all CO_2-emitting industries and utilities in Europe.

When integrated into new plants, or retrofitted into existing plants that are fired with biomass or waste using current best practice, the total CO_2 emissions would be reduced by more than 85 per cent compared to conventional fossil fuel-fired lime and cement plants, without significant operating issues, energy or capital penalty.

When the LEILAC project is completed in 2021, a cement and lime industry CCS Roadmap will be developed. It would be amazing if the world, which pours gigatons of concrete per year, were to switch to this type of technology.

Alternatives to Portland cement

"Limestone-free cements can be achieved through chemical 'activation' of by-product materials or by producing an array of cementitious compounds based on magnesium oxide," according to Jenny Burridge, the head of structural engineering at The Concrete Centre, UK.[19] There are several candidates. The use of magnesium silicates eliminates the CO_2 emissions from raw materials processing. In addition, the low temperatures required allow the use of fuels with low energy content or carbon intensity (i.e. biomass), thus potentially further reducing carbon emissions.

Commercial processes now exist which capture raw flue CO_2 gas from industrial sources and convert it into calcium carbonate cement-based building materials, permanently sequestering the CO_2 and yielding a high-strength material that can be used without any other cement or binder system to make concrete products.

Alkali-activated and 'geo-polymer' cements gain their strength from chemical reactions between a source of alkali (soluble base activator) and aluminate-rich materials, which are waste products: fly ash, municipal solid waste incinerator ash (MSWIA), metakaolin, blast furnace slag, steel slag or other slags, or other alumina-rich materials. They tend to have lower embodied energy/carbon footprints than Portland cements (up to 80–90 per cent but this is dependent on the source of the aluminate-rich material). Production is covered by a standard: PAS 8820:2016 Construction materials.

Ferrock, another cement substitute, is composed partly of iron dust reclaimed from steel mills that also sequesters carbon dioxide in the form of iron carbonate, and has a greater compressive strength than mortar made with Portland cement.

Another substitute for concrete is hempcrete, which is made from hemp and lime. Hemp, when growing, absorbs atmospheric carbon dioxide. Lime, when applied this way, also absorbs atmospheric carbon dioxide, making material carbon negative. While not having the structural strength of concrete (its typical compressive strength is around 1 megapascal (MPa), over 20 times lower than low-grade concrete, and its density is 15 per cent that of traditional concrete), with a k-value (a measure of heat conductivity) of between 0.12 and 0.13 watts per metre-Kelvin (W/mK), it offers some insulation value.

Hempcrete can be used in many situations where concrete is currently used and is of interest also because of its breathability. This lends it to use with other natural building materials to create buildings that have a pleasant internal atmosphere that do not suffer from damp or condensation.

The final piece

I think that industry may well be the final sector that one day becomes completely decarbonised and meets the requirements of the 'one planet' city. In that future, waste and pollution will become distant memories. They will be viewed in the same way that we look upon the practice of throwing rubbish and excrement into the streets – with shame and disgust.

The points made in Chapter 4 about reducing the length of supply lines and making social and environmental criteria part of procurement processes and purchasing decisions, are relevant here. Governments are recommended to investigate the example of Sweden and its PRINCE project described in that chapter.

Notes

1 See: www.eceee.org/ecodesign/
2 Shipley, A. and Elliott, R. N., *Ripe for the Picking*, American Council for an Energy-Efficient Economy, Washington, DC, 2006.
3 *UN Environment Programme (UNEP) Year Book 2014*.
4 Neufeld, L., et al. (Eds), *The New Plastics Economy: Rethinking the future of plastics*, World Economic Forum, 2016. See: www3.weforum.org/docs/WEF_The_New_Plastics_Economy.pdf
5 See: www.recyclingtoday.com/article/acc-plastic-recycling-diversion-100-percent-goal-usa/
6 Scott, M., "Companies Need To Design In Sustainability To Products – And Investors Need To Encourage Them", *Forbes*, August 2018. See: www.forbes.com/sites/mikescott/2018/07/30/companies-need-to-design-in-sustainability-to-products-and-investors-need-to-encourage-them/#93ee3025c6fa
7 Russell, E.H., "Ellen MacArthur: Why I believe in the 'circular economy'", *Spears*, August 2018. See: www.spearswms.com/ellen-mcarthur-why-believe-circular-economy/
8 Guertzgen, S., *Circular Reasoning: Why Chemical Manufacturers Are Engaging In The Circular Economy*, 8 September 2018. See: www.manufacturing.net/article/2018/08/circular-reasoning-why-chemical-manufacturers-are-engaging-circular-economy
9 Xhang, C., et al., "Implementation of industrial waste heat to power in Southeast Asia: An outlook from the perspective of market potentials, opportunities and success catalysts", *Energy Policy*, July 2017, 106. DOI: 10.1016/j.enpol.2017.03.041. See: www.researchgate.net/publication/316119645_Implementation_of_industrial_waste_heat_to_power_in_Southeast_Asia_an_outlook_from_the_perspective_of_market_potentials_opportunities_and_success_catalysts
10 International Energy Agency, *Tracking Industrial Energy Efficiency and CO2 Emissions*, 2007. Available at: www.iea.org/publications/freepublications/publication/tracking_emissions.pdf (accessed 25 August 2018).

11 Such as: Oung, K., Thorpe, D. and Fawkes, S., *Best Practices and Case Studies for Industrial Energy Efficiency Improvement*, UNEP, 2016.
12 Shan, Y., et al., "City-level climate change mitigation in China", *Science Advances*, 2018.
13 Lavery, G., et al., *Next Manufacturing Revolution*, Lavery Pennell, 2degrees, Institute for Manufacturing, 2013.
14 According to the 2018 Forbes Global 2000, a list of all publicly traded companies from 60 countries which collectively accounted for a market value of $56.8 trillion.
15 InfluenceMap, "Corporate lobbying: How companies really impact progress on climate", 2018. See: https://influencemap.org/climate-lobbying
16 See: https://influencemap.org/evidence/Supporting-renewable-energy-legislation-4630 5f7c6bbfeeb43305ae8382a4b843
17 See: www.theguardian.com/environment/2015/aug/20/bp-lobbied-against-eu-support-clean-energy-favour-gas-documents-reveal
18 International Energy Agency, 2007, op. cit.
19 What is required from design codes and specification for concrete – barriers and challenges for new materials, presentation by Jenny Burridge, Head of Structural Engineering, The Concrete Centre, 2016.

9

MANAGING WATER IN THE AGE OF CHANGE

We need water, but it can also be a threat to life. Whole ecosystems, our health, our food supplies and our energy security all rely on its sensible management; and water-related disasters such as floods, droughts and storms account for almost 90 per cent of the 1,000 most disastrous events that have occurred since 1990. In 2019 Mozambique's port city of Beira became the first city in the world to be destroyed by climate change.

The 'one planet' city must therefore be able to ensure a reliable supply of sufficient clean water for all of its citizens' essential needs, the proper management of sanitation to protect health, and resilient protection from extreme weather events – whether drought or flood.

The sanitation and shortage challenges

This is currently not the case everywhere in the world. Around 2 billion people presently have no choice but to drink contaminated water, causing the death of one child every minute that passes. By 2025, two-thirds of the world's population may face water shortages. The economic loss risk associated with water-related disasters such as floods and droughts is also increasing in all regions. The impacts of water-related disasters are felt most harshly in developing countries such as Bangladesh and Haiti. A third of groundwater supplies are already in distress, while most rivers or sites that could be dammed have already been developed. Los Angeles, Beijing and Seoul are among the world's most water-stressed mega-cities, the top five being Chennai, Istanbul, Tehran, Hyderabad and Kolkata.[1]

Then there is the impact on the biosphere. Although they cover less than 1 per cent of the Earth's surface, freshwater habitats are home to more than 10 per cent of known animals and about one-third of all known vertebrate species. Around the world, wetland areas have more than halved since 1900. Rivers have become increasingly disconnected by dams and other infrastructure. Reservoirs have changed natural flows and trapped more than a quarter of the total sediment load globally that used to reach the ocean. Eutrophication (starvation of oxygen) and toxic pollution due to poor agricultural and industrial practices are now major sources of water quality degradation.

To tackle these problems, water is the subject of several Sustainable Development Goals, particularly Goal 6, which calls for adequate and equitable sanitation and hygiene for all and the reduction of water pollution by 2030. Water also features in SDGs 3, 11, 12 and 15. The central importance of water is why the UN General Assembly declared the period from 2018 to 2028 the International Decade for Action, 'Water for Sustainable Development'. One of its reports states that "if the world continues its current path, projections suggest that we may face a 40 per cent shortfall in water availability by 2030".[2] Pause a moment and consider the implications.

The world can no longer take water for granted. From individuals to countries, we need to better understand, value, and manage water with the respect it deserves.

SDG 6 opens a path for progress on water-related issues that is complemented by the adoption of the Paris Agreement on Climate Change. At its heart is Integrated Water Resources Management. The key word is 'integrated'. It involves all interested parties together taking a holistic approach to risk reduction and building resilience. This may be problematic for cities because 60 per cent of all watercourses cross borders, and are therefore outside their jurisdiction. But what they do have potential control over is the management of the water that arrives within their boundaries.

This comes in the form of rainfall, watercourses, sea flooding, and the import of goods that contain water, such as food. Whether water supply is used by industry or individuals requires enforcement of legislation to control pollution.

The need for public ownership

The municipality of Paris took over control of its own water supply and treatment from the private sector with a body called Eau de Paris in 2008. Its prime motivation was that it was in the public interest, with the belief that access to water is a right, defined by a United Nations declaration, and that being able to manage it directly improves efficiency and democracy – a form of governance based on the common interest. Other cities have followed, with Barcelona looking like the latest to do so at the time of writing.

This principle has led to an international struggle against the sale and privatisation of water services. Canada's Blue Communities Project encourages municipalities and Indigenous communities to support the idea of a water commons framework and a campaign to phase out the sale of bottled water in municipal facilities and at municipal events.

The Blue Communities movement, begun in Paris and Bern in Switzerland, has spread to other municipalities around the world, from Cambuquira in Brazil to Thessaloniki in Greece, besides being adopted by schools, religious communities and faith-based groups. All have agreed to treat water as a common good that is shared by everyone and is the responsibility of all. But these signees still represent just a tiny fraction of all the world's administrations.

Cape Town – the first city drought caused by climate change

It is doubtful whether Cape Town in South Africa would have been able to avert disaster in 2015–2018 had its water supply not been in public ownership. As climate change bites and areas experience more severe and prolonged droughts, other administrations are going to have to plan to take the same kind of precautions as this city's authorities did when faced with their crisis.

Cape Town was the world's first metropolis to be forced to confront severe water scarcity in the age of climate change. 'Day Zero' was the Armageddon-style name given to the date on which water was predicted to run out for the inhabitants of this metropolitan area with a population of almost 4.5 million, should nothing be done.

The area's largest source of water is Theewaterskloof dam, which was down to nearly a tenth of its 480 billion-litre capacity. Rainfall had dropped from an average of 1,100 mm in 2013 to just 500 mm in 2017. A drastic civic water conservation campaign was put into practice. Over three years, residents more than halved their water use from 1.2 billion litres a day in 2015, to just over 500 million litres at the start of 2018. "We're going to become one of the world's water-resilient cities," Helen Zille, the Western Cape's provincial premier and a former mayor of Cape Town, said at the time.[3]

What techniques did the authorities use? Suburban restrictions were enforced at 50 litres per day per person, which compares to the global average use of 185 litres. Sectors such as tourism and agriculture bore the brunt of this. The campaign used behavioural nudges, public praise, surveillance and dire warnings.

To keep the public informed of dam levels and water usage, a colour-coded map showed adherence to water restrictions street by street online. Escalating tariffs on the rich were put in place to pay for basic water for the poor. The drought had a levelling effect, with all sectors of society affected, so the campaign, in this society notoriously divided between the rich white and the poor black, painted water scarcity as a common problem. Parsimonious use of water in the rich suburbs still seemed luxurious, however, when compared to the facilities available to the inhabitants of the informal settlements, where 25 litres a day was always close to their actual use.

Expensive restaurants served special 'dry' menus on paper plates. Discounts were given to customers in hair salons who washed their hair prior to their appointments. Organisers of any event provided their own water to attendees. Golf courses became parched.

Neil Armitage, a professor of civil water engineering at the University of Cape Town, said that the campaign was successful. "If we have another dry year as bad as the last, it looks as though we could survive it again. People didn't like being treated like naughty children. But it worked."[4]

If Day Zero had been reached, defined as when overall dam levels reached 13.5 per cent (the lowest was 19 per cent), residents would have been rationed to 25 litres per day. Despite this, farming became the biggest casualty. Harvests and crop yields have been reduced, impacting upon agricultural employment.

The richest households could afford to install private boreholes and storage tanks. Those in the middle stopped using a washing machine. The crisis made everybody realise that, although the universal right to water was in the democratic constitution of the country, it was very far from being put into practice.

Water supply in the region is managed by a publicly owned water department, which had already done much to reduce wastage in the system, keeping overall consumption rates at the same level for the previous decade despite a 30 per cent rise in population over that period. As part of the effort to achieve this, water meters were given to the heaviest users, who also had the highest tariffs.

But the region is still over-exploiting most of its river systems, according to the country's Institute for Security Studies. The water department internally collapsed as controversy raged over who was to blame for the drought. In an attempt to adapt to the ongoing shortfall, desalination plants, wastewater treatment units and groundwater extraction projects have been built, but these are not yet anywhere near sufficient to meet the demand. Desalination plants are still prohibitively expensive.

"Conservation is going to be the big game in town for the foreseeable future – and Cape Town has shown how to do it," says Neil Armitage.

Other cities in the world are beginning to face similar crises, such as Isfahan in Iran and Jakarta in Indonesia. The latter is sinking by 10 cm per year as a result of the over-extraction of groundwater and because of urban development (impermeable surfaces), which has prevented aquifers from recharging. As a result, piped water supplies only a small amount of the city's population of 10 million and it is often polluted.

A holistic approach

So, a holistic approach is required to manage water consumption: controlling supplies, treatment, increasing water-use efficiency by curtailing waste, protecting against water scarcity, and restoring water-related ecosystems. It can imply the use of rainwater harvesting, desalination, sustainable wastewater treatment, and recycling and reuse technologies.

A lack of investment in water infrastructure leads to multiple problems, but awareness of how to manage such a multi-faceted challenge at a governance level is often lacking. Putting Integrated Water Resources Management into practice requires leadership well beyond the water industry itself, in many governmental departments, so as to align objectives. Local communities need to be brought into the conversation to participate in improving water and sanitation management.

As with energy, the provision of good-quality data is vital; what gets measured gets saved. There is therefore a prior need to invest in water-related data and systems in order to share, analyse and take decisions on these data. Then there needs to be a way to engage on a permanent basis the full range of stakeholders to take responsibility for water resources in an inclusive manner.

Several international standards are available to help with this by giving expert step-by-step guidance. ISO 14046 is a life-cycle assessment standard covering the measurement and reporting of a 'water footprint', setting an international benchmark for estimating the impact of water use, and ways to improve efficiency and reduce overall consumption. In addition, ISO/TR 14073, Environmental management – Water footprint, contains examples on how to apply ISO 14046, with extra guidance.

Flood and supply management: working with nature

The local and regional natural environment's innate characteristics can be harnessed to both secure supplies and protect from floods. Any decisions taken on investment in infrastructure should look at this approach first. In the words of WWF's *Living Planet Report*:

> It is, in the end, nature that replenishes the freshwater that underpins all economic activity. Rainforests pump moisture into the atmosphere and the 'sky rivers' that flow from them water crops thousands of kilometres away from where they stand. Wetlands purify water and recharge the aquifers from which springs flow. Natural systems also contribute to water security through the role they play in maintaining climatic stability.[5]

It is a sad fact that the management of watercourses is in general in a terrible state, with around 80 per cent of all industrial and municipal wastewater being released to the environment without any prior treatment.[6]

In the local catchment areas, building natural infrastructure can involve any of the following: planting trees to minimise run-off and curb flooding; changing farming practices so that the soil retains more organic material which will hold more water in times of flood or drought; using natural wetlands to control flooding; and using managed reed beds for ecological sanitation treatment and nutrient reclamation. These are examples of 'nature-based solutions'.

> Nature-based solutions are inspired and supported by nature and use, or mimic, natural processes to contribute to the improved management of water. [They] can involve conserving or rehabilitating natural ecosystems and/or the enhancement or creation of natural processes in modified or artificial ecosystems. They can be applied at micro- (e.g. a dry toilet) or macro- (e.g. landscape) scales.[7]

The water cycle is an endlessly recirculating global system that underpins this regeneration of water supply. Nature-based systems of water management therefore become part of this process and an example of the circular economy. They are intrinsically restorative and regenerative by design, in contrast to the prevalent linear approach, which is a 'take, use, dispose' model of water consumption. The regenerative, 'one planet' city will restore water resources to a pristine level.

The water cycle consists of physical, chemical and biological processes that take place in a landscape, and especially in soils. These influence the quality of water as it moves through an ecosystem, as well as influencing soil formation, erosion, and sediment transport and deposition. Large energy fluxes affect this cycle, such as the latent heat of condensation and evaporation. This can exert a cooling effect and be used to help to regulate urban climates; so there are many associated benefits of this approach besides just maintaining the quality and supply of water.

Other benefits include products that are directly obtained from ecosystems, such as food, fibre and energy, from their processes, such as good air quality, flood control, and climate regulation, nutrient cycling and soil formation, and cultural services such as recreation.

Planting trees and other vegetation, returning organic material to the soil to help it retain and process water, and the conservation and the creation of new wetlands, which are the liver and kidneys of the world just as forests are its lungs – these are all practices that are part of the process of repairing the damage we have done to the water system and the environment in general so that it can support us better and improve biodiversity. They are practices that can only be successfully managed strategically at a governmental level and by cooperation between all parties in a given catchment area.

The following practices all help with the regulation of water supply, including safeguarding against drought, but they also offer the following additional benefits:

- Wetlands restoration/conservation: water purification, biological control, water temperature control, riverine flood control.
- Constructing wetlands: water purification, biological control, water temperature control, riverine flood control.
- Providing green spaces (bioretention and infiltration): water purification, water temperature control, reducing urban stormwater run-off.
- Permeable pavements: water purification, reducing urban stormwater run-off.
- Water harvesting: reducing urban stormwater run-off.

Industry can be a very thirsty water user and therefore has an interest in protecting the quality and security of its sources. The World Business Council for Sustainable Development has published various case studies showing excellent examples of how companies are using nature-based solutions.[8] For example, in Mexico's water-starved Puebla Tlaxcala Valley, a production plant belonging to the Volkswagen Group partnered with the Comisión Nacional de Áreas Naturales Protegidas (National Commission for Protected Natural Areas) to plant trees, dig pits and construct earthen banks to prevent run-off caused by earlier deforestation. This resulted in over 1.3 million m^3 per year of additional water being returned to aquifers.

Permeable surfaces

In urban areas, the provision of water-permeable surfaces instead of concrete and Tarmac, of ponds, reservoirs and bunds to absorb high levels of rainfall, and of 'green' roofs, all help to both replenish groundwater and give protection from extreme rainfall events.

Sustainable urban drainage systems (known as SUDS) use permeable pavements, roads and car parks, and create natural habitats such as ponds and lakes that both clear the air and water, provide pleasant environments and habitat, and provide space for excess water during rainstorms.

SUDS incorporate natural processes like evaporation, infiltration, and plant transpiration, and the biological action of bacteria to process pollutants around the plants' roots, to complement traditional 'grey' infrastructure, and reduce the overall amount of water entering local storm sewers or surface waters.

In the City of Philadelphia, USA, a US$2.4 billion plan for public infrastructure called Green City, Clean Waters, calculated that the use of SUDS would give benefits worth almost $3 billion compared with benefits of under $100 million deriving from the traditional piped alternative. The additional benefits include: changes to property values, green jobs created, a reduction in greenhouse gas emissions, and reduced crime.

In the city of Birmingham, UK, the city-wide benefits of retrofitting SUDS through the creation of green streets (not including the benefits relating to a reduction in the number of heat-related deaths, and enhanced biodiversity and health) were quantified at £1.5 billion over 40 years. On one street, a benefit of over £906,000 or 7.5 times the site costs of £121,000 was calculated. If water reuse infrastructure was added, to store and recycle run-off locally for irrigation and toilet flushing, the benefits would increase dramatically to nearly £8.3 billion across the city and nearly £3 million at the site scale. An added benefit of ponds and reservoirs is beautiful leisure spots, attracting plenty of wildlife.

Reducing water consumption

With water becoming an increasingly scarce and valuable resource, local authorities, institutions and companies can lead by example in reducing wasteful consumption. Just as with energy, the monitoring of water use is the place to start, followed by implementing a water-management system. This will help to identify very simple ways to easily capture and reuse water without extensive capital investment. It is helpful to follow the Water Minimisation Hierarchy; in a nutshell this consists of the following priorities, in order:

1. Eliminate wasted water;
2. Improve efficiency and use alternative sources;
3. Reuse water;
4. Recycle water.

Any organisation can adopt a water use policy. A good one should have the following elements:

1. Set a target for reducing water use in buildings, with further, more stringent requirements for all new-build and major refurbishment projects;
2. Embed the targets within corporate policy and processes;
3. Set corresponding requirements in project procurement and the supply chain;
4. Measure performance at a building level relative to a corporate baseline; and report annually on overall corporate performance.

The policy could incorporate any of the following metrics or key performance indicators:

- The volume of potable water consumed: gallons or cubic metres per year, with reference to the number of full-time-equivalent occupants (commercial spaces), number of visitors (retail, leisure or public spaces), or number of residents (housing); or with reference to
- [Square feet or square metres of net lettable area per year]; or to
- [Percentage reduction in use relative to a given year].

As an example, the supermarket chain Sainsbury's in the UK saved half of their water use, the equivalent of 393 Olympic-sized swimming pools each year, with strategies that have included eradicating underground leaks, fitting pre-rinse spray taps and low-flush toilets in all their stores, and investing in rainwater harvesting for all new stores as standard, as well as retrofitting these units in existing stores; even reclaiming water from car washes. This has resulted in savings for individual stores of hundreds of thousands of pounds each year, with a fairly short return on their investments. Using rainwater for toilet flushing in one city store (Swansea) achieved an annual mains water consumption saving of 1,300 m^3.

Saving water also saves carbon emissions because of the energy used in processing and pumping it. In the UK, mains water contains an average of 1.47 kg of embodied carbon dioxide equivalent per 100 litres, through sourcing, transportation and heating. The embodied carbon contained in water supplies elsewhere will depend upon the proportion of fossil fuels used in grid-supplied electricity in the local area.

Opportunities exist in almost all industrial plants to save and reuse water. Water has many uses in industry, such as steam used during processing, cleaning, in evaporative cooling towers, distillation, pumps, and boilers. Potential savings are hard to estimate as they are process specific, but they can be as high as 90 per cent. The Water Technology List is a list compiled in the UK of certified products that are amongst the most water efficient available.[9] Tax relief is available on their purchase. In the USA, the Watersense label fulfils a similar function.

Waterless urinals are now common, and mains water is not needed to flush the waste away. Composting toilets are less common but besides possessing the same advantage, additionally, the waste is composted and later used as a fertiliser, retaining the nutrients in the waste. Compact models of composting toilets are available off the shelf that look almost like the toilets we are used to (although avoid models that use electricity to heat the waste to accelerate decomposition).

Rainwater harvesting

Rainwater may be collected, stored and used for toilets, washing machines, gardening and other purposes. It does not need disinfecting, merely filtering. The larger the roof, the more cost-effective the measure. In order to calculate the benefit, the annual rainfall for the location should be known, together with the roof area. This is then multiplied by a drainage factor which is dependent upon the roof type. The higher the roof factor, the greater the proportion of rain falling on the roof will reach the gutter and be collected. Since rainfall is sporadic, storage may be needed. The size of the tank should be sufficient to hold about 18 days' worth of demand, or 5 per cent of annual yield, whichever is the lowest.

Rainwater harvesting can be integrated with SUDS. Provided it is properly treated if pollutants are likely to be present, all surface water on a site can then potentially be collected and reused. Surface water drainage systems would lead into a tank, from which the water can be pumped for reuse. The sizing of this tank would be calculated using a method that takes into account extreme weather conditions, such as once-every-50-years occurrences, or provision for overflow into balancing ponds/swales. These would be open to visual inspection and periodic cleansing. Integration of the two systems, SUDS and rainwater collection, makes it a double win.

Sewage treatment

As with transport infrastructure, new communities do not need to repeat the past mistakes of developed countries. It is wasteful to use expensively cleaned water to transport waste when waterless urinals work perfectly well, and sewerage can be separated into black water and greywater, which are each then managed differently.

Nutrients in wastewater must not be dumped into watercourses, which causes pollution such as algal blooms by depleting the oxygen in the water. Instead they can be retained and recycled in keeping with the closed-loop approach. As mentioned above, doing so can yield a financial revenue from sales to fund maintenance. Nutrients are retained in different ways depending on the sewage treatment system.

There are many ways of processing sewage. A comparison of them will take into account the full life-cycle energy and carbon costs versus benefits, the land area, the availability of supply and markets for products, and health and safety. Most conventional systems and anaerobic digestion systems will generate methane gas and sludge. This biogas can be retained and used within an on-site combined heat and power plant to generate heat and electricity for local consumption, or purified and injected into a local gas network if there is one. A Thermal Hydrolysis Plant can generate even more biogas.

Anaerobic digestion can be combined with other forms of suitable organic waste, which a business would contract from suppliers who can ensure a reliable and regular supply. Digesters process the waste to produce methane and sludge. The sludge is composted and returned to the land. The methane is collected and either used directly as a fuel for cooking, heating or transport, or burnt in a combined heat and power plant. There are advantages in scale.

Constructed wetlands

These are designed to mimic the functionality of natural wetlands and are now commonly used for treating domestic and even some industrial effluent. Wetland vegetation such as reeds is planted within a protected area lined with clay or plastic, and the effluent is allowed to flow in. Around their roots, populations of microbes naturally arise and work hard to remove pollutants in the water such as excess nitrogen, phosphorus, potassium and organic pollutants.

As they grow, these wetlands soon look indistinguishable from natural ones and may also serve to increase the quality of reclaimed water by recharging underground aquifers when the water percolates down through the soil and rocks.

Humans produce about 4.5 kg of nitrogen and 0.6 kg of phosphorus per person per year.[10] If this is treated in a constructed wetland, the water leaving it can have relatively high levels of these nutrients, making it perfect for irrigating cropland.

Constructed wetlands are highly productive. Clean water is often allowed to flow out of the end of these wetlands into a plantation of short-rotation coppiced willow or other fast-growing trees. Harvesting and composting the biomass of reed bed systems and any downstream willow plantations, allows the compost to be sold. Alternatively, the branches of the willow or other trees, harvested on a short-rotation coppice basis, may be processed into furniture, paper or packaging, or chipped and seasoned for fuel for sale. Research is showing the potential for the production of biogas too, where one giant reed bed can generate methane yields greater than that of corn or sorghum.[11]

Both natural and constructed wetlands can also biodegrade or render harmless a range of pollutants. One piece of research found that almost half of a total of 118 pharmaceuticals that were monitored in conventional wastewater treatment were removed only partially but constructed wetlands did at least as good a job, as shown in one study in Ukraine[12] and other pilot studies.[13]

The business opportunities for utilities here should be self-evident. Profits generated from the products of the treatment of effluent help to reduce costs to customers, as well as reducing pollution and recycling nutrients.

In this way sewage treatment becomes a closed-loop business, offering employment and producing sellable products; quite a different business model from the currently dominant approach, but one with multiple benefits.

Aquaculture

Domestic sewage can be used on fish farms. Around Kolkata in India, farmers developed a technique of using domestic sewage for fish culture and other agricultural purposes, 80 years ago. They supply vast amounts of fish to customers in this densely populated Indian city. It is the largest operational system in the world to convert waste into consumable products.

At its peak, 600 million litres of sewage per day was diverted into stabilised sewage ponds, which were originally used as a source of water for growing vegetables, and developed into the large-scale expansion of a sewage-fed fish culture system. The downside is that it does use a large amount of land surface, some of which more recently has been appropriated for urbanisation.

Water reuse

According to Alexandros Makarigakis, programme specialist at UNESCO's Water for Human Settlements section, "Everyone should look at wastewater use – it's a low-lying fruit that should be picked up first".[14]

In many parts of the world treated municipal wastewater is directly returned to pipes downstream for drinking purposes. London is a case in point, but Singapore and parts of Australia and Namibia, California, Virginia and New Mexico do this.

Indirect reuse happens when treated wastewater is returned to aquifers or surface sources and eventually ends up again as drinking water where it may be consumed again downstream.

Many industrial and commercial units can reuse untreated wastewater if it is of sufficient quality, for example in industrial processes, such as water that has been used for cooling or heating. It can also be used for the irrigation of public parks and gardens, or golf courses. In these contexts, wastewater may be treated by an existing effluent treatment plant, using a dissolved air flotation unit to clarify the wastewater, followed by a submerged biological filter.

Simply by investing in membrane technology (microfiltration and reverse osmosis plant) to further treat the effluent, the quality of the treated water is such that it can be reused on site for certain applications, such as boiler feed water, top-up feed for cooling towers, washing of non-sensitive items, etc. The treated water offers a high level of purification and a relatively low energy and carbon footprint.[15]

In industrial parks, it is possible to use closed-loop systems – a form of industrial symbiosis described in the chapter on industry. For example, in Kalundborg, Denmark, the Asnæs Power Station receives 700,000 m^3 of cooling water from a Statoil facility each year, which it treats to use as boiler feed water, plus about 200,000 m^3 of Statoil's treated wastewater for cleaning each year. The cooling water becomes steam that is piped back to Statoil, as well as to other businesses, such as a local fish farm. This saves local water resources nearly 3 million m^3 of groundwater and 1 million m^3 of surface water per year.[16] The power plant uses saltwater from a nearby fjord for some of its cooling needs, which reduces the withdrawal of freshwater from a nearby lake. This results in a lot of hot saltwater, some of which goes to the fish farm's 57 ponds. Asnæs also supplies the city, Novo Nordisk, and Statoil with steam for a district heating system which replaces the equivalent of about 3,500 oil-fired boilers (furnaces).[17]

Greywater reuse systems for individual buildings reclaim water from baths, washbasins and showers, or other places where water is not likely to be too contaminated, store it and most usually reuse it in flushing toilets. Savings vary from 5 to 36 per cent. Treatment is necessary because warm, nutrient-rich grey water incubates bacteria when stored. This means that greywater should only be stored temporarily and systems require more frequent attention. Chemical disinfectants such as chlorine or bromine compounds may be added.

In larger systems, the greywater may be treated in small sewage treatment plants, using membrane filtration or ultraviolet technology. This is normally not cost-effective unless such a system already is in place.

With both building-level rainwater and greywater reuse, the carbon cost of creating and installing the system, plus running it using the pumps, should be taken into account before a decision is taken. Any pumps used should be of the minimum power specification for the job and not over-sized. If possible, gravity-fed systems should be deployed to minimise or remove the need for a pump. In both cases as well, using the water for irrigation purposes is the simplest end use.

Treating rainwater run-off

Where rainwater runs off surfaces that are polluted, such as in industrial facilities or roads, it can contain hydrocarbons and other chemicals. Some form of retention and filtration or biological treatment is necessary to avoid contamination of agricultural land or watercourses.

Advice for city leaders on regenerative sanitation

The proper strategy for cities depends upon whether they already have universal water supply and sanitation. Planning and financing need tight control in either case. Adequate sanitation cannot be achieved by the private sector and NGO investment alone, and grants are necessary to make it affordable. There will be economic benefits which compensate the cost. Any administration planning expensive sewered sanitation and treatment plants should think again, and opt instead for much cheaper decentralised systems, faecal sludge management and infrastructure for container-based sanitation. For all developing country cities, universal sewered sanitation would not be affordable, and is impractical in dense urban slums; new types of on-site and container-based sanitation technology render that over-expensive option obsolete.

Containers remove material for district-level treatment and provide the opportunities for nutrient reclamation and resale – an income stream – described above. City leaders can benefit therefore from taking a city-wide, holistic and inclusive sanitation approach, and giving contracts for on-site or container-based sanitation that cover large enough geographical areas to be financially viable. Local standards and regulations must be set to ensure the quality and sale of the products made from treated toilet resources. Public bodies, by being a customer themselves, can provide and encourage markets for these products and supply appropriate enforcement of standards.

Here are four examples:

Bogota, Colombia

For cities like Bogota with urban slums, to be more sustainable the authorities plan to focus on the treatment and reuse of toilet resources. The Water Service, Sewerage and Sanitation Company of Bogota is owned by the municipality and there are plans for it to generate revenues from selling the products of its sewage treatment plants, specifically fertiliser, biodiesel, chemicals, and biomass for feed. All of these come within the concept of the 'circular sanitation economy'. They are seen as a source of plant nutrients and energy and not as waste to be disposed of.

Senegal

Senegal has not completed universal water supply or treatment. Its urban sanitation is handled by a branch of national government, therefore cities have no direct control over its management. Although the concept of the circular sanitation economy is officially recognised, it is very little practised to date.

Two-thirds of the capital Dakar's homes are served by on-site sanitation and one-third by sewered sanitation, but the national policy places equal emphasis on, and has future plans for, both. Faecal waste is collected by licensed contractors and taken to treatment centres which will turn it into commercial products, notably treated water, fertiliser and energy.

At the scale of the whole city, this would be a significant circular economy innovation. The authorities will increase tariffs to fully cover the operational expenditure. They view improving the legal and regulatory environment as a way of encouraging commercial investors.

Durban, South Africa

The concept of the circular sanitation economy is well understood in South Africa where there are several examples of it in operation. With only half the city's 3.7 million people connected to a sewerage system, and a further quarter with no sanitation at all, the municipal council-run water and sanitation company (EWS) has the job of managing all types of sanitation services.

For capital projects, EWS is half-funded from national government grants and half from commercial loans repaid from operating revenues. With a good credit rating, it pays low interest rates. For operational expenditure, its annual $140 million budget is derived 84 per cent from tariffs and 16 per cent from government grants.

Its treatment plants generate revenues from selling fertiliser, biodiesel, chemicals, and biomass for feed. Durban recognises that the human right to sanitation means it must be affordable to all, and so financial management incorporates an overt tariff subsidy from rich to poor households.

Jodhpur, India

Not every city has had success with selling products made from waste to generate revenue. In Jodhpur, 70 per cent of its population of 1.1 million inhabitants are connected to an ageing sewerage system – although its treatment capacity covers far less. This is managed by the municipality and financed out of general taxation.

Opposition to buying the products came from the perception that products made in the public sector should be free. From now on the city government plans to ring-fence sanitation accounting in the budget, and to encourage large institutions such as hospitals, hotels, educational institutions to build their own wastewater treatment plants. These could be run by private-sector contractors, who would apply circular-economy concepts and sell the products.

Regulatory changes support investment

"Improving the capacity of local government institutions and water utilities for the delivery of services, and prioritising household use while regulating water use for large scale agriculture or industry, are essential to address water stress," says Vincent Casey, the senior water, sanitation and hygiene manager for the charity WaterAid. "Raising awareness about sensible water use, securing additional water sources and increasing storage capacity while at the same time reducing water loss through leakages and illegal connections are more practical measures that help tackle water stress in cities."[18]

UNESCO's Alexandros Makarigakis agrees: "As we have seen in California and in Cape Town, a three- or four-year drought can change things drastically. The way that you go about your day can change very quickly once climate change hits. It's a reality. We have to live with it and plan for it."[19]

Local authorities find it necessary and desirable to adjust their regulatory requirements to permit and even encourage the use of green and water-friendly infrastructure in urban areas. In certain cities traditional regulations discourage the use of urban agriculture and food gardens, and even the provision of green infrastructure such as planting trees, green roofs and green walls. In such circumstances they should be changed to align with sustainability goals. New benchmarks and technical standards are currently being introduced; UNESCAP, for example, provides an e-learning course for policymakers on Shifting Towards Water-Resilient Infrastructure and Sustainable Cities.[20]

Urban, regional or national authorities can engage in discussions with land managers on higher land in their catchment areas, and offer incentives so that they can become stewards, and manage their land in such a way as to conserve and recharge underground aquifers, and take other measures to prevent floods downstream where the urban areas lie. In such a way they could be rewarded for contributing to freshwater reserves that can be drawn upon by the urban areas, and contributing to flood prevention.

New York City provides a good example of this, where $300 million per year of water treatment operation and maintenance costs have been avoided using 'payment for environmental services' schemes for watershed protection.

Its Green Infrastructure Plan, in place since 2010, integrates nature-based and traditional 'grey' approaches to the capture and treatment of urban run-off, following calculations that showed that investment in green infrastructure was cheaper than in conventional stormwater management. Rain gardens, bioswales, green roofs, constructed wetlands and other nature-based approaches were implemented by the Department of Environmental Protection using funds from water rate payers and capital infrastructure investments made by other city agencies. The Department offers grants to private property owners to provide green infrastructure on their land.

Publicly funded stormwater green infrastructure systems such as bioswales (ponds and trenches along contours designed to capture rainwater run-off) and 'stormwater capture greenstreets' are sized to take all run-off generated within their tributary areas during around 90 per cent of all wet weather events per year. Significant co-benefits were then noticed, such as improved biodiversity, a reduction in air temperature through shading, more beautiful communities, and opportunities for ecological stewardship and a reduction of flood risks.

Latin America also has much experience of this; there, countries have worked together to form an association of water and sanitation regulatory entities to rationalise investments resulting in payments to landowners for their environmental services. And in Vietnam, water users, operators and utilities collectively paid $54 million to communities planning in upland forests for the watershed services that they were providing, per year.

It is integrated water policy and management that has been responsible for the success of the massive task of cleaning up the Rhine river, which used to be the most polluted in Europe as it passed through an area of intensive, dirty industries. This approach integrated political and regulatory strategies to reducing pollution, emissions reduction, the restoration of ecologies, and flood prevention and mitigation. It resulted in the Water Framework Directive, an extremely important EU-wide measure for the protection of watercourses across administrative boundaries. It can serve as a model for pursuing the same approach in industrialised, highly polluting areas of developing countries.

The latest United Nations *World Water Development Report* suggests that "cities, companies and water utilities could invest much more in nature-based services. There is mounting evidence that such investments are cost-effective and make good business sense, while at the same time generating co-benefits such as biodiversity conservation, community benefits, climate change adaptation, and jobs and training."[21]

Such a holistic approach to problem-solving and investment that provides multiple benefits like this is a characteristic of excellent 'one planet' governance. It is a challenge to business-as-usual, fragmented, ineffective policies. As with many of the other solutions in this book, success involves the participation of many different stakeholder groups, from governments to NGOs, citizen groups, and landowners. It offers a means to encourage change through consensus-building on overall system objectives and by identifying win-win outcomes that satisfy multiple interests.

Financing the change

The UN Report also advises that these measures do not necessarily require additional financial resources to make them happen. Instead they usually involve redirecting and making more effective the use of existing financing.

Nevertheless, some extra investment can often be required. To provide confidence for investors in the climate and green bonds market that it is worth investing in this area, a series of criteria that score the climate-adaptive potential and the environmental impact of 'water bonds' has been developed. These bring the awareness and knowledge of the world of technical water management to the attention of the worlds of finance and investment.[22]

In 2016, the first bond scored against the standard was issued, and within a year over $1 billion had been raised against the standard, including the first African issuance from Cape Town, with help from the accountancy firm KPMG.

UNEP has estimated the amount of water-related investment required in the range of $6.7 trillion by 2030 to $22.6 trillion by 2050.[23]

The World Water Council, based in Marseille, France, works to position water at the top of the global political agenda to support this agenda. It brings together over 300 member organisations from more than 50 different countries (NGOs, construction companies, governments, academics, local governments) at the triennial World Water Forums. At the last such Forum, held in Brasilia in March 2018, it laid out ways to increase investment in sanitation and policy recommendations for ministers and local authorities, encouraging them to be bold and embrace change, although critics accuse it of promoting the privatisation of water supply.[24]

It is complemented by the International Water Association, headquartered in London, with a global secretariat based in The Hague and offices in Beijing, Bangkok, Nairobi and Singapore, which has members primarily amongst water professionals in more than 130 countries. It also is engaged in concerted action to advance standards and best practices in sustainable water management over issues covering the full water cycle. It has three main programmes: Basins of the Future, Cities of the Future, Water and Sanitation Services.

Sponge cities

China's central government has taken this even further and come up with the idea of the 'sponge city' to improve and manage the availability of water. I rather like this idea as an example of regeneration in cities. The 'sponge city' uses a mixture of nature-based solutions – such as constructed wetlands, parks, revitalised lakes and bioremediation – and built infrastructure – such as permeable pavements, green roofs and walls.

The objective is for 70 per cent of rainwater to be absorbed and reused. China aims for one-fifth of all urban areas in sponge cities to meet this goal by the year 2020, rising to four-fifths by 2030. Some 16 pilot sponge cities are to be constructed across an area of over 450 km^2, with more than 3,000 planned construction projects costing about $1.25 billion.

The concept has been integrated in urban regulatory planning and ecological restoration at city and district levels in Shenzen and Guangdong provinces. Rain gardens and bioswales are used to collect run-off and remove some of the pollutants with some of this water being returned to the natural system and some being used.[25]

Associated standards

For water supply, the ISO 4064 series of standards define the requirements for water meters that monitor both cold drinking water and hot water, while ISO 24510, ISO 24511 and ISO 24512 deal with activities relating to drinking water. At the opposite end of the pipes, others deal with sludge recovery, recycling, treatment and use in most circumstances. In particular the ISO 16075 series governs treated wastewater reuse: for irrigation, in urban areas, and industry.

Principles for Water-Wise Cities

The multiple challenges presented by water are an opportunity for administrations to develop systems that protect water resources and are resilient to changing and uncertain conditions. The International Water Association has concluded that many cities today lack governance, infrastructure and institutions to provide even basic water services. In these cities, radical new strategies are needed to benefit from this opportunity. The ways of the last century are no longer appropriate. Disruptive interventions that enable reuse, recovery and recycling of resources can and should be developed and implemented, it says.

To help city administrations it has come up with the following Principles for Water-Wise Cities.[26] These present a framework to assist urban leaders and water professionals to develop and implement regenerative and resilient urban water services, through working together on a shared vision:

1. Regenerative water services
    - Replenish water bodies and their ecosystems
    - Reduce the amount of water and energy used

- Reuse, recover, recycle
- Use a systemic approach integrated with other services
- Increase the modularity of systems and ensure multiple options

2. Water-sensitive urban design

    - Enable regenerative water services
    - Design urban spaces to reduce flood risks
    - Enhance liveability with visible water
    - Modify and adapt urban materials to minimise environmental impact

3. Basin-connected cities

    - Plan to secure water resources and mitigate drought
    - Protect the quality of water resources
    - Prepare for extreme events

4. Water-wise communities

    - Empowered citizens
    - Professionals aware of water co-benefits
    - Transdisciplinary planning teams
    - Policymakers enabling water-wise action
    - Leaders that engage and engender trust.

Water is a magical, pure thing. It is the basis of all life. Managed properly, and under normal circumstances, it is delightful to be close to. We seek it out because it makes us feel good. It is no coincidence that in many religions it has symbolic, spiritual significance. For all these reasons, the 'one planet' city should not take it for granted, but cherish it as never before.

Notes

1. WWF.
2. WWAP (United Nations World Water Assessment Programme), *The United Nations World Water Development Report 2016: Water and Jobs*, UNESCO, Paris, 2017.
3. Cotterill, J., "How Cape Town beat the drought", *Financial Times*, May 2, 2018. See: www.ft.com/content/b9bac89a-4a49-11e8-8ee8-cae73aab7ccb
4. Cotterill, 2018, op. cit.
5. Grooten, M. and Almond, R. E. A. (Eds.), *Living Planet Report – 2018: Aiming Higher*, WWF, Gland, Switzerland, 2018.
6. WWAP (United Nations World Water Assessment Programme), *The United Nations World Water Development Report 2017. Wastewater: The Untapped Resource*, UNESCO, Paris, 2017.
7. WWAP (United Nations World Water Assessment Programme)/UN-Water, *The United Nations World Water Development Report 2018: Nature-Based Solutions for Water*, UNESCO, Paris, 2018.
8. See: www.wbcsd.org/Programs/Food-Land-Water/Water/Circular-water-management/Resources/Case-studies
9. See: www.eca-water.gov.uk

10 Mateo-Sagasta, J., Raschid-Sally, L. and Thebo, A., "Global wastewater and sludge production: Treatment and use", pp. 15–38 in *Wastewater: Economic Asset in an Urbanizing World*, P. Drechsel, M. Qadir and D. Wichelns (Eds), Springer, Dordrecht, The Netherlands, 2015.
11 Morrison, E. H. J., Banzaert, A., Upton, C., Pacini, N., Pokorný, J. and Harper, D. M., "Biomass briquettes: A novel incentive for managing papyrus wetlands sustainably?" *Wetlands Ecology and Management*, 2014, 22 (2), 129–141. doi.org/10.1007/s11273-013-9310-x.
12 Vystavna, Y., Frkova, Z., Marchand, L., Vergeles, Y. and Stolberg, F., "Removal efficiency of pharmaceuticals in a full scale constructed wetland in East Ukraine", *Ecological Engineering*, 2017, 108 (Part A), 50–58. doi.org/10.1016/j.ecoleng.2017.08.009.
13 UNESCO/HELCOM (United Nations Educational, Scientific and Cultural Organization/Baltic Marine Environment Protection Commission – Helsinki Commission), *Pharmaceuticals in the Aquatic Environment in the Baltic Sea Region: A Status Report*, UNESCO Emerging Pollutants in Water Series, Vol. 1, UNESCO, Paris, 2017. See: unesdoc.unesco.org/images/0024/002478/247889E.pdf
14 Kirk, A., "Cities in the Face of Drought", *Daily Telegraph*, 8 August 2018. See: www.telegraph.co.uk/news/cities-in-the-face-of-drought/
15 See my book *Energy Management in Industry* (Routledge, 2014) for a chapter on water management and efficiency in industry.
16 Domenech, T. and Davies, M., "Structure and morphology of industrial symbiosis networks: The case of Kalundborg", *Procedia – Social and Behavioral Sciences*, 2011, 10, 79–89.
17 WWAP, 2017, op. cit.
18 Kirk, 2018, op. cit.
19 Kirk, 2018, op. cit.
20 UNESCAP (United Nations Economic and Social Commission for Asia and the Pacific), *Shifting towards Water-Resilient Infrastructure and Sustainable Cities*, ESCAP Knowledge Hub for Sustainable Development, 2017. E-learning course. sustdev.unescap.org/course/detail/9 (accessed July 2017).
21 WWAP, 2018, op. cit.
22 Michell, N., "How to Plug the Gap in Water Investments", *Development Finance*, 2016. See: news.devfinance.net/how-to-plug-the-gap-inwater-investments?utm_source=160613&utm_medium=newsletter&utm_campaign=devfinance.
23 UNEP, *Transboundary River Basins: Status and Trends; The Transboundary Water Assessment Programme*, 2016. See: http://twap-rivers.org/
24 World Water Council, *8th World Water Forum Highlights*, Marseille, France, 2018. See: www.worldwatercouncil.org/sites/default/files/World_Water_Forum_08/Outcomes-of-8th-WWForum_WEB.pdf
25 Embassy of the Kingdom of the Netherlands in China, *Factsheet Sponge City Construction in China*, Beijing, Kingdom of the Netherlands, 2016. See: www.nederlandenu.nl/binaries/nl-netherlandsandyou/documenten/publicaties/2016/12/06/2016-factsheet-sponge-cities-pilot-project-china.pdf/2016-factsheet-sponge-cities-pilot-project-china.pdf. Xu, H. and Horn, O., *China's Sponge City concept: Restoring the Urban Water Cycle through Nature-Based Solutions*, ICLEI Briefing Sheet, ICLEI, Bonn, Germany, 2017. See: www.iclei.org/fileadmin/PUBLICATIONS/Briefing_Sheets/Nature_Based_Solutions/ICLEI_Sponge_City_ENG.pdf
26 International Water Association, *Principles for Water-Wise Cities at the 2016 World Water Congress and Exhibition in Brisbane*, London, England, 2016. See: www.iwa-network.org/wp-content/uploads/2016/08/IWA_Principles_Water_Wise_Cities.pdf

10

'ONE PLANET' NEIGHBOURHOODS

The places we live in define our quality of life and the opportunities for us, both personal and within our community. At present too many people are either homeless, suffer inadequately serviced informal settlements, find themselves in housing developments not designed to foster community cohesion, or in badly designed neighbourhoods. An important factor is frequently the affordability of housing, which is linked to land prices.

The aspiring 'one planet' city may decide to improve all of these circumstances. To do so would require city authorities to adapt their approaches and expectations, listen to their citizens, and alter their planning regimes. The development of 'one planet' cities is likely to happen by the staged adaptation of existing neighbourhoods alongside the creation of new ones. This chapter examines some of their features that we haven't yet touched on, while Chapter 11 focuses on the buildings themselves.

Informal settlements and homelessness

There are no global statistics for homelessness, according to Leilani Farha, UN special rapporteur on housing, but she says that around 6 billion people are inadequately housed across the world, and around 900 million are living in informal settlements and encampments. According to one estimate, by 2050, nearly one-third of all humanity is projected to live in informal settlements, since population growth is greatest in urbanising developing countries.

Before then, by 2030, at the current rate of expansion, over 700 cities will have populations of more than 1 million. "Without access to adequate, secure and affordable housing there is no equality, there is no end to poverty, to health and well-being, to sustained access to education, to employment," Farha says.[1] Lack of affordability means that over 55 per cent of households in sub-Saharan Africa spend more than 30 per cent of their income on housing costs.

Sustainable Development Goal 11 pledges to make cities and human settlements safe, inclusive, resilient and sustainable by 2030. Local and national authorities are making uneven progress towards achieving that goal. The UN has created a platform called Local2030[2] for cities and local governments to work towards achieving it. But what do citizens themselves in the cities of the developing world – those who live in the informal settlements that are often dismissively termed shanty towns – think about what they need?

In South America informal settlements are known as *favelas*. Theresa Williamson, an urban planner and executive director of NGO Catalytic Communities, works in Rio de Janeiro, where 1,000 favelas exist, some over 50 years old, mapping and surveying them, and, with her team, finding out what residents are doing to support their communities and what they need from the authorities, who often see them as a problem.

As a result of this work, Williamson has come to the conclusion that community land trusts (CLTs) are the proper legal vehicle through which these families can obtain official recognition and self-determination. This raises crucial questions of fairness, ownership, and principally the affordability of housing.

Tackling affordability of housing

In non-capitalist societies, all land has the same value and therefore housing in one location will cost the same as in another – often just the cost of building the house. In a capitalist society, a piece of land in a given location can rise in value over time from nearly nothing to millions, without costing the owner anything or producing anything of value to society.

Property rights in land in this instance are therefore a form of monopoly; only one entity can own a block of land. Owners can withhold their land from the market to extract the maximum price from those who need it for housing. So housing will always be relatively expensive for those who have to pay the market price for land. Prices will tend to rise out of the reach of those on lower incomes.

To tackle this barrier to those on low incomes needing to buy or rent a home, subsidies are often given by local governments as part of a policy to make housing more affordable. But the end result is typically that the subsidy ends up in the pockets of landowners, without making access to land and housing cheaper for new buyers and renters.[3]

There are two ways around this paradox:

1 The land rent scheme

The Australian Capital Territory's land rent scheme (LRS) in south-eastern Australia is a publicly run system that provides people with a means of accessing homes at a below-market price. The regional government initially retains ownership of the land beneath the new house that's supplied (and therefore the security of its rising value). It gives occupants a perpetual lease, in return for an annual land rental payment at a discounted rate (in this case it was 2 per cent of the land value: it could be a fixed annual payment, a share of household income, or a lower rate of market value).

Residents pay a much lower than market rent for the land in return for renouncing the right to profit from any future land price increases. The agreement includes a clause stating that when the household income of a resident exceeds the qualifying income threshold, they must exit the scheme within two years "... by purchasing the land outright or by moving out and transferring the lease to another applicant who meets the eligibility criteria for the discounted land rent".[4]

A government could choose to offer loans to pay for the building construction to residents since many may not afford to pay for it themselves (this is not the case in the LRS).

The scheme has been successful: many LRS residents who left the scheme but stayed in their home used the savings they made from it to buy their house at market price, including the land beneath it. The disadvantage of selling the land to them is that fewer dwellings remain in the scheme, and so for it to continue, the Land Agency must keep finding new land lots to offer.

2 Community land trusts

CLTs are incorporated legal bodies that operate in parallel to housing co-operatives and co-housing groups. Their advantage is that they can be controlled by the larger community as well as acting as a vehicle for shared housing. CLTs can also provide freehold housing for sale, for cross subsidy of the affordable homes or covenanted freehold homes, which must be sold back to the CLT, in order to be resold to local people.

CLTs can reduce the cost of housing compared to the market price by more than a quarter. For example, over its 25 years of operation, the Champlain CLT (Burlington, Vermont) offered housing 26 per cent below market prices on average. A CLT owner can save 52 per cent of their housing costs over ten years compared to renting.

A CLT sets a geographical boundary for itself, defining the extent of the community it serves. The general rule about eligibility for housing is that it is for local people within that area. If there are still vacancies after all of those people have been housed, then eligibility is progressively extended to neighbouring communities. Generally, the CLT's community boundaries are set to be the same as administrative boundaries whether those of community councils, towns, or parishes, etc.

The trusts have to obtain land and subsequently act in a management capacity in perpetuity (or subcontract to a Registered Provider). According to Ian Crawley, a training officer from the British Community Land Trust Network, "A big part of our work is confidence building. Confidence needs to be given not just to the members of the CLT but to councillors and local officials that what is being proposed is achievable". He says it is generally best to supply a mix of housing tenures from rented to ownership.

There remains the challenge of acquiring the land at a below-market price. Dr. Murray offers several routes: public land could be transferred at little or no cost by local or state governments; the CLT could purchase an agricultural or industrial site that is re-zoned for residential use; it might acquire part of a larger site for free as a condition put on a development for a large land developer; it might obtain crowd-sourced or philanthropic funding to acquire a site, or existing building, at market value; it could even be a community organisation that already owns a site.

Crawley says that the CLT always charges ground rent (about £250 per property per year in England) on the freehold, which is let on a 125-year lease, in return for the CLT acting as stewards of the whole site. "In some cases this ground rent can be paid in advance for the whole period of the lease which can be useful finance." He says that rent levels "are best set at the level that enables the CLT to house those it is targeting. Rents are unlikely to be less than 60 per cent and more likely to be 80 per cent (of market rates), but will also be determined by the business case for the scheme".

CLT residents have the right to sell their CLT property, or to sublet it, but this must be approved by the CLT Board and the constitution will ensure that this is conditional on the sale being at below-market prices, to households meeting the specific criteria set by the CLT in the constitution. Residents have the right to occupy as long as they wish, but may recover the value of any investment in dwelling improvements in the event of resale; therefore incentives to maintain the home are the same as for those of private homeowners.

To keep housing affordable to new residents, the CLT's constitution sets a formula for resale price: this is either a discount – say, 50 per cent – on the relative market price of the land component (the house being the same as the market price), or an index-based formula relative to a desirable affordability metric, like median local incomes. For example, the council-run Citywide Inclusionary Affordable Housing Program in San Francisco sets a limit of no more than 33 per cent of purchasers' income being used at first to pay for the housing; and, for resale, to this is added the percentage change in the median income from the date of purchase, the cost of approved capital improvements, and the real estate agent fees equal to 5 per cent of their total.[5]

Local authorities can support CLTs in their development plans. Crawley gives the example of a Health & Wellbeing park in the UK, "where the Local Planning Authority helpfully agreed to a specific policy in the Development Plan for the 65 acre site". City authorities have set inclusionary zoning policies in several US cities: Chicago, San Francisco, Davis in California, Montgomery County in Maryland, Burlington in Vermont (known for many progressive affordable housing initiatives), and Boston. North of the border in Canada, Toronto, Montreal and Vancouver have policies such as Vancouver's 'non-market housing policy'.

CLTs are growing in popularity. In the UK the number of CLTs has grown from two to 280 since the 2008 financial crisis. There are just 250 in the USA; Bernie Sanders started one of the first when mayor of Burlington, Vermont. This now manages over 2,200 dwellings.[6] The 27-year-old Citywide Inclusionary

Affordable Housing Program in San Francisco manages over 800 properties and over the last 50 sales, the average buyer obtained a 47 per cent discount on their home's market price, saving about US$12 million.[7]

Favela community land trusts

Back in Rio, this is why Williamson supports CLTs. It must be recognised that land serves a social function. Williamson says:

> There are three main reasons a household might choose to participate in a Favela Community Land Trust rather than seek individual title to the land:
> 1 *Permanence*. The residents' main concern is being able to stay in their homes and maintain their community, rather than being able to sell their homes at full market price.
> 2 *Affordability*. They require subsidies because they cannot afford full property tax, utilities, and other market-rate costs of living associated with the 'formal city', such as businesses likely to operate in a speculative setting.
> 3 *Community Management*. They prefer their community to manage its own development rather than relying on government agencies, which are often absent or ineffective.[8]

She believes that this route could be deeply transformative.

> Communities would ensure their tenure security through cycles of economic growth and decline, amid gentrification and eviction. They would also build on the legacies of resilience and resistance in favelas, preserving the unique characteristics of individual neighborhoods and their residents. They would use their collective, formal status to lobby for cultural recognition, subsidized utilities, and other amenities, and material improvements.

I have visited favelas in Brazil and informal communities in Tanzania and can vouch for the fierce sense of pride and determination residents feel, and their entrepreneurship in the informal economy as a way of bootstrapping themselves out of a precarious, poverty-stricken existence. Favelas start as squatted public land, consisting of structures that are thrown together out of any material to hand. As wealth is generated they evolve, and buildings are constantly repurposed with more durable materials, with extra storeys added.

A strong community sense develops, with informal support groups formed. There can be crime, but then, crime is everywhere. What matters for the authorities is that this natural self-improving process is supported and encouraged rather than meeting it with blanket obstruction. It is much cheaper for the authorities to work with these communities to provide water, sewerage services and power, than to bulldoze them and rehouse everyone, thereby destroying the community spirit. It is the height of madness to bulldoze them and do nothing for the inhabitants.

Williamson is helping to develop a Sustainable Favela Indicator in Rio de Janeiro. It is based on three sets of measurements: ecological footprint, basic services, and well-being. She says:

> The idea is to have both a simple qualitative questionnaire and an urban database quantitative set of indicators for each of these three areas that we can draw on in assessing the 'full sustainability' of an informal settlement. By 'full sustainability' I mean that we will not only include typical footprint/service indicators, but truly measure the value the settlement offers its residents (many of these will fall in the well-being realm, such as solidarity networks and a sense of security from proximate relations, etc.). The indicator will assess both the settlement's environmental/social assets and its environmental/social challenges and the primary goal is for the settlement's leaders and residents to use the results in self-organizing for further improvements, by building on their assets to address their challenges.[9]

She is going on to host workshops for community leaders and land rights activists in partnership with the Brazilian Urban Law Laboratory at the Federal University of Rio and the Caño Martín Peña settlements in Puerto Rico.

Caño Martín Peña

Caño has been widely studied and shown to successfully demonstrate the potential of favela CLTs to effectively provide formal, titled ownership without the risk of gentrification, while building on a community's existing social attributes. The eight Caño Martín Peña communities in Puerto Rico adapted the CLT model to support existing informal settlements in 2008. They were typical informal settlements begun in the 1930s.

The 26,000 residents were supported by the passing of a law that created the Martín Peña Canal Special Planning District, a public corporation called ENLACE, and provided for the future incorporation of the CLT. Residents established a non-profit entity to promote economic, social and community development that could liaise with ENLACE. Some 2,000 of the 6,500 or so families opted to join the CLT and became entitled to pay a lower property tax in exchange for forgoing their right to speculate, and guaranteeing permanently affordable housing.

Residents were granted surface right deeds with the right to inherit and maintain ownership of their home. ENLACE retained title to the land beneath. This protects residents from rising real estate values. The General Regulations state the CLT's role as a "mechanism of collective possession in order to solve the problem of the lack of ownership titles" and to "avoid involuntary displacement" of residents.

In 2015, the CLT won a Building and Social Housing Foundation World Habitat Award recognising it as a model for other informal communities. Initiatives around the world are inspired by how residents organised themselves to decide the best model, craft legislation, and develop the community affordably. When households are a part of a favela CLT, they continue building on the collective assets of the community, rather than lapsing into the more self-interested thinking associated with full individual titles.

So, when Hurricane Maria hit Puerto Rico, the Caño was globally linked and within months was able to galvanise supporters including other CLTs around the world to raise hundreds of thousands of dollars to aid their rebuilding efforts. This is true resilience in action. CLTs can be economic engines coordinated by collective community priorities.

The sustainable characteristics of informal communities

Favelas in Rio have developed the following attributes that we commonly associate with sustainable communities:

- High degrees of collective action
- Cultural incubation
- High rate of entrepreneurship
- Affordable housing in central areas
- Housing near work
- Low-rise, high-density and highly sociable neighbourhoods
- Flexible use-based architecture
- Mixed-use, pedestrian-centred streetscape
- High use of bicycles and public transport.

As this book attempts to show, these are highly desirable attributes of successful communities anywhere. Of course, favelas have many problems. Amongst them are poor access to adequate health care, reliable transport, safe public spaces, and more hazardous environments where municipal solid waste collection services are unavailable. But there is nothing special about favelas that produces criminal activity. This is more to do with the criminalisation and stigmatisation of poverty and lack of economic opportunity.

Neglect by the authorities of education, infrastructure and other amenities happens for budgetary reasons. By giving residents titles to their property, they are automatically given addresses, and so can be brought into the taxable economy. Taxes provide an income stream with which to fund public services. It is often the simple lack of addresses that prohibits this in many cities, and perpetuates citizens' exclusion from public services – because the authorities cannot afford them.

Back in Rio, Williamson says, "If only a quarter of the community forms a [favela] CLT, the fact they have done so will limit the speculative potential of their community's real estate permanently, because there will not be large tracts of land available for speculation." But she does admit that this "will depend on heavy investments in existing community organizing efforts, to inform residents about the risks and opportunities."

The opposite temptation

City administrators in developing countries such as sub-Saharan Africa are sometimes tempted by plans by architects and speculators to revamp neighbourhoods using buzzwords like 'eco-cities', 'smart cities', and 'new urbanism'. These terms can mask a terrifying prospect of mass evictions of millions from their homes to furnish the greed of a new kind of capitalist imperialism.

Proposals exist to renew, extend or replace cities in Angola, the Democratic Republic of the Congo, Ghana, Kenya, Nigeria, Rwanda and Tanzania. Professor Vanessa Watson from Cape Town has made a special study of these urban master plans that can be found on the websites of international architectural, engineering and property development firms, and condemns the architects who came up with them.[10] She says they betray a desire to emulate the developments of Dubai, Singapore or Shanghai, with iconic building shapes thrown in, glass tower buildings and manicured landscapes with motorways that are a throwback to 20th-century modernism, since exposed as disastrous.

If they are serious then there is a worrying contradiction here. If they are not, then the greenwash propaganda needs to be exposed. These architects want to project the idea that Africa is on the up. Economists point to a ready and willing workforce, thousands of square miles of space to be developed and exploited, huge natural resources ready to fire the furnaces of industry in the rest of the world. They capitalise on the psychology of a continent looking wistfully at Hong Kong or São Paolo and thinking "me too?"

But, Watson, believes, to go down this route would be a disaster because the picture ignores the fact that the bulk of the population is extremely poor and living in informal settlements. Wherever such fantasy plans are being put into practice already, vulnerable, low-income groups are being shoved aside brutally.

Michael Goldman has called these processes "speculative urbanism".[11] He cites the example of Bangalore, where the main business of government has become that of land speculation and the dispossession of populations who happen to get in the way. Goldman suggests that transforming rural economies into urban real estate is "… the principal tension running through the urban periphery of much of Asia today".

Watson observes: "Newly created parastatal organisations designed to fast-track particular large projects are externally funded and have little or no local oversight, and hence local government has been carved up into the older bureaucracies left in charge of small maintenance budgets and the new autonomous agencies fed by international loans but also large obligations of risky debt finance."

This is beginning to happen in sub-Saharan Africa. International property companies are vying with each other to carve up the territory. They see the chance for big profits in a new version of colonial imperialism: capitalist imperialism. It's unrelated to the needs of most of Africa's inhabitants.

Watson found a coterie of property developers, designers, engineering and infrastructure companies, finance and IT firms – and even companies promoting urban sustainability. But this is sustainability without the social equity. It's a complete misuse of the term.

Frequently, you will find – and not just in sub-Saharan Africa – that when the above buzzwords are used, there is no assessment of the life-cycle impact of the proposed developments. You will find the names of big engineering consultancies. You will not find these proposals compared to others with lower impact, on a democratic, inclusive, smaller scale. You will find no references to urban growing, agro-ecology, urban ecology. Nor will you find that these cities are planned around local communities where people don't have to travel far to get to work.

Watson concludes by observing that "these visions and 'master plans' may or may not materialise or may be implemented in part", but what happens will partly "depend on the extent to which the various urban groupings disadvantaged by these processes are able to collaborate and resist. There is no doubt that the scale and extent of change envisioned in these plans might be sufficient to mobilise shack dwellers, unemployed youth, local informal and formal business and the NGO sector at a citywide scale, to effectively counter these interventions."

The challenge of public space

Speculation doesn't just affect the affordability of homes but the provision of genuine public space. By 'genuine' I mean space available to the public, whether privately or publicly owned, which is designed to be attractive and to be used by people. This is a question that affects both property law and design. Charles Montgomery, in his book *The Happy City*,[12] is amongst those who have documented the creeping corporate domination of plazas, streets and shopping centres, and how they can implement designs that discourage people from stopping in them and even make them feel depressed.

Public space includes parks and squares, but also the streets, pavements and street corners; even the shopping centres that replace outdoor shopping streets and public markets. Streets are vital to the quality of life, the most important and immediate type of public space. Street venders lend vibrancy, variety, cheap produce and a semi-formal economy to public life and should not be sacrificed for the sake of some notion of modernity or tidiness as they have been in some cities such as Johannesburg or Bangkok.

'Fluid' or 'agile' planning can specify multiple uses for the same spaces. Most cities designate a single purpose for a piece of land. An agile approach would allow that land to be used in different ways as the need arose, said a World Economic Forum report.[13]

The Australian city of Melbourne has seized on this opportunity, the report details, turning more than 90 ha of pavement into parks and pedestrian plazas in the last 30 years. "Land use is possibly the underlying element that allows most of the other phenomena to unfold." Zoning should be able to respond to changing needs. But the price of land is a big challenge too.

The sustainable and equitable cities of the future must include spaces for all people to congregate for all kinds of reasons, but especially just to be spontaneous.

The main reason why there has been a trend towards the privatisation of public space has been an economic one: cash-strapped public purses become vulnerable to the demands of private equity. This trend has been found across the world, in Europe, North American, Commonwealth and post-communist countries. While public-private partnerships and the contracting out of public services are often essential for any development to occur, it should still be possible for city authorities to place restrictions, as part of the conditions of planning permission, on what private owners can and cannot do, and the amount of genuine public space that is provided.

People need to feel safe in public spaces, but what might be safe for teenagers in one city can be the opposite in another: teenagers enjoy shopping centres in some cities for this reason, whereas in others they are excluded for loitering. People should not feel that they have to be consumers and spend money in order to remain in a public space. All spaces do not need to be designed to appeal to everyone at all times, or they may appeal to nobody. Large, diverse cities need a diversity of public space provision.

We do not want "urban spaces that are … spaces of order and control, of aesthetic homogeneity and uniformity".[14] Instead, public space should deliver a culturally local sense of place, giving a sense of belonging and identity.[15] Local authorities should plan in advance sufficient public space well distributed throughout the city.

In many cities, especially in Asia, governments have failed to provide or maintain public spaces – gardens and parks, sporting facilities and community health centres. In some cases, such as Eastwood City, Manila, the private sector has responded, because they have noticed that by providing attractive places to live and work, they have a better chance to sell and rent accommodation and business space.

The concept of the "right to the city" has been developed, which tackles all of these issues. It incorporates the notion that ownership of urban sites goes hand in hand with an acknowledgement of citizen rights.

The right to the city

UN-Habitat's New Urban Agenda – the framework adopted in 2016 for managing urbanisation across the world – states:

> Pillar 11. Human Rights (HR)
> We share a vision of cities for all, referring to the equal use and enjoyment of cities and human settlements, seeking to promote inclusivity and ensure that all inhabitants, of present and future generations, without discrimination of any kind, are able to inhabit and produce just, safe, healthy, accessible, affordable, resilient and sustainable cities and human settlements to foster prosperity and quality of life for all. We note the efforts of some national and local governments to enshrine this vision, referred to as 'right to the city', in their legislation, political declarations and charters.[16]

In the negotiations leading up to the Habitat III conference in Ecuador in October 2016, which produced the above document, the 'right to the city' was controversial, with civic activists arguing that it should be the broad principle for shaping urban growth because it would bring about cities that are just, equitable and accessible to everyone. But many governments resisted this, fearing unrealistic and unaffordable claims by citizens and its ambiguous meaning.

So in the final text quoted above the vague slogan 'cities for all' was adopted and 'the right to the city' is mentioned only once. Activists continue to state that genuinely sustainable cities of the future must incorporate the right to the city.

What does this entail? It means the "right of all inhabitants, present and future, permanent and temporary to use, occupy and produce just, inclusive and sustainable cities, defined as a common good essential to a full and decent life".[17] This implies that they should be democratically managed; provide adequate housing; and balance urban and rural needs. Access is seen as crucial. Issues such as land tenure and urban land use are recognised as potential battlegrounds between competing interests, namely the rich and the poor.

Various global bodies are championing this cause. The United Cities and Local Governments Committee on Social Inclusion, Participatory Democracy and Human Rights has drafted the Global Charter-Agenda for Human Rights in the City, which includes an action plan for making cities more democratic and solidarity-based through dialogue with citizens. A World Charter for the Right to the City emphasises the rights of all people to live with dignity in urban areas.

The inclusion of the Right to the City concept in the New Urban Agenda was also supported at Habitat III by some local and regional governments as well as some Latin American and European countries. Countries as far apart as Belgium, Kenya and New Zealand provide examples that include community land trusts created to democratically manage and administer urban zones. Colombia has legislation to redistribute 'socially created land value'. Brazil's 2001 City Statute, regulating urban policy, is supposed to ensure that the right to the city has legal value.

However, in practice this is not the case. As happens too often in states with weak independent judiciaries, the 'right' is often conveniently forgotten. According to Theresa Williamson, and the watchdog Rio On Watch, it is one of those laws "for the English to see" ("*para inglês ver*"). The tradition of enacting such laws has continued since the ineffectual Feijó's Law to end the slave trade (which did nothing of the sort) in 1831. The notion of "*para inglês ver*" therefore, "suggested the signing of the law was purely Brazil's measure to placate Great Britain, when really there was no intention of carrying out the promises".[18]

Nevertheless, the concept also has been supported by various other charters, including the:

- European Charter for the Safeguarding of Human Rights in the City (Saint-Denis – France, 2000)
- World Charter on the Right to the City, the result of discussion on the occasion of the World Social Forum in Porto Alegre, Brazil (2001)

- Charter of Rights and Responsibilities of Montreal (Canada, 2006)
- Mexico City Charter for the Right to the City (Mexico, 2010)
- Global Charter-Agenda for Human Rights in the City (Florence, Italy, 2011)
- Gwangju Human Rights Charter (South Korea, 2012).

At least if it is enshrined in a statute, citizens may refer to it in their struggles. The Global Taskforce of Local and Regional Governments supports linking the New Urban Agenda, the 2030 Agenda on Sustainable Development and the Paris Agreement on climate change to reinforce the relationship between SDG 11 and its financing and monitoring. The belief is that this will enable local governments to develop integrated strategies.

Back to housing

Now we can see that when we talk about housing, we're not just talking about buildings, even net zero carbon buildings. Housing is about communities and affordability, and is a fundamental aspect of human rights in cities. Every human being deserves decent, secure shelter. Homelessness in a 'one planet' city should be as rare as possible – preferably illegal.

The problem is that despite the fact that everybody agrees with this statement, discussions about it are often abstract and theoretical. There is no mandate to implement the aim in practice and no accepted principle of what the right to 'adequate' housing actually means – nor has there been except in past and present communist countries. Does 'adequate' include the 'micro-home' dwellings found in alleyways in fast-expanding Ho Chi Minh City that are the size of postage stamps and occupied by families? Definitions are needed.

Despite the fact that South Africa's 1996 constitution included the right to 'adequate' housing as a fundamental principle, this has not produced the hoped-for benefits, partly because the right was expressed in general terms, with no target indicators to specify place, minimum standards or time limits. The national housing programme proved to be too narrow, prescriptive and unrealistic given the budgetary constrictions. A flexible array of appropriate responses would have been more suitable. Instead the policy has ironically produced sterile dormitory settlements that have perpetuated segregation and exclusion.

This is why communities need to be listened to and included in conversations about housing policies. These policies and the resulting legislation need to make reference to indicators and targets and be subject to regular independent reviews. Only in this context can we then talk about the type of buildings that we will live in, the subject of the next chapter.

Neighbourhood design and green space

These topics are partly covered in the chapters on regenerative cities and transport, but are worth reiterating in this chapter. Green space is often public space – parks, areas where there are trees and open space, play areas for children and sports facilities. Research on the proximity of green space has shown that it is a good idea for every citizen to live within 400–500 m of a green space. If not, then for city dwellers to be able to pass areas of green space on their walk to a transport hub or work/shopping area; in other words not to have to make a special journey further than this, otherwise that journey may well not be taken.

Why is this a good idea? Well, for mental and physical health for a start. According to a British Parliamentary study[19] of the evidence, "people who live within 500 metres of accessible green space are 24 per cent more likely to meet 30 minutes of exercise levels of physical activity", although "some researchers suggest that it is 'perceived' access rather than measured proximity that influences activity levels". In addition, "a correlation has been observed between those living closest to greener areas and reduced levels of mortality, obesity and obesity-related illnesses".

In terms of neighbourhood design, the Parliamentarians continue: "insufficient footpaths or the presence of busy and dangerous roads prevent easy access and deter use, particularly for children." Addressing the complaint that green-space provision is expensive, they offer the counter-evidence that "Local businesses and property developers benefit from additional green space through job creation, visitor spending and house prices. For example, it is estimated that living within 600 metres of a park in London adds 1.9 to 2.9 per cent to property value, while a high quality park could add 3–5 per cent".

The model for new neighbourhoods could perhaps be a modern adaptation of the traditional 'village' design (i.e., with a greater density) around a 'village green' or multifunctional common, central, open space, which facilitates social interaction.

How to achieve the desired result? A working group called the Sustainable Urban Neighbourhoods Network, which looked into this in England, advised: "Developers, planning authorities and local communities need to work together."[20] It advocated strong civic leadership to help improve design standards, with design codes written in such a way as to ensure that later phases are not 'dumbed down', i.e., the sustainability aspects left out; and the use of development agreements rather than master plans since they provide a greater degree of flexibility.

Risks and returns should be fairly shared between the public and private partners. The sustainability aspects of a new development will particularly improve the affordability of the housing, because it reduces costs for residents. They also said that to build up a sense of community that leads people themselves to take care of their neighbourhood, development agreements might cover the way a new community is managed, and it needs to be budgeted for.

Ideally, neighbourhoods should contain a mix of land uses and housing. Experience shows that if affordable housing is sprinkled amongst other types of housing, this reduces the risk of ghettoisation; it fosters tolerance and avoids the build-up of resentments or intolerance. Gated communities are the opposite of this inclusive approach, fostering a rich–poor divide. Mixing retail, light industry and entertainment venues amongst housing, together with greening the neighbourhood, makes it more resilient, interesting and lively.

Above all, it's people that make neighbourhoods and communities. How they get on together is vital to their success. The 'one planet' city should be a happy city, as far as possible. One good way to foster a sense of community in advance of a new development is to involve them in designing it. This is exemplified by the Baugruppen (German for 'Build Group') movement, which is so exciting that I've made it a case study in Chapter 15. This chapter also goes further on the subject of private-public development partnerships.

Now, it remains to consider buildings themselves. This is the subject of the next chapter.

Notes

1 Oral statement by Leilani Farha, UN Special Rapporteur on the right to adequate housing, 11 July 2018 at the High Level Political Forum 2018 – Thematic Review of SDGs implementation: SDG 11: Make cities and human settlements inclusive, safe, resilient and sustainable. See: www.ohchr.org/Documents/Issues/Housing/HighLevelPoliticalForum SustainableDevelopment.pdf
2 See: www.local2030.org
3 Murray, Dr C.K., *Unspoken Alternatives To Expensive Housing: How to deliver affordable housing without inflating landlord profits*, The Australia Institute and Prosper Australia, September 2018. See: https://bit.ly/2wQobrN (accessed September 2018).
4 Ibid.
5 Crabtree, L., et al., *The Australian community land trust manual*, 2013. https://researchdir ect.westernsydney.edu.au/islandora/object/uws:26922/datastream/PDF/view. *San Francisco CA: Inclusionary Affordable Housing Program*, Wellesley Institute, 2010. Available at: www.wellesleyinstitute.com/housing/inclusionary-housing-case-studies/
6 Temkin, K., et al., *Shared Equity Homeownership Evaluation: Case Study of Champlain Housing Trust*, The Urban Institute, 2010, p.3. See: http://cltnetwork.org/wp-content/uploads/2014/01/2010-Case-Study-of-CHT.pdf
7 Temkin, K., et al., *Shared Equity Homeownership Evaluation: Case Study of the San Francisco Citywide Inclusionary Affordable Housing Program*, The Urban Institute, 2010, p.6. See: www.urban.org/sites/default/files/publication/29266/412239-Shared-Equity-Homeow nership-Evaluation-Case-Studyof-the-San-Francisco-Citywide-Inclusionary-Affordable-Housing-Program.PDF
8 *LandLines* magazine, July 2018, 30 (3), The Lincoln Institute of Land Policy.
9 By email.
10 Watson, V., "African urban fantasies: Dreams or nightmares?" *Environment and Urbanization*, 1 April 2014, 26 (1), 215–231. See: https://doi.org/10.1177/0956247813513705
11 Goldman, M., "Speculative Urbanism and the Making of the Next World City", *International Journal of Urban and Regional Research*, Blackwell Publishing Ltd, 1 December 2010. https://doi.org/10.1111/j.1468-2427.2010.01001.x
12 Montgomery, C., *The Happy City: Transforming our lives through urban design*, Penguin, 2013.
13 *Agile Cities: Preparing for the Fourth Industrial Revolution*, World Economic Forum, 2018. See: www3.weforum.org/docs/WP_Global_Future_Council_Cities_Urbanization_rep ort_2018.pdf
14 Tran, H. A., "Urban space production in transition: The cases of the new urban areas of Hanoi", *Urban Policy and Research*, 2015, 33 (1), 79–97.
15 UN Habitat, *Tracking Progress Towards Inclusive, Safe, Resilient and Sustainable Cities and Human Settlements SDG 11 Synthesis Report*, 2018.
16 *The New Urban Agenda*, United Nations, A/RES/71/256, 2017, ISBN: 978-92-1-132731-1. See: http://habitat3.org/the-new-urban-agenda/

17 De Paula, N., *The "Right to the City" and the New Urban Agenda*. Policy Brief, IISD Reporting Services, SDG Reporting Hub, July 2016. She is quoting from the Global Platform for the Right to the City, a civil society initiative established at the International Meeting on the Right to the City, in São Paulo, Brazil, in November 2014. See: http://sdg.iisd.org/commentary/policy-briefs/the-right-to-the-city-and-the-new-urban-agenda/
18 Ashcroft, P., "Two Centuries of Conning the 'British': The History of the Expression 'É Para Inglês Ver,' or 'It's for the English to See' and Its Modern Offshoots", *Rio On Watch*, May 2015. See: www.rioonwatch.org/?p=21847 (accessed 9 August 2018).
19 Wentworth, Dr J., "Green Space and Health, British Houses of Parliament Office of Science and Technology", *POSTnote*, October 2016, 538.
20 Falk, N. and Carley, M., *Sustainable Urban Neighbourhoods – Building communities that last*, URBED and Joseph Rowntree Foundation, 2012, ISBN: 978-1-85935-899-3.

11

BUILDINGS IN A 'ONE PLANET' CITY

In the 'one planet' city, buildings will form an intrinsic part of the urban ecosystem, helping to regenerate it. Pleasant to occupy, they will not rely on fossil fuels for heating, cooling or power supply, with many of buildings' present impacts 'designed out'.

The building challenge

Buildings around the world now are responsible for about 30 per cent of all greenhouse gas emissions. This is why there is now a big drive for buildings both new and already existing to become 'net zero' or 'net positive' in terms of the emissions for which they are responsible. In other words, to either be responsible for no greenhouse gas emissions or actually absorb carbon dioxide from the atmosphere and store it, or generate even more renewable energy than they consume.

But this task is getting harder and harder, because the equivalent of a city the size of Paris is being completed every week,[1] such is the frantic rate of construction – of urbanisation – going on around the world. According to Andrew Steer, president and CEO of the World Resources Institute:

> By 2050, the global floor area in buildings is expected to double to more than 415 billion square metres, and buildings' energy demand may increase by as much as 50 per cent, but the buildings sector is a laggard in climate action with average energy consumption per person remaining practically unchanged since 1990. The IEA 2-Degree Scenario[2] requires that building-related CO_2 emissions drop by 85 per cent from current levels by 2060. This means all new buildings must be zero carbon by 2030, and all new and existing buildings must be net zero carbon by 2050.[3]

If you can't picture 415 billion m^2, it's a little more than the area of England and Wales combined, or about 2.5 per cent of all the world's land surface. And this doesn't include roads.

Net zero buildings

So how do we get to net zero? The World Green Building Council (World GBC) has issued a challenge to the global building industry to set a target to eliminate carbon emissions arising from new building use by 2030. It says this can be achieved by employing a combination of efficiency measures and renewable energy.[4] Any company signing up will have to track, verify and report publicly on their buildings' energy and carbon metrics, and encourage their supply chains and partners to make a similar commitment. The council also has a programme to promote and support the target that all buildings be net zero carbon by 2050. This would take a massive amount of retrofitting.

The World GBC is a global network of over 70 Green Building Councils and their 32,000 member companies. It defines a net zero building (NZB) as: "a highly energy-efficient building with all remaining operational energy use from renewable energy, preferably on-site but also off-site production, to achieve net zero carbon emissions annually in operation." Terri Wills, CEO of the World GBC, says he knows the vision is ambitious, "but we know that the building industry has the knowledge, the technologies and the capability to deliver".[5]

Kevin Hydes, the CEO of a global network of building design professionals, Integral Group, says[6] his company believes that there are two key ingredients to designing net zero buildings: positive people and simple engineering. Construction industry workers require training in a whole new set of techniques.

What some cities are doing

Although a new market, NZB is growing fast with a sevenfold increase since 2010 in the USA and Canada, amounting to 45 million square feet (ft^2) of commercial building space, according to the New Buildings Institute, one of the North American bodies which verifies buildings.[7] Looked at another way, the percentage of commercial office space certified by LEED or Energy Star – two relevant standards – has risen to 38 per cent in 2017 from less than 5 per cent in 2005.

California is already implementing a plan that all new residential buildings be zero net energy by 2020 and all new commercial buildings be so by 2030. Additionally, 50 per cent of existing commercial buildings will need to be retrofitted to this standard by 2030. Though it admits that this is a huge shift, it believes that recent advances in energy efficiency best practices, solar photovoltaic and energy storage position the building community for success. The California Public Utilities Commission is using a legislative tool to achieve this market shift: the California Long-Term Energy Efficiency Strategy Plan.

Within California, Santa Monica's updated Green Building Code already requires all new low-rise residential buildings to be designed to use 15 per cent less energy than allowed in the 2016 California Energy Code and achieve an Energy Design Rating of Zero. The city believes itself to be the first city in the world to enact a net zero building code for their municipality. All new high-rise residential and non-residential buildings, hotels and motels will be designed to surpass the 2016 California Energy Code's energy requirements by using 10 per cent less energy.

Andrew Steer says that the building sector has a big advantage – "the technology and design techniques for zero carbon buildings are both available and cost-competitive". These include passive design principles, energy-efficient equipment and storage, carbon-negative materials and a combination of on-site and off-site production of clean energy. But 'permit-by-permit' change is not fast enough, and instead we must "shift market expectations – in prominent, high-growth places – of what acceptable building assets look like".

As examples of good practice, Steer cites Shanghai's Changning District, which recently put in place an energy monitoring platform that now tracks 160 of the district's 165 public buildings. This project is supported by the World Bank.

Mexico City, and Medellin, Colombia, are among cities which are using any mechanisms they can as a part of their business model to achieve this goal, including energy tariff levies, municipal bonds, concessional finance, and some public budgets such as facilities maintenance funds. In Colombia, growth is spurred by financial markets, with Bancolombia having issued its first green bond (see Chapter 14 on finance), which raised 350 billion Colombian pesos to expand financial services for private-sector investments that address climate change.

Between 2008 and 2016, 1.6 million m^2 of building space has been certified green in Colombia, and in 2017, it boasted 384 green building activities. Medellin hosts a Global Innovation Centre, one of the first NZBs in South America, whose design prototype achieved Net Zero Carbon at full operational energy densities (including laboratories) and Net Zero Energy as office space only, both without the purchase of green energy or carbon offsets. (These tactics – purchasing green energy to compensate for the emissions made by the building – are seen by some as 'cheating' – that is, achieving zero carbon without having designed a zero carbon building.)

Colombia's President Juan Manuel Santos has praised the efforts of his country's construction sector to tackle climate emissions: "We are the first country in Latin America that is already imposing a series of conditions on environmental sustainability, meaning that constructions have to be more environmentally friendly. The contribution of the construction sector to reduce the factors causing climate change is a very important step."[8]

This shows the value of strong leadership. If a poor country like Colombia can do this, surely the richer ones can?

Passive solar and passive house building design

Technically, how will the above be achieved? Passive solar and passive house architecture are separate but overlapping approaches. The goal of passive solar architecture is to optimise the use of naturally available light and heat by designing the building with respect to the exterior conditions and location in such a way as to capture and keep the sun's energy within the building (if heating is required); or to prevent the heat from gaining access (if cooling is required). A fundamental element is how the building faces the sun.

The goal of passive house design is slightly different: to manage thermal energy gain and loss, regardless of the source (it includes heat from people and technology), to minimise the amount of energy needed to run the building. This entails much insulation, controlling the number of air changes per hour, and eliminating anything that conducts heat or cold through the building's skin.

'Passive' solar is distinct from 'active' solar, which refers to solar thermal or electric technology.

Clearly the two concepts – passive solar and passive house – overlap, but at their extremes, depending on the climate, can give rise to different building forms:

- The ideal form for a passive solar building in a temperate climate would be longer on the north–south-facing side than the east–west-facing sides, with double or triple-glazed windows on the sun-facing side and very small or no windows on the opposite side.
- The ideal form for a passive house design would be a cube. This is because it minimises the surface area to volume ratio, limiting heat loss.

Traditional passive solar architecture is exemplified by the 'Black Forest' passive solar 'earthship' in Colorado Springs, Colorado. Made of rammed earth and reused tyres, its south-facing conservatory traps solar heat and stores it overnight in the thermally massive structure. It functions well in this location and climate, but would be inappropriate in, say, a hot, dry climate or a city, so for cities we have to turn to a different type of design.

Since not every building can face south or be unshaded, passive house design is more widely applicable than traditional passive solar. In principle, a passive house building can be built anywhere. The same cannot be said of passive solar buildings.

In both cases, a comfortable interior climate is assured by using a large amount of insulation. Ventilation is based on convection, controlled air ingress and, where appropriate, heat exchangers or heat dumps, amongst other techniques. Natural light is allowed in to the extent that it balances the needs to avoid glare and use artificial lighting. Daylight should become the primary light source for health, productivity and sustainability. In the health-care sector, for example, the use of daylighting has health benefits.

These principles will save energy and running costs over the lifetime of the building and provide comfort in all seasons and climates. They also represent a safe investment into the future, because there is constant added value through decreased operation costs.

Experience over the last 30 years has helped to determine successful design in different climates throughout the world and to reduce build costs to the same or only slightly higher than conventional buildings. Lifetime operational costs are considerably cheaper – yielding on average savings of 85 per cent.

Design tools and modelling can establish the needs and requirements of all aspects of a building and their inter-relationships. This process may also reveal opportunities for multiple functions to share space and reduce the footprint of the building. These techniques have now been refined to such an extent that modern passive house buildings can be visually indistinguishable from conventional buildings – or can also be beautiful. The point is that passive solar buildings don't need to have a particular visual style.

Passivhaus

Passivhaus is a way of ensuring that the twin aims of passive solar building – to reduce fossil fuel energy use and increase comfort – are met empirically: it is a strict, performance-related standard. The key distinguishing factor is the rigour applied in the design process, and measuring a specific maximum level of energy use by a certification stage. These buildings may need no conventional heating or cooling system, making them very cheap to run.

The first Passivhaus row or terraced houses were built in Darmstadt, Germany in 1990. In 1996 the first Passivhaus Institut was founded to promote and regulate the standard. Since then, tens of thousands of Passivhaus structures have been built, and many existing buildings have been renovated to Passivhaus standard.

Passivhaus institutes have since been founded in many countries, including the USA. The principle continues to spread and be adapted to the differing cultural and climatic conditions. As a result, the techniques to achieve the standard have expanded considerably. Passive houses are now found as far apart in latitude as Qatar and Scandinavia. It is possible to construct them for the same cost as for normal building standards in most countries.

The Passivhaus Standard stipulates a way of analysing all the heat gains and losses within a building. It requires that the space-heating or -cooling requirement must not exceed 15 kWh/m^2 of floor space per year, as well as other technical tests.[9]

Passivhaus in the Middle East and North Africa countries is presently in an early phase. In 2013, to test the concept, two villas in Qatar were constructed side by side, one according to the Passivhaus standard and the other according to conventional construction practices in the country. They were then monitored. The results were not quite perfect but did indicate that the Passivhaus achieved the required comfort criteria, with indoor temperatures maintained below the 25°C limit, and performed much better than the standard villa. The PV panels on the roof generated sufficient electricity to power the remaining cooling load. The conclusion was that the technique works in the region's hot, arid climate.[10]

In Eastern Europe, the residential and commercial building in Kaufbeuren Bavaria was the first to fulfil the criteria for the certificate Passivhaus Premium, which goes beyond Passivhaus! With a heating demand of only 8 kWh/m^2 per year it has massive energy efficiency, but also generates renewable energy using a 250 m^2 PV system on the roof, making it carbon positive.

In the USA, the PHIUS+2015 Passive Building Standard released in 2015 is spurring new growth in passive buildings and is based upon climate-specific comfort and performance criteria.[11] Passivhaus is being developed in California, including in Los Angeles.

New York City Council passed a law in 2016 prescribing that most new city buildings and major retrofits will need to achieve LEED Gold Standard and cut energy use in half, and mandated that from 2019 energy performance design targets are to rise to Passive House-like levels by 2022. The Manhattan Borough Board passed a resolution in 2016 supporting the Passive House Classic and Passive House Plus (net zero) and Passive House Premium (net positive) standards for potential application to new construction and renovation. It is doing what administrations should do everywhere: leading by example with city-owned new buildings and substantial renovations as proofs of concept.

Cornell Tech in New York City is the world's tallest passive house building in 2017. At 270 ft, with 350 residential units, it forms a template for Green Design in NYC. The tower is designed to use 60–70 per cent less energy than a similar construction with a conventional design. The architects, Handel Architects LLP, and developers, The Hudson Companies & The Related Companies, regard it as "a new paradigm for affordable, high performance buildings".

Another good example is a Passivhaus apartment block built by construction company Rhomberg in the residential park Sandgrubenweg in Bregenz, Austria. Interestingly, two almost identical blocks were built: one to Passivhaus standard and the other to a 'low carbon' standard, and were monitored for energy use. The Passivhaus was supposed to have an annual heating requirement of 9.03 kWh/m^2 and the low-energy house 33.23 kWh/m^2. In practice, the occupants in the passive house consumed 39.9 kWh/m^2 and those in the low-energy house 36.3 kWh/m^2. The reason was that the occupants in the Passivhaus had the heating over 2°C higher than the designers had planned for (22.1°C instead of 20°C). This shows how occupancy behaviour can frustrate the best intentions of designers and why post-occupancy evaluation is a recommended practice for architects and developers.

Another example is 425 Grand Concourse Development, Mott Haven, Bronx, New York, a Passivhaus-certified 26-storey residential tower containing 277 units of affordable housing for low- and moderate-income families. It includes a charter school, medical and community facilities and, visually, looks almost identical to any other tower block.

'Passive solar' refers to the 'in use' phase of a building. But it would be self-defeating to make this phase zero carbon and ignore the other three phases: material sourcing, construction, and eventual dismantling. Buildings should be at least zero carbon when totalling the impacts of all four phases. Any unavoidable emissions may be balanced by actions which remove carbon dioxide from the atmosphere and store it in the building fabric, or which generate renewable energy.

Negative carbon buildings

A building which generates more renewable energy than it consumes over its entire lifetime is termed 'negative carbon'.[12] In summary, this entails:

- Minimising the use of fossil fuel energy during the supply chain and process of construction;
- Encouraging the use of materials which store atmospheric carbon in the fabric of the building, such as timber products;
- Constructing and managing the building in such a way that it minimises the emission of greenhouse gases during its lifetime and eventual demolition;
- Encouraging the capture, generation and even export of renewable energy.

The Solcer House, a prototype designed by Cardiff University's Welsh School of Architecture, is a negative carbon building. Its foundations contain locally produced low-carbon cement, and its walls are of a structural insulated panel system made from expanded polystyrene or extruded polystyrene foam. A 17 m^2 transpired solar collector on the south-facing wall sits below a 40 m^2 integrated photovoltaic roof which is claimed to produce a surplus of electricity. Windows are high performance, there is a ventilation heat-recovery system and a heat pump operating on the heated air from the transpired solar collector during cold weather. Surplus electricity is stored in lithium-ion batteries, and surplus heat in a water store.

Eco-minimalism

Personally, I am an eco-minimalist when it comes to building design. This means I tend to prefer solutions that avoid the use of technology, because not only can technology go wrong or not be used correctly, it also comes with its own life-cycle ecological footprint. As a general rule, if you can achieve the same goal without it, why use it? This does mean the application of a little ingenuity. It is too easy for architects or developers to rely upon technology to achieve zero carbon or negative carbon results if all that is being measured is the in-use stage.

I accept that large, well-integrated buildings have to incorporate a network of controls, connected independently or centrally, operated automatically or manually, that measure and respond to both external and internal stimuli, perhaps operating on a local area Wi-Fi network. Basic functions often include monitoring, controlling and optimising energy usage. Advanced systems can measure rainfall, wind speed and occupancy density. The aim is to manage the building environment so that the supply of energy perfectly balances with the use made of it. In such cases the use of technology with a management contract can be justified.

The same eco-minimalist rule applies to the materials from which a building is constructed. It is a great idea to use materials which lock up atmospheric carbon dioxide, where possible. Natural cellulose-based materials do this: usually timber products, because trees and other plants, while growing, absorb carbon dioxide. When harvested and used to make furniture or buildings, that carbon is kept out of the atmosphere, reducing the greenhouse effect.[13]

Steel is a possible exception to this rule. Despite its high embodied carbon from its manufacturing process, it is long-lasting and can be recycled. In Chapter 8, on industry, I take a look at alternatives to traditional concrete that also absorb atmospheric carbon.

Timber skyscrapers

Many large and high-rise structures are now being constructed using frames made of timber products. These 'lock up' atmospheric carbon, have varying degrees of insulation ability, and are easy to work with. They are also biodegradable or easily recycled at the end of the building's life, and may support local agro-forestry.

There are now many examples of tall buildings made out of timber products such as glulam and cross-laminated timber – layers of laminating stock timber bonded with adhesives to form large, strong, structural members.

The world's tallest timber building, at 85.4 m, is in Brumunddal, in the Ringsaker municipality of Hedmark, Norway. At 18 storeys high, it includes apartments, a hotel, offices, restaurants and common areas, and a swimming hall. The building was designed by Voll Arkitekter. Its architect, Bård S. Solem, says: "The Mjøstårnet is a pilot project that can pave the way for other sustainable projects that explore new boundary-breaking solutions around the use of materials. The environmental effects of using timber in tall buildings are substantial; it can reduce emissions in the production of materials for the load-bearing structures by 35 to 85 percent."[14] He adds: "In the longer term, we need changes in our society. We must think completely differently. The Mjøstårnet shows that you can build big, tall buildings in timber, with sustainable materials that release low greenhouse gas emissions during production."

The timber product supplier, Moelven, claims to have noticed a significant shift towards the use of timber in large constructions. Its CEO, Rune Abrahamsen, says: "In recent years, we've seen an increasing demand for renewable building materials, with their potential for storing carbon. The international climate agreements have led to a greater focus on choice of materials. We don't think the Mjøstårnet will be the record holder for very long. But we, and especially the climate, can live with that."

Design and location

Different layouts for building developments are appropriate for different climates. In hot, windy areas, organic, non-grid layouts provide shade and can be designed to prevent wind funnelling. Grid layouts with orientations borrowed from other climates, and wide spacing of buildings, do not provide shade or wind shelter. Orientation is important only for solar thermal and PV.

The form of a building also has an effect on its energy use. A low surface area-to-volume ratio is better for a building that wishes to conserve energy for heating. This is the ratio between the external surface area and the internal volume. It is a measure of compactness.

Size is also a factor: a smaller building will have a smaller ecological footprint. And a small building with the same form as a larger one will have a higher surface area-to-volume ratio. Buildings with the same level of insulation, air-change rates and orientations but differing ratios and/or sizes may have significantly different heating/cooling demands. Therefore:

- Small, detached buildings should approach square or circular floor shapes;
- Larger buildings may have more complex shapes but require thicker insulation to achieve the same level of insulation.

High-ceiling buildings have a higher heating need but lower cooling needs. Lower ceilings reduce the resources required and the cost. As buildings increase in size, besides permitting greater design flexibility, this lets designers use daylighting and natural cooling strategies to reduce energy demand.

Traditional cooling methods reinvented

Per Heiselberg, of Aalborg University, Denmark, has spent much of his professional life looking at the opportunities for net zero carbon buildings and natural cooling, which might seem odd for somebody from a climate most people think of as cold already. He is interested in particular in night-time ventilation.

Many of his principles are ancient and traditional in hot and dry parts of the world. There, buildings often have thick walls and small windows, with shutters on the windows in the daytime. Some also use wind towers, solar chimneys, underground pipes and evaporative cooling. "My project is about developing ICT [information communications technology] solutions for occupants in houses and offices to illustrate the usability of indoor environments," Heiselberg told me. "We are transferring these approaches from warmer areas of the earth to colder areas."

It involves the use, at night-time when the air is cooler outside, of openings at the ground level facing prevailing breezes, to direct them into the building, relying on the stack effect (warm air rising) to drive the day's heated air inside, up and into the sky through openings in the roof.

"We stress the importance of different strategies for different climates," he says. The approach is also applicable in retrofitting existing buildings.

> In lots of cases we just use the windows and doors that are already present, but with the addition of automatic control strategies and activators, which is more clever than having to open and close windows manually. It is relatively cheap

to install. It might be necessary to create new roof openings and use solar chimneys, and install automatically opening louvres in the downstairs windows where, for security reasons, you would not want to open the windows at night.

Our experience is that in many countries where there is a movement to eco-retrofit existing buildings they always try to improve energy efficiency for the cold season, but this creates a larger demand for cooling in the summer. So we are advocating that when you apply measures to a building during a retrofit to reduce the winter heating needs, you should also apply measures to reduce the summer cooling needs.

There's one type of climate where it is not appropriate, and that is a hot, humid environment, because the temperature and humidity do not change at night-time. Here, new strategies are still needed based on traditional architectures.

Buildings which mimic nature

Mike Rainbow is another engineer interested in a solution to the rising demand for cooling. He worked with architect Mick Pearce on one of the world's most famous passively cooled commercial buildings, the Eastgate Centre in Harare, Zimbabwe, constructed in 1997, which has since become a beacon for elegantly solving this problem with minimal energy use – just with some low-power fans.

Its design principles are based on biomimicry, drawing on lessons learned from a study of how termite mounds stay cool inside. A typical termite mound is 30 ft high, and is full of holes which exhale and inhale air like a giant lung as temperatures rise and fall throughout the day. Whereas the surface temperature may be as high as 40°C, below the ground it will be up to 11°C lower, so the mound draws air up from below ground to cool it.

The seven-storey Eastgate Centre is constructed from a double layer of concrete slabs and bricks which have a high thermal mass and thermal diffusivity – like the materials which compose the termite mound – meaning that they can absorb much heat without changing temperature. This, together with design details, helps with amplitude dampening – a measure of resistance to temperature penetration – and phase shifting – delaying the time taken for an extreme external temperature to reach the interior, and then intercepting it on the way through. The exterior of the building is 'prickly' like a cactus, which increases the surface area. This encourages more heat loss during the night. Small windows mean that little heat is allowed in during the day. Extended overhangs provide additional shade.

Inside, at the base of the building, low-powered fans pull in cool air at night from the outside through cavities in the floor slabs, and let it rise up through the building. In the daytime, warm air rises up the building and out through the chimneys, which are tall to increase the pressure differences between the top and bottom and encourage the stack effect. The result is that the average interior temperature is 28°C during the day and 14°C during the night, leaving it to use around 35 per cent less energy than other similar buildings in Zimbabwe.

The holistic, low-tech solution

A low-tech solution for both heating and cooling buildings – which the designer hopes will become a standard technology in regions of the world where both winter heating and summer cooling are required – has been developed by Dr Peter Holzer in Vienna, Austria, where changes in the climate over the last ten years have led him to consider the need for this approach. "We have changed here from a climate of moist with cold winters to moist temperate with mild winters, so cooling in our area is a new thing. Before climate change and urbanisation we did not need it," he told me.

"Building with thermal mass and using night cooling in the summer has worked in central Europe for hundreds of years. But these changes have led me to think about how to solve the problem of the necessity of summer cooling without having to use a new bundle of energy-consuming equipment." The technique his team has developed uses reversible heat pumps, boreholes and water as a carrier, delivering heating or cooling to or from absorption or radiative surfaces – floors or ceilings in the buildings.

Heat pumps are simple – working on the same principle as a refrigerator. They are very efficient since for every unit of electricity you put in you can get three units of heat out. The boreholes – polypropylene pipes set in concrete in holes up to 150 m deep, one servicing each apartment block – can probably last up to 60 or 70 years.

Their research into their performance discovered something neat: "Putting heat back into the ground in the summer makes the winter heating more efficient. Where better to take this heat from than the building – which would otherwise overheat in the summer?" This solves two problems with one switch and a heat exchanger. It's unsurprising that Holzer is also a big fan of low-tech, eco-minimalist approaches. "I'm a mechanical engineer and at university I was fascinated by thermodynamics. My teachers wanted me to study power plants and atomic power while I was always interested in smaller technologies: solar, biomass, renewable energy. Later I moved on into building physics and building simulation."

"My current fascination has to do with houses for 'good enough' comfort using as little technology as possible. My ideal is the unplugged house! If we pulled the plug would it be able to be run by only the inhabitants with quality of life?" Holzer is a fan of the concept of 'adaptive comfort'. "I think it is a big mistake to set excessive aims, instead of sufficient ones. There's no sustainability without sufficiency. For example, in the case of mechanical ventilation with heat recovery, in a usual residential circumstance in my climate I would do without it, relying on insulation. These systems need extra investment, extra energy and professional servicing to change filters and so on."

Holzer says that his dual use of surface heating/cooling systems are "booming" in Austria. "New projects are under construction. Due to the summer recharging of the boreholes, the heat pumps are operating at very high efficiency, reaching performance coefficients in the range of 4 and above." Note: 3 is the average; 4 is exceptional.

His hope is that because of its ease of application and cost effectiveness this system can become a standard technology to apply in future almost anywhere in the world except where there is high humidity, where it is unsuitable.

Solar cooling

In these hot, humid climates, there is another solution – although it will work anywhere there is sufficient sunshine: solar thermal collectors are a particularly good match for powering space cooling and air conditioning because the sun's heat is always available when it is most needed.

Solar heat collectors similar to those used to provide heating can also be used to power heat engines that provide cooling. This becomes even more cost-effective when the solar collector is used for space or water heating when not needed for cooling. Solar thermal is a well-developed technology but the commercial deployment of solar-driven cooling lags behind solar photovoltaics simply because not many people know about it. This elegant technology really needs government support similar to that which solar photovoltaic panels have received, to bring the price down through mass production. I describe this approach in another book, *Solar Energy Pocket Reference*. [15]

A simplified evaluation tool called Easy Solar Cooling is available to help assess the cost performance of different technologies and system designs under different operating conditions.[16]

Living buildings

Margaret Robertson, author of *Sustainability Principles and Practice*, [17] and a member of the American Society of Landscape Architects, believes that the flagship LEED standard for energy efficiency in buildings is being superseded by a new standard, the Living Building Challenge. This is a rating system that she says is regenerative; a green building certification programme and sustainable design framework. Living buildings produce more energy than they use, and collect and treat all water on-site.

"LEED still means using resources, which means poisoning ourselves – if only a little bit more slowly," she told me in conversation.

> So what if instead we thought of the building as being like a tree or a flower where it only uses the water and energy that come from the sky and does not create any more waste than can be absorbed locally? All waste from one part of the system becomes food for another part of the system. That's the idea behind the Living Building Challenge. It goes beyond the idea of net zero energy and net zero water. We begin to think about doing some repair. What if our building could be a good citizen in the natural world around it and begin to give back more than it takes?

There are various grades for the standard and even a standard for a living community that is regenerative. In total, seven communities in the USA are taking this challenge and have been certified, and one in Mexico City.

Post-occupancy evaluation

The designers of buildings often make predictions about the amount of energy they will consume when they are occupied. But it's a very good question as to whether in practice these predictions come true, and often they do not. This difference has come to be known as the 'performance gap'. Post-Occupancy Evaluation is a methodology that is designed to test this and find out why. Different countries have different methodologies.

In Britain, it is defined by BRE.[18] For them, metering is the standard way to do it, coupled with normalisation for weather. In New Zealand, a proposed evaluation framework[19] is much more comprehensive. It includes both quantitative measurements and questionnaires to residents. This framework covers not just buildings but neighbourhoods, and captures data on energy use, greenhouse gas emissions, transportation, land use, water, ecology, energy production, equity and health: almost everything a 'one planet' neighbourhood could monitor.

It is almost as comprehensive as Wales' 'one planet' development reporting (see Chapter 16).

The buildings of the future

By now, you should have realised that this chapter is not about what the houses and other buildings of the future will look like – they could look like anything. What concerns the 'one planet' city are the materials they are made of, how they are made and how they perform, and finally what will happen to them when they are eventually decommissioned – to be recycled, their parts refashioned into something else.

Who knows what materials will be invented or mass produced in the future? Houses are already being printed in sections in factories and assembled on-site. They may be built out of forms of carbon not yet imagined, materialised from atmospheric carbon dioxide or methane.

Dwellings are likely to become smaller, to match those found in some Asian cities, to house the greater human population in a more compact urban area, which is inherently more sustainable. This is because in the 'one planet' future, people may eat out, use local laundries, and so have no need for large kitchens, freezers and washing machines. They will require less storage space because entertainment is virtual, and as they reduce their ecological footprints people will not need to own so many products.

I can imagine such compact, modular homes stacked in layers with green terraces for roofs and green walls. I picture apartments combined with offices for groups of home-workers. Some may connect with indoor farms or smaller growing areas such as conservatories, and gardens or aquaculture farms. These urban growers will reprocess carbon dioxide, nutrients and water, turning them into food, reducing the dependency of cities on farmland hundreds of miles away. Green buildings will help to keep the area clean and the temperature down, and provide food for insects and birds.

Some neighbourhoods may have floating homes, or homes on stilts. In the Sun Belt region, homes may even be underground, with light tubes taking them daylight, while on the surface are greenhouses or factories, or even deserts that are too hot to venture into in the daytime.

All things are possible. But the fundamental truth is that for humanity to survive, the places we occupy will have to have no dependency on fossil fuels as they keep their occupants safe and healthy, and support far more biodiversity than they do now.

Notes

1 Abergel, T., Dean, B. and Dulac, J., *The Global Status Report 2017 International Energy Agency (IEA) for the Global Alliance for Buildings and Construction (GABC)*, United Nations Environment Programme, 2017, ISBN 978-92-807-3686-1. See: www.worldgbc.org/sites/default/files/UNEP%20188_GABC_en%20(web).pdf
2 As explained in the International Energy Agency's report *Energy Technology Perspectives 2017 – Catalysing Energy Technology Transformations*. See: www.iea.org/etp2017/
3 Steer, A., *Q&A: What Is the Future of Green Building?* World Resources Institute, 4 June 2018. See: www.wri.org/blog/2018/06/qa-what-future-green-building
4 In Toronto, Canada, 6 June 2018. See: www.worldgbc.org/news-media/world-green-building-council-calls-companies-across-world-make-their-buildings-net-zero.
5 A report on the page of the above link describes what business, governments and non-governmental organisations can do to achieve net zero buildings.
6 In a press release, 6 July 2018: www.integralgroup.com/news/integral-group-is-one-of-three-founding-signatories-to-the-wgbcs-net-zero-carbon-buildings-commitment/
7 This claim is made along with a whole suite of tools on their website: https://newbuildings.org/hubs/zero-net-energy/
8 *Greening Latin America – One Building at a Time*, International Finance Corporation, World Bank, May 2016. See: www.ifc.org/wps/wcm/connect/news_ext_content/ifc_external_corporate_site/news+and+events/news/greeninglam_one+building+at+a+time
9 More information on this and all of the other subjects in this chapter are available in my book: *Passive Solar Architecture Pocket Reference*, Routledge, 2017.
10 Khalfan, M. and Sharples, S., "Thermal Comfort Analysis for the First Passivhaus Project In Qatar", *Proceedings of SBE16 Dubai*, 17–19 January 2016, Dubai-UAE. Available at: http://repository.liv.ac.uk/2047759/
11 Projects are recorded on the website passivehousecal.org
12 On average the carbon impact of the construction of conventional buildings is between 10 per cent and 20 per cent of their use during their lifetime.
13 An average new house made with conventional materials contains the equivalent of 50 tonnes of CO_2 as 'embodied carbon' that could be reduced to 38 tonnes with greater use of timber and modern methods of construction, or even to approximately 25 tonnes by using cellulose-based materials, such as hempcrete. See: Monahan, J. and Powell, J. C., "An embodied carbon and energy analysis of modern methods of construction in housing: A case study using a lifecycle assessment framework", *Energy and Buildings*, 2011, 43, 179–188.
14 In a press release dated 26 September 2018. See: www.moelven.com/news/news-archive2/2018/mjostarnet-a-sustainable-pilot-project/

15 Thorpe, D., *Solar Energy Pocket Reference*, Routledge, 2018, ISBN: 9781138501201.
16 For this and plenty of case studies see: www.solair-project.eu/218.0.html
17 Robertson, M., *Sustainability Principles and Practice*, Routledge, 2017, ISBN: 9781138650244.
18 BRE, *GN32: BREEAM New Construction Guidance Note –*
   Energy Prediction and Post Occupancy Assessment, Guidance Note GN32 V1.0 March 2018. See: https://bregroup.com/brebreeam/wp-content/uploads/sites/3/2018/04/GN32_BREEAM_UKNC_2018_Energy_prediction_and_post_occupancy_assessment_v0.0.pdf
19 Boarin, P., et al., *Post-Occupancy Evaluation of Neighbourhoods: a review of the literature, University of Auckland, Building Better Homes Towns and Cities*, National Science Challenge Working Paper 18-01, 2018. See: www.buildingbetter.nz/publications/SRA4/Boarin_Besen_Haarhoff_2018_POE-Neigbourhoods_WP18-01.pdf

12

MOBILITY IN A 'ONE PLANET' CITY

Cheap and quick connectivity is vital for prosperity, development and social cohesion. It is a prerequisite for integrating people to places, opportunities, markets and jobs – the reasons why people live in cities. However, adequate transport is unavailable for the world's poorest and most vulnerable people.

'One planet' mobility is about eliminating or reducing the air pollution and carbon emissions from travel. This requires green integrated transport systems that are inclusive, safe, accessible and affordable, and ample provision for walking and cycling. These don't just help alleviate pressure on road systems and curb climate emissions, they are also essential for addressing poverty, health and social exclusion … and greening cities.

Like water, transport cuts across many other areas, from health to energy via planning. To tackle this holistically, then, requires improved understanding, collaboration and public engagement from administrations.

Vehicles are presently responsible for terrible air pollution in cities, leading to premature deaths and severe health problems, not to mention uncountable casualties and 1.3 million fatalities per year from traffic accidents: a slow massacre. On a larger scale, aviation is responsible for 3 per cent of global emissions, and shipping for another 3 per cent; it goes almost without saying that both must be decarbonised.

What is a car? "10 tons of resources are turned into one ton of car which transports about 100 kg of humans at an average speed of 15 km/h for an average distance of 1 km per trip."[1] How sustainable is that?

What do cars mean for streets? In many cities a great deal of public space is reserved for parking privately owned cars. For example, in Vienna, nearly 70 per cent of street space is given to motor vehicles, despite the fact that privately owned cars remain inactive, on average, 95 per cent of the time. There is therefore great potential for citizens to reclaim urban space and improve liveability and health.

The central role of planning

Urban transport is a function of planning. Unplanned urban expansion and the historical preference for suburbia have increased reliance upon the automobile. Cars and traffic are not the same as urbanity: they are its opposite. They have caused a growth in transport-related emissions, an epidemic of traffic fatalities and casualties, and the loss of agricultural land, lengthened travel times, and increased the costs of servicing urban residents – all of which are characteristics of unsustainable urban settlements.

Efficient land use is paramount for the sustainable development of cities, yet only half of all cities assessed by the United Nations consult their citizens on proposals for major roads or highways or alterations in zoning. So the challenge in practice for cities in the post-carbon age is to serve the mobility needs of people by redesigning streets, formal parking areas and street networks so that the greatest proportion of urban dwellers can have a good quality of life.

The design of streets decides what happens on them: how people behave and enjoy themselves. Multiple uses need to be catered for: cyclists, pedestrians, goods vehicles, public transport, taxis. Urban ecosystems are dynamic; requirements for what they can provide change according to the time of day, time of week and time of year, and over decades. This is why sustainability in street design is about resilience (to climate extremes), flexibility (in order to adapt to changing needs), and responsiveness (to citizens' needs, to invite public life).

As with waste management, traffic management involves a hierarchy. For waste it is, in decreasing priority: eliminate, minimise, reuse, recycle. For traffic it is similar: first eliminate the need to travel; then reduce the remaining travel distances; facilitate walking, cycling and accessibility; have good, cheap, frequent and reliable public transport, both local and national; and then allow shared vehicles, such as car-shares or taxis; with private vehicles coming at the bottom of the hierarchy, preferably electric or low carbon.

New or quickly expanding cities can do much worse than to emulate the famous example of Curitiba in southern Brazil, which designed its expansion around a two-tier bus system. In this city it is possible to get to anywhere from anywhere very easily because the frequent and cheap buses don't just travel radially to and from the centre, but in concentric circles at varying distances from the centre. There are rapid transit buses which have their own bus lanes and less frequent stops so that travellers may journey at speed over longer distances, and there are more local bus services with their own bus lanes and more frequent stops.

In general, buses are cheaper to introduce than trains, metros or trams because they require less infrastructure and are more flexible. Cheaper still is the provision of cheap or free walking and cycling facilities that are separate from car traffic. This will include cycle hire facilities, cycle parks and cycle lanes. Increasingly, electric bikes are being used, but there should be measures to alleviate safety concerns since they can travel remarkably fast and silently!

In the interests of accessibility it must be remembered that the very young, the infirm and disabled may not be able to cycle or walk, so pavements and access to public transport must be made fully accessible to those with wheelchairs and buggies. As urbanist David Harvey says: "The right to the city is far more than individual liberty to access and other resources: it is a right to change ourselves by changing the city."[2]

The widespread provision of electric vehicle charging services will speed up the transition of private cars and vans away from carbon-based transport, reducing noise and air pollution. All of this should make streets much more pleasant places to be.

Top cities

Singapore topped the overall Urban Mobility Index in 2018[3] with a score of 59.3 per cent out of 100, followed by Stockholm in second with 57.1 per cent and Amsterdam in third at 56.7 per cent. The index scores cities on 27 indicators according to factors such as different ways of getting about, transport infrastructure, initiatives that aim to improve mobility, car-sharing schemes, mobility-as-a-service platforms, autonomous vehicle initiatives, air quality, accident rates, and the length of the commute. At the bottom of the index, alongside Baghdad, came … Atlanta, Georgia, USA, a city defined by sprawl.

The Urban Mobility Index report notes that urban mobility is one of the toughest system-level changes for cities, and is actively hostile to innovation. It identifies that what is needed is "system level collaboration between all stakeholders of the mobility ecosystem to come up with innovative and integrated business models". Many of the poorly performing cities lack a clear vision and strategy on how their mobility systems should look in the future.

Different strategies would be appropriate to different cities according to their level of maturity:

Cities in mature countries with a high proportion of motorised individual transport need to become more public transport and sustainability oriented. Mature cities with a high share of sustainable transport modes need to fully integrate the travel value chain to foster seamless, multimodal mobility. At the other end of the scale, those cities in emerging countries with partly underdeveloped mobility systems must establish one or more sustainable mobility cores with the ability to satisfy short-term demand at reasonable cost and avoid replicating mistakes from the early history of cities in Europe and America.

These cities have the opportunity to take advantage of the best planning developments in the world, even cities in developing countries, such as Curitiba. Imaginative solutions are required to generate revenue, where ticket sales are not enough, such as sponsorship and the aggregation of third-party services that appreciate from the indirect benefits of public transport and cycle routes, such as improvements in air quality and health. Reductions in journey time have a knock-on effect in economic performance and so economic models should be designed to reflect this.

Walking

Minimising the need to use the car involves making walking more attractive. Walking allows people to notice their neighbourhoods and stop to talk to others or pop into a shop. It's also great for health. Many cities are instituting their own programmes to encourage walking, the first step of which is to make it feel safer for citizens to walk to shops, cafés, schools and work. Typically, it needs to be made possible through traffic-calming methods and road signage for pedestrians to have priority at road crossings and intersections.

For example, the WALK Friendly Ontario programme encourages municipalities to create and improve the conditions for walking by awarding Bronze, Silver, Gold or Platinum designations. Annie Matan, a researcher and lecturer at Curtin University Sustainability Policy (CUSP) Institute, Curtin University, Perth, Western Australia, at a conference in Sydney, Australia called Walk21, told delegates: "The health benefits and subsequent economic benefits of creating walkable urban areas are measurable."[4] Part of the problem, she said, is that walking and cycling are often not factored into decision-making, because the people who manage the budgets for health are not usually the same as those who manage the budgets for transport and planning.

There is therefore an urgent need for city leaders – mayors and senior administrators – to be made fully aware of the need to factor in these benefits – and how to do it. As Lucy Saunders told the same conference, "The health benefits of more walking in the city go beyond the substantial physical activity benefits, including reduced harms from noise and road traffic collisions and greater social cohesion". She explained that in London, the organisation Transport for London "have 10 new ways of working to ensure that health benefits are considered across all strategic transport planning in the city from option appraisal through business case development to evaluation", and this is detailed in a Health Action Plan, which contains a framework, 'Healthy Streets', highlighting ten evidence-based indicators that can be used to assess how 'healthy' a street is and identify what needs to change to make it healthier. Pedestrians are at the centre of this with 'Pedestrians from all walks of life' at the top of the list.

Pedestrians need to connect with the mass transit systems at transport hubs, described in more detail below. David Mepham, an urban access consultant/academic from Melbourne, argues that park-and-ride and light rail are more pedestrian-friendly than bus rapid transit (BRT), but this is based on Australian experience.[5] Different countries will be more culturally or geographically suited to different approaches. Whatever modes are chosen, Sonia Lavadinho, a Swiss urbanist, argues that "placing walking at the core of the multimodal [mobility] system" is vital, and is "a paradigm shift" – a way to "both optimize the geometry of the transit network and reinforce the degree of attractiveness of the major multimodal hubs".[6]

This is being tried out by Transdev/Semitag, the public transport operator of the French city of Grenoble, where the hubs are known as 'Carrefours de Mobilité', or 'Crossroads of Mobility'. There, the transit network has been substantially modified to include new tramway and BRT lines, accompanied by a new philosophy that puts walking at the core. While valuing all mobility solutions (tram, bus, train, bicycle and car-sharing) it is also "repositioning the pedestrian at the heart of concerns and providing a comfortable experience and intuitive journey".

Cycling

The same issues face measures to promote cycling, which has similar health benefits but even more potential to displace the need to travel by car and van. It is sometimes easy to justify the provision of cycle lanes when they are cheaper than roads, but not when they are in addition to roads – unless the associated benefits of cycling are factored in. In places where roads are too narrow for both a cycle lane and vehicular traffic, cities are advised to look at Barcelona's superblock solution, described below, or a similar solution which involves alternating streets prioritised for cars and vans, with streets prioritised for cycles and walking.

Cities like Copenhagen and Amsterdam are often talked about as bike-friendly cities; however, they may be exceptional because of the already existing bike-friendly culture and the fact that they are flat. For an example of a different kind of city that is also promoting cycling I turned to Vienna, capital of Austria. A new Viennese government in 2010 set a goal of doubling the number of cyclists by 2015. It now has over 1,379 km of cycle paths, cycle lanes and cycle routes through low-traffic zones. To achieve this the city funded a new Vienna Mobility Agency to support walking and cycling with a publicity campaign. In 2013 it hosted the VeloCity conference at which it launched the Vienna Cycling Manifesto.

I asked Florian Lorenz, a resident biking evangelist, how the campaign motivated people to cycle more. "It had the slogan 'putting enjoy on the move' to make people understand the enjoyment while cycling and all of its positive aspects," he said. The manifesto called for a "redefinition and redistribution of public space" for cycling. What did this mean? "When you think about it, it is happening in all the cities which are promoting active transport," he said. "It is about taking public space and assigning [it] into different transport modes, even turning a parking lot into a summer restaurant. In Vienna it is also relevant to the city's mission statement to make better use of public space especially in new housing developments on the fringes of the city in old industrial land and to encourage cycling. In Vienna we have several municipal departments involved in the planning of urban cycling. The planning, development, and streets departments, and a few others, have all been involved in creating the manifesto from the start."

I wondered if there had been any objection from businesses to the pedestrianisation of streets. Lorenz said: "It is always controversial at first because businesses will panic about losing money, but in streets in central Vienna that were pedestrianised in the 1970s, they are now amongst the most highly sought after shopping streets by top end retailers. Nobody from any interest group will tell you that they want to put cars back in there. It does need a lot of time to get used to it."

In Vienna, an integrated mobility card allows holders to access not just buses, trams and metro, but also cycle-hire and car-share schemes. The Citybike network allows anyone to register at a Citybike terminal with a credit card, choose a bike and pedal away. For a one-off registration fee of €1, users have access to 1,500 bicycles distributed between 121 stations. The first hour is free, after which a usage fee is charged. It has over 200,000 users per year. It does not make a profit, but is subsidised and sponsored. Vienna also has two car-sharing systems. Lorenz told me that some businesses also use cargo bikes to deliver food, as in Holland and Copenhagen, where families also use such bikes to transport their families. Although it's hilly, some residents have electric cargo bikes to give their legs a little extra help.

Bike City is another ambitious Viennese project, which Lorenz says, "is about making a truly bike-friendly city, with bicycle parking slots provided in new apartment blocks and fewer car parking slots than was standard before, plus installing large elevators in the blocks capable of holding three bikes and three people."

This is fine for developed countries. Elsewhere in the world, bike sharing in African cities has lagged due to lack of dedicated lanes and the perception that cycling is for the poor, worsening traffic jams and pollution. Morocco's Marrakech was the first African city to get a bike-share programme in 2016 with the support of the United Nations as it prepared to host a climate conference. "The attitude towards cycling is still bad across Africa. It is seen as a poor man's form of transport and has been pushed aside by planners," says Amanda Ngabirano, an urban planner at Uganda's Makerere University.

Experts are keen to increase cycling across the continent as it could reduce traffic jams and air pollution and enable more people to get around their cities, improving job opportunities. "Simultaneous to building infrastructure is building a bicycle culture, by raising awareness on the benefits that cycling can bring to people's lives," says Stefanie Holzwarth, a mobility expert at UN-Habitat, an urban development agency.

Car-sharing

To reduce the number of vehicles on the roads, car-sharing schemes are springing up everywhere, often with the use of a simple phone app. The Mobility Factory SCE is a European co-operative enterprise born out of a collaboration between three citizens' co-operatives in Europe to successfully share electric cars with their members, who share experiences and ideas. They have developed together a shared IT platform that is part of The Mobility Factory, a European co-operative society that will soon allow co-operatives from across Europe to offer an e-car sharing services to their members.

Alterna is another new co-operative for mobility services such as car-sharing or bike-sharing that was set up in 2017 in Valencia with the aims of accelerating the transition towards a clean and sustainable mobility, reclaiming public space in cities, and creating an e-mobility movement. Alterna is collaborating with other mobility cooperatives in Europe such as Partago and Som Mobilitat in order to built a network to share knowledge and resources together.

Making cities denser and converting the suburbs

At present, the urban population and the simultaneous de-densification of urban areas are both increasing as cities sprawl and suburbs splurge over the countryside. This is completely unsustainable; if this trend were to continue the area of the planet covered by urban settlements would reach over 3 million km^2 by 2050.

Suburbs are a function of cheap oil. Detroit stands as an example of a post-industrial city whose urban core collapsed after it sprawled, and therefore could not manage the economic crash following the nosedive of its car-manufacturing industry. No one wanted to live there and property became worthless. It is usually the middle class that wish to move to the suburbs, often when they start a family, but if the right kind of high-rise dense neighbourhoods are provided, with appropriate, family-friendly facilities and transport links, they will not feel the need to do so.

A report from Smart Growth America and the Metropolitan Research Center, *Measuring Sprawl*, found that inhabitants of cities with less sprawl have higher quality of life. In general, people have greater economic opportunity in compact and connected metropolitan areas. They spend less of their household income on the combined cost of housing and transportation, have a better sense of community, have a greater number of transportation options open to them, and those who live in compliant, connected metropolitan areas tend to be safer, healthier and have longer lives than those in suburbia.[7]

The *Measuring Sprawl* report notes that: "As residents and their elected leaders recognize the health, safety and economic benefits of better development strategies, many decision makers are re-examining their traditional zoning, economic development incentives, transportation decisions and other policies that have helped to create sprawling development patterns."

Moreover, suburbs are responsible for double the climate emissions of urban centres. Suburbs alone account for half of total US household carbon footprints. Lower Manhattan has an average household carbon footprint of 32.5 tCO_2e (tonnes of carbon dioxide equivalent) compared to Montclair Essex County, New Jersey, where it rises to 68.3 tCO_2e, more than double.

Christopher Jones and Daniel M. Kammen of the Goldman School of Public Policy in California, used national household surveys and developed econometric models of demand for energy, transportation, food, goods and services to derive their figures. These were assigned to US zip codes, cities, counties and metropolitan areas. They write in the report that "We find consistently lower household carbon footprints in urban core cities (40 tCO_2e) and higher carbon footprints in outlying suburbs (50 tCO_2e)."[8]

Ending the suburbs requires a new model of design – especially for larger cities. This more sustainable model involves abandoning the idea of a single core or city centre. Instead, a series of centres, like pearls on a necklace or in a matrix, are connected by faster urban public transport systems.

Mobility nodes

These centres are mobility nodes where travellers can connect from rapid transit buses, trams or train stations to local modes of transport: buses, taxis, cycles, walking. At and around these nodes are places of employment, housing, shopping, entertainment and other services. The nodes might have greater population or building density than the spaces between them.

Johannesburg is an example: much of the population was living in informal townships, and the authorities took the decision to invest in mass transit services to link them together, rather than trying to re-engineer the whole city along traditional lines with one centre. Job and residential densities are being intensified in these development hotspots, thereby increasing the proportion of the population able to access the subsidised transport system.

Therefore it is much better to upgrade informal or older, low-rise settlements, increasing the level of density. It also makes it cheaper for the municipality or utilities to provide services, over a smaller area of land. Large roads transporting citizens from suburbs to the centre divide communities and make dysfunctional cities, such as Los Angeles.

When Seoul's mayor opted to demolish the eight-lane highway that ploughed through the city centre, he announced: "Seoul is for people, not cars."

Low-emission zones

Air quality is a driver for policies to tackle mobility. To this end, many cities are now introducing low-emission zones (LEZs). These are schemes that cover specific areas to reduce air pollution and discourage certain types of vehicles from entering. Vehicles may be charged for doing so especially if they do not reach a minimum standard for emissions to enter the zone. At the time of writing over 220 are in place throughout Europe. Few examples exist elsewhere, however, although Singapore was the first city in the world to implement an electronic road toll collection system for congestion pricing, and Dubai, Ontario and Hong Kong have similar, but limited systems. New York City has considered one but not implemented it yet.

London introduced its LEZ in 2003, with a congestion charge. It has been shown to have radically improved bus services, made journey times more consistent for drivers, and increased the efficiency for distributing goods and services throughout the city. Since 2006, the scheme's administrators, Transport for London, have reported a drop in traffic of 15 per cent and congestion by 30 per cent. Traffic volumes in the charging zone are now nearly a quarter below a decade previously, allowing central London road space to be given over to cyclists and pedestrians. Air quality has improved in the centre but not in the area surrounding the zone, however, which has repeatedly breached legal levels.

How it works: if a vehicle enters the 21 km^2 zone between 7:00am and 6:00pm on a weekday, the owner will pay a flat daily rate which has risen from £5 in 2003 to £11.50 in 2018. Residents receive a 90 per cent discount and registered disabled people can travel for free. Emergency services, motorcycles, taxis, minicabs and zero-emission vehicles are exempt.

It succeeded because it was part of larger efforts to improve travel across all modes of transport in the city and had the clear and convincing reason that it would reduce traffic in the centre and generate funds to reinvest in improving public transport services. On the day it was introduced, 300 extra buses were added to the central London network. One year later, the mayor, Ken Livingstone, reported that 29,000 more passengers were entering the charging zone by bus during the morning rush hour compared to the previous year, and over the following decade the number of private cars entering the zone dropped by 39 per cent.

Recently, however, with the advent of Uber, many more private-hire vehicles have been entering the zone, increasing congestion and lengthening bus journey times. These types of services have become a serious problem for many cities, especially in cases where it is cheaper and quicker to use them than to use public transport.

London has looked to other cities for solutions to this problem, such as Stockholm where the 35 km^2 zone encompasses two-thirds of the city's residents, and the charge varies according to the time of day, up to a maximum limit. London consequently decided to extend the congestion charging zone and replace the daily flat rate with a charging structure sensitive to when and where vehicles enter it and the length of time they spend within it. This new extended ultra-low emission zone covers an area 18 times larger than the Central London Ultra Low Emission Zone; from 25 October 2021 it will include the inner London area bounded by the North and South Circular Roads. Affected are motorbikes that do not meet Euro 3 tailpipe emission standards, petrol cars and vans that do not meet Euro 4 standards (roughly the equivalent to not being more than 15 years old for cars in 2021), and diesel cars and vans that do not meet Euro 6 standards (roughly the equivalent to not being more than six years old for cars in 2021).

A further idea being considered is devolving the national vehicle excise duty (an annual charge that increases with the pollution level of the vehicle) from the national level to the Mayor of London's office to give city leaders another means to encourage sustainable travel.

In Spain, since the city of Barcelona introduced its temporary environmental zone in 2017, other zones are banning the Euro vehicle emission classes 0–3 from traffic whenever there is a peak in air pollution. Madrid has introduced three environmental zones. Since the beginning of November 2018, only vehicles displaying a Spanish environmental badge 'Distintivo Ambiental' are able to enter the Madrid Central zero emissions zone. Within the larger M30 zone, vehicles without a badge will already be excluded on the second day of higher levels of air pollution. Parking bans have been put in place for certain vehicles, and taxis may only drive if they display another badge. If an air pollution peak lasts three days, the driving ban can be extended to a larger environmental zone. A free Green-Zones app is provided to help drivers. Foreigners visiting cannot yet buy an environmental badge for the Spanish environmental zones so they have to leave their vehicles outside the permanent environmental zone of Madrid Central.

LEZs in developing countries

Cities in developing countries find it difficult to implement LEZs because they require much investment and are considered to be unfair to the poor. Jakarta exemplifies this. There, a BRT system (see below) was first introduced; now the city is constructing mass rapid transit and light rail transit systems to increase public transport. In August 2016, a traffic control policy was implemented that allowed only even-numbered licence plate vehicles to enter certain corridors during even days, and vice versa. In 2018, the first airport rail link opened and an Electronic Road Pricing system like Singapore's is scheduled for 2019.

A system has been considered for Delhi, home to some of the worst congestion and air pollution in the world, but is problematic because the city is so large. Nalin Sinha, programme director of the Initiative for Transportation and Development Programmes, says that the problem is fairness. "With a system like this, the rich Mercedes owner will be able to pay the toll, but the poor man in the Maruti will be denied access. It is far better to employ total restrictions like car-free zones or high automobile taxes."[9] Nothing has yet been decided, and meanwhile millions suffer from very poor air and congestion.

Beijing implemented China's first low-emission zone in 2017 with the aim of reducing its notorious air pollution and improving residents' health. It targets heavy-duty vehicles and was expected to prevent the emission of 11 metric tons of particulate matter and nitrogen oxides each day in its first two years. Shanghai, Hangzhou and Suzhou are considering similar moves.

A survey by the World Resources Institute recommends the following for any municipal government considering introducing such zones:

- Setting clear and strong objectives.
- Conducting comprehensive studies on implementation, such as charging fees and targets.
- Allocating revenues to transportation system improvements in a transparent manner.
- Using proven and appropriate technologies.
- Effective communication strategies that respond to public feedback.
- Providing viable alternative travel options and mitigating potentially unwelcome impacts before implementation.[10]

Superblocks: a modular solution

A solution to those cities finding it difficult to ban traffic from certain areas is offered by the concept of the superblock. This is being trialled successfully by Barcelona. The strategy is to permit traffic on certain roads, and on other streets leave the space open for green space, cycling and walking.

The 'superblock' is Barcelona's exciting system of limiting vehicle access, designed partly to tackle the 3,500 premature deaths and thousands of cases of chronic adult bronchitis, cardiovascular events and asthma attacks caused by air pollution in the city. The concept derives from Ildefons Cerdà, an engineer influenced by Utopian Socialism, and is based on the idea of the green city, full of fresh air and light.

A superblock is formed by nine square existing city blocks, of around 400 by 400 m each, inside of which, in the first phase of implementation, traffic is reduced to one lane, and public transport traffic diverted to the perimeter roads. Inside them the speed limit is 10 km/h and it is not possible to drive through. This frees up most of the area for cyclists and pedestrians. On the roads on the edges of all the superblocks the speed limit is 50 km/h.

In the second phase citizens have the right to make more use of the freed-up space for market trading, parks (green space), games, sports and parties, as well as cycling and walking. Tree plantings, rainwater management, green roofs and façades are installed, increasing the proportion of green and open spaces by up to 35 per cent. The new green spaces will connect a citywide network – properly regenerative and helping to improve biodiversity and the urban climate.

Under the Barcelona Sustainable Urban Mobility Plan, the city will also introduce measures to reduce the number of vehicles through extending its orthogonal bus network, expanding the network of bike lanes, reducing the number of free parking spaces and increasing the price of short-term parking.

Eventually the project should cover an area of 2.7 square miles and contain 503 superblocks, almost half of the city's current traffic area. Public transport, especially bus lanes and the bicycle network, will be expanded massively, with the aim to reduce vehicle traffic by 21 per cent.

It is all part of a wider programme to build a stronger, resilient city, and is a remarkable attempt to put human beings at the heart of the urban ecosystem. In the words of landscape architect Sigrid Ehrman, it "stresses the importance of citizens' relationships with each other and the city itself". The model is scalable and so it can be applied to new developments and the urban regeneration of compact inner-city suburbs and low-density settlements anywhere.

There has been some opposition, but this has melted away over time. One fear was that the remaining traffic-permitted roads would become more congested, but in fact traffic has decreased as people walk and cycle more.

Ehrman says that the project is based on an Ecosystemic Urbanism model, and that because of its neighbourhood scale it has the potential to foster a bottom-up approach with local citizen engagement.

She adds: "Cities are key players in addressing the challenges of climate change and a growing population. However, this is neither reflected in governance and regulatory frameworks nor institutional and economic structures, where cities play a subordinate role to countries or regions. Giving cities the control – and the budget – is required for them to fully implement strategies like the Ecosystemic Urbanism model, and to drive the urban transformation."[11]

Freight

Cities additionally have to make it possible for goods and services to be delivered, and this typically means plenty of lorries. What are the alternatives?

Wherever railways and waterways exist, they could be used more efficiently, providing capacity to move more freight. For example, the EU has a target of shifting 30 per cent of long-distance road freight to rail and waterways by 2030 and 50 per cent by 2050 to take more traffic from the roads. It recognises that this requires a long-term commitment to improving service quality, lowering costs (to create a level playing field between modes) and increasing capacity. More non-motorised forms of distribution may be encouraged, and the remaining ones eventually driven by hydrogen fuel cells (as some already are), as it's the most appropriate zero-emission technology for long-distance heavy vehicles.

To keep freight out of town, districts may have consolidation centres, with freight transferred to cleaner modes on the first/last part of their journey. Encouraging non-motorised vehicles for the first and last miles of delivery can make a great deal of difference to road design and use, air quality, and place-making, removing polluting, dangerous and noisy vans and lorries.

In order that goods may be transported with minimal delays, separate lanes, back streets and parking provision for goods and service vehicles may be provided. In the absence of this, access for such vehicles could be timetabled to avoid commuter rush hours, for example, reducing the risk of traffic jams. Removing subsidies for road freight and aviation and making the alternatives more viable, would help shift to rail and water and allow for them to work together more.

Ending traffic fatalities

This is an epidemic that makes no headlines: at least 1.3 million people around the world die every year on the roads (that's twice as many as die of malaria). People in the future will look back on this as barbaric.

One man determined to end this mass slaughter is Claes Tingvall. In 1997, sitting at his desk in an office in Stockholm while looking at traffic fatality statistics, Tingvall suddenly thought: "What if nobody ever had to die in an auto accident?" A crazy idea, right? But sometimes it takes one person to question what everybody else accepts. Claes went on to found Vision Zero and has been able to put his ideas into practice as the director of traffic road safety at the Swedish National Road Administration in Stockholm.

In April 2014 New York City became the first US city to adopt the Vision Zero policy, which has also been adopted by Los Angeles, Paris, Berlin, Stockholm and state of Victoria in Australia. New York targeted 50 roads and junctions with the highest fatality levels, introducing slow zones, speed cameras, street lighting, traffic signals and community outreach. By 2017 road deaths were down 28 per cent and pedestrian deaths 45 per cent.

But Vision Zero is not successful everywhere, because of campaigns by drivers against restrictions. So in the course of a conversation with Claes, I asked him why he thought this 'slow massacre' doesn't attract more outrage. He put it down to "a culture of blaming the victim because they did something wrong and of not taking responsibility". His response? To use economic arguments to show that it is worth investing more in traffic safety and to appeal to ethics. "In other man-machine systems we arrange it that even the most vulnerable are protected, why not traffic systems?"

In the beginning people laughed at him, "or indeed were very upset and frightened," he said. "Many people said it was impossible, but they were only looking at the figure zero. You probably can't get to zero deaths but you might get to within 90 per cent." Nowadays there is an ISO standard for road safety, and Claes says that 43 of the European states have a towards-zero ambition.

Many of the gains in Sweden were made by creating "a more forgiving way of designing roads and intersections," as Claes puts it. Deaths dropped as a result by over 70 per cent in five years. In rural areas it was done with good management of speed cameras and traffic calming together with divided narrow roads. Some 90 per cent of traffic fatalities occur in low- and middle-income countries, and of those, 70 per cent involve vulnerable road users: pedestrians, cyclists, motorcyclists.

Claudia Adriazola-Steil shares Claes's vision. When working for the government of Peru she designed that country's first road safety policy. As the Health & Road Safety Programme director for EMBARQ, the Center for Sustainable Transport and Urban Development at the World Resources Institute in New York City, she is responsible for rolling out similar policies across the world.

EMBARQ's Health and Road Safety team assists cities, particularly emerging mega-cities, in their urban planning by recommending street designs that calm traffic, reducing block sizes, and finding ways to improve walking environments for pedestrians and promote cycling. Claudia passionately advocates that "focusing on traffic safety leads to multiple wins: cities can improve air quality, increase the opportunities for physical activity, reduce stress and alleviate inequalities. Shifting the method of transport to high quality, sustainable options is a significant part of the puzzle in making roads safer, but other tactics are necessary as well."

Chief amongst these modes is cycling, which is gaining in popularity almost everywhere. "Cyclists bring a lot of benefits, because they are happier, will make fewer hospital visits and take fewer days off work." Claudia says that in addition to shifting to sustainable transport, encouraging high-density development helps to reduce the number of traffic injuries. Together with mixed land use, it enables people to avoid transport, reducing risk. Using World Bank data, she points to a direct correlation between traffic fatalities and vehicle miles travelled.

You and I might believe that some countries have a culture of aggressive driving that must make them harder to tackle. Claudia sees it differently. "If you have a city like Lima you can be stuck in traffic for two or three hours, which makes you reluctant to stop at a traffic light or a stop sign because you are in despair. Maybe the city is not functioning any more. To solve this you must think how people live in the city and find ways to move facilities such as hospitals closer to where people need them to cut down the need for long journeys. You need to look at the entire system."

She believes the same approach is valid for tackling dangerous motorcycling. "It's very sad that in countries like Brazil and India you get a lot of young men who die in traffic accidents or suffer severe injuries that leave them disabled for life because they are riding a motorcycle," she says. "Again, we need to look at what is happening behind all this. If distances are too big to bike or walk then they must take their motorbike. A lot of countries are making a mistake in providing subsidies for motorcycles. The disadvantages of motorbikes are huge and these are not factored into these decisions."

Speed is a big factor in traffic safety, of course. "If you cut down the speed people are allowed to travel at they become more aware of other road users. Unfortunately, some cities – when they develop roads – do not think clearly about how pedestrians are going to use the area."

In some places – counterintuitively – they even take away all of the road signs, having slowed down the traffic. By doing this you force people to negotiate between themselves how they are going to navigate the shared space. "Humans actually function better in a social environment than in a regulated environment," Claes says. "When we understand that we need to negotiate we can become more clever and more aware of what is going on and we become more dynamic. It has been tried in Sweden and we do it really well. Some older people get really upset and they want rules, but the point is rules are unnecessary. The only rule should be that you don't want to crash with anyone else."

Bus rapid transit

Claudia is a passionate advocate of BRT services, a faster than conventional bus, public transport system designed to improve capacity and reliability, with dedicated lanes. "Mexico City has four BRT lines and they are building a fifth. BRT is expanding to further cities in Mexico and around the world, with many representatives from other cities coming to Mexico City to see what is happening there," she says, giving the examples of Rio de Janeiro, Istanbul, Bangalore amongst the 160 cities that have already implemented BRT systems.

Another persuasive factor is that besides saving lives, with BRT "you can also move more people. For example in a lane that can carry 1,000 people you can carry many more thousands using buses for the same amount of road. This will cut congestion and increase mobility as well as traffic safety. People who don't use private transport will also be more physically active, cutting health costs and cutting pollution."

Self-drive cars

What about automated transport? Does Claes think that this is a realistic proposition to cut road injuries and fatalities? "For sure. Technology is much more reliable than humans. The new generation of self-driving vehicles will be programmed by engineers who will make sure as best they can that they anticipate everything that can go wrong so they will force the car to drive in the most cautious manner possible, something that perhaps we never do after we have passed our driving test!"

The question is, will we be patient enough to tolerate this? Perhaps we won't have any choice.

Financing new transport systems

Many cities can't afford the investment in infrastructure needed to implement sustainable transport policies. What is the solution? "It's not always about spending a lot of money but about changing attitudes and priorities," says Claes. "The question really for any city is whether it can survive without going in that direction. In parts of South America and Asia they are more radical than in America or Europe. The World Bank as a lender in the past has been a disaster but nowadays it's not going to finance building a major road through a city without thinking really seriously about it. Instead they think more about sustainability and liveability."

Claes agrees that bulldozing roads through is still happening a lot in African cities, putting this down to consultants and planners who need to get rid of their old ideas and be retrained or replaced.

As cities and countries incrementally approach Vision Zero, mopping up the few remaining injuries and fatalities requires lateral thinking. In Sweden this has moved on to thinking about making road and pavement services softer. "Human beings," Claes says, "weren't designed to walk on stones and concrete, so sometimes even a fall from standing can do a great deal of damage and even cause death. This is insane. We are researching new, softer types of surface that can help to minimise injury". Personally, I quite fancy the idea of bouncing along a bungee surface.

Let's leave the last word on this topic to Claudia: "Mobility is very important, but not at the cost of health and your life and the quality of life".

That should be the prime motto for any 'one planet' city.

Notes

1 Zbicinski, I., *Product Design and Lifecycle Assessment*, 2006, quoted in *Futurama Redux: Urban mobility after cars and oil*, exhibition catalogue, Vienna, 2018.
2 Harvey, D., "The Right to The City", *New Left Review*, 2008, 53, 23–40. See also: Harvey, D., *Rebel Cities: From the Right to the City to the Urban Revolution*, Verso, 2013, ISBN: 9781781680742.
3 Little, Arthur D., *Urban Mobility Index 3.0*, March 2018. See: www.adlittle.com/sites/default/files/adl_urban_mobility_index_3.0_2018_ranking.pdf
4 Matan, A., Talk: Walkable urban forms: Modelling the potential human health impacts of transport options using urban development assessment models, in Walk21 (XV International Conference on Walking and Liveable Communities, Luna Park, Sydney), Abstracts, 21 October 2014.
5 Mepham, D., Why is light rail more pedestrian friendly than bus rapid transit? A review of Australian urban transit accessibility, in Walk21, 2014, op. cit.
6 Lavadinho, S., Enhancing transit for walking: Gearing towards the multimodal city, in Walk21, 2014, op. cit.

7 Ewing, R. and Hamidi, S., *Measuring Sprawl 2014*, Smart Growth America and Metropolitan Research Center. See: www.smartgrowthamerica.org/app/legacy/documents/measuring-sprawl-2014.pdf
8 Jones, C. and Kamman, D., "Spatial Distribution of U.S. Household Carbon Footprints Reveals Suburbanization Undermines Greenhouse Gas Benefits of Urban Population Density", *Environ. Sci. Technol.*, 2014, 48 (2), 895–902. See: https://pubs.acs.org/doi/abs/10.1021/es4034364
9 "Congestion Pricing", *Outlook India Magazine*, 1 March 2008. See: www.outlookindia.com/website/story/congestion-pricing/236827
10 Wang, Y., et al., *Study On International Practices For Low Emission Zone And Congestion Charging*, World Resources Institute Working Paper, January 2017. https://wriorg.s3.amazonaws.com/s3fs-public/Study_on_International_Practices_for_Low_Emission_Zone_and_Congestion_Charging.pdf?_ga=2.60420521.529340390.1542105438-328231776.1542105438
11 Ehrman, S., "Here Come The Superblocks", *La Pinya* web blog, 28 September 2018. https://lapinyabarcelona.com/blog-archive/superblocks

13

HOW SMART IS A 'SMART CITY'?

To what extent will the 'one planet' city rely on technology? And isn't a 'one planet' city smart by definition?

Science fiction stories about highly automated futures with cities run by computers frequently contain dire predictions. *The Matrix* trilogy has humans completely taken over by machines. Stories in Charlie Brooker's *Black Mirror* television series features control systems breaking down with unfortunate results. Kurt Vonnegut's *Player Piano* is a variant on the Luddite theme of workers smashing machines that have put them out of jobs.

But perhaps the template for this kind of dystopia was E.M. Forster's *The Machine Stops* published as long ago as 1909, in which people live in little boxes and rarely go outside, communicating via a kind of instant messaging/video conferencing network and worshipping a deity called the Machine. They have forgotten that humans created the Machine, and its self-repair mechanism has begun to fail, a development regarded as heretical because of the supposed omnipotence of the Machine. It was the inspiration for a David Bowie song, 'Saviour Machine', whose chorus line cautions civilians against staking their lives upon it.

India's drive for smart cities

Nevertheless, that seems to be exactly what we are doing. All over the world, cities are being wired up for all kinds of supposedly good reasons. Perhaps the most ambitious programme is the 100 smart cities mission launched by Indian Prime Minister Narendra Modi, estimated by consultancy Sustainability Outlook to cost US$45–50 billion over the next five years. The idea is to put cities such as Pune, Jaipur, Surat, Kochi, Ahmedbad, New Delhi, Chennai, Visakhapatnam, Ludhiana and Bhopal at the forefront of the 21st-century urban experience.

There is a parallel project for 50 'solar cities' to enable at least 10 per cent of electricity to be supplied from solar panels over five years. Among the technologies being suggested by Western technology companies such as Cisco and Ericsson, hungry for the business of fitting out these cities in the second most populous nation on the planet, are high-resolution cameras for city surveillance; sensors for managing parking; smart meters for measuring, billing and reducing energy use; solutions for managing waste and renewable energy provision; intelligent street lights and traffic controls.

Indian Chief Minister Devendra Fadnavis said of the projects slated for his home town Nagpur, "We are trying to ensure that infrastructure development in the city keeps pace with its growth".[1] This growth includes transportation, water, sanitation, health care and unauthorised development. Transport projects include a metro and the Nagpur–Mumbai Super Expressway. While the latter is not particularly sustainable, other projects, such as those in Pune, incorporate the redesign of streets, footpath retrofitting, making space on roads for social activity, rainwater harvesting, low-income skill development and health care, LED lighting, solar rooftops, and an e-governance centre including grievance redressal and smart customer services through online apps. Quite a list.

A main focus is on public transport, with the installation of GPS and real-time tracking of buses through a mobile app, a vehicle health monitoring system, intelligent road-asset management, traffic mapping using mobile GPS and e-challan (an official electronic document) for payment of traffic fines.

While this will be very impressive for Indian cities, Western cities have learned that smart cities are about much more than technology. The overall aim of the smart city project for them is to help government perform better and the public to receive improved services.

The need for open data

It's a given that open data, in other words transparency, is an absolute necessity for projects to succeed in a democracy. The inclusion of the populace – tapping into the wisdom of crowds – is the ability given by the technology to enable information sharing and communication to be not just one-way (top down) or two-way (up and down) but every-way. It is crucial for success – or 'citizen buy-in'.

According to Dario Hidalgo, director for integrated transport at EMBARQ, 'smart' does not necessarily mean 'sustainable'. "In order to be sustainable, a city needs the sustainability parts first as a design prerequisite before implementing 'smart' components," he said. In a webinar I ran on this topic, Melanie Nutter, a former director of the San Francisco Department of Environment, agreed. She said that policies should come first; other elements follow. Sensors, for example, can be an enabling form of assistive technology.

For her, the original idea of smart cities, which we can call Smart Cities 1.0, focused on technology and engineering. But the concept behind Smart Cities 2.0 is about putting people first. "You can't manage what you can't measure," she said.

"Smart by itself is not a measurable goal or strategy. We must answer questions like 'Do we want smart sprawl or revitalisation?' Technology is about data but community revitalisation is about wisdom."

Mobility

Borrowing technology derived from the smart energy metering of commercial buildings and the use of closed-circuit television (CCTV) cameras and sensors for traffic control, mobility has been an early focus for Smart Cities 1.0. Apps are assisting the sustainable transport agenda by pulling in transport data from all aspects and aggregating them. For example, traffic sensors are used to control traffic lights to speed up the flow of traffic dynamically and unblock hot spots of congestion.

One main cause of air pollution and congestion is drivers circulating in the hunt for parking space. Milton Keynes, a city in England, has tackled this with sensors to manage the use of 20,000 parking spaces, allowing information to be provided on roadside displays and smartphone apps to guide vehicles towards available parking spaces. As well as optimising the use of existing parking infrastructure, 'smart parking' systems of this type potentially reduce fuel consumption and emissions from vehicles. After detecting an arrival or departure, the sensors send information wirelessly to lamp post-mounted solar-powered repeaters which transmit the data to a hub where it is processed and displayed on the council's public information dashboard and smartphones, with parking bay status displayed as red (occupied) or green (free) on Google Maps.

Smart traffic management can be used to ensure that service deliveries are made at night time when they interfere less with other traffic. Storm Cunningham, publisher of *Revitalization News*, cites the widespread prevalence of Uber, Google Transit and car-sharing apps. He believes that the next thing will be "the wider use of congestion pricing [as in London] and parking management and traffic management which is green since it reduces emissions".

But for Dario Hidalgo, the only fully sustainable vehicle is the bicycle, and in many cities people are going back to it. It is the main mode of transport in some cities such as Copenhagen. Its connection to the 'smart' agenda is about efficiency, cheapness and health. He does not think that driverless cars are particularly sustainable, although they do hold the promise of being able to slash fatality and injury figures from traffic accidents. "Cycling," he says, "can be helped by smart technology such as bike sharing projects."

Storm adds that walking, too, is properly sustainable. "Pedestrians need to be formally incorporated into transit plans and we need a better analysis of where the blocks are to encourage walking. Revitalisation of community districts to make them more attractive for walking through should be high on the agenda."

The smart city agenda has huge advantages for compact, connected cities such as Barcelona, which has rushed to adopt it. Contrast this with a city as sprawled and car-dependent as Atlanta. For Dario, the only way for this type of city to reduce its traffic impacts is "to become less centred and more dependent upon nearby local hubs for services and jobs. Properly sustainable programmes will encourage people to shift to more sustainable forms of transport or not travel at all by creating homes next to jobs and services rather than to drive their cars more efficiently."

Melanie Nutter agrees: "Real-time data will reach everywhere. The 'smart cities' hype does not represent a silver bullet solution to all the problems cities face. Cities thinking about implementing the technology need to engage with providers properly and be fully informed before engaging in procurement and talking to the private sector about the best way it can fulfil the aims of the sustainability agenda."

Smart cities need a new type of management

While the smart city idea works on a project basis, it does not work so well at a systemic level, Storm Cunningham believes.

> The basic decision-making rules of civic government need changing to accommodate the cross-departmental and cross-sectoral opportunities of the technology available. The places where the smart city agenda is working is where it focuses on renewal, re-purposing existing infrastructure and reconnecting existing assets. Humanity itself needs to adapt as it moves into cities and is destroying the world's ecosystems. Up until now, humanity has been in adaptive conquest mode. Now we need to move into 'adaptive renewal' mode.

Melanie Nutter cited a survey of 300 urban sustainability directors which found that they had to take a step back in order to examine how to best implement a smart city strategy. They needed to examine the way their city was governed, managed change and integrated new technology. For restructuring of city governance in order to successfully implement smart technology, the order of implementation should be: vision first, then strategy/ies, then plan(s) or programmes, then projects. To achieve buy-in by both city officials and citizens, their confidence in the agenda needs to be raised first by pilot projects.

"It needs a lot of ground work to be done," Melanie advised. A task force to define a digital strategy and how the city is to use data needs to be created as a next stage, followed by the appointment of data officers. Included in the strategy should be a means of helping the private sector better understand the administration's needs.

The WCCD is attempting to standardise city data so cities can learn from each other, share and compare experiences, and improve the quality of life of citizens. It encourages the use of open city data and provides a consistent and comprehensive platform for standardised metrics, using ISO 37120 Sustainable Development of Communities: Indicators for City Services and Quality of Life, the international standard discussed in Chapter 3. Their website publishes the results of the open data collection for certified cities ranging from Mexico City to Makati (Philippines), via Minna (Nigeria) and Makkah (Saudi Arabia). Remember from Chapter 3, the standard is focused on meeting the urban Sustainable Development Goals but not sustainability within planetary boundaries.

The idea of a smart city presents a big opportunity for medium-sized cities, which can change faster than huge metropolises that are more complicated in management. Kielce (population 200,000) was one of the first cities in Poland the WCCD assisted to obtain ISO 37120 Platinum Certification. Data are collected using the standard about all aspects of the city and published for all to see using an online map as the interface. Its mayor, Wojciech Lubawski, sees one advantage of certification being that data on the city's performance "is so reliable that we do not question it at all. The goal of each mayor, the goal of local government, is to raise all these standards to a higher level. Making the wrong decision can affect the quality of people's lives. And that's why all this data, knowledge and statistics are extremely necessary."[2]

Often, data collection using the ISO means that unexpected things are discovered. "We might discover that we should allocate more money on something that is needed and expected by residents, or discover through those external entities what our strengths and weaknesses are that should be dealt with more thoroughly," said Lubawski. "People often follow the path of least resistance, but they should be aware that they may have poor outcomes if they make decisions that do not reflect research."

Ms. Jadwiga Skrobacka, Kielce's manager of the Office for Intelligent Management of Sustainable Development, added some other advantages:

> Everyone looks at decisions from their point of view and sees only a fragment of reality. But describing the context in more detail and showing it on the map, as we try to do, seems to create much more opportunity to include residents in the process. Including residents is more common, and in the smart city idea, data sharing is crucial. Everyone learns, both the employees who have their responsibilities and the residents who are informed about the city's activities. It becomes possible to agree on common directions ... on certain key values, according to which we imagine the future of our city, and these values guide the various activities that we will undertake. It becomes easier to prioritise investment decisions. I also think that the educational role of city-to-city learning is very important.[3]

The 'wisdom of crowds'

Cities can use social networks too, to tap into the 'wisdom of crowds', but this needs to be conducted sensitively. For example, in a referendum in Vancouver on transit strategies, citizens were asked to post in their ballots, which reduced participation; responses should also have been possible via email or text message. In the interests of total inclusivity of the population, off-line approaches are just as important as online ones. No ballots should be self-selecting.

Self-selecting groups do frequently use technology to get momentum behind a project, however. The High Line is often given as an example of an iconic piece of urban, walkable green space. It is a 1.45-mile-long elevated linear park, greenway and rail trail created on a former New York Central Railroad spur on the west side of Manhattan. It would not have happened without substantial public campaigning. The organiser used the crowdsourcing fundraising website Kickstarter to establish support and raise $100,000, making an offer to the administration that it did not refuse, and allowing the project to bypass the restrictions of the planning system. Although a desirable outcome was reached, the High Line has been criticised as a development for educated, well-off, middle-class people – and all views need to be heard.

The democratic idea of citizen-led social innovation is central to the URBACT Smart Cities network, a European-funded initiative. In a 2012 paper on Social Innovation For Smart Cities. Peter Ramsden observes that

> the smart cities of the future will be innovating in how they deliver social policies as much as they do in organising transport networks or reducing their carbon footprint. Smart cities will be innovating to prevent problems happening rather than trying to solve problems when it is too late. Most of all, smart cities will organise their innovation effort to focus on priorities and to look for new solutions. Citizens should be in the driving seat. They should be invited to participate and respected for their views. They should be able to share their needs and speak truth to power while trusting that working with officials and experts will lead to new solutions.[4]

Five years later Rasha Elgazzar and Rania El-Gazzar argued that, "ICT is a means of achieving intelligent resource management in a city, but it does not necessarily presuppose that the city is successful in being smart and sustainable."[5]

Measurement and verification

In Smart City 2.0+ then, applying the approach of using standards to check and compare progress, smart technology provides the opportunity for cities to measure their sustainability and include citizens' views.

It's also possible to use modelling with feedback from sensors around the city. In Taiwan, the g0v Air Pollution Observation Network involves distributing hundreds of air-quality sensor 'airboxes' which, using IoT (Internet of Things – the implementation of ubiquitous interconnectivity) technologies, permits citizens to provide real-time air-quality information from wherever they are. Thousands of citizen contributors have accumulated a massive database, revealing the air quality in the places where they are active. This is connected to the Civil IoT programme, which has a four-year budget of TWD4.9 billion. It collects an enormous amount of environmental data on air quality, meteorology, water resources, earthquakes, disaster relief and more, integrating them into a high-speed computing environment. This permits collaborators to discover correlations between social activities and environmental phenomena more quickly.

I agree with Storm Cunningham that while 'smart' infrastructure projects can be beneficial, their overall impacts need to be estimated and measured as part of the decision-making process on whether projects should go ahead. City administrators are often bedazzled by salespeople from technology companies proclaiming the wonders that digitisation and the IoT can achieve. They are touting for business and their own interests are paramount in their minds. Administrators need to conduct their own cost-benefit analyses; these should include a full life-cycle analysis of the impacts of installing the technology, including the embodied energy.

A project cannot be sustainable if the environmental cost of installing and maintaining it is greater than the benefit. Rather than mounting sensors expensively on infrastructure around the city – technology which can go out of date remarkably quickly – it may be better to, as Taiwan has done, employ mobile apps and citizen engagement.

Additionally, keeping the data open (as ISO 37120 requires) rather than as intellectual property owned by the commissioned companies (which may be a condition of their contracts) will also multiply the uses to which the data can be put by the city.

Different cities will go different ways according to their culture. Singapore is an extreme example of technology taking over a city; Singaporeans accept a level of intrusion into privacy that would not be acceptable in other countries. Its top-down solution is to put sensors all over the city to create a 3D X-ray map, monitoring everything from temperature, climate and pedestrian movements, to toilet flushes and appliance behaviours inside apartments, 80 per cent of which are public housing. The motivation, accepted by citizens, is that it enables the city authorities better to look after its occupants. This might be the closest city on the planet to some of the dystopias imagined by science-fiction writers.

Hackathons

Contrast this to, say, Amsterdam, an example of a city that runs hackathons. Hackathons are events where anyone with coding ability is given access to all sorts of sources of city data and invited to come up with creative ways of using it to improve quality of life. In that city, winners of a competition included projects to adapt the existing cameras of an arena into a smart facial recognition system for better security, safety and experience; a stadium app solution alerting fans for availabilities of bathroom, food and beverages in a venue; and a crowdsourcing app helping citizens fix the city by engaging and initiating projects with fellow citizens.

A hackathon in Stockholm came up with the idea of installing cheap noise and air-pollution sensors around the city to measure and visualise noise levels and air pollution in different areas, in order to help control excess noise and reduce pollution. In Germany a company called Indalyz Monitoring & Prognostics (IM&P) has developed a unique piece of prognostic software that uses artificially intelligent algorithms to predict when which component of a machine will break down and have to be replaced.

Cynics might say that the only thing needed to make E.M. Forster's prediction come true is that the machine then repairs itself and takes over the running of a city like Singapore. But optimists would argue that although human ingenuity is constantly coming up with new ways to use the IoT to make improvements supposedly for our benefit, international standards are being developed to make sure that all of these talk the same language and provide guidelines to help cities (and their officials) plan with greater wisdom.

In order to measure the performance of smart community infrastructures, ISO/TS 37151 (Smart community infrastructures – Principles and requirements for performance metrics) outlines 14 categories of basic community needs from the perspectives of residents, city managers and the environment. This standard helps with the assessment of the abilities and qualifications of city planners and consultants to guarantee the safety and security of residents.

Ultimately, it's up to citizens themselves to ensure that their governments adopt this type of practice, and safeguard themselves from misuse of information.

Cities and national governments need monitoring systems that produce accurate and close to real-time data and information, to design and inform their actions, and policies aimed at uplifting urban services and the quality of life of their residents. The appropriate disaggregation of data at different levels and forms is necessary in order to report consistently on performance at the urban, subnational and national levels, but privacy rights must be built in.

Financing data collection is an issue for many countries. Those with many cities, and those with limited human resources and funds, can adopt various strategies to cope with big demands for data (e.g. to satisfy UN and other international requirements, like the Sustainable Development Goals). One of these is a national 'sample of cities' approach, developed and piloted by UN-Habitat. It is an important means of aggregating national urban performances in a consistent and systematic manner. Using a sample to represent the whole offers a low-cost option of monitoring fewer representative sets of cities consistently, rather than all cities, and being able to report national-level performances.

Ultimately the level of digitisation in service of the smart agenda is going to be a trade-off between cost versus benefits, the degree of open citizen engagement that can be tolerated by a government, and the overall social and environmental impact. To refer back to the webinar I conducted on this topic, the conclusion was: "a city is only as smart as it citizens" – and its leaders. Information, evidence, equity and education are the pillars of a smart, 'one planet' city. Technology is its servant, not its master.

Notes

1 Khapre, S., "Smart Cities project: With just a month to go for second round, govt pushes Nagpur development", *Indian Express*, Mumbai, 28 April 2016. See: https://indianexpress.com/article/cities/mumbai/smart-cities-project-with-just-a-month-to-go-for-second-round-govt-pushes-nagpur-development-2773492/
2 Mayor of Kielce, "Poland Speaks on WCCD ISO 37120 and Smart Cities", *WCCD*, October 2018. See: http://news.dataforcities.org/2018/10/mayor-of-kielce-poland-speaks-on-wccd.html
3 "Mayor of Kielce, Poland Speaks on WCCD ISO 37120 and Smart Cities", *Data for Cities News*, 1 October 2018. See: http://news.dataforcities.org/2018/10/mayor-of-kielce-poland-speaks-on-wccd.html

4 Ramsden, P., *Social Innovation For Smart Cities: The state of the art*, URBACT, 2012.
5 Elgazzar, R.F. and El-Gazzar, R.F., *Smart Cities, Sustainable Cities, or Both? A Critical Review and Synthesis of Success and Failure Factors*, SMARTGREENS 2017 Conference. www.researchgate.net/publication/316623082_Smart_Cities_Sustainable_Cities_or_Both_A_Critical_Review_and_Synthesis_of_Success_and_Failure_Factors

14

FINANCING THE WAY TO A 'ONE PLANET' FUTURE

It won't be cheap to shift the way we live to within the limits of what the planet can provide, although it will provide great financial benefits also. Who will pay – and how?

The UN has estimated that US$6 trillion a year will have to be invested in sustainable infrastructure globally over the next 15 years. Of this, it says $6 billion needs spending on climate resilience in the form of disaster risk management.[1] This would generate total benefits in terms of risk reduction of $360 billion, equivalent to a 20 per cent reduction of new and additional annual economic losses,[2] according to the Global Commission on the Economy and Climate.

They say that we have run out of time for incremental steps, generic proposals, or statements of broad principle. Governments should put a price on carbon of at least $40–80 per ton and move towards mandatory climate risk disclosure for major investors and companies. They also point to the huge rewards if we do the right thing: "More compact, connected, and coordinated cities are worth up to US$17 trillion in economic savings by 2050"; "Sustainable agriculture and forest protection together could deliver over US$2 trillion each year in economic benefits"; and "low-carbon growth could deliver economic benefits of US$26 trillion to 2030 — and this is a conservative estimate."

And it's not just about money: they emphasise that a people-centred approach is needed to ensure lasting, equitable growth and a just transition.

The problem facing cities

Cities are at a disadvantage in that they cannot usually tax carbon and other polluting activities, with the possible exception of congestion charges to improve air quality (although states can). Any attempt by a city to tax polluting businesses risks losing to another city the investment that business can bring.

Imagine much higher taxation on 'luxury' goods that go beyond satisfying basic needs and on goods with higher impacts, with the proceeds being directed towards investment in the circular, zero carbon economy. At present, any attempt by cities to tax excessive consumption – over and above the existing taxes on larger, and second or empty homes – could drive highly qualified and well-paid workers away, or be democratically unpopular. It might be possible for national governments to do this, but there are not many precedents.

Capitalism has traditionally outsourced the environmental cost of its reckless appropriation of natural resources and services to beyond the balance sheet. The reason why it is difficult to finance activities which are proven to be environmentally beneficial is that the economy has been used to paying a low price for the services nature provides.

This goes some way to explain why it is hard to find a market for goods made from recycled materials that stimulates recycling to happen. It is why it is hard to find a business model for anaerobic digestion to deal with organic waste, because of the high upfront costs, despite the multiple, sellable products that result. And it is why it is hard to finance the energy-efficient deep refurbishment of existing buildings, despite the obvious long-term savings and improvements in the internal comfort that result.

Brave attempts are being made, in the form of subsidies, natural capital accounting, and investor confidence-building with respect to energy-efficiency projects. But until at national and global levels subsidies cease to be given to activities such as fossil fuel extraction, deforestation, and industrial processes that do not treat their effluent, and the tax regime shifts in favour of regenerative land management, and of closed-loop, zero-pollution manufacturing, we are fighting an uphill battle.

So to fight this battle where can the money come from? How to pay for the new sustainable infrastructure required? This is the subject of much activity around the world in financial institutions. Whether we like it or not, investors must see a way to benefit from sustainable investment. And to prevent the money being misspent, it needs to be overseen in a fair, transparent and verifiable manner. At the same time, not-for-profit and community-led profit-sharing schemes are increasingly being seen as a viable alternative to conventional investment. This chapter examines several possibilities being deployed by or within cities around the world.

Green bonds

Green, Social and Sustainability Bonds are seen as one solution. They are any type of bond instrument where the proceeds will be exclusively applied to eligible environmental and social projects.

2017 saw over 1,500 green bonds in 37 countries issued with a total value of $155.5 billion, a 78 per cent growth on 2016. The largest had a value of $10.7 billion and there were three sovereign Green Bonds from France, Fiji, Nigeria.

According to the International Capital Market Association, there is increasing evidence that green bonds can outperform non-green bonds and appear less volatile in stressed markets. They also are traded somewhat less than non-green bonds.

For example, Caja Rural de Navarra is a cooperative, regional and retail-focused bank that provides banking and financial services in northern Spain: Basque Country, Navarre and La Rioja. It has issued a Sustainability Bond Framework under which it offers multiple bonds and uses the proceeds to finance and/or refinance expenditures that promote rural socioeconomic development in Spain. It addresses several UN goals: sustainable cities and communities (SDG 11) and consumption and production (SDG 12), in addition to affordable and clean energy (SDG 7).

Eligible projects include almost all aspects of 'one planet' living: sustainable agriculture, renewable energy, energy efficiency, sustainable forest management, waste management, affordable housing, social inclusion, education, and economic inclusion. Caja Rural de Navarra has allowed itself to be independently investigated by Sustainalytics, an independent environmental, social and governance and corporate governance research, ratings and analysis firm. The bond scheme in itself incorporates a number of indicators that include, for example, the number of disadvantaged families housed in social housing, and the area of previously non-forested land that is forested.

Any number of similar schemes are being set up around the world. Municipalities themselves are issuing green bonds to finance green infrastructure. However, there is a feeling amongst some that green bonds are too restrictive because the cost of due diligence is high; why not just issue ordinary bonds? To which the answer is: to provide confidence and transparency; as always, there will be a trade-off between what it is desired to do, and the ease and cost of doing so.

Green bonds operate in a context of numerous factors affecting investment, governance and development, such as globalisation, changing political landscapes, ecosystem depletion, urbanisation, resource utilisation, demographics, climatic patterns and climate change, employee attitudes and consumer preferences. Their potential application is correspondingly manifold. A range of currencies, geographies and maturities now feature in the market, with new issuers and new types of green bonds such as asset-backed securities from Toyota linked to hybrid and electric vehicle loans, and Indian Railways, IDBI Bank and the International Finance Corporation. These bonds are seen as vital to support sustainable development in developing nations, for example Latin American countries looking to increase their renewable energy capacity.

To secure trust, the International Capital Market Association's Green Bond Principles and Social Bond Principles, as well as the Sustainability Bond Guidelines, have become leading frameworks for the issuance of green, social and sustainability bonds.[3] But others exist. The Climate Bonds Initiative, a registered charity in England and Wales working to increase bond markets for financing climate change solutions, has developed the Climate Bonds Standard and Certification Scheme, a

Fairtrade-like labelling scheme for bonds. Its Climate Bonds Taxonomy is seen as a vital backbone, defining investments that are part of a low-carbon economy, overseen by a board reporting to the governors. Its members are other not-for-profit organisations, such as California State Treasurer John Chiang and the Investor Network on Climate Risk, represented by Peter Ellsworth at Ceres. It played a key role in advising China on the set-up of its green bonds market and manages a bond certification scheme.

Europe, China and India are currently issuing the greatest value of bonds. India's Securities and Exchange Board Green Disclosure Norms attempted in 2016 to design a separate set of standards to demarcate the issuance of green bonds from regular corporate bonds which are in line with the Green Bond Principles. The big advantage seen by the Indian regulator is that they are playing an increasingly important role in attracting capital for the development of infrastructure projects in India, in other words much-needed inward investment.

The world leader is China, which issues green bonds to finance low-carbon and other environmental projects. China has rocketed from not being involved in the market at all pre-2015 to becoming the planet's second largest issuer in 2017 with $37.1 billion.[4] This is because of the size of the challenges the country faces, with the People's Bank of China estimating that China will need hundreds of billions of dollars a year to fund climate change projects ranging from renewable energy to mass public transport, pollution abatement and climate change mitigation.

Green bond standards

The International Organization for Standardization is currently developing a Green Bonds Standard, ISO 14030, that expands the Green Bond Principles. "Different definitions for green bonds have confused and deterred investors," says Dr John Shideler, chair of technical committee ISO/TC 207, which is developing the standard. "We hope to clear that up."[5]

Other standards exist to support investors and investment decisions. ISO 14007 will enable organisations to determine and communicate the costs and benefits associated with their environmental aspects, impacts and dependencies on natural resources. ISO 14008 describes a set of tools for assigning monetary values to environmental impacts.

"There is a growing drive towards valuing natural capital, as well as a need to undertake a monetary assessment of an organisation's environmental aspects and impacts," observes Martin Baxter, chair of ISO/TC 207's subcommittee SC 1, Environmental management systems. "Therefore, having a set of standardised, harmonised methods becomes important."[6]

Finance for adaptation, resilience and the transition to a low-carbon sustainable economy is also supported by ISO 14097 for the assessment and disclosure of climate-change risks of investments.[7]

Competition for finance

Although China's $37.1 billion seems a large figure, this is but a trickle compared to the $90 trillion required to make the global economy less dependent upon fossil fuels.[8] So whereas green bonds are on the up, to have a chance of making any headway in steering the world to a 'one planet' future, they will have to outcompete and overtake the growth in conventional investment for unsustainable infrastructure.

Also, there is evidence that green bonds are being used to finance projects that are questionably sustainable. While praise may go to Mexico for issuing $6 billion of green bonds, and to Mexico City for being one of the first to sign the Green Bond Pledge – which seeks to have cities, public authorities and world's largest corporations commit to increased use of green bond finance to pay for new, sustainable infrastructure – the bonds are to finance Mexico City Airport. Meanwhile Mexico has a simultaneous aim of producing 2.6 million barrels of oil per day, an increase of 30 per cent.

Orlando International Airport, financed with green bonds, was awarded a high score on S&P's Global Ratings Green Evaluation for its ongoing works on a new South Terminal Complex project. How can using green bonds in this way be right when the aviation industry accounts for between 4 and 9 per cent of the total climate change impact of human activity, and airports themselves are a source of considerable greenhouse gas emissions, harmful air pollutants, water pollution and contribute to a loss of agricultural land through their daily operations?

In its defence, the airport says it will run totally on renewable energy and achieve reductions of 30 per cent in water and 40 per cent in energy consumption compared to the existing airport. Some 70 per cent of the water consumed in the airport will come from on-site water and wastewater management projects. Design firm Foster + Partners also claim that for much of the year, temperatures will be maintained almost completely by outside air, with little or no additional heating or cooling, and the roof will be covered with solar panels and collect rainwater. Nevertheless, the project remains intensely controversial.

This illustrates a consistent problem found in the world of conventional business and governance: a desire to 'have one's cake and eat it'. To have a chance of making a meaningful difference, green bonds should be used only for totally sustainable projects, including all impacts; and countries and cities need to end their reliance on fossil fuels.

Other large-scale financing schemes

But green bonds aren't the only large-scale game in town.

In Europe, the European Investment Bank, the European Bank for Reconstruction and Development and the World Bank Group are behind a fledgling project called Global Urbis, which is an ambitious global initiative to provide cities with financing and technical assistance to mobilise private capital. URBIS is a dedicated advisory platform for investment support to cities.

In the USA, progressive states, including Michigan, Hawaii, New York and Rhode Island, are bypassing the Trump administration's lack of interest in clean energy by setting up their own green banks. New York State has set up, with $1 billion backing, the NY Green Bank to help finance clean energy projects in the state. One of these is targeted at solar energy and the other at providing finance to commercial and multi-family building owners so that they can install energy-efficient retrofits and renewable energy. The bank has been working with other green loan providers across the USA to standardise the loan application process so that solar and energy-efficiency equipment becomes as easy for companies to finance as industrial machinery or vehicles.

On the other side of the continent, in California, the Green Bank function is held by the State Treasurer's Office. This invests a portion of funds from the state's Pooled Money Investment Account in bonds that finance green projects throughout the world. The office operates two authorities that finance and administer programmes and projects that promote green jobs and green California industries, protect air and water quality, and encourage recycling, the conservation of natural resources and the use of renewable energy. These are the California Alternative Energy and Advanced Transportation Financing Authority, and the California Pollution Control Financing Authority.

Australia's Clean Energy Finance Corporation was established by the Australian Government under the Clean Energy Finance Corporation Act 2012. Its investment policies include a mission to "accelerate Australia's transformation towards a more competitive economy in a carbon constrained world, by acting as a catalyst to increase investment in emissions reduction".[9] By particularly backing solar power it has helped to progress Australia to a position approaching the top ten of solar power-supporting countries around the world.

In South-East Asia, the Malaysian Green Technology Corporation is effectively a green bank under the purview of the Ministry of Energy, Green Technology and Water. To date it has invested £629 million in clean energy, water, construction and transportation projects in the form of soft loans. These are loans supported by the government, which bears 2 per cent of the total interest/profit charged by the financial institution, to make the loans cheaper, while carefully controlling the type of project that can be backed.

Nevertheless, sustainable investment is not happening fast enough. The biggest policy change that could stimulate the change fast enough is to tax polluting activities; in other words to increase the global economy-wide carbon price to a level adequate to achieve emission-reduction targets consistent with the Paris Agreement. This is between $40 and $80 per metric ton of carbon dioxide. However, the average carbon price across the Group of 20 (G20) largest economies at the time of writing is $16 per metric ton.[10]

Impact investing

Impact investing is yet another channel for investors to put their money into socially and environmentally sustainable projects. Investors who have done so, and who participated in a survey, reported in 2018 an average compound annual growth rate of 13 per cent, with their collective assets growing from $30.8 billion in 2013 to $50.8 billion in 2017.[11] They include fund managers, banks, foundations, development finance institutions, pension funds, insurance companies, and family offices. Interestingly, two-thirds of them said that their investment choices are made not by ethics, but primarily by risk-adjusted, market-rate returns.

Impact investing is about supporting all of the UN Sustainable Development Goals. The Global Impact Investing Network (GIIN) puts the bill to reach all of the goals by 2030 at a staggering $5–7 trillion, which is the biggest investment challenge in the history of the planet (bigger than the one for sustainable infrastructure alone).

Barriers to the spread of impact investing were identified in GIIN's survey, but it says these are reducing year on year. In particular there is a perceived lack of appropriate capital across the risk/return spectrum, and a lack of a common understanding or definition and segmentation of the impact investing market. Also, the range and availability of exit options are sometimes less than in conventional markets. GIIN itself is working to reduce these barriers so that more investors can fund solutions to the world's challenges.

Again, at present, we find that there is a plethora of different criteria and standards that often confuse investors. "If we want impact management to become the norm for every enterprise and investor, as the SDGs demand, we need shared principles, reporting standards and benchmarking methods for impact," says Clara Barby, who is the lead facilitator of the Impact Management Project.[12]

To tackle this problem, the World Benchmarking Alliance, a group of nine organisations with stakeholders in over 50 countries, have already cooperated to agree on norms for impact management and declared an ambition "to rank businesses on their sustainability performance and contribution of leading companies in key industries that will highlight the best and worst performers through benchmarking. The benchmarks will provide financial institutions, companies, governments, and civil society with information that can be used to challenge businesses to do better".

Public and private finance

Involving the private sector is usually necessary for providing infrastructure and investment. But there must be effective management in the public sector (see the cautionary case study of a London borough council in the next chapter). Once a city has been built, its fundamental fabric often changes very little. Decisions taken at the expansion stage, like opting for denser, more compact urban forms because they are less energy-consuming and more manageable,[13] are crucial in determining sustainability levels and quality of life. Letting the private sector take these decisions alone is not advisable, as companies bring their own agendas and problems.

For example, Kathmandu, the capital of Nepal, has recently been expanding quickly. When residents were asked about what was attracting people to the city, 86 per cent said they thought it was centralised services such as educational institutions, equipped hospitals and industrial activities. But when asked how this affected their quality of life they listed the negative consequences: more air-pollution, insufficient sanitary landfill sites, comparatively low water supply, loss of scenic beauty, and the deterioration of agricultural land.[14]

Top of their list for improvements were: the provision of an integrated solid waste management system; the prioritisation of non-motorised vehicles for transportation; recreational sites and open space; rainwater harvesting and groundwater protection; a well-designed land use plan; sustainable energy; and 'green' buildings. If you add ecological sewerage treatment, these are common features of eco-cities everywhere. We saw many of these in Chapter 3, as metrics for the standard for sustainable communities. These need investment coupled with skilled oversight by the city government, which has not yet happened.

The types of projects presently most in need of private-sector involvement are energy efficiency/retrofits of existing or new buildings, the provision of renewable energy, and the creation of transport infrastructure. African cities have the strongest demand for private-sector involvement. In terms of climate adaptation, over half (27) of the 50 largest Association of Southeast Asian Nations (ASEAN) cities are, like Manila, located on the coast and thus at risk of flooding due to sea-level rise and land subsidence. Investment is therefore required to protect these areas and can only come from the private sector.

In the ASEAN region, fast-expanding cities like the Indonesian capital, Jakarta, and its second largest city Surabaya, have the worst traffic congestion and air quality in the region, and poor public transport provision. Local governments like these could also benefit from private-sector involvement, but only if they are confident that they possess the necessary expertise to negotiate and manage such partnerships. Unfortunately, local governments like these are frequently underfunded, insufficiently trained and institutionally weak, so they are unable to facilitate the building of appropriate infrastructure. This is witnessed by the slow progress of the Jakarta monorail project, which has been in the making for 15 years. However, to be fair, the city has succeeded in installing the main network of its Bus Rapid Transit system, the TransJakarta, a lower-cost alternative to rail.

In Bangkok, public-private partnerships have proceeded fairly well. They are responsible for a number of public transport projects including the extension of lines for the Skytrain, railway, metro and monorail networks, all of which are due to be completed over the next 10–15 years. Ten new river bridges are also scheduled to be built within the next ten years, connecting the two sides of the city.

So, private investment, however welcome, must not be seen as a magic bullet. There are ways to get small things done without the involvement of big business. We will look at some case studies in the next chapter.

Meanwhile, a global body tackling this issue is the Blended Finance Taskforce. This aims to mobilise large-scale private capital for investment in the SDGs, especially for sustainable infrastructure. Blended finance is thought to be one of the best ways to attract the extra cash from the private sector to meet the bill. The word 'blended' comes from the notion of combining private finance with public money by development banks. Currently the ratio between the two finance sources for investment by development banks is on average half and half. Substantially more private finance is needed if the world's problems are to be tackled successfully and the SDGs met, according to its new report, *Better Finance, Better World*.[15]

The majority of this money will be needed for sustainable infrastructure projects in emerging markets, which are often seen as too risky for mainstream investors. Using public funds can attract private investment by mitigating these risks. Blending public and private capital could be a win-win for both investors and global development. But the blended finance market needs to scale in order to get from billions of dollars of aid or public funds from the development banks, to trillions of dollars of private investment.

Green mortgages

Moving down the scale, there is a need for green mortgages to stimulate the market for sustainable building and renovation. At the current renovation rate of just 1 per cent annually, it would take a century to decarbonise the entire European building stock. Currently, 39 European Banks[16] including BNP Paribas, ING Bank, Nordea Bank and Société Générale, are currently testing a set of standards proposed under the Energy Efficient Mortgages Action Plan (EeMAP). Coordinated by the European Mortgage Federation – European Covered Bond Council, the scheme is aimed at making loans to energy-efficient homes cheaper.

Consumer research discovered a positive customer reaction to the idea of an energy-efficient mortgage, especially in Italy (80 per cent very/quite appealing), followed by the UK (66 per cent very/quite appealing), with very low outright rejection of the idea.[17] According to E.ON's B2C [business to consumer] market manager, Marco Marijewycz, "The single most important reason in all markets for finding the product appealing is the prospect of cutting the cost of the mortgage."

The World Green Building Council's Europe network has issued a call to action addressed at banks, investors, the building sector, energy companies and governments, to develop partnerships and to work to understand how policy changes can support the spread of green mortgages. Making energy efficiency become part of every mortgage conversation with borrowers is vital to stimulate awareness and demand. They also called for 'building renovation passports' to support phased, deep energy renovations of a building by improving the availability of data for valuers and lenders, and to ensure that any renovation works are planned and implemented in a technically sound manner.[18]

Financial regulators have the scope to relax capital requirements for energy-efficient mortgages, on the basis of their lower-risk profile. This concerns the minimum amount of capital that a bank must hold in reserve in relation to the size of its loan portfolio, intended to keep the bank solvent in the event of a crisis and directly related to the risk associated with the bank's assets. A relaxing of these requirements for lower-risk, energy-efficient mortgages would allow banks to do more lending.

According to Ursula Hartenberger, global head of sustainability at the Royal Institution of Chartered Surveyors:

> It is essential that valuers and lenders take a wider view beyond energy efficiency and traditional value drivers, looking at more 'intangible' aspects such as: the potential impacts of – and on – climate change; configuration and design, including use of materials and concepts increasingly associated with health and 'wellness'; accessibility and adaptability; building 'intelligence' and other 'costs in use'. Access to consistent and comparable building information data, as captured by the proposed building passports, is essential for valuation and property professionals in order to provide banks with robust advice and facilitate informed lending decisions.[19]

Social businesses

Moving further down the financing scale involves supporting millions, if not billions more than the above schemes, however worthwhile they are. The challenge for finding the 'safe and just space' of Kate Raworth's doughnut within environmental limits involves bringing up the living standards of the majority of people on the planet, as well as reducing the consumption levels of the richest.

Muhammad Yunus revolutionised finance for the poorest when he started providing tiny loans to Bangladeshi villagers at market interest rates without requiring collateral, in so doing becoming the founder of the microcredit movement and eventually winning the 2006 Nobel Peace Prize. He saw a very large need for very small loans to the type of people that banks would never normally touch, but whose lives could be transformed by being given the opportunity to buy small pieces of equipment and enabling them to start a business.

The Grameen Bank was born with the ambition to provide zero-collateral micro-loans to primarily low-income rural Bangladeshis (usually women – who make up 97 per cent of the typical loan portfolio). Grameen has equity/convergence at its heart, seeing credit 'as a human right'. Loans are usually small, in the order of 100–1,000 Taki (a few dollars to tens of dollars), to buy equipment. Lenders are supported through peer pressure to abide by the principles of solidarity lending, and by a set of values known as the 16 Decisions (which include prescriptions about environmental protection and promoting social justice). In reciting this code, which they do, they are taking a pledge to follow these rules.

Escaping from poverty may mean that the ecological footprints of Grameen borrowers increase, but the principle of 'equity within planetary limits' requires a decrease in the environmental footprints of some citizens while a growth in others, so that, globally, we all 'contract and converge'. The 16 Decisions cover related issues such as limiting family size, keeping the environment clean, and the use of disease-limiting sanitation. The initiative explicitly seeks to empower the low-income fraction of the population it works with according to the principles and practice of social justice. The principle is also present because borrowers pledge to work with each other in a democratic and ethical manner towards common goals.

The idea of microcredit was hugely successful and has spread throughout the world. Yunus believes that social businesses are the way to solve environmental and social problems. He is on record as saying that "whenever I see one such problem I see it as an opportunity to start a business". He recognises that everybody has innate entrepreneurial skills but these do not have to be at the service of profit for its own sake but that the profits can be ploughed back in to serving the social or environmental aims of the business once the basic needs of other business have been met.[20]

In his book, *A World of Three Zeroes*, he gives countless examples from all over the world of how social businesses and the new economics can tackle the problems of inequality, climate change, joblessness and pollution by unlocking the entrepreneurial potential of individuals and groups and their desire to better themselves to work towards a world of zero poverty, zero unemployment, and zero carbon emissions.

He has set up the Yunus Centre to promote social businesses as a way to fight poverty and to help academic institutions develop programmes focusing on the sector. Recently he called for the creation of more 'social business cities', which he says have sprung up from Germany to Japan, and could help to solve social issues in our rapidly urbanising world.

"We can activate the creativity of individuals and corporates together to solve the problems of cities. If you can convert some welfare programmes into social business, then the money comes back to you, it becomes very powerful," he said at a recent Philanthropy for Better Cities Forum in Hong Kong. Yunus said that inviting social businesses, which aim to solve social problems while turning a profit, to invest in areas like health and housing, could free up money for cash-strapped cities to spend on development. He believes that learning about social businesses should become part of "normal growing up" through education in schools.

The Yunus Social Business Centre at the University of Florence and ARCO (Action Research for CO-development) have created the Social Business City Program, to create an enabling ecosystem for social businesses. It believes that "a city-focused approach allows us to identify and tackle social problems in a more effective and efficient way, since each territory has distinctive features and socio-economic dynamics". The programme involves a wide range of stakeholders, addressing different components of the local community, with a participatory approach.

Taiwan

Taiwan is a country that is fully supportive of social enterprises in order to smooth and support its path to democracy and development, because it believes that they harness the wisdom of crowds and the energy of the people. Its government has created a Social Innovation Lab, both a one-stop venue for accessing government services but also containing the fruit of collective wisdom from more than 100 social entrepreneurs. It has become a springboard for social innovation, and similar bases have been set up in Taoyuan and Taichung.

Agoood is one social enterprise to have emerged. In November 2017, Agoood launched a crowd-funding programme named Endowed City – Assist Street Sellers Rise to Their Feet. Through a novel design of mobile stations – illustrated by painters with Down's syndrome – they changed the role of street sellers from receivers of charity to providers of diversified services, including Wi-Fi hotspots, phone charging, tourist guidance, and selling fair-trade goods; in other words, instead of being perceived as a problem, as we saw in Chapter 10 in Cape Town, South Africa, they were seen as a route to solving several social problems at once. Agoood began with a goal of raising TWD 800,000 through crowd-funding, but greatly exceeded that. This success shows that such programmes can be persuasive, for they aim to solve real social issues.

Taiwan has even democratised the process of drafting legislation, with a Regulations Sandbox Platform, on which law and regulation clarification consultation services are offered to innovation practitioners. The government believes in collaborating with the civic sector, to build "a robust environment suitable for social entrepreneurship to grow, where the power of civil society could be brought into full play" in the words of Audrey Tang, the digital minister for Taiwan.[21]

She adds: "If we say that the concept of social enterprise focuses on the change of business ideas, which is from being profit-oriented to social-impact driven, then social innovation is to change the relationship between social groups through scientific innovations and technological applications, in order to devise new solutions to social issues."

We saw another such project, The g0v Air Pollution Observation Network, in the previous chapter on smart cities. Tang says that the Taiwanese government is committed to support social entrepreneurship and social innovation to link more tightly Taiwan's development with the UN's 17 Sustainable Development Goals.

As we have seen before in this book, it is when grass-roots citizens' energies and experience are yoked to the power and motivation of government leaders, rather than being pitted against each other, that seeming miracles can be achieved.

Local multiplier effect

Here is another thing that cities can support. We have mentioned the 'multiplier effect' of spending locally elsewhere in this book. Whenever you spend cash, you can either do so in locally owned businesses, or in businesses owned outside of the city area, via the Internet or through chains and multinationals. If the latter, once it leaves the area, it is less likely to return. But by supporting locally owned businesses, the same money will be spent again, and possibly again. Each time it is spent, it benefits somebody in the area – supporting a business or a job. This is called the multiplier effect, and it has been given a value, which varies according to the location, but typically between 1.4 and 2.0 times the value of the amount spent.[22]

The New Economics Foundation has developed a tool, Local Multiplier 3,[23] as a simple and understandable way for cities to measure local economic impact.

If city authorities and other organisations choose to procure their goods and services within the city, enormous benefits can therefore result depending upon the scale of their spending.

A recent example is the city of Preston in the north of England, which bucked the trend during the recent decade of austerity that has seen living standards plummet, by changing its procurement strategies along these lines. It followed the example of Cleveland, Ohio. Cleveland has a variety of local procurement programmes designed to drive contracting and purchasing to locally owned businesses. As a result, in 2014, the city spent 39 per cent of its total $147 million in contracting with businesses that are either local and small, or local and minority- or female-owned.

When Preston Council chose to emulate this as a way to challenge austerity, it found that local businesses were often too small to bid for them. The Council overcame this by splitting large tenders into smaller jobs that enabled small local companies to put in offers. For example, instead of having a tender to supply school meals in their entirety, they were broken down into puddings, main meals, fruit and snacks.

One study of the effect of this concluded: "Since 2013, over £70 million has been redirected back into the Preston economy; £200 million invested into the Lancashire economy; spending behaviour within public bodies has been transformed; and, new tools for a fairer economy have been developed."[24] In 2018 Preston rose to the top of a list of 'top improvers' British cities in a 'Good Growth Index' from consultancy PricewaterhouseCoopers.[25] The study found an improvement in employment, workers' pay, house prices, transport, the environment, work-life balance, and equality. Unemployment more than halved, from 6.5 per cent in 2014 to 3.1 per cent in 2017.

Another study concluded, about the benefits of community wealth-building:

> [It] is a local economic development strategy focused on building collaborative, inclusive, sustainable, and democratically controlled local economies. Instead of traditional economic development through public-private partnerships and private finance initiatives, which waste billions to subsidise the extraction of profits by footloose corporations with no loyalty to local communities, community wealth building supports democratic collective ownership of – and participation in – the economy through a range of institutional forms and initiatives.[26]

It sounds like a win-win strategy. And it has the additional benefit of shortening supply chains and so the carbon footprint of transport associated with the supply of goods and services. Why seek answers elsewhere when they could be right on your doorstep?

Local currencies

Another way of capitalising upon this effect is to issue a local currency or to manage a local exchange trading system. There are now many cities around the world which have their own currency. It provides an alternative economy alongside the mainstream one, and enables much greater local control. As a self-contained system, money can circulate endlessly, continuing to create economic benefits.

Amongst the most famous local currency is the Bristol Pound.[27] Bristol is a thriving city, well familiar with 'green' initiatives, in the south-west of England. With this currency you can buy a house, travel on buses and eat in cafés. In fact, it is possible to purchase just about anything using the currency, since it is accepted by over 800 businesses. Ciaran Mundy is manager of the scheme and when I interviewed him he told me that the advantages for members are that "it is a chance to invest in their own community. For the businesses, they can build a very strong and loyal customer base in their community. These businesses are often at the heart of their communities and help them realise how important they are".

How does the Bristol Pound work? Mundy explains: "Once you join, you can change pounds sterling into Bristol Pounds on a one-to-one basis. For a while, the mayor, George Ferguson, had his £51,000 salary paid entirely in Bristol Pounds." The system functions electronically but also on paper; there is a credit card and a mobile phone app. "You can make deposits in sterling and withdraw in Bristol Pounds, and vice versa. It is very easy," Mundy said.

There are, at the time of writing, around 1 million Bristol Pounds in circulation, "and most of them stay in that currency," says Mundy. Taxes and business rates can be paid in the currency, meaning that Bristol City Council can be directly supported. In return, they, alongside other organisations in the city, support it directly themselves by offering their employees the chance to take a proportion of their salaries in Bristol Pounds. Energy bills can even be paid in Bristol Pounds to the local renewable energy provider, Good Energy.

Greening the entire financial system

Much of the above is about green finance, not greening finance. To reach the global 'one planet' target, the entire financial system must be greened. Some countries such as Australia are working out how to do this. The European Union's High-Level Expert Group on Sustainable Finance, published its final report last year.[28]

It contained these recommendations:

1. A classification system, or taxonomy, to provide market clarity on what is sustainable
2. Clarifying the duties of investors' when it comes to achieving a more sustainable financial system
3. Improving disclosure by financial institutions and companies on how sustainability is factored into their decision-making
4. A label for green investment funds
5. Making sustainability part of the mandates of the Supervisory Authorities
6. A standard for green bonds.

According to Catherine Howarth, the chief executive at ShareAction, a non-profit organisation working to make the investment system a force for good, the EC is following up well on the recommendations but she has identified a significant fault line in acting on the recommendations: she says it's relatively easy to influence the "sustainable" segment of the market, like green bonds, by introducing cross-cutting anti-greenwashing tools – as the UK has confined itself to doing – but "achieving a systemic shift of capital from brown assets to green is much harder".[29]

To integrate sustainability risks in investment decision-making in order to mitigate financially material risks within a "business as usual" time horizon is difficult enough, but to actually design out negative investment-driven externalities that affect the lives of millions of people whose savings drive capital markets is the ultimate challenge.

Conclusion

All of the above initiatives are inspiring. However, they need to scale up and spread quickly in order to meet the urgent global challenges presented by climate change and the Sustainable Development Goals. Governments and businesses must change their procurement and investment strategies to support only environmental, sustainable, socially constructive and well-regulated initiatives. A great deal of learning on the part of city managers and investors is therefore needed before the dream of a 'one planet' world is realised: one where all investment is sustainable.

Notes

1 *The Sustainable Infrastructure Imperative: Financing For Better Growth And Development*, The Global Commission on the Economy and Climate, 2016, United Nations. See: www.un.org/pga/71/wp-content/uploads/sites/40/2017/02/New-Climate-Economy-Report-2016-Executive-Summary.pdf
2 *New Climate Economy Report*, Global Commission on the Economy and Climate, World Resources Institute, 2018. See: https://newclimateeconomy.report/
3 Bond Market Contact Group, *Green & Social Bond Market Update*, European Central Bank, Frankfurt, 2018, and International Capital Market Association. www.ecb.europa.eu/paym/groups/pdf/bmcg/180206/2018-02-06_-_BMCG_-_Item_3_-_Update_on_the_green_and_social_bond_market_-_ICMA.pdf
4 Climate Finance Initiative, *China Green Bond Market 2017*.
5 Gould, R., "The secret to unlocking green finance", *ISO News*, 8 May 2018. See: www.iso.org/news/ref2287.html
6 Gould, 2018, op. cit.
7 Gould, 2018, op. cit.
8 *New Climate Economy Report*, op. cit.
9 See: https://annualreport2018.cefc.com.au/governance/our-purpose/
10 IHS Markit press release, September 2018: https://news.ihsmarkit.com/press-release/energy/paris-we-have-problem-worlds-largest-economies-not-getting-policy-mix-right-avo
11 *Annual Impact Investor Survey*, Global Impact Investing Network, 2018. See: https://thegiin.org/research/publication/annualsurvey2018

12 See: www.worldbenchmarkingalliance.org/the-start-of-the-impact-management-project-network/
13 Some experts advocate floor area ratio (FAR) as a useful measure, but it is an unreliable density control metric in residential areas since floor area per person varies (usually increasing with income). In order to measure and satisfy population demands, particularly for planning mass rapid transit services, but also water and energy, the index of floor area per person is more useful.
14 Thapa, M., et al., "Assessment of Element of Sustainable City for Urban Environmental Management of Kathmandu Valley", *Conference Proceedings*, ADNEC, Abu Dhabi, UAE: Ecocities in Challenging Environments, 2015.
15 *Better Finance, Better World*, Blended Finance Task Force, Business & Sustainable Development Commission, 2018. See: http://report.businesscommission.org/
16 For full list see: http://eemap.energyefficientmortgages.eu/pioneers/
17 Marco Marijewycz, M., et al., *Creating an Energy Efficient Mortgage for Europe: Consumer Research Insights*, E.ON, EeMAP, 2018. See: http://eemap.energyefficientmortgages.eu/wp-content/uploads/2018/04/EeMAP_D2.7_E.ON_Final.pdf
18 Richardson, S., *Creating An Energy Efficient Mortgage For Europe: Towards A New Market Standard*, WorldGBC Europe, 2018. See: www.worldgbc.org/news-media/creating-energy-efficient-mortgage-europe-towards-new-market-standard
19 Ibid.
20 Yunus, M., *A World of Three Zeroes: The new economics of zero poverty, zero unemployment, and zero carbon emissions*, Scribe, 2017.
21 Tang, A., "Social Innovation in Taiwan", *Medium*, February 2018. See: https://medium.com/@audrey.tang/social-enterprise-in-taiwan-3eb96d4dc8a7
22 Miller, P., *Economic Multipliers: How Communities Can Use Them for Planning Wayne*, University of Arkansas, Division of Agriculture, Community & Economic Development, 2014. See: www.uaex.edu/business-communities/economic-development/FSCED6.pdf
23 See: www.nefconsulting.com/our-services/evaluation-impact-assessment/prove-and-improve-toolkits/local-multiplier-3/
24 *The Preston Model*, The Centre for Local Economic Strategies. See: cles.org.uk/the-preston-model
25 Goode, V., et al., *Good Growth for Cities 2018: A report on urban economic wellbeing*, PwC and Demos, 2018. See: pwc.co.uk/government-public-sector/good-growth/assets/pdf/good-growth-for-gities-2018.pdf
26 Hanna, T., et al., "The 'Preston Model' and the modern politics of municipal socialism", *Open Democracy*, 12 June 2018. See: www.opendemocracy.net/neweconomics/preston-model-modern-politics-municipal-socialism
27 See: bristolpound.org
28 See: https://ec.europa.eu/info/publications/180131-sustainable-finance-report_en
29 See: https://www.euractiv.com/section/energy-environment/opinion/the-beacon-of-sustainable-finance-in-europe-must-not-lose-its-flame/

15

A MENU OF CASE STUDIES

There are no places in the world that I am aware of which are measurably 'one planet'.

I'm sure that there are many clusters of people living very good lives within planetary boundaries, primarily because they consume little and what they do consume comes mainly from their local environment, which they care for. Usually they will be in fairly remote areas, and not urban; their impacts are unlikely to be monitored.

Therefore in choosing the case studies to include in this chapter I had to be selective. Each one of them has some qualities and features that would be desirable for a 'one planet' neighbourhood, town or city. Since every community is different, they will want to choose different aspects from amongst the following. It's possible to think of it as a menu, although there's no implication that it is exhaustive or exclusive. For each example I have endeavoured to highlight the chief features which I regard as useful for our purposes.

I thought it worth ending with one which demonstrates the kind of slips that are possible, because we need to learn from the mistakes of others as well as their successes.

The Baugruppen model

This case study focuses on the provision of affordable social housing.

Useful aspects: The involvement of residents in the design of their neighbourhood and housing, which fosters community cohesion; the zero-profit housing model; the support of local authorities; affordable housing; the use of non-financial criteria in selecting contractors to work with.

Baugruppen means "group build" in German. The model originated in Germany. It is a process that enables individuals to group together to become their own designer and developer. They deliver custom-built and individually designed homes and communities. The future residents design and develop the community, according to their long-term needs, rather than investors doing it, who typically prioritise their own economic benefits. The process of working together in advance of construction helps to create a sense of community, as members collaborate on identifying their own needs and designing their homes and shared spaces.

In the last decade, Germany has seen over 1,800 Baugruppen developments. With their increasing popularity, the city of Hamburg has set aside 25 per cent of its land for Baugruppen developments. Other countries are fast copying this model because of its success in solving multiple social and housing problems.

Baugruppen is a 'zero profit' housing model that has the potential to deliver higher-quality, more sustainable homes, designed for long-term needs rather than profit. Traditional developer building costs include:

- Land: 15–20 per cent
- Construction: 45–50 per cent
- Finance and holding: 8–10 per cent
- Fees and marketing: 6–8 per cent
- Developer profit: 15–20 per cent.

With Baugruppen there is no developer, so no marketing costs or developer profits. This allows for up to 28 per cent savings, which makes it possible to develop higher-quality accommodation with the same or a smaller budget. This was proven in a study of six self-build projects in southern England, which all resulted in significant financial benefits.

Local authorities can support individual Baugruppen projects by offering access to cheap land, and this model can be applied anywhere, regardless of the financial scope of a local authority. The latest research shows that if local governments want to solve their housing crises they must take a more proactive, participatory role and engage not just in house building, but community building.

By stipulating in tenders for architects and contractors to provide the homes that ecological and social criteria are valued over profit helps to build a market for companies that are motivated by such concerns. The contracts limit profits for builders to 15 per cent.

Three examples:

So.vie.so

The So.vie.so development in Vienna[1] consists of 111 subsidised rented apartments, communal facilities of different size, shared green space with neighbouring housing schemes and spaces for small businesses, on former (brownfield) railway land. A housing co-op was formed for the residents to support the planning process and on-going management.

In the pre-build stage residents collaborated on planning their communal spaces. Different task groups were formed, and, over a year of planning, the facilitators provided by the housing trust gradually withdrew. This allowed the now-experienced group members to take over tasks such as maintaining the communication processes, and organising regular meetings to decide upon the allocation and use of funds or maintenance activities.

This is unlike many new developments, in which the inhabitants find themselves thrust alongside strangers with whom it may take years to form a sense of community. Here, the community is formed from the start, and with it a sense of responsibility for looking after the neighbourhood. In this development it is possible to choose from a number of different types of apartments, all of which are constructed according to Passivhaus principles.

To finance the scheme, those joining pay an initial, one-off deposit. This goes towards the land and construction costs, usually between €15,000 and €30,000 for a medium-sized to large flat. The higher the initial contribution, the lower the monthly rent. At the end of the tenancy, this deposit is paid back. With Baugruppen generally, some co-op schemes offer buy-out options to tenants. High earners are discouraged, with a ceiling for tenant salaries. The rents are fixed for ten years and existing rent contracts can be extended beyond then.

Vauban

As described in my previous book, *The One Planet Life*,[2] in the city of Freiburg, Germany, the city council made a conscious decision that developmental rights in the Vauban district (affordable and sustainable housing) would be preferentially given to Baugruppen rather than developers. The city and working group felt that prioritising affordability (through collaboratively built projects), would make it attractive and feasible financially for families to live there.

Rather than bidding wars, lots were awarded to future inhabitants who met certain sustainability criteria. These included: social diversity, most ecologically sound. The city council provided facilitators to help a Baugruppe procure legal and financial representation for their project. The development has a population density of around 5,300 inhabitants over 38 ha, but it has an open feel.

Vienna Wild Garden Housing Project

Vienna Wild Garden Housing Project is for "people who want to shape their living and living situation" – single, family or senior.[3] As usual, future inhabitants got to know each other in the process of participatory planning to develop and design, together with a planning team, a cross-generational, mixed community.

The planning team consisted of a single architect, a reality lab to test building management, a landscape architect and a developer. A common garden, meadow and wild hedge are included for the benefit of the inhabitants. It is located about 30 minutes from Vienna city centre, but cars are only permitted to remain at the edge of the neighbourhood and park in collective garages under the buildings, as in Vauban. Electric car-sharing, cycle parks and new public transport stations are provided.

There are approximately 1,100 dwelling units altogether over 26.5 acres, and 11.37 acres of green and open spaces. The buildings include semi-detached houses to a multi-storey residential building. There is a neighbourhood centre, and an outdoor nature kindergarten, approximately 200 rented properties, a local energy supply, 80 owner-occupied apartments, and 50 self-financed apartment buildings.

Vienna eschews conventional developers for housing developments: housing co-operatives, combined with state-owned apartments, make up an amazing 60 per cent of Vienna's housing today. This and other case studies and model policies from more than 80 cities in 35 countries can be found in *Sharing Cities: Activating the Urban Commons*. [4]

Heathcott Road, Leicester, UK

This case study focuses on the provision of an affordable, Passivhaus-standard, social housing estate with gardens for growing food.

Useful aspects: The necessity for key driving individual(s); comprehensive community consultation; asset transfer from the local government; 100 per cent affordable housing; leasing of land to a developer; training; and the link to food production.

This development contains 68 Passivhaus-standard semi-detached homes, all at an affordable rent that is set at about 20 per cent less than the assessed local market value. These homes can be heated for as little as £13 per year. There are four one-bedroom flats and 23 two-bedroom, 20 three-bedroom and three four-bedroom houses. The first residents moved in in June 2017.

The project is the brainchild of Neil Hodgkin, head of development for the resource centre Saffron Lane Neighbourhood Council (SLNC), a charity with aims including the relief of poverty, the environment, conservation and heritage, and economic, community development and employment.

The area is one of the most deprived in the city of Leicester. In 2007 Hodgkin had an idea for an urban community farm to grow vegetables for SLNC's day-care service users. From this, Saffron Acres was born: allotments and a community garden that provide education and volunteering opportunities. Now, fruit grown on Saffron Acres is turned into preserves to be sold as part of a project providing skills training for local unemployed people and adults with learning difficulties.

Determined to rejuvenate the area, Hodgkin identified housing as a key issue. SLNC embarked on a lengthy process of consultation with hundreds of local residents about the area's housing needs. It acquired 22 acres of former derelict allotment land as an asset transfer from Leicester City Council for £1. Asset transfers of their land like this are an extremely useful way in which local governments can help regenerate areas in partnership with the community. This land is covenanted so that it can never be used for private profit. It will always be held in trust by SLNC.

SLNC oversaw the project, leasing the land to a developer, and engaged the architects and builders. The housing area is 13.3 acres. The cost was £9 million. Construction took 70 weeks with 40 builders, four of whom were solely dedicated to achieving the Passivhaus credentials. The project has won several awards.

Next to the houses is a permaculture farm, intended to provide education on food growing, cooking and healthy eating, an allotment, beehives, a flower meadow, rejuvenation of field ponds, reinstating of hedgerows and fruit tree planting. Residents are encouraged to work and grow their own fruit, vegetables and supplies. Existing community gardens are next door. This makes the development close to 'one planet' without a specific policy mandating it, or verification measurement for ecological footprinting.

The income generated by the development pays for a full-time debt and welfare support officer who is also on-site and plays a key role in advising and liaising with the community. Buoyed by the success of the development, SLNC has a further £1.6 million project to build 20 more housing units. The income from these will help to pay for a further two staff to do more work in the community.

Resident Claire says that the fact that the Resource Centre is so firmly integrated with the development is a real benefit. "The Centre has lots of services we can use; my next plan is to see more of what's in the area and become part of the community. I can start working. I feel like I've really landed on my feet. Being here has changed our lives and opened up everything."

"This project shows that communities can plan, deliver and manage their own housing and address specific wider social needs," said Hodgkin. "Retaining money within the community to directly deliver services within the community to help solve local social issues can offer longer-term solutions towards sustainable regeneration of neighbourhoods."[5]

The Cannery, Sacramento, USA

This case study focuses on the provision of a larger, low-carbon, mixed-use development with a farm and agricultural college, plus sustainable transport.

Useful aspects: An inclusive multigenerational approach to residential development built around a farm, agricultural training college, community-supported agriculture, some affordable housing, jobs, renewable energy, energy efficiency, low-carbon transport, with the support of the city government. The financial approach is conventional: developer investment recouped from high-value sales and retail outlet rent.

We met The Cannery in Chapter 5. Here is a little more detail about this project which stops short of 'one planet' status because it has no monitoring of residents' ecological footprints, nor are they incentivised to reduce these.

The Cannery is on the site of a former tomato cannery (brownfield site) in Davis, on the outskirts of Sacramento near San Francisco, California. It is similar to a garden village. Covering approximately 100 acres of land, it contains a mix of land uses consisting of low-, medium- and high-density residential; a mixed-use business park; stormwater drainage retention; greenbelts, agricultural buffers, an urban farm, parks; and a neighbourhood centre. There are 583 residences, with an average density of 9.5 units per acre; with many sizes, types, densities and styles of housing, including ownership and rental, detached and attached homes in low, medium and high densities ranging from three to 30 units per acre.

Together, these sites could accommodate employment opportunities for approximately 600 to 850 jobs, reducing the need for travel by the occupants. The Cannery combines environmental engineering and landscape architecture elements into a neighbourhood plan. It contains five districts:

1. The Cannery Farm District
2. The Cannery Commerce District
3. The Urban Residential District
4. The Traditional Neighborhoods District
5. The Neighborhood Park District.

A 7.42 acre urban farm is included as a community asset and as a transition between urban uses and adjacent agricultural land. This is part of 20.8 acres of open space consisting of the open space/bioswale, agricultural buffer on the north edge, agricultural buffer/urban farm on the east edge and greenbelts. It is an adaptive reuse and redevelopment of a former industrial site located within the city limits.

It is owned by the New Home Company Inc., a fully commercial private company. New Home plans to deed the land to the City of Davis, which will lease it to the Center for Land-Based Learning, a not-for-profit.

Water and sewerage services are provided by the City of Davis. Planning the development was achieved in full cooperation with the planning department under normal processes, but was favoured by City Hall's policy approach. All the planning documents are on the city's website.[6]

The sale of market-price houses supplemented the affordable housing. There are 110 affordable homes (16 per cent), including 45 units suitable for rental to very low-, low- and moderate-income households.

For the energy needs, the neighbourhood design includes street layouts, building orientation and landscaping designed to accommodate passive and active solar energy systems, and to capture natural cooling and heating opportunities. Design treatments for passive solar are balanced with the neighbourhood's overall objective of reducing heating and cooling demands and providing solar-ready rooftops on south-facing roofs.

Energy efficiency measures increase the buildings' performance, liveability and comfort well beyond the city's minimum requirement of the 2010 California Green Building Standards (Cal Green) Tier 1 requirement. Residential uses exceed California's 2008 Title 24 Energy Code by 40 per cent, which is equivalent to 33 per cent greater than 2010 Cal Green Tier II requirements. The mixed-use site will exceed California's 2008 Title 24 Energy Code by 15–20 per cent.

All single-family detached and attached homes will have a 1.5 kW system installed at initial construction with the option to upgrade if desired, upgrading to 'net zero living'.

On transport, 9.9 miles of on-site bicycle and pedestrian improvements that are connected to existing cycle and pedestrian links into the city centre have been built. All places are no more than a ten-minute walk or a five-minute bicycle ride from one another. Every residence is within approximately 300 ft of a trail, park, greenbelt or open space area.

As if this was not all excellent enough, the district is planned around an agricultural college that teaches young residents how to grow food on the land adjacent. The college's mission is "to inspire, educate, and cultivate future generations of farmers, agricultural leaders and natural resource stewards". It is managed by the Center for Land-Based Learning, which "cultivates opportunity for youth, for agriculture, for business, for the environment" and runs the California Farm Academy to help those wanting to break into a career in agriculture. There are two farming businesses and three farmers at Cannery Farm. Residents can buy food in a veggie box Community Supported Agriculture scheme. Food is also sold in a local market.

Garden cities

This case study focuses on the provision of large-scale regenerative developments.

Useful aspects: Building regenerative, zero carbon communities around gardens and food production.

Ebenezer Howard was a British socialist reformer and the founder of the Garden City movement, whose idea was to provide "a web of solutions as to how we could live, from energy and local food to access to green space and health care". In the 19th century, Howard founded the Garden City Association, which became the Town and Country Planning Association (TCPA). Finance for the cities came from rich investors who received interest from rents. He tried to obtain financial support from working-class co-operative organisations but didn't succeed. As a result he had to ditch his plans for a co-operative ownership scheme with no landlords, and introduce short-term rent increases.

This lack of control over ownership meant that later these lovely homes in desirable places became priced out of the reach of the working class. Nowadays, the formation of Community Land Trusts to manage these ventures is advocated. These are able to take advantage of the rise in land value caused by the development, while being able, in principle at least, to attach conditions that mean that the homes they rent out cannot be sold later at a vast profit.

When the time came to build the forerunner of garden cities, Welwyn Garden City in England, a novel aspect of it was that its residents, via a community fund, would share in the profits of the development process, rather than just the developers. This meant that there would always be finance available to look after the town. Residents also shared in the governance and development process, and could join workers' co-operatives (such as a general store and a building and joinery firm) which gave them employment, and cultural, artistic and health-supporting activities. With a huge amount of effort these were made an intrinsic part of the development process so that the city would be a rewarding place to live on all levels.

Welwyn Garden City ended up quite different from Howard's original ideal for many reasons. Now a city of over 100,000 people, it is a Garden City in form but not practice; unfortunately its agricultural belt never happened. Howard thought that the Garden City should be socially, economically and ecologically sustainable. For example, land should be bought at low, agricultural values and all houses and factories should be leaseholds. Investors would be paid back from the rent, and any increase in land values would benefit the whole community. Extra money from the rent could then finance pension funds and community services.

The TCPA draws six recommendations, or lessons to be learned, from the above experiences:

1. the need for a dedicated, local, planning consent process;
2. the up-front development cost was in every case repaid by the value of the towns, however they were initially financed;
3. correctly composed partnerships can deliver the towns quickly;
4. the importance of good, clear design and master planning;
5. the need for a long-term plan for stewardship;
6. the need for public support and involvement.

Fast forward to 2007, and the UK Government proposed a series of 'eco-towns'. Unlike eco-villages, these were top-down, not bottom-up proposals. Although many failed to materialise (mainly because lessons 3 and 6 above were not heeded), they did result in a set of eco-town standards[7] and some developments such as Bicester,[8] Whitehill and Borden.

The standards stated that eco-villages should be: zero carbon, resilient to climate change, 30 per cent of homes should be 'affordable', have low-carbon transport with local jobs, that 40 per cent of the land should be green space and increase biodiversity, including: "particular attention should be given to land to allow the local production of food from community, allotment and/or commercial gardens." They should also be water-efficient and flood-resistant and manage their waste efficiently. Finally, they should "develop their master plan" with "a high level of engagement and consultation with prospective and neighbouring communities". Tragically, the Conservative-led coalition government scrapped these standards in 2010.

Modern garden cities and 'one planet' developments

The principles adopted for modern garden cities are similar and build on what has gone before, but for the TCPA the food and water supply and treatment elements appear to be lacking the emphasis they are due. However, this concept can be combined with 'one planet developments'. These typically use agro-ecological cultivation and agro-forestry to provide many of the land-based businesses, supplying local communities with their produce and services. This automatically improves biodiversity, soil condition and sequesters carbon in the soil.

Net carbon sequestration could be a vital function of regenerative garden cities, if this were adopted as a principle. In 2014 a plan to give garden city status to up to 40 English towns won the £250,000 Wolfson Economics Prize. Winner David Rudlin imagined a fictional town called Uxcester, and applied that concept to Oxford (2011 population: 150,000). Rudlin's financial model shows that for every plot developed, the same area again could be allocated for parks and gardens which are publicly accessible to the whole community rather than kept in private hands; 20 per cent of new homes would be affordable housing.[9]

Several garden towns are now in development in the UK. In 2016 Oxfordshire Council proposed a 'garden village' extension to Eynsham, which received initial government backing. It adopted some of Rudlin's ideas, plus the idea that it could generate much of its energy from solar power and that half of its homes be 'affordable'. The council proposed: "There is significant scope for community ownership/involvement in managing new community infrastructure such as allotments, green infrastructure, community energy, community facilities etc. The establishment of a community management company can provide a key vehicle to achieve this."[10]

In the south-east of England, Braintree District Council, Colchester Borough Council, Tendring District Council and Essex County Council have all agreed to work together to create three new 'Garden Communities' of 30,000–40,000 new homes alongside significant new infrastructure, jobs and associated facilities. These will be based on the ethos of garden cities being promoted by the TCPA. An overarching wholly council-owned body, North Essex Garden Communities Limited, will coordinate the development of the sites and the strategic approach to growth, funding and infrastructure delivery. It will establish further companies (called Local Delivery Vehicles) for each proposed Garden Community, working with landowners and local communities to bring forward schemes that truly deliver on garden city principles. Capacity funding has come from central government.

The councils have also agreed to provide funding of up to £460 million to "pay for delivery of infrastructure in a more timely and co-ordinated way than could be achieved via traditional development models".[11] Specific sites and boundaries have not yet been determined but will be refined through the Local Plan decision-making process. However, already, transport plans are being drawn up with an aim that 24 per cent of trips will be through active means (walking, cycling), with 38 per cent by bus rapid transit and the same proportion by private vehicle.

Housing association Bournville Village Trust (BVT), situated at Lightmoor garden village, is developing another modern urban village in Telford, where the Industrial Revolution began. The new village is a joint venture between BVT and the Homes and Communities Agency (HCA). It will eventually comprise around 1,000 homes, a primary school, community centre, health centre, nursery and shops, as well as parks and numerous green open spaces. BVT provides services to 8,000 homes of mixed tenure and 25,000 people. The Trust also manages agricultural estates totalling around 2,500 acres, so is able to provide food for the village. It uses a system of covenants to preserve its stewardship, which include a maintenance charge. Charmingly, the management committee plants a tree in the community orchard every time a child is born in the village – an idea first thought of by George Cadbury, the founder of Cadbury's chocolate company, which he housed in the 'Bournville factory in a garden', the Bournville garden village for its workers, in 1895, and the Trust itself, to manage it.[12]

The delivery body for 'one planet' garden towns can therefore be anything from a large development corporation to a CLT owned by the community. Whatever they are, they should:

- Have consistent leadership, ownership, and monitoring provision, and be in place (or set up a not-for-profit stewardship body to be in place) in perpetuity;
- Have commitment to the garden city principles;
- secure the funding to deliver the business plan;
- Work in a transparent and inclusive manner, communicating and involving actively everyone affected;
- Be able to broker agreements and cajole delivery by its partners, including through the use of statutory powers where appropriate (e.g. planning and compulsory purchase).

In the absence of public finance for a garden city, it is still possible to use Howard's model today. A study by Wei Yang and Partners and Peter Freeman[13] for the Wolfson Economics Prize showed that over 15 years the sale of some land acquired at near agricultural land prices for private housing in the garden village could permit the provision of land for 30 per cent affordable housing plus parks, leisure, transport networks, utilities, school and other community facilities, at no cost.

Clear targets for social-rented homes and other forms of sub-market housing should be set. These targets should be dictated by local circumstances, revealed by analysis (e.g. homelessness, waiting lists). As a benchmark, new garden cities should aspire to 30 per cent of homes available for social rent; other forms of sub-market housing, such as shared-equity and low-cost or discounted ownership, should form a further 30 per cent of homes; and there should be a clear mechanism to ensure that such housing is made available in perpetuity.

This is a demanding standard, but, as the new towns demonstrated, it can be delivered if bodies are able to offer cheap land to social-rent providers. Between 69 per cent (at Basildon) and 97 per cent (at Peterlee) of housing in the Mark One new towns, built just after the Second World War, was for social rent, although it is worth noting that both these examples were large-scale, new developments.

New garden cities present a rare opportunity to provide a diverse range of housing types and tenures, delivered by a range of providers – from garden city development corporations or local authority housing companies in partnerships with housing associations, private-sector house builders and small and medium-sized enterprises through to smaller, citizen-led models such as co-operatives and CLTs.

Community co-housing and CLTs are both models that could be adopted. Letchworth and Welwyn both included co-partnership housing models that provided a unique form of tenure combining features of a tenant co-operative with a limited-dividend company, whilst in continental Europe many such schemes have been developed in recent years. While co-housing is not yet mainstream in Britain, it is growing in popularity and success, as schemes such as Cannock Mill in Colchester demonstrate.

Barcelona: the world's most radical city?

This case study focuses on an entire city, from a people-centred angle.

Spain's Barcelona hosts many examples of new ways of running cities and of citizen involvement in civic and economic activities, and is actively promoting itself as an inspiration for other cities to follow. The city has a long, radical tradition going back to the anarchist collectives documented by George Orwell in *Homage to Catalonia*, his book about his experiences fighting alongside the anarchists against the fascist forces of General Franco. It is unsurprising, then, that, following the particularly severe effect upon Spain of the banking crisis of ten years ago, creative grass-roots responses to austerity have emerged.

Barcelona is home to a radical citizen-led movement that coalesced in June 2014 under the 'Yes we can' (*Podemos*) slogan into the platform *Barcelona en Comú*, an organisational structure for individuals, activist groups and political parties. This linked networks of local assemblies allowing people to engage in policy decisions. Ada Colau, a former housing activist, astonished everyone when in June 2015, as part of *Barcelona en Comú*, she was elected mayor, the first woman to hold the office. "Democracy was born at local level, and that's where we can win it back," she declared.

She had been a founder of the *Plataforma de Afectados por la Hipoteca* (Platform for People Affected by Mortgages) that was set up in 2009 in response to the rise in evictions caused by unpaid mortgage loans and the collapse of the Spanish property market. (She co-wrote a book, *Mortgaged Lives*, based on her experiences.) In one of her first speeches Colau called for "an end of the political class removed from the people". She was not alone: the same year saw radical mayors elected in Madrid, Valencia, Zaragoza and La Coruña, and together they announced the 'Rebel Cities' network: a group of cities confronting central government, devising their own policies, and making a worldwide plea for other cities to join. A handbook is available for other cities to follow.[14]

From the same movement that gave birth to Colau came the Catalan Integral Co-operative (CIC) ('Integral' is perhaps best translated as 'holistic'). Its goal is to build an anti-capitalist co-operative structure not just for the benefit of its own fee-paying members but for the Commons as a whole. "The main objective of the CIC is nothing less than to build an alternative economy capable of satisfying the needs of the local community more effectively than the existing system, thereby creating the conditions for the transition to a post-capitalist mode of organization of social and economic life," writes George Dafermos, author of a report[15] about it. It has been actively involved in developing infrastructures as diverse as barter markets, a network of common stores, an alternative currency called the 'eco', a 'Cooperative Social Fund' for financing community projects and a 'basic income programme' for paying members for their work. Its activities are not confined to Barcelona, but extend across Catalonia.

The CIC is a collection of about ten committees with responsibilities for different topics. For example, the Economic Management Committee is responsible for the economic management of the co-operative, the Legal Committee manages legal matters, the IT Committee deals with the IT infrastructure and so on. Each works largely autonomously but to coordinate their activities, the co-op holds 'permanent assemblies' once a month where members make collective decisions based on consensus.

It has about 600 'self-employed members'. There are also 20 'self-managed pantries' run by local consumer groups wishing to purchase products made locally or by producers associated in other parts of Catalonia, chosen through an online list of over 1,000 items supplied by currently 70 producers and distributed by vans. According to Dafermos, the co-op is "based on direct exchange and the use of alternative community currencies. The way this ecosystem operates represents the model of the autonomous public market envisioned as a means of satisfying the needs of the local community … a model for the transition to a post-capitalist economy".

A minimum income scheme

This radicalism extends to the official level. The city is one of several places in the world that are trialling a 'minimum income' scheme – B-MINCOME – in two of the city's poorest neighbourhoods. Here, citizens receive a guaranteed minimum level of income. Receipt for some of them is conditional upon agreeing to some level of community work, by volunteering. Others have other conditions, or none at all, and the results of the trial will be evaluated to determine the most successful model.

The designers of the scheme – which is supported by a grant from Urban Innovative Actions, a European Commission initiative that supports projects investigating "innovative and creative solutions" in urban areas – took experience from the governments of Finland, the Canadian province of Ontario, and the Dutch municipality of Utrecht, all of which have designed guaranteed income experiments in their own areas.

Smart city

Barcelona is also smart in the digital and ecological senses of the word. As one of the leading smart cities worldwide,[16] 50 per cent of street lighting is LEDs fitted with sensors to switch on when they detect motion and dim when streets are empty, saving 30 per cent of previous energy. Some 19,500 smart meters monitoring and optimising energy consumption have been installed across the city, including a sensor system helping drivers to locate available parking spaces, reducing congestion and emissions.

There is a Bicing app, providing updated information on the location of public bike stations and bike availability, and the city has one of the biggest free public Wi-Fi networks in Europe. Smart technology is also used to improve the speed and efficiency of the city's new 'orthogonal bus network', and digital bus shelters are also in place. All of this is helping with the superblock project in the city, described in more detail in Chapter 12.

Now calling itself a 'fearless city', Barcelona is positioning itself as a focus for a movement. According to Dr Bertie Russell, research fellow at the Urban Institute in the University of Sheffield, Barcelona's and Madrid's 'Decidim' process of citizen involvement in decision-making is good because it allows citizens to set the policy agenda, not just react to it.[17]

Barcelona – home of Antoni Gaudí – is continuing to be every bit as revolutionary as that unique man's architectural style, pioneering 21st-century solutions that address the kind of citizen disillusionment with power that has fuelled reactionary movements elsewhere in the world in the last two or three years. But by positioning itself within an alternative movement, it is determined that its ideas can be replicated and supported elsewhere.

Examples from developing countries

Across three continents, citizen-led struggles are forcing municipalities to work with them to build more sustainable futures for themselves.

The housing category of the 2018 Transformative Cities award[18] was won by mass housing projects in Solapur, India, that have provided affordable housing to thousands of workers previously living in slums. By forming co-operatives and by persuading the federal and state governments to dedicate funds for the purpose, these workers have helped build over 15,000 houses since 2001. Another 30,000 houses began construction in January 2018 to be completed in four years.

Many of these homes – typical size around 555 ft^2 (around 50 m^2) – are for beedi workers, cigarette-rolling women who are often the sole bread-winners for their families, low-paid, and previously renting tiny shanties in slums. The land-purchase cost was shared equally by the worker, the central government and the state government, but the workers had to wage a long struggle to win their demand and have previous debts cancelled.

The awards prove that the sheer strength of workers' sustained movements, with the cooperation of governments, can deliver results. On the other side of the world, in Bolivia, every day the residents of San Pedro Magisterio village used to have to fetch water from springs near their polluted river. The grass-roots community organisation founded a water co-operative, drilled wells and built the basic infrastructure to bring water to their homes. The funding to solve all these problems came from contributions made by the community itself, who did all the work too.

They followed this with a long campaign to build a wastewater treatment plant to clean up the highly polluted river. The community ended up planting 500 reeds to set up a reed bed ecological sewage treatment system serving 4,000 people. Resident Doña Magui says they are now trying to replace the reeds with arum lilies because they perform the same function and will help keep the treatment plant going, because residents will sell the lilies and plough the profit back into maintenance. "As far as the state is concerned, we don't exist," says Doña Magui, adding that it was the residents themselves who built the first school, the church and the first roads.

The Spanish city of Cádiz also won an award for its action plan against energy poverty. This campaign featured active cooperation between local government leaders and ordinary citizens. A group of unemployed citizens was trained as energy advisers and given an eight-month contract by the city council, tackling unemployment, energy poverty and climate change at once. The energy advice team gives families in Cádiz advice on how to optimise their energy contracts so they pay as little as possible. In just three months, the team ran 60 workshops, gave 640 people training on energy issues, and advised 70 families in their homes, reducing their electricity bills by 20–50 per cent.

Villages in transition

This short case study focuses on community participation, agriculture and energy for town regeneration.

Luzy in France is part of another network, the Village in Transition[19] movement. It boasts a farmers' corner, associative café, a donation shed, all run by citizens, with municipality support. Luzy is also a member of POTEs (Ordinary Energy Transition Pioneers), fostered by pan-European initiative Energy Cities. Its spokesperson, Carine Dartiguepeyrou, says POTEs "are every-day-life innovators and visionaries working in areas related to the energy transition that Energy Cities, the Bourgogne Franche-Comté Region and ADEME, the French national energy conservation and environment agency, are forming into a network". She adds: "POTEs are efficient and innovative ... they collaborate and take care of others by helping them make progress in their project and overcome difficulties."[20]

She explains that, "A good example of this is the 'hold-up' method, a collective brainstorming process which resulted in the café and the farmers' market at Luzy. Starting with a challenge faced by each project, the participants put forward solutions to help find a new business model and the farmers' market to perpetuate its activity. In order to solve challenges faced by project leaders (social entrepreneurs, researchers, engineers, government, NGOs ...), the actors who participate discuss and exchange ideas until they arrive at solutions."

Eco-cities compared

How do so-called eco-cities compare? Could any be said to be anywhere near the same as 'one planet' cities? Researchers at the Singapore Centre for Sustainable Asian Cities conducted a comparative study in 2015. [21]

The researchers compared the properties of 200 different eco-cities around the world and found that they fell into three basic categories depending on the motivations for creating them, whether they were social, economic, to do with identity marketing, or 'professed corporate social responsibility'. The three categories were:

1. Social and community focused, emphasising the provision of public or social housing, community facilities and amenities (social infrastructure), which are typically government-initiated projects such as those in Singapore and the Hong Kong townships.
2. Environmentally focused, with declared intentions to achieve specific goals of environmental sustainability, for example, low carbon footprint and optimisation of resource use and consumption, tackling the 'closed-loop metabolism' that we have looked at in chapters on resource efficiency and industry. Falling into this group: the Chenjia eco-community, Hammarby Sjöstad, Wuxi Eco-City, Yuelai Eco-City, and the Changxindian Low Carbon Community.
3. Technology focused, featuring inclusion of technological innovations such as renewable energy, waste management and transport systems with design and investment motivated by a wish to testbed and implement new technologies. Examples include: Masdar City, Tianjin Eco-City, Nanjing High Tech Island, Dongtan Eco-City, Hongqiao New Central Business District, and Tangshan Bay Eco-City.

No one city is entirely focused on one category and in reality they may possess the features of all three. The Singapore researchers looked in detail at three of them:

- Hammarby Sjöstad in Stockholm, Sweden, a harbour brownfield site, redeveloped according to a resource flow model with an integrated, closed-loop system of flows of energy, water and waste.
- Wuxi Eco-City, China, a demonstrator low-carbon eco-city, developed in partnership with Sweden to implement clean-tech solutions in the built environment. Waste is sorted and recycled and non-recyclable waste incinerated in a combined cooling, heating and power plant (CCHP) to produce electricity and heat.
- Punggol Eco-Town, Singapore, a high-rise public housing project of Singapore's Housing and Development Board: the "first eco-town in Singapore", integrated with 'ecological' features, a waterfront concept, and green areas, hoping to balance three components: environmental (clean, healthy and comfortable environment, greener transportation, enhanced urban greenery

and biodiversity, reduced waste, water sufficiency, etc.); economic ("sustainable and inclusive growth"); and social.

Here's how they compare on six different categories and seven performance indicators:

1 Resource conservation and management

Hammarby has an ambitious waste-reduction target, and included extensive development strategies of waste-to-energy systems. Wuxi's targets are not as clearly expressed, but nonetheless included definitive efforts for conversion of non-recyclable waste. Punggol's waste targets are pegged to the national recycling rates, but included pragmatic and implementable design proposals to achieve them.

Performance indicator: Reduce waste to landfill

Hammarby Sjöstad: Reduce waste to landfill by 90 per cent.

Wuxi Eco-City: Aiming for all household waste to be sorted.

Punggol Eco-Town: National goal of attaining 70 per cent recycling rate by 2030. (In 2013, the recycling rate nationwide was 61 per cent, approximately 20 per cent of which was household waste.)

Relevant planning strategies:
- Converting organic waste into biosolids for fuels and fertiliser;
- Converting combustible waste for district heating, cooling and electricity;
- Using automated waste disposal systems;
- Providing area-based collection points for hazardous waste;
- Building a centre for environmental information to encourage eco-friendly lifestyles;
- Half of waste that is non-recyclable to be incinerated by CCHP for a local network;
- All organic waste and sludge from wastewater treatment processed and converted to biogas to fuel bus services;
- Provision of separate chutes for waste recycling in common areas;
- Equipping the town with pneumatic waste conveyance systems to increase efficiency and convenience;
- Providing incentives to reduce food waste.

Performance indicator: Household water consumption

Hammarby Sjöstad: 100 litres/capita/day. (In 2007 water consumption was 150 litres/capita/day.)

Wuxi Eco-City: 120 litres/capita/day.

Punggol Eco-Town: Nationwide target of 140 litres/capita/day. (In 2013, water consumption was 151 litres/capita/day.)

Relevant planning strategies:
- Eco-friendly installations;
- Low-flush toilets;
- Recycling greywater and stormwater;

- Reusing greywater and rainwater harvesting;
- Improved efficiency in water usage;
- Smart water meters.

2 Protecting the natural environment

Performance indicator: Provision of public green space
   Hammarby Sjöstad: 15 m^2 of green courtyard space per apartment unit.
   Wuxi Eco-City: 15 m^2 of green courtyard space per apartment unit.
   Punggol Eco-Town: Park provision ratio of 8 m^2 per capita.
   Relevant planning strategies:
- Enhancing the existing natural environments;
- Providing neighbourhood parks and common greens within the residential precincts;
- Redeveloping, reusing and transforming old brownfield sites into green public spaces;
- Involving local community in the development of the green commons;
- Recycling greywater and stormwater;
- Providing better connections between parks through the park connector network;
- Provision of habitats and routes for birds and butterflies.

3 Climate change mitigation and adaptation

All have energy consumption reduction targets. Hammarby invested in renewable energy; Wuxi its closed-loop waste-to-energy plant; Punggol prioritises energy efficiency.
   Performance indicator: Building energy consumption
   Hammarby Sjöstad: 50 kWh/m^2/annum.
   Wuxi Eco-City: ≤85 kWh/m^2/annum for residential buildings; ≤100–125 kWh/m^2/annum for public buildings.
   Punggol Eco-Town: National standard of 20 per cent energy reduction in the new estates and 30 per cent energy reduction in existing estates.
   Relevant planning strategies:
- Apply a 100 per cent Renewable Energy System, e.g., solar cells, natural gas, hydropower and biofuels;
- Using combustible waste as a biofuel for district heating and electricity; heat from wastewater treatment for heating and cooling;
- Waste from processed sewage into natural gas as fuel for cars and buses;
- A smart system in households for residents to monitor energy and water consumption;
- Applying the closed-loop resource optimisation;
- CCHP to convert combustible waste to district heating and electricity;
- Surplus electrical energy heats water for buildings;
- Enforcing energy-efficiency labelling for household appliances.

4 Sustainable transport

Performance indicator: Public transport share

Hammarby Sjöstad: 80 per cent public transport share. (In 2015, 52 per cent of journeys were by public transport and 27 per cent by walking/bicycle.)

Wuxi Eco-City: 80 per cent green transport rate (including public transport, cycling and walking).

Punggol Eco-Town: Modal share of journey using public transport at 75 per cent of the total journeys in 2030.

Relevant planning strategies:
- Provision of comprehensive public transport routes;
- Different mode choice of transport;
- Frequent public transport services;
- Networks of bicycle routes and facilities;
- Car-pooling system;
- Limitation of private car-parking places;
- Extending the rail network;
- Intensifying land use around public transport stations;
- Piloting an electric car-sharing scheme;
- Increase the quantity of bus services;
- Integration of various transportation modes;
- Development of a well-connected pedestrian network.

5 Support the local economy

All towns aim to provide local jobs.

Performance indicator: Resident population employed within township

Hammarby Sjöstad: No specific targets. In 2010, 6,544 people were already employed in the area.

Wuxi Eco-City: No specific targets.

Punggol Eco-Town: No specific targets.

- Hammarby had a large provision of 250,000 m^2 commercial facilities to support local employment;
- Wuxi provided a commercial and mixed-use area in close proximity of under 300 m from a residential area;
- Punggol's strategy is to encourage investments through collaboration in research and development on new technologies and innovations in urban sustainability through collaboration with local research institutions.

6 Enhance social well-being

All three towns made provision for social amenities and their accessibility in various ways in their master plans.

Performance indicator: Resident population employed within township
Hammarby Sjöstad: No specific targets.
Wuxi Eco-City: Kindergarten within 300 m; primary school within 500 m; high school within 1,000 m; walking distance to residential area.
Punggol Eco-Town: No specific targets.
Relevant planning strategies:

- Provision of social services and special dwellings: 400 student flats, 59 homes for the elderly and dwellings for assisted residential care; educational and cultural facilities, including 8 schools, 17 pre-schools, two high schools, library, theatre, chapel and an environmental centre.
- Comprehensive range of public facilities which is regulated by the planning standards that Singapore new towns have adopted; for example, one primary school (around 1.8 ha) for every 2,300 dwelling units, and one college (around 6 ha) per new town.

The above are just a few of many attempts around the world to reduce the impact of city districts and provide a decent quality of life for all citizens. While unfortunately none measure their progress against absolute planetary boundaries, I hope that they provide inspiration, ideas, and the ability to critically interrogate any planned or existing project against 'one planet' standards.

To finish this chapter, here is a more cautionary case study:

Elephant Park, London, UK

A case study focusing on urban renewal and the relationship between the local government and a large developer.

Useful aspects: For a city authority involved in a district-level regeneration project, care over contract negotiation, oversight and community involvement.

Elephant Park is known as a leading sustainable urban district development in the borough of Southwark, south London.[22] The £3 billion, 9.71 ha development was begun in 2013 and is due to complete by 2025. It involves the regeneration of the Heygate Estate, a post-war social housing area.

It is one of 19 projects in the world in C40's Climate Positive Network accredited to 'partner' status, whose purpose is to support the development of neighbourhoods and districts that seek to meet an emissions target of net-negative operational greenhouse gas emissions. C40 Cities Climate Leadership Group is an organisation set up to "help cities replicate, improve and accelerate climate action", that connects 90 cities, and whose president is former New York City Mayor and property billionaire Michael Bloomberg.

A C40 Climate Positive Development is expected to be a district-scale, mixed-use development that includes:

- Walk- and bike-ability;
- Highly efficient buildings;

- Low-carbon energy sources;
- Recycling, landfill diversion programmes and, ideally, utilising waste as a resource;
- Close proximity to high-quality mass public transport;
- Compact, mixed-use development; and
- Expansion of the positive impact into the surrounding community.

In order to go "beyond carbon neutral", projects can earn Climate Positive Credits by sequestering emissions on-site and abating emissions from surrounding communities, perhaps by creating a park, or investing in LED streetlight upgrades in a neighbouring area, energy retrofits for existing housing, and connecting neighbouring buildings into a low-carbon district heating network.

Local administrations hosting these projects are encouraged to "utilise holistic planning, to create policy and regulatory frameworks to encourage Climate Positive development in regeneration projects, and use their procurement powers to promote Climate Positive transformation of industrial areas". C40 Cities claims to independently monitor and verify the claims of developers and local authority partners in its Climate Positive programme, according to its Good Practice Guide.[23]

In the case of Elephant Park some of these outcomes, promised at the outset when Southwark Council gave it initial planning permission, have not materialised. A consequence is that the development, due to be Climate Positive by 2025, will need significant carbon offsetting. It also cannot be said to provide genuinely affordable housing.[24]

In its favour, its Zone 1 homes in central London have been certified to Passivhaus standards; yet each of these 15 three- or four-bedroom townhouses in the Victorian style, featuring cross-laminated timber, costs at least £1.495 million.

Developer LendLease's Master Regeneration Plan[25] originally proposed that a woodchip-powered combined heat and power plant would generate 100 per cent renewable energy that would supply the new Aylesbury estate development next door,[26] and two other existing nearby housing estates, making it the largest such district heating network in London. But the plans were dropped[27] in favour of a smaller gas-fired district heating network that does not comply with the council's 20 per cent on-site minimum renewable energy aspiration set out in its Strategic Policy 13.[28] Other commitments for encouraging public transport and green space have been watered down too.[29]

The private-public dilemma and affordable homes

The dilemma for Southwark Council, which faces all local governments in London and elsewhere, was how to finance such massive regeneration schemes when central government does not offer sufficient support and land prices are so high. Councils usually turn to the private sector, but the trade-off is typically the loss forever of publicly owned land to the private sector and the loss of homes for social rent. Social-rent homes, with the lowest rent levels, are disappearing in favour of so-called 'affordable-rent' homes, where tenants can be charged up to 80 per cent of what the private market demands.

Writer Catherine Slessor reported in the *Architects Journal*:[30] "Despite assurances, out of the proposed 2704 new dwellings, the number of socially rented units has shrunk to 82 from 1194 when the estate was operational. The last residents left in 2013, evicted, decanted and dispersed miles from their original neighbourhood." According to Transparency International,[31] all of the 51 Elephant Park's South Gardens units were sold abroad, at prices of £790,000–£1,500,000 to international investors.

It's been calculated that so-called 'affordable' houses in the Elephant and Castle area can only be occupied by people earning £60,000–£90,000 a year. This excludes the average Southwark resident, who cannot afford it. Southwark Council Councillor Mark Williams, the cabinet member for regeneration, has defended their policy by saying, "All homes sold by LendLease help pay for the affordable homes. It is not within the council's interest to manage regeneration schemes which do not offer the best deal for residents – everything we do is for the benefit of residents."[32]

LendLease and this project have been praised by the Greater London Authority,[33] the UK Green Building Council[34] and other bodies and won several prestigious architectural prizes.[35] There are many less green, social and scrupulous developers than LendLease. But local governments elsewhere seeking development and wishing to avoid a scenario like this may prefer not to work with big developers; to form a not-for-profit legal entity for delivery; to listen to and heed all of their existing community; to provide genuinely affordable housing; to have design schemes, not master plans; to have qualified negotiators and lawyers scrutinise the contracts; and to spread the risks.

Notes

1 For more information see the website: www.sovieso.at
2 Thorpe, D., *The One Planet Life*, Routledge, 2015.
3 See: www.wildgarten.wien
4 Cicero, S., et al., *Sharing Cities: Activating the Urban Commons*, Shareable, 2018, ISBN: 978-0-9992440-0-5. See: www.shareable.net/node/89997/done?sid=11941
5 Interviewed by the author for: www.smartcitiesdive.com/ex/sustainablecitiescollective/work-starts-europes-largest-permaculture-passivhaus-eco-community/1082106/
6 See: http://cityofdavis.org/city-hall/community-development-and-sustainability/development-projects/the-cannery/environmental-review
7 UK Ministry of Housing, Communities & Local Government, *Planning Policy Statement: Eco-towns – A supplement to Planning Policy Statement 1*: ISBN 9781409816836. This Planning Policy Statement (PPS) provided the standards any eco-town had to adhere to before it was cancelled for all areas excluding north-west Bicester on 5 March 2015. See: www.gov.uk/government/publications/eco-towns-planning-policy-statement-1-supplement
8 See: nwbicester.co.uk
9 See: http://urbed.coop/sites/default/files/20140815%20URBED%20Wolfson%20Stage%202_low%20res3.pdf
10 See: www.westoxon.gov.uk/media/1483153/West-Oxon-Garden-Village-EoI-July-2016.pdf
11 See: http://hyas.co.uk/finance-delivery/delivery-models-approved-garden-communities/
12 See: www.cadbury.co.uk/about-bournville

13 Wolfson Economics Prize, Finalists' Submissions, Compendium of Non-Technical Summaries, Policy Exchange, 2014. See: policyexchange.org.uk/wp-content/uploads/2016/10/wep-2014-compendium-of-finalists.pdf
14 *How To Win Back The City En Comú: Guide To Building A Citizen Municipal Platform*, Barcelona En Comú, 2016. See:
   https://barcelonaencomu.cat/sites/default/files/win-the-city-guide.pdf
15 Dafermos, G., *The Catalan Integral Cooperative an organizational study of a postcapitalist cooperative*, P2P Foundation and Robin Hood Coop, 2017. See:
   http://commonstransition.org/the-catalan-integral-co-operative-an-organizational-study-of-a-post-capitalist-co-operative/
16 See: www.europarl.europa.eu/RegData/etudes/BRIE/2018/614701/EPRS_BRI(2018) 614701_EN.pdf
17 Stringer, J., "Analysis: Bristol's Next Move Should Look To Barcelona", *The Bristol Cable*, 2017. See: https://thebristolcable.org/2017/10/analysis-bristols-next-move-look-barcelona/
18 See: http://transformativecities.org/
19 See: www.energy-cities.eu/Community-voices-in-Luzy-village-of-the-future-mates-of-the-future
20 See: www.energy-cities.eu/Community-voices-in-Luzy-village-of-the-future-mates-of-the-future-5146
21 Malone-Lee, L.-C., Heng, C.K., Nasution, I., *Designing Beyond Compliance: Aspirational Performance of Eco-Cities*, Centre for Sustainable Asian Cities, Department of Architecture, School of Design and Environment, National University of Singapore, Conference Proceedings, ADNEC, Abu Dhabi, UAE: Ecocities in Challenging Environments, 2015.
22 See: www.southwark.gov.uk/regeneration/elephant-and-castle?chapter=6
23 See: http://c40-production-images.s3.amazonaws.com/good_practice_briefings/images/3_C40_GPG_Climate_Positive_3_Mar.original.pdf?1457019377
24 See: https://southwarknotes.wordpress.com/2012/07/24/lend-lease-outline-planning-application-responses-and-objections/
25 See: www.southwark.gov.uk/download/downloads/id/4889/elephant_and_castle_regeneration_agreement_appendix_6
26 See: http://moderngov.southwark.gov.uk/mgConvert2PDF.aspx?ID=16241
27 See: www.london-se1.co.uk/news/view/5052
28 See: www.southwark.gov.uk/planning-and-building-control/planning-policy-and-transport-policy/development-plan/local-plan
29 See: http://35percent.org/sustainable-development/
30 Slessor, C., "dRMM's Trafalgar Place is the first trumpet blast of the world to come", *Architects Journal*, 5 October 2016. See: www.architectsjournal.co.uk/buildings/drmms-trafalgar-place-is-the-first-trumpet-blast-of-the-world-to-come/10012478.article
31 Cowdock, B., "Faulty Towers: Understanding the impact of overseas corruption on the London property market", Transparency International UK, March 2017. See: www.transparency.org.uk/publications/faulty-towers-understanding-the-impact-of-overseas-corruption-on-the-london-property-market/#.W5KbrpNKiHo
32 See: www.vice.com/en_uk/article/qkq4bx/every-flat-in-a-new-south-london-development-has-been-sold-to-foreign-investors
33 See: www.london.gov.uk/sites/default/files/17._environmental_standards_study_0.pdf
34 See: www.ukgbc.org/wp-content/uploads/2018/03/UK-GBC-Leading-the-way.pdf
35 See: https://nextgeneration-initiative.co.uk/

16

WALES AND 'ONE PLANET' DEVELOPMENT

What is 'one planet' governance? What are the implications for a local or national government in taking on board the full responsibility of having to change its decision-making style and processes so that its policies and their implementation fall within the requirements of planetary boundaries?

To answer this question, the current chapter will take a deep dive into the experience of the Welsh Government in implementing its Well-being of Future Generations Act, which passed into force in 2016. This is one of the world's leading examples of attempts to transform governance so that all decisions are measurably sustainable.

Wales' Future Generations Act

Wales is one of the four nations that make up the United Kingdom, with limited devolved powers and a population of around 3 million. It has adopted a definition of sustainable development that means "the process of improving the economic, social, environmental and cultural well-being of the country by taking action, in accordance with the sustainable development principle". This in turn is defined as meaning "that any publicly funded body must act in a manner which seeks to ensure that the needs of the present are met without compromising the ability of future generations to meet their own needs".

The Welsh Government realised that in order to achieve this it needed a revolution in the way that the country was managed. Rather than pass a 'sustainable development act', a decision was taken to give the legislation the name 'The Well-being of Future Generations Act'. The rationale was that citizens could more readily connect with the idea of creating a better world for their children and their descendants than with an abstract jargon term like sustainable development. A Bill was drafted and, over a year or so, a number of public meetings and online consultations were held around the country under the banner 'the Wales we Want' to find out how people felt.

Eventually, the Act was passed in 2015 at a celebration held in the Wales Millennium Centre in Cardiff Bay, attended by Nikhil Seth, director of the Division for Sustainable Development, Department of Economic and Social Affairs, United Nations, who said: "The Wales Future Generations Act captures the spirit and essence of two decades of United Nations work in the area of sustainable development and serves as a model for other regions and countries. 'One Wales, One Planet' captures it all. We hope that what Wales is doing today the world will do tomorrow."

Any private citizen or community group can now challenge decisions by public bodies on the grounds that they do not take account of the needs of future generations.

The Act is designed to achieve the following seven goals, chosen as the seven chief aspects of – and requirements for – human well-being:

1. A prosperous Wales: an innovative, productive and low-carbon society which recognises the limits of the global environment and therefore uses resources efficiently and proportionately (including acting on climate change), and which develops a skilled and well-educated population in an economy which generates wealth and provides employment opportunities, allowing people to take advantage of the wealth generated through securing decent work.
2. A resilient Wales: a nation which maintains and enhances a biodiverse natural environment with healthy functioning ecosystems that support social, economic and ecological resilience and the capacity to adapt to change (for example climate change).
3. A healthier Wales: a society in which people's physical and mental well-being is maximised and in which choices and behaviours that benefit future health are understood.
4. A more equal Wales: a society that enables people to fulfil their potential no matter what their background or circumstances (including their socio-economic background and circumstances).
5. A Wales of cohesive communities: attractive, viable, safe and well-connected communities.
6. A Wales of vibrant culture and thriving Welsh language: a society that promotes and protects culture, heritage and the Welsh language, and which encourages people to participate in the arts, sports and recreation.
7. A globally responsible Wales: a nation which, when doing anything to improve the economic, social, environmental and cultural well-being of Wales, takes account of whether doing such a thing may make a positive contribution to global well-being.

These are now legislative requirements as the goals are enshrined in the Act. The Act doesn't stop there. It also requires government departments to conduct their business in a completely different way from before – and from the way in which most other governments operate. This is also why the Act is revolutionary. The new way is defined by "the five ways of working". These are:

1. Thinking long-term: balancing short-term needs with the needs to safeguard the ability to also meet long-term needs.
2. Prevention: proactive behaviour to prevent problems occurring or getting worse that may help public bodies meet their objectives.
3. Integration: considering how the public body's well-being objectives may impact upon each of the well-being goals, on their objectives, or on the objectives of other public bodies.
4. Collaboration: with any other person (or different parts of the body itself) that could help the body to meet its well-being objectives.
5. Involvement: the importance of involving people with an interest in achieving the well-being goals, and ensuring that those people reflect the diversity of the area which the body serves.

A former civil servant, Gretel Leeb, who worked on the implementation of the Act, describes it as "a multi-faceted piece of legislation that cuts right across government". According to her, "it's about good governance. Who wouldn't want to be thinking long-term about actions if executing good government? The five ways and the sustainable development principle are common sense, as Carl Sargeant [the cabinet secretary responsible for the Act at the time] said. He called it the 'common sense act'. If we apply it full-bloodedly it can lead us into a different way in how we live our lives".

How the Act came about

Jane Davidson was the Labour Party education minister (2000–2007) and environment, sustainability and housing minister (2007–2011) in the Welsh Assembly Government who oversaw the introduction of the concept of the Act. She explained to me how it came about, and it is interesting to quote her in full.

> The Act essentially had its roots in 1999 when the National Assembly for Wales was set up with a clause in the Government of Wales Act 1996 which established it, for the new National Assembly for Wales to promote sustainable development in everything it did. Because I spent my first year in the Assembly as Deputy Presiding Officer, I got excited about the fact that the Assembly had been given this duty – which did not apply in any other part of the UK. The Assembly was then a corporate body, and the Executive was required to prepare a Scheme in which we had to say what we should do to make our actions comply with sustainable development. Following an election, the incoming Executive would then be required to review that and make its own Scheme.
>
> The first Scheme in 2003 was 'Ten actions for a better planet' and the second in 2006 perhaps doubled its length. When, in 2007, I became Minister for Environment, Sustainability and Housing, I saw that the two Schemes

were full of good ideas but not a manifesto for delivering sustainable development for Wales, nor were they influencing everything that the then Welsh Assembly Government was doing. (The Government title was established following the second Government of Wales Act in 2006). If a body has a duty to promote sustainable development, there has to be a tangible difference in the actions it takes from other bodies. I didn't feel there was. I drew strongly upon the Sustainable Development Commission (SDC) – an independent group of experts set up by the UK Government to monitor its own actions. In 2005 the SDC had been given clear support from all four nations of the United Kingdom – governments of different political colours – about what a sustainable set of policies would look like, and I was keen to be in the vanguard.

I asked Rhodri Morgan, the First Minister, to give me responsibility for sustainable development on his behalf, i.e. a super-responsibility over and above my portfolio At the time we [The Labour Party] were negotiating a coalition government with Plaid Cymru [the National Party of Wales] and I needed to be able to ask any minister to do something about sustainable development on behalf of the First Minister. By agreeing this, he opened the door to enable me to begin to align the actions of the whole of government with the principles of sustainable development.

The first step was to commission the Stockholm Institute at the University of York to review the manifesto with which we came into power, including the emission reduction targets. From their report it was clear that we wouldn't increase nor reduce Wales' emissions with this manifesto. I was worried about the ecological footprint of Wales, that we were using at least three planets' worth of resources. We had to reduce this and be able to demonstrate this, but clearly our current plans would not be enough. Cabinet colleagues were supportive that change was needed but we were all working within a set coalition agreement that all ministers had to deliver. There was little they could do other than be supportive. So a government-wide Scheme wouldn't be deliverable without collective agreement. We needed a big idea to engage within and without government.

This was 'One Wales One Planet', which we published in 2009. [This document set an aspiration for Wales to reduce its ecological footprint to 'one planet' by 2050.] We saw climate change as the biggest threat so I established the Welsh Committee on Climate Change (CCC) with all party representation and external experts; we set statutory municipal recycling targets – and are now third or fourth in the world's good recycling nations league as a result – and we consulted widely. 'One Wales One Planet' was a vision of what a sustainable Wales would look like and it relied on our new CCC, civil servants and the Stockholm Institute to detail the actions we should be taking. This was a hugely exciting development, which I hoped would lead to more evidence based decisions, more collaborative working with other agencies and better longer term outcomes. I think it would be fair to say that One Wales One Planet was seen as a very positive initiative but still not influencing all cabinet delivery.

Three things happened next: firstly, WWF-Cymru [the Welsh branch of WWF] conducted an independent policy review with Netherwood Sustainable Futures (a part of Cardiff University) of all ministerial delivery which judged clearly that the Welsh Assembly Government was actively supporting unsustainable decisions. A second independent review by the Welsh Audit Office found that the commitment to sustainable development, even though it was in legislation, had not been cascaded through the civil service. In practice, the top two tiers of officials only used the language of sustainable development with ministers but did not cascade it down to the vast majority.

So if neither the business nor political components of government were delivering on sustainable development, when the principle was supported, something else had to be done. Although people liked 'One Wales One Planet', no legislation existed to make it happen. The third key factor was the tenth anniversary conference of the SDC in England at which I'd been invited to speak. At the time, the Coalition Government between the Liberal Democrat and the Conservative parties had just been established. The Secretary of State for Environment didn't attend. It was clear that it was not a priority and soon after that the SDC was disbanded by the UK Government which had committed to reducing non-departmental public bodies. This for me was the ultimate irony – that the independent expert body established to advise all UK nation governments on delivering sustainably and for the long term in the interests of current and future generations should be disbanded overnight. Legislation was the only answer!

I have always been proud that from 1999 the new National Assembly for Wales had a duty on sustainable development laid down at the start, but what had become clear as we tried to implement it, was there were still legislative and cultural barriers. The duty was not defined; there was no notion of what sustainable delivery would look like. On the way home between Bristol and Cardiff I wrote the four main framework points of the now Well-being of Future Generations (Wales) Act: that it would apply to all the devolved public sector in Wales; that there would be a Commissioner with sufficient powers to change behaviour; that as public services are already audited by the Welsh Audit Office, the same office would audit under a new sustainability framework and that the Welsh Government itself would be subject to the Act.

I took my proposals through the political route within the Labour Party for inclusion in the Labour Party Manifesto for the 2011 election. When Labour came into government in 2011, the proposed act was included in the Programme for Government and was on its way with all the key framework components in place. This was critical, so that any future government of any political colour would deliver sustainable development because it would be at the heart of the legislative programme with a focus on responsible management for the benefit of current and future generations. Importantly, the legislation should both be permission to think differently for the organisations involved – and others wanting to join the journey, should align with the developing Sustainable Development Goals and be challengeable in the courts.

Developing the Act

Jane Davidson left politics, but the work she began continued. Following the publication of proposals for legislation in a Sustainable Development White Paper in 2012, the Welsh Government sought the views of the Welsh public on what mattered to them for the future in a national conversation called 'The Wales We Want', which coincided with the United Nations' global conversation on 'The World We Want'. Wales' Conversation was kick-started with a publication in 2014 called 'The Wales We Want by 2050', with a view to discovering the long-term sustainable development goals that people thought were the most important for them and their children and grandchildren.

Over 7,000 people joined this 'Welsh Conversation' as individuals and through community groups, public meetings and the recruitment of "future champions" to act as advocates of communities and organisations. There were also champions recruited to represent the views of young people. The Future Generations Act was then drafted, containing the seven goals and the five ways of working. The first minister for Wales observed that the Act could also be seen as a global model of how the SDGs are translated into action at the regional and local levels.

One planet development

Prior to this process, Wales had also enshrined the concept of One Planet Development in planning law. This came within the context of its Sustainable Development Scheme, "One Wales: One Planet", that includes the objective that within the lifetime of a generation, Wales should use only its fair share of the Earth's resources, and "our ecological footprint be reduced to the global average availability of resources – 1.88 global hectares per person in 2003". This aim was carried forward into the Act.

One planet development is defined as "development that through its low impact either enhances or does not significantly diminish environmental quality". It "can either be single homes, co-operative communities or larger settlements" that "may be located within or adjacent to existing settlements, or be situated in the open countryside".[1]

Strict conditions are applied: they "should initially achieve an ecological footprint of 2.4 global hectares per person or less in terms of consumption and demonstrate clear potential to move towards [a] 1.88 global hectare target over time. They should also be zero carbon in both construction and use". If located in the open countryside, "they should, over a reasonable length of time (no more than five years), provide for the minimum needs of the inhabitants in terms of income, food, energy and waste assimilation". Further guidance for rural 'one planet' developments stipulate zero waste, nutrient reclamation, zero carbon transport, improving biodiversity, respect for Welsh culture, and so on.

The Welsh Government has provided a useful online tool for calculating the ecological footprint per person of households applying for planning permission and living on one planet developments, using expenditure as a proxy.

This subject was covered extensively in my 2014 book, *The One Planet Life*,[2] to which this current title is considered a sequel. At the time of writing, around 35 households have received planning permission for one planet developments in the open countryside, but none have been proposed yet for urban areas. Defining the criteria for these is the dominant subject of this book, and has been considered by The One Planet Council, of which I am a patron.

In an important sense, the existing one planet developments have acted as pioneers and test cases of what one planet living everywhere might be, and have highlighted certain advantages and shortcomings. What has not been tested is one planet development at scale. The Well-being of Future Generations Act and One Planet Development, taken together, offer a glimpse of how it might play out in Wales.

Developing the indicators

The Act stated that the indicators "... must be applied for the purpose of measuring progress towards the achievement of the well-being goals". To develop the indicators that might measure progress towards the seven goals, advice was sought from the Public Policy Institute for Wales. It advocated that indicators should measure outcomes which should "resonate with and matter to the public". The total number of indicators should be limited – no more than 40 – including five or six headline ones. They should form a coherent set, which can be justified and framed by a narrative about what progress means for Wales and lead to significant progress over the long term.

In October 2015 there was another national consultation, "How do you measure a nation's progress?" This involved a series of events across Wales and ran for 12 weeks, seeking views on a set of 40 national well-being indicators for Wales. Some 175 responses were received, more by far than any previous statistical consultation exercise, demonstrating the range of interest in the national indicators. In March 2016 the final indicator set was published, expanded to 46 indicators.

Since the goals are inter-related, it was seen as vital that the national indicators were aligned with this aim. Therefore there is no single, exclusive indicator for each goal; instead, a 'key' is provided to show which indicators made a significant contribution to which goals. Similarly, there is an indicative map of the Welsh national indicators against the 17 Sustainable Development Goals. Furthermore, it was seen as critical that the national indicators provided a collective set of outcomes that were seen as desirable in Wales and that were not the responsibility of any single public body.

A paper prepared by the Welsh statisticians said:

> For example, we wanted to measure the overall health outcomes of the nation not just a measure of the National Health Service. Better performance in unscheduled care (however that is measured) would contribute to better population health outcomes. But so would lowering air pollution, delivering a

higher skilled population, or improving economic activity. We wanted to ensure we were measuring the collective outcomes, in this case overall population health, so that all public bodies would recognise their contributory role.[3]

There was much discussion about the balance of subjective and objective measures. Some aspects of well-being cannot be measured by objective administrative data or survey responses, such as whether people consider themselves to be 'safe' or living in 'cohesive communities'.

Consequently it was decided to develop new data sources to capture individuals' feelings about personal well-being. The National Survey for Wales, an annual household survey of over 11,000 people, already existed to provide an understanding of a range of topics including the performance of public services and personal well-being, and so new questions were added to it to provide more subjective indicators; the Survey is now the source for 14 of the indicators.

While these subjective, qualitative indicators are important, unless the wider, objective quantitative indicators which refer to planetary boundaries – such as ecological footprint and carbon footprint – are given due weight, there will be no sufficient natural resources in the future on which all the others depend.

In the end 46 national indicators were decided upon, listed in the Act's associated 'Technical document'.[4] These include versions of the Sustainable Development Goals' indicators, the ecological footprint of Wales, plus indicators for carbon emissions and social and cultural goals such as promoting the Welsh language.

BOX 16.1 THE NATIONAL INDICATORS FOR WALES' WELL-BEING OF FUTURE GENERATIONS ACT

1. Percentage of live single births with a birth weight of under 2,500 g.
2. Healthy life expectancy at birth including the gap between the least and most deprived.
3. Percentage of adults who have fewer than two healthy lifestyle behaviours (not smoking, healthy weight, eat five fruit or vegetables a day, not drinking above guideline levels and meet the physical activity guidelines).
4. Levels of nitrogen dioxide (NO_2) pollution in the air.
5. Percentage of children who have fewer than two healthy lifestyle behaviours (not smoking, eat fruit/vegetables daily, never/rarely drink and meet the physical activity guidelines).
6. Measurement of development of young children.
7. Percentage of pupils who have achieved the "Level 2 threshold" including English or Welsh first language and mathematics, including the gap between those who are eligible or are not eligible for free school meals. (Replaced from 2017 by the average capped points score of pupils.)
8. Percentage of adults with qualifications at the different levels of the National Qualifications Framework.

9. Gross value added (GVA) per hour worked (relative to UK average).
10. Gross disposable household income per head.
11. Percentage of businesses which are innovation-active.
12. Capacity (in MW) of renewable energy equipment installed.
13. Concentration of carbon and organic matter in soil.
14. The ecological footprint of Wales.
15. Amount of waste generated that is not recycled, per person.
16. Percentage of people in employment, who are on permanent contracts (or on temporary contracts and not seeking permanent employment), and who earn more than two-thirds of the UK median wage.
17. Gender pay difference.
18. Percentage of people living in households in income poverty relative to the UK median: measured for children, working age and those of pension age.
19. Percentage of people living in households in material deprivation.
20. Percentage of people moderately or very satisfied with their jobs.
21. Percentage of people in employment.
22. Percentage of people in education, employment or training, measured for different age groups.
23. Percentage who feel able to influence decisions affecting their local area.
24. Percentage of people satisfied with their ability to get to/access the facilities and services they need.
25. Percentage of people feeling safe at home, walking in the local area, and when travelling.
26. Percentage of people satisfied with the local area as a place to live.
27. Percentage of people agreeing that they belong to the area; that people from different backgrounds get on well together; and that people treat each other with respect.
28. Percentage of people who volunteer.
29. Mean mental well-being score for people.
30. Percentage of people who are lonely.
31. Percentage of dwellings which are free from hazards.
32. Number of properties (homes and businesses) at medium or high risk of flooding from rivers and the sea.
33. Percentage of dwellings with adequate energy performance.
34. Number of households successfully prevented from becoming homeless per 10,000 households.
35. Percentage of people attending or participating in arts, culture or heritage activities at least three times a year.
36. Percentage of people who speak Welsh daily and can speak more than just a few words of Welsh.
37. Percentage of people who can speak Welsh.
38. Percentage of people participating in sporting activities three or more times a week.

39. Percentage of museums and archives holding archival/heritage collections meeting UK accreditation standards.
40. Percentage of designated historic environment assets that are in stable or improved conditions.
41. Emissions of greenhouse gases within Wales.
42. Emissions of greenhouse gases attributed to the consumption of global goods and services in Wales.
43. Areas of healthy ecosystems in Wales.
44. Status of biological diversity in Wales.
45. Percentage of surface-water bodies, and groundwater bodies, achieving good or high overall status.
46. The social return on investment of Welsh partnerships within Wales and outside the UK that are working towards the United Nations Sustainable Development Goals.

Reporting on progress

The Act also requires Welsh ministers to, "in respect of each financial year beginning after the date on which national indicators are published ... publish a report (an 'annual well-being report') on the progress made towards the achievement of the well-being goals by reference to the national indicators and milestones".

These reports are published in line with the UK Code of Practice for Statistics, with the content and methods being independent of ministers. They are therefore official statistics products, which ensures that there is clarity and separation between them and the Welsh Government's own reports on its performance against its well-being objectives. This reinforces the fact that the goals require collective action. The national indicator reports are compiled using Microsoft Power BI reporting tools, which were chosen because the software has the advantage that reports can be produced in-house and it results in an interactive, multi-functional report that costs no more than the time and effort of government statisticians.

Ministers must also publish a future trends report, and take account of any action taken by the United Nations in relation to the Sustainable Development Goals, and the UK's assessment of the risks of the current and predicted impact of climate change.

The Act also requires Welsh ministers to set National Milestones to help with determining whether progress is being made. At the time of writing these were about to be developed. All public bodies including health boards and trusts must also report annually on their progress towards meeting their well-being objectives.

How it is working in practice

Since this transformation required of government is so huge, it will take many years to have the complete effect envisaged. As I write it is only into the third year.

There are 44 public bodies covered by the Act that are required to carry out sustainable development. There are also around 144 divisions and 5,000 civil servants in the Welsh Government, all working on policies.

This means that mutual awareness, corporate oversight, effective governance and policy integration are a big challenge. And, as Gretel Leeb observes,

> It is not just about public bodies, but all sectors and civic society. For example, it is in the nature of the third sector that organisations tend to be very single interest. You could say that's as much a problem as it is with government departments. Just as government departments now have to think in an integrated way, so too should single interest organisations; collective focus on the goals requires us all to think in a more integrated way and to work together. So the Act has the potential to change not only the way that government works, but all organisations and people in Wales.

Public Services Boards

The Act required the setting up of 19 regional Public Services Boards based on administrative boundaries. The members of these county-wide boards include representatives of all publicly funded bodies, plus representatives from civil society. Community and town councils are also covered if, for each of the preceding three financial years, their gross income or expenditure was at least £200,000.

Each Public Services Board is charged to "improve the economic, social, environmental and cultural well-being of its area by contributing to the achievement of the well-being goals by making an assessment of well-being in its area and setting objectives to maximise achieving those goals, and by members taking all reasonable steps to meet those objectives".

Each Public Services Board must come up with its own local well-being plan, targeted at the specific needs of its area. Its assessment must refer to the national indicators and future trends reports.

Local Planning Authorities must pay attention to the local well-being plans and understand what their communities need. They must engage the public so that it can influence development proposals at an early stage.

The state of the environment is incorporated into the local plans and the overall schema for the whole country in the following way. A public body called Natural Resources Wales is a member of the independent Future Generations Commissioner for Wales' Advisory Panel, and a statutory member of all the Public Services Boards. It provides the evidence base on natural resources to inform their assessments of well-being and their local well-being plans. The body also advises planning departments to help them prioritise and improve infrastructure.

The state of the natural environment is also intended to be protected and enhanced by the Environment (Wales) Act and the Planning (Wales) Act, both published around the same time as the Well-being of Future Generations Act, and by a National Development Framework, when a two-year consultation is completed on the Welsh Government's land-use priorities.

The role of the independent commissioner

An independent future generations commissioner for Wales was set up to check whether the Act is being implemented correctly. The office was established by the Act itself to be a "guardian for future generations" – challenging short-term, narrow policymaking, and reporting on how effectively public bodies are acting and thinking and can think and plan for the future. The office has 21 staff, an Advisory Panel, and a budget of £1.4 million a year. The first *Future Generations Report* is due by 2020.

Since April 2017 it has been engaged in dialogues with the country's councils, health boards, national parks, fire and rescue services, the Welsh Government itself and national organisations, such as Natural Resources Wales and Sport Wales, and the 19 Public Services Boards, on their work towards a collective 345 objectives.

Sophie Howe is the current commissioner. Her background prior to accepting this role was as a local councillor, then adviser to the government on legal matters, equal rights and human rights. She was attracted to the role of commissioner because she saw as a result of her earlier work a need for reform of public service and that policy and practice were insufficiently holistic. She told me that she recognises fully the connections between the social and the environmental sides of the Act although her emphasis is on the social and economic.

The Commission's budget does not permit it to scrutinise every single decision taken by government departments and public bodies, so it has to take strategic choices about what to call the government to account on. One of its first critiques was of a proposal to spend £1.5 billion driving a motorway relief road at Newport through a nature reserve. Howe said: "This report was well received. The narrative is now shifting. It seemed a great deal to spend and a lot to lose for an average of four minutes' travel time-saving. The guidance under which the decision is taken is now reformed in favour of the Future Generations Act rather than road building, although a final decision is yet to be taken. It is a good sign."

She calls the act

> brilliant legislation, a challenge to the old system. There are people who do not want to change. A million pieces of legislation still perpetuate that system and we are trying to reform them to ally with the Act concentrating at the moment on reforming planning and housing policy.
>
> This legislation is the biggest culture change in public service ever seen in Wales and government has not resourced this sufficiently. I am writing to the Cabinet Secretary to ask what resources she is allocating to planning officers under the act. My team can't resource all of that.

"People are passionate about it," she continues. She speaks regularly to organisations to encourage them to use it.

> For example I spoke recently at the National Association for Educational Leadership – head teachers across Wales – because it is important on the future of education for kids. They need to learn how to cope with artificial intelligence and skills to create a low carbon society. Only a handful of head teachers had heard about the Act and all of them went away saying that it is amazing. They can see all the interconnections.

I asked her how she saw the conflict between politicians' short-term aims and the long-term aspirations of the Act. Ministers own decision-making after all and may have quite different ideological priorities (from each other and from the Act's priorities).

> I am optimistic. There is a genuine understanding by politicians of the need to take long term decisions. However, having the tools to do that is a different matter, and some things are beyond politicians' control: the annual budgets and the grants to Wales from [the UK Government in] Westminster, and departments' budgets for example. We spend half of the total government budget on the National Health Service, and Local Authorities, which includes education and social services. That's a big chunk.

I wondered whether she thought social and environmental criteria could trump economic ones when public bodies chose suppliers for goods and services. She cited an example of a Local Health Board, when considering whether it should sell off some surplus land adjacent to a hospital but instead choosing to maintain it as a community facility, providing allotments.

Most people I spoke to in researching this chapter said that the Act empowers citizens to challenge the decisions made by publicly funded bodies if they feel the decisions do not comply with the requirements of the Act to protect future generations. Because the Act is so new this is only just beginning to be tested in the courts. In the week that Sophie Howe and I spoke, the first case was being brought by lawyers against the closure of a comprehensive school, Cymer Afan Comprehensive, in the Upper Afan Valley, which was ordered to close at the end of the current school year by Neath Port Talbot County councillors. Parents of all of the 229 pupils were fighting the consequent need to travel to a new 'super school' around 40 minutes' bus ride away. At the time of writing a Judicial Review over the decision, focusing on issues of equality and the Well-being of Future Generations Act, was being prepared.

"It is slowly filtering into consciousness," said Howe. "There are issues with statutory guidance for lawyers and reaching a robust position in the obligations on public bodies, tests they have to apply. We are applying and testing a decision flowchart. There will probably be a test case. Who can enforce the Act is something yet to be decided."

Howe noted that of the 46 indicators, some Public Services Boards can take action on individual ones, some have set objectives.

The way they are planning to measure is not yet very good. The first annual review, which I am preparing now, will give feedback on the best and worst indicators, and my office is talking to Cardiff University about the best way to measure them and linking them to the Sustainable Development Goals. We aim to help P[ublic] S[ervices] B[oards] to consider all aspects – economic, social and environmental – and we have to come down to individual decisions and try to influence them to do the right thing. Some people criticise that there is not sufficient focus on nature or biodiversity. I think that there is. We pick the topics of planning, housing, transport, and social care as being especially of higher priority. If there is to be a public enquiry we are on all four sustainable development pillars, the low carbon pathway and reflecting all the goals.

"Procurement is not being used effectively," she says.

I have challenged the government because it has a complex web of procurement – no one can quite fathom it. Those who handle procurement are mystical beings who speak a different language. Policy reflects these decisions rather than the other way round and procurement policy has not been updated since the Act came in. This is a missed opportunity. Following my challenge the government is working out how the system needs to change to reflect the Act. While we're waiting for this we are pulling in Cardiff Business School and experts into figuring out what would it look like. Opportunities exist in specific regions, for example the City Deal, and some small scale schemes already exist, e.g. Public Health Wales, where, when they were combining offices, the estates lead intervened and they gave a grant to a community organisation to take care of a building for people with learning difficulties etc. There are many 'frustrated champions' in the public sector, and the Act gives these people permission to challenge the system.

Finally, Howe said that she believes the £6 billion annual budget in Wales should be used "not just to buy from the existing market but to create the world we want to see, such as school dinners from local producers. We need to change the way we procure. I'm confident it will drive this change".

Self-critiquing progress

Two years after the Act had been passed, the government commissioned independent research to see if anything had changed in the way it was working. The researchers, from Cardiff University, discovered that, as might be expected, civil servants had largely continued doing things the way they always had. The Welsh Government, in considering the findings of Cardiff researchers on how the Act was working, took the unprecedented step of convening a series of three incremental workshops for leaders of NGOs in the country and senior civil servants from different government departments. Their joint purpose was to see what could be done to expedite change.

I had previously attended earlier public consultations, feeding into the process leading into the final Act on behalf of the One Planet Council and others, so I attended all three workshops early in 2018 and it was a fascinating process. I witnessed first-hand the genuine commitment of all players to make the changes that were required and an acknowledgement of how difficult it was.

The feeling was that there were high expectations of the Act to begin with, but a lack of preparation for its introduction. This led to a wide variety of attitudes and expectations of and within government, which had given rise to inconsistency. In the country at large there remained ignorance of the Act and its implications. Different stakeholders tended only to interpret it in relation to their immediate concerns.

So participants at the workshops overwhelmingly felt a need to increase clarity about the Act's meaning in order to overcome the differing interpretations and to avoid the 'cherry picking' by departments of some aspects of the Act while neglecting other aspects; also, to overcome the vested interests and the 'silo mentality' of departmental working, that can lead public employees to interpret the Act's goals as it suits them, while ignoring the Act's mandate to integrate all seven aspects and the five ways of working. All of this clarification has to be done, but with scarce resources to support it.

One way in which workshop participants felt this could be achieved was to emphasise that the Act may tackle multiple problems and obtain multiple wins with a single integrated strategy: a concept that is not widely understood because it involves formally recognising benefits to a strategy which fall outside a particular departmental remit. Taking this approach sets a problem for the government's accountants, since it makes it harder to attribute benefits to a given cost centre. A methodology still needs to be developed to enable costs and benefits to be shared between departments. Data gaps will necessarily be identified, particularly in the environmental area, and there will be a lack of resources to patch these holes.

Questions from participants included: Who should do the assessments? How far should they go, and are they done early enough? Do the assessments change anything? How do we know? How are the data shared? All of these yet require some work to be answered.

Politicians and civil servants are unused to thinking in the time frames which the Act assumes, i.e. up to 60 years. This challenge affects procurement and funding. However, a benefit of long-term thinking is that it can provide long-term security for citizens and business. It was reported that the Assembly Government's new permanent secretary appreciated very well the importance of, and the difficulties associated with, decisions whose effects span beyond political cycles and spending term plans.

Maximising the involvement of all stakeholders in decision-making was identified at the workshops as a major area of concern. Although achieving this will give better outcomes, in reality this practice is patchy, and often omitted or done too late in the process. Involving stakeholders in decision-making is not the same thing as consultation once a policy has been formulated, and is rarely done at the early stage of identifying problems and discussing potential solutions. Consultation respondees often find themselves faced with a *fait accompli*. This type of open democracy was seen as a function of the culture prevailing in national and local government, which are both risk averse and therefore reluctant to be open, sharing and seeking views.

This led workshop co-convener Anne Meikle, CEO of WWF-Cymru, to observe that "new activity needs to be shaped by the goals not just retrofitted. The five ways of working need to be integrated into the process. Reframing existing policies or retrofitting them onto the Act are not good enough".

There were also questions from the workshop participants about whether administrations had the personnel capacity and the necessary skills to engage the public/stakeholders.

At this early stage in the implementation of the Act there was still much work going on to establish baseline positions to help measure progress, and to understand and use future trends in the development of policies. "Nothing will be done properly unless hearts and minds are behind this, so how people think demands a culture change," said Anne Meikle. "The new Permanent Secretary is emphasising the importance of this in the form of a 'future proofing' programme. Leadership and performance management are seen as being vital but there are no illusions about how far there is to go."

The barriers identified included culture change, mindset, and a tick-box mentality. To address 'silo working' a number of budget allocation suggestions were made:

- A radical idea to achieve better integration would be a new approach in departmental budget allocation, requiring integrated and collaborative plans that are only funded when joint policy and planning are clearly evidenced. Jointly developed proposals would be funded across departments.
- The management of budgets and funding allocations in Welsh Government can make collaborative, strategic procurement difficult – more actual or notional pooled budgets could possibly be a solution.
- There will be a need for budgets in different portfolios to work together to deliver the projected impacts, so a way will need to be found of accounting for wider, shared benefits.
- It was suggested that civil servants could be moved around more on secondment, also bringing in people from outside to change the decision-making process and human resource opportunities.

In this way, project leads could then pick teams from across government, depending on the task at hand. In time, recruitment policy might change to facilitate this and the practice could be built into performance reviews.

The following challenges were also identified and would be faced by any administration pursing a similar path:

- *Policy and integration*: to address the opaque 'black box' of decision-making where inputs are not tracked and unexpected decisions emerge, more transparency is required over this process. This would reduce political risk and increase the confidence of stakeholders, the public and the third sector. Shared outcomes are the goal to be aimed at. Overview is required with responsibility for assessing alignment with the goals and the objectives and meeting the indicators, but more work is needed on clarifying the indicators. A communications effort is needed to better explain the workings of Welsh Government to citizens and stakeholder organisations.
- *Procurement*: the development of a system for 'valuing' multiple benefits across the well-being goals. Sustainable procurement implies a need for awareness training for buyers and supply chains. As an example, in the procurement of a 'Warm Homes' service to address fuel poverty, contractors are asked to show how their proposals include the wide benefits of the goals. There is work to develop a similar approach consistent with this across government. Too often contracts go to the same narrow list of large companies. Smaller, local companies should be permitted to bid. Therefore framework agreements need to be recalibrated to the objectives of the goals. More collaboration across cost centres and finance heads is needed, and a way needs to be found to properly and fully value multiple benefits in the procurement process. Continuous professional development will be needed to make cross-departmental procurement more joined up. The National Procurement Service needs to support local-level procurement too.
- *Capacity and capability*: lack of shared understanding was seen as a barrier. Recognising and rewarding the five ways of working with an awards scheme was thought to be something that might help. A skills gap and a cultural deficit in the ability of people to think about and factor in long-term impacts need to be addressed. Better ways of showing progress and being accountable need to be explored.

"It's complicated and challenging," concluded Meikle. "We need a common view and to know we are getting better. The annual reports and indicators require government to report on its progress but do we have the right evidence? What would it look like? The KPIs [key performance indicators] need to change and be questioned. Do we have a common view? Do the annual reports chart progress? They can't be qualitative, we must know what success looks like. Do you have to report on every goal for everything?"

Developing the process

Following these workshops, I asked some key officials in the administration their views, some of whom were not able to speak on the record because of privacy requirements of the civil service, which might seem a little odd, given the condition of transparency expressed in the five ways of working, but it is simply because civil servants are wary of being accused of expressing a political opinion, which they are not supposed to do.

The process of reporting on the indicators is in fact self-critical and iterative.[5] I was directed to a quality report on the national indicators. This explains the origins of the data used to show whether the indicators are being met. For example, for the indicator concerning the concentration of carbon and organic matter within the soil, 300 sites of 1 km^2 across Wales are surveyed. These are selected at random from 26 land classes, except for brownfield sites.

Other reports detail the fact that "natural resources are being depleted faster than they can be replenished", "biodiversity is declining", and that "no ecosystems in Wales can be said to have all the features needed for resilience".[6]

Glyn Jones, chief statistician, responded to my enquiries by saying: "The national indicators are used as a basis for the annual Well-being of Wales report, which I publish as Chief Statistician independently of Ministers under the Code of Practice for Statistics.[7] The Well-being of Wales report provides a report on our progress as a nation against the goals, as required by the Act. As well as the national indicators it uses other data sources that are relevant to each goal to provide a narrative on progress."

In addition, the Environment (Wales) Act sets out a statutory approach to reporting on Wales' greenhouse gas emissions with a series of emissions targets defining a pathway to a 2050 goal of reducing Wales' emissions by at least 80 per cent below 1990 baseline levels. This is the same as the UK as a whole. Like the rest of the UK it also introduces five-yearly carbon budgets which limit cumulative emissions over this period. These targets and budgets are based upon production emissions in Wales and are reported in line with international guidelines from the United Nations Framework Convention on Climate Change.

In recognition of the wider impact Wales can have on global greenhouse gas emissions the Environment (Wales) Act includes a further requirement to report upon the emissions associated with the consumption and use of goods and services in Wales, whether produced in Wales or elsewhere. Wales' carbon footprint will be reported alongside the reports detailing its performance for each five-yearly carbon budgetary period.

Glyn Jones said that "Combining all these approaches provides a comprehensive view of how Wales' greenhouse gas emissions are changing over time and allows us to effectively monitor our progress towards reducing Wales' emissions".

How are these results shared?

Glyn Jones testified that the reports he produces "are shared widely through our stakeholder networks which include other public sector bodies and Public Service Boards" and other committees. He said that "for local authorities, an additional spreadsheet is being prepared that brings together any national indicator data in one place to help them access the data relevant to their area in as easy a manner as possible".

He issued a caveat, that the data are to register outcomes, and not for measuring performance, a distinction that seems odd to outsiders: "It's important to note that the national indicators are designed to help us understand national progress against the goals. They are not performance indicators for Welsh Government or any other public body."

Perhaps the National Milestones yet to be set will fulfil this function. Jones added that:

> there are no targets for the national indicators. The national indicators are intended to help us understand whether we are making progress towards the goals. This is a critical measure of the progress we are making as a nation and the collective outcomes to which the Act is contributing. The indicators also highlight areas we need to consider further, and should lead to further research questions to help us understand how we are doing.
>
> The narrative around the indicators, and the Well-being of Wales report, is an important one for all public bodies, Welsh Government included, to consider. Public bodies under the Act are expected to contribute towards this national progress. However, the importance of separating this out from measuring performance came out strongly in our consultation on the national indicators. In discussions with stakeholders they felt it would potentially undermine collective action if they were perceived to be performance indicators of individual public bodies.

Reducing the ecological footprint of Wales

Data on the ecological footprint of Wales are commissioned regularly from the Stockholm Environment Institute. Their last report[8] concludes: "The Welsh Government has a goal to achieve 'One Planet Living' within the lifetime of a generation. This will require a reduction of the impacts of Welsh consumption by approximately two thirds."

That is quite a challenge.

The methodology for the footprint calculation is in line with that used by the Global Footprint Network – i.e., it uses UN collected data. It is not the same as for calculating a material footprint. As mentioned in Chapter 3, this relates to a calculation of the amount of primary materials required to support demand in a country or city. Glyn Jones observes: "We did, however, in the national indicators technical document, reference the Material Footprint and Material Footprint per capita as an example of contextual information which may aid analysis."

The Stockholm Environment Institute's 2015 report goes on: "Whilst currently unsustainable, consumption patterns change, and resource and energy efficiency can improve, and it is Welsh Government's role to develop and implement policies that facilitate the transition to One Planet Living in Wales."

This is clear about where the responsibility to reduce the footprint lies. It continues: "The consumption of food, housing, transport, consumer items, private services and public services together accounts for 85 per cent of the ecological footprint. The specific consumption categories that contribute the most to the ecological and carbon footprints are household energy use, transport, construction and consumption of meat, fruit and vegetables."

This is a message which has barely travelled outside the obscure document in which it sits. I have never heard an official or politician relay this message. Yet the report repeats the message when discussing the carbon footprint of Wales:

> The carbon footprint of Wales has been calculated at 11.1 tonnes of carbon dioxide equivalent per capita ... Most of the greenhouse gas emissions associated with Wales' carbon footprint occurs from purchases of domestic goods and services. Therefore the priority should be to improve both the efficiency of production and size and type of consumption within Wales. It is also important to reduce emissions released outside Wales but that contribute to products consumed within Wales, along international supply chains.

This is making the same point about the impact of imports and exports which we discussed in relation to Sweden in Chapter 4. It adds: "The primary land type dominating the footprint is the land required to sequester emissions of carbon dioxide, which arise from the burning of fossil fuels."

If burning of fossil fuels ceased, this would lead Wales to become carbon positive. Regional authorities are named in the report: "The top three local authorities with the highest carbon footprint are Ceredigion, Powys and Isle of Anglesey. Those with the lowest carbon footprint are Merthyr Tydfil, Blaenau Gwent and Caerphilly." The three highest are largely rural, and therefore have a higher reliance on private transport.

Finally, the report issues a challenge:

> In order to achieve the One Planet Living goal a significant degree of change is required. This will need all sectors and stakeholders to contribute. The Welsh Government's current policies are aimed at reducing the impacts of Welsh consumption in a number of ways; the most significant of these target reductions in greenhouse gas emissions in Wales. Further policy development and implementation will be needed, and targeting steeper reductions in early years will increase the probability of meeting both carbon and One Planet Living goals in the longer term.

Conclusion

There is no doubt that this is a ground-breaking piece of legislation whose effect will be felt for decades to come, provided it does not get repealed by a future government. The fact that it gives a licence to citizens and community groups to challenge the decisions of public bodies in the name of future generations is potentially massively empowering. It is also potentially game-changing over decades, in that it could transform the mindsets of both elected representatives and officials in national and local government and public bodies, leaching out into the country at large.

In theory, the Act is emphatic on the equal importance of the goals for the well-being of Wales as well as the well-being of individuals; and the goal to be a globally responsible nation ensures commitment to reducing climate change and other transnational challenges to sustainability. Yet there remains a risk that in the interests of pleasing citizens, the wider picture might become neglected. A presentation by Welsh statisticians[9] observes that their annual reports under the requirements of the Act should primarily answer the question: "Are we becoming more prosperous, healthier, a more equal nation?"

Equality and health are laudable aims. Prosperity is usually defined in terms of wealth, and we know that since wealth gets spent and consumption is the biggest part of any country's ecological footprint, this could have counter-productive effects. While we obviously want to avoid people living in hardship, a sustainable society would ensure that people had sufficient to meet their needs but not too much; perhaps it needs a different definition of prosperity which refers to a feeling of contentment, linked to reducing the income gap between the richest and poorest.

This is the meaning of 'contraction and convergence' as described in Chapter 2: the reduction of the overall ecological footprint while removing social and economic inequalities.

While many of the subjective, qualitative indicators are important, unless the wider, objective quantitative indicators which refer to planetary boundaries – such as ecological footprint and carbon footprint – are given due weight, there will be no sufficient natural resources in the future on which all the others, along with society itself, depend. The requirement for Wales to reduce its ecological footprint to one planet within one generation, is fundamental to genuine sustainability, and unless it is highlighted, it will not be factored into decision-making.

As we learnt, again in Chapter 2, from the work on 'The Real Doughnut' by Daniel O'Neill and his colleagues, it is a massive challenge to provide a good life for all citizens within planetary boundaries. The width of Kate Raworth's 'Doughnut' is currently zero, and creating space for it depends ultimately on reducing consumption and creating a totally circular zero carbon economy. Reducing consumption is seen as counter to the prevalent model of economic growth, which is widely believed to be necessary to support a population.

Yet while sufficient economic activity is of course necessary, any definition of what constitutes sufficient economic activity must make reference to taking consumption levels to a sustainable level, i.e. fairly distributed and within planetary boundaries: contraction and convergence.

One of the relatively easy things to do to tackle both the ecological footprint and the carbon footprint and meet the requirements of the Well-being of Future Generations Act and the Environment (Wales) Act is to shorten supply chains and make those supply chains gradually more circular: to source locally as far as possible, and reuse everything, including nutrients in food. Those goods which cannot be provided nationally should be sourced ethically and with the footprint of production in mind.

By sourcing more goods and services within Wales, authorities would have much more control over their impacts and be able to make production more circular and resource efficient.

The 'five ways of working' strongly imply using the procurement policies of public bodies to achieve this. Procuring goods and services locally within Wales will result in more intensive use of the countryside and greater employment, leading to the sustainable regeneration of rural areas as they service the urban areas.

Public procurement policy is already changing along these lines. The Welsh Government and local government are expected to lead by example by procuring from local firms/suppliers using social and environmental criteria, as well as economic criteria, in the choice of suppliers as part of that process. Smaller firms will be able to apply for contracts to supply hospitals, schools, prisons, colleges, care homes, and so on, instead of contracts going only to large companies. They will be encouraged to cooperate with each other in consortia.

This approach is intended to widen the market. It will encourage sustainable practices right down the supply chains. The end goal is the same as 'one planet' developments in rural Wales but on a countrywide scale. Processing facilities, and logistics, will need to be linked to production on a more local scale, not that different from how they used to be 100 years ago – before the era of cheap fossil fuels.

With this in mind, for the benefit of public administrators and their students, I reproduce here the advice of Gretel Leeb, enthusiastic former civil servant:

1. Always look outward and ahead, 360 degrees around an issue or a proposition.
2. Pay careful and constant attention to the truly long term and to Future Trends reporting.
3. Be interested in, and understand what is happening in other policy areas.
4. Avoid thinking and acting in rigidly linear and sequential ways – the world and its circumstances are continually changing around you and your objectives – develop the art of intelligence-led, adaptive management and keep discussing changes with the people affected by them and with your ministers.
5. Place integrated impact assessment at the heart of policy development and delivery.
6. Listen carefully, pay close attention to and learn from the real-life experiences of the citizens you are trying to help and the organisations who work in their interests.
7. Keep on checking your path and your assumptions about impacts – once is never enough – the world is changing even as you move …
8. The way to keep on checking is to keep on talking – establish and 'look after' a running conversation with civil society and stakeholder organisations.
9. Never be afraid of needing to change course!

This could be great advice for public servants and politicians everywhere.

Notes

1 Welsh Government, "Technical Advice Note 6 Planning for Sustainable Rural Communities", July 2010. See: https://gov.wales/topics/planning/policy/tans/tan6/?lang=en
2 Thorpe, D., *The One Planet Life*, Routledge, 2014.
3 Further information: Jones, J., Leake, S. and Charles, A., *How Welsh statisticians are helping to measure the progress of a nation*, Paper prepared for the 16th Conference of IAOS, OECD Headquarters, Paris, France, 19–21 September 2018. See: www.oecd.org/iaos2018/programme/IAOS-OECD2018_Jones-Leake-Charles.pdf
4 Available at: https://gov.wales/docs/desh/publications/161115-national-indicators-for-wales-technical-document-en.pdf; see also: https://gov.wales/docs/desh/publications/160316-national-indicators-to-be-laid-before-nafw-en.pdf
5 Welsh Government Statistics for Wales, Progress reports against the well-being goals. See: https://gov.wales/statistics-and-research/well-being-wales/?tab=data&lang=en.
6 Welsh Government Statistics for Wales, *Well-being of Wales 2017–18: Measuring progress towards the national well-being goals*, Slide Show. See: www.slideshare.net/StatisticsWales/wellbeing-of-wales-201718
7 See: www.statisticsauthority.gov.uk/code-of-practice/
8 Lungley, H., with Dawkins, E. and Croft, S., *Ecological and Carbon Footprints of Wales, Update to 2011*, Stockholm Environment Institute and GHD, July 2015. See: https://gov.wales/docs/desh/publications/150724-ecological-footprint-of-wales-report-en.pdf
9 Jones, G., Leake, S. and Charles, A., *How Welsh statisticians are helping to measure the progress of a nation*, Welsh Government Paper prepared for the 16th Conference of IAOS, OECD Headquarters, Paris, France, 19–21 September 2018. See: www.oecd.org/iaos2018/programme/IAOS-OECD2018_Jones-Leake-Charles.pdf

17

SIX STEPS TO A 'ONE PLANET' CITY

In this final round-up chapter is just one suggested blueprint for how a community from a neighbourhood up to a mega-city or nation in size may decide and set a pathway to 'one planet' status. These 'one planeteers' would take on responsibility for making a better world. They could choose to guarantee their town's security and resilience into the future, and that it would pay back the debt to nature that civilisation has built up over the last few decades. Here it is, in summary form, followed by some simple suggestions for individual topics.

The basic 'one planet' requirements

1. That to aim towards 'one planet' living should become an underlying principle of all planning and official policy as de facto the only objectively verifiable regenerative strategy.
2. That the same set of social and environmental criteria should be used to assess all planning applications and procurements.
3. That these criteria should be informed by appropriate indicators including life-cycle and ecological footprint analysis, to enable all potential and actual projects to be compared and evaluated for their impacts.
4. Official policy should support all areas and sectors to use 'one planet' living principles and methods to become more productive and more biodiverse – regenerative.
5. Reduction of excessive consumption should be the next social revolution: this entails doing more with less, and believing that meeting basic needs is sufficient for everybody.

The six-step path towards 'one planet' cities and communities

1. **Obtain community buy-in and feedback at all levels:**
   Hold a series of public meetings and online and off-line consultations to explain the context and aims in order to obtain feedback and community buy-in. This might take a year.

2. **Decide which standards and objectives to use:**
   These will include a methodology and accounting system and be applicable to all sectors such as soils, biodiversity, water, energy, buildings, transport, well-being, etc. They must include ecological footprinting.

3. **Set a baseline – the current situation:**
   Use data and surveys to ascertain the starting point from which goals will be set: on the supply side, the productivity of its ecological assets (green space and water bodies); on the demand side, the ecological assets/resources required to produce the natural resources and services it consumes.

4. **Set targets for each sector over realistic timescales:**
   A system similar to that applied by the UK Climate Change Commission could be adopted, along with a version of NPV+ to test the results of different scenarios. A set of five-year plans may result, each with a budget and a set of targets. The overall target could be, say, 40 or 50 years away, to meet everybody's basic needs within planetary limits. Each short-term target will be a step closer to the overall one. Each sector (biocapacity, water, energy, buildings, transport, industry, etc.) will have its own schedule.

5. **Set in place ways to measure them:**
   This should be based on what data are easy and cost-effective to gather, and relate to the baseline situation, chosen metrics and sector targets. The data should be transparent and publicly available. Everybody should be able to view the progress being made.

6. **Ratchet down consumption over one or two generations:**
   Each five-year plan will have its own evaluation period to check that all expected benefits are resulting, to share experiences, to accommodate criticisms, to potentially revise plans, and to celebrate successes.

Principles

Planning conditions, procurement policies and all governance procedures should adopt the following principles and approaches:

Working with people

- Respect and include all views and all citizens
- Public bodies to use the Five Ways of Working
- Recognise and award or celebrate achievers in all areas and sectors
- Actively solicit ideas and solutions
- Foster virtuous feedback loops.

How will we improve biocapacity?

- Planning conditions to specify net gains in biodiversity for all development
- Regenerative cities (green walls, roofs and other infrastructure)
- Parks: lots of biodiversity
- Trees and bushes: make them all productive – nuts and fruit
- Indoor, vertical, rooftop, food growing, including mushrooms, aquaculture
- Vertical farms: can be all scales from large warehouses to inside homes, shops and cafés, and can include aquaculture and poultry
- Allotments, community gardens, therapeutic gardens
- 'Incredible edible' plants at every opportunity
- Plenty of opportunity for land-based jobs and jobs in food processing
- Procurement along supply lines to include conditions to yield net gains in biodiversity.

Use a low-tech, cost-effective approach for infrastructure/development

- Life-cycle analysis to reduce overall biocapacity demand and cost
- Make it human scale
- Deploy low-impact materials that sequester carbon where possible
- Keep it simple: eco-minimalism – do more with less
- Keep it local: support the local economy by procuring locally
- Plan for easy maintenance and end-of-life ease of removal, recycling
- Design out the possibility of less sustainable behaviour.

Water

- Conserve water use
- Reuse water
- Prevent water pollution
- Rainwater collection from roofs
- Replenish underground aquifers.

Zero or net-positive carbon

- Energy efficiency, reducing demand and need
- Passive design new buildings; older buildings deep retrofitted for energy efficiency
- Decarbonise travel
- 100 per cent renewable energy supplies

- Supply (local): heating, cooling, electricity
- Investigate feasibility of district heat mains
- Distributed generation: fourth-generation solar will be almost invisible, often building integrated
- Some wind turbines
- Combined heat and power
- Anaerobic digestion for biogas and compost
- Heating: geothermal, ground- and water-source heat pumps, solar thermal, passive solar, biogas
- Inter-seasonal storage
- Demand-management technology
- Building Information Management systems in larger buildings
- Incinerate only absolute waste in CHP plants
- Climate-appropriate street layout and vegetation planting to either reduce heat island effect and therefore cooling energy demand, and/or reduce heat demand, depending on location.

How will we get around?

- Minimise need to travel far by locating housing near jobs, services, shops etc.
- Separate vehicular traffic from pedestrians and cyclists
- Networked, latticed town and neighbourhood centres planned around transport hubs
- Efficient, cheap public transport: metro, subway, buses, trams, light rail and intercity
- Promotion of walking, cycling by design of streets and neighbourhoods
- Separation of local and longer-distance traffic
- Lots of bikes and bike hire schemes
- Rickshaws, cargo bikes for people and goods delivery
- Electric vehicle hire and share schemes.

Closed-loop systems

- Waste designed out
- Plastic designed out
- Products designed to be reused and recycled
- Reward waste minimisation and sorting of waste, especially plastics
- Food, agricultural waste and sewage have nutrients reclaimed
- Repair and upcycle workshops in every neighbourhood
- Culture of exchange, swop, repair and reuse to replace throwaway culture
- Local industrial symbiosis systems.

Climate change adaption

- Extreme weather preparedness
- Storm swales and underground lagoons
- Raised areas in regions of potential floods

- Sustainable urban drainage
- Protection from overheating
- Vegetation to tackle heat island effect
- In areas of high winds, buildings close together to prevent funnelling.

How will our neighbourhood serve us?

- Human-scale, people-centred design
- Markets – formal and informal, street and covered, valued for small traders and social interaction
- Conserve amenity value and embodied energy of old buildings while retro-fitting for efficiency
- Preserve communities through provision of affordable housing and full mix of housing and land use
- Respect and support all cultures
- Work, education ideally located near to home
- Public transport and parks within walking distance
- Leisure, shops, entertainment and museums nearby.

How will we get healthy?

- Neighbourhood-level drop-in health centres and hospitals
- Some consultations may be distance tech-based
- All people to have access to cheap, fresh, healthy fruit and vegetables
- Cooking workshops and education
- All-ages-friendly cities
- Mental health support
- Accessibility everywhere: ramps, not steps
- Walking, cycling, pathways, wayfinding.

How will we complain and get justice?

- Implement 'The Right To The City'
- Root out corruption
- Independent watchdogs and transparency
- Use technology to maximise participation in decisions, campaigns
- Smart sensors, feedback loops and algorithms to help manage some basic services
- Smart involvement uses the wisdom of crowds, open data, etc.
- Sensitive, anonymous surveillance is part of data collection, linked to algorithms which attempt to model and predict needs so they can be met efficiently.

Standards

Existing legal frameworks, tools and standards can help save reinventing the wheel. Standards assist with setting up processes and monitoring progress. I would recommend exploring the following, which have been examined earlier within these pages:

- *General*: the UN Sustainable Development Goals
- *For governance*: Wales' Well-being of Future Generations Act (Chapter 16)
- *For baselining*: Quantify the metabolism of the city using material flows (Chapter 3)
- *For reducing the impact of consumption*: the ecological footprint (Chapters 3 and 4), and One Planet Development (Chapter 16)
- *For carbon footprinting*: PAS 2070 (Chapter 3)
- *For reducing the impact of imports and exports*: the PRINCE model in Sweden, and EEMRIO ('environmentally extended input-output modelling') (Chapter 4)
- *For promoting energy efficiency in organisations and industry*: ISO 50001 (Chapters 3 and 6)
- *For smart and sustainable communities*: ISO 37120 (Chapter 3)
- *For environmental management*: ISO 14001 (Chapter 3)
- *For promoting the circular economy*: ISO 14040 Life Cycle Analysis (Chapter 3)
- *For measuring the long-term value of local investments in infrastructure and natural capital*: the Global Footprint Network's Net Primary Productivity (Chapter 3) and NPV+ tool (Chapter 4)
- A natural resources and biodiversity survey is also recommended.

In a nutshell

> You cannot go on destroying your environment because then your environment will destroy you. So you really must realise what are your limits. But even then you can still grow by making much more sophisticated elements. You can grow in economic terms even if you don't grow in physical terms, by making more high-value activities, so wealth and well-being can increase even if size does not.
> – *Pedro B. Ortiz, in conversation with me*

Ortiz is a former deputy mayor and director-general for metropolitan planning at Madrid's regional government, and author of *The Art of Shaping the Metropolis*. In 1996 he created a Regional Plan that was a revolutionary method of metropolitan planning called the Metro-Matrix Method, one way of addressing the challenge of the explosion of urban growth and mega-cities.

Post-script: support for city leaders

City leaders must use all possible channels to reduce their cities' environmental impacts to at least 'one planet' if not better, while fulfilling the Sustainable Development Goals. They are able to respond more directly to the concerns of their citizens than national governments. If unsupported by national governments, they may attempt negotiation to devolve more power – particularly over budgets – from them, to be able to shape their own destinies.

City leaders need support to achieve all of the above. Support networks at global level already exist, such as Global Footprint Network, ICLEI, the C40 Cities Climate Leadership Group and the UCLG. These are primarily about information-sharing and facilitating, not formulating agreements. But could mayors and other city leaders rule the world? Many think they could do a better job than their host nation leaders. The Global Parliament of Mayors is a governance body of, by and for mayors from all continents, set up to make this pitch. In October 2018 they met in Bristol, a former European Green City, and issued a declaration that opened like this:

> From migration to climate change, from inequality to population health, global and national institutions are not delivering to the fullest. Citizens of the world need global governance to evolve, seeing cities and international networks sit alongside national and international leaders as equal partners in shaping global policy.[1]

At present the Global Parliament of Mayors is not primarily focused on climate change or planetary limits, more on issues like health, security and migration, but this could – and needs to – change.

It is ultimately up to citizens to demand a 'one planet' city. They may well need to educate their leaders in how to imagine a better world, and to make it so.

Note

1 *Global Parliament of Mayors Annual Summit 2018 Declaration*, 23 October 2018. See: https://globalparliamentofmayors.org

18

ONE DAY IN A 'ONE PLANET' CITY

A short story

As I described in Chapter 1, films, television and literature are filled with images of dark futures for humanity. To make the future we want I suggest that we need to imagine it first. You, yourself, could draw or write your own vision of what a regenerative 'one planet' city of the future might be like, in which humanity is living within the boundaries of what the planet can provide, but which might bring its own dilemmas. Here is one of mine: a snapshot of a few minutes in the life of a woman and her granddaughter in a place with many of the features described in these pages.

When I looked up I could no longer see Tara. The glint of the sunlight on the surface of the slide must have blinded me for a moment. At first I thought nothing of it; she had probably seen a friend and run to her, or perhaps was enjoying the Rainforest Roundabout.

I struggled to see her red top, quite distinctive, amongst the other children and adults in the play-park. I hurried amongst the playground furniture, all of which were in the shape of animals. Many of them are now extinct: an elephant, a lion, a gorilla, a polar bear. Why do they make them like this? To remind children of what they are missing, I often think. To keep them alive in the popular imagination because they are no longer real …

Peering under the rope climbing frame, strung between two giraffes, I was becoming more frantic by the second. Where was Tara?

She and I were accustomed to coming here on the two mornings a week that both of her parents worked – her father Pietr in the vertical farm on the south side of our block, where he produced pak choi and rocket, and her mother Kes as a teacher in the local high school. Neither of them had to work more than four days a week, so that meant there were two days when they overlapped, and two days when they both had time off to spend together.

That morning, as usual, I had collected Tara from their apartment along the garden roof terrace from mine. Despite the heat outside it was cool inside their well-insulated flat. Before I left, Kes wanted my opinion about the new city vote. "They want everybody to take home the same amount of wages no matter what job they do."

"What? Doctors paid the same as cleaners?" I asked as I pulled Tara's jumper over her head.

"No, doctors would still be paid more, but after a certain amount – the same as for everyone – their wages would go into the community kitty. Like with the profits of a social enterprise. People will have enough money from their wages to spend on the things they need, but anything more than that will go to improve the community, like the school, hospital or the farm."

"What's wrong with that?" I asked. "The smarter people will feel better about themselves and have more status even if they don't keep more money for themselves because they're putting more into the community."

Kes frowned. "Pietr wants to keep more for himself to spend on his hobbies or travel. I told him that wouldn't be fair on everybody else. It's not right that some people should be rich, and in the end all jobs are necessary aren't they, really?"

"Yes. Why would people want more than they need?" asked Ricard, Kes's seven year old, who was standing there ready to cycle to school with his mother. "That's like the bad old days."

"Hasn't he heard about contraction and convergence?" I asked. Before I could get an answer Tara was pulling my arm off.

"Come on!" she was yelling. "I want to go to the play-park!"

"See you later!" Kes waved to me as we had disappeared through the front door. It was good to see how much happier she looked now she was working fewer days a week.

There had been a torrential downpour in the night, but all of it had drained away into our lovely lake at the bottom of the slope. This lake, surrounded by rushes and reeds and buzzing with dragonflies, was part of our sewage nutrient-reclaiming system and also held the coils of the heat pumps that heated and cooled the apartments, and it even gave us a few carp for our dinners and eggs from the ducks. We could hear the bullfrogs croaking from the path.

"Nana?" Tara asked me as she chewed some fine strands from a plant that we passed on the pathway: "Why does fennel taste minty?"

"Because you haven't tasted mint properly," I replied. "Here, taste this." I handed her a leaf from another plant in the herb bed that we were passing. I often tended this bed, so was quite entitled to do this. "This is real mint. Notice the difference?"

A smile lit her face. "Oh, yes! This one is sweeter!"

"So the first one is not really minty," I said. "If you had tasted aniseed, you would recognise a hint of that flavour."

"It kind of is minty though," she replied, determined to have her way.

Often I worked in these gardens, mostly just weeding these days, and showing the younger ones what was what. My back was not good enough any more to do the heavy work but I could still be useful. The kids were fed up with me telling them how this area used to be an industrial park and there used to be chemical works and other nasty things before it became home to 5,000 households. They didn't believe me when I said that in the past older people like me were put away in special buildings and not allowed to see children and teenagers. "Those buildings used to be called care homes," I would say. "I'm glad I'm here and not in one of those."

"So am I," children sometimes told me. "Who else would tell us how to look after the plants – and teach us the old songs?"

"Everything alright?" asked my neighbour Miko, who was watching her son play on an armadillo-shaped ride.

"It's Tara. She's vanished. I'm sure she is safe …" I checked myself. "Actually I'm getting worried. She was here a minute ago. I only turned away for a second and when I looked back there was no sign of her." I gave Miko a description of what she was wearing.

"Have you alerted Sentinel?" she asked.

"Not yet. I didn't want to trouble it …"

She looked around. "I can't see her anywhere." Suddenly, she jumped. "Careful!"

Miko pulled me back out of the cycle lane just in time as a troop of cyclists whizzed by, some of them on electric bikes. I watched them disappear over the rise like a flock of birds. I really think they go too fast.

"Sorry. I forgot to look where I'm going!"

"I really think you should tell Sentinel. If the cameras have recognised her face anywhere … or anyone's spotted her … they'll tell you … You know. Just in case."

"But then her parents will find out and they'll think I wasn't paying attention, and they won't let me bring my granddaughter here again." As soon as I said this I realised I was being selfish. "No, you're right. Tell it."

Miko spoke into her device. "Missing child. Tara Serrado. What's her date of birth?" she asked me, and relayed the answer.

While we waited for a response, the bus rapid transit clattered by beneath a cloud of parakeets squabbling amongst themselves as they travelled from a group of bladdernut trees planted at one end of our 'village green' to one of huckleberries in another corner.

I had two seconds to remember how nice it was that we all looked out onto the village green area and could watch the children, and how, since there were so many wild birds and animals in our local ecosystem, not to mention domesticated chickens and ducks that we kept for eggs, that nobody needed to keep pets any more. The children could play without fear of catching diseases from dog poo. In the past so much land and resources was used to feed and look after pets; it was ridiculous when so many people went hungry as productive land grew more scarce because of climate change.

Miko's device pinged.

"Where is she?" I craned my neck to look at the map where a yellow light was flashing.

Every child wore a bracelet holding a GPS chip so that at any time the city knew where it was. Sentinel had picked this up and was showing us her location.

Any vulnerable person also wore such a bracelet. There had been a big debate recently about whether everyone should have to wear one. Some people objected that they needed privacy. Others said this was like the people who protested against laws to bring in seat belts in vehicles, CCTV, and self-driving cars, when the result of all of these had been that nowadays hardly anybody died in a traffic accident and crime had gone down. Another vast improvement on the barbarism of the past. It seems inconceivable nowadays that people tolerated the deaths of millions of people a year just for the sake of their personal freedom to drive anywhere they liked.

Miko pointed towards the herb beds. "Over there!"

I don't think I had run faster in two decades. But I still couldn't see Tara anywhere. The light was blinking on Miko's map where the raised beds were. I thrust aside bushes of herbs. And there she was. Tara was crouched in a raised bed surrounded by big plants, tasting every single one.

She looked at me with a big smile on her face. "I like this one. But I don't like that one." She was pointing at an aniseed tree.

I couldn't help laughing.

"What's funny, Nana?" she said. "I just wanted to find the plant you were talking about. It doesn't taste a bit like fennel."

"Never mind." I took her hand and she jumped down from the bed onto the path. "Just tell me next time before you run off, please."

"Why?" She looked puzzled. "Sentinel can always find me."

INDEX OF PEOPLE

Abrahamsen, Rune 191
Adler, Stephen 107
Adriazola-Steil, Claudia
    211–212, 213
Armitage, Neil 152

Barby, Clara 230
Barrios-Paoli, Lilliam 74–75
Barrett, John 34
Baxter, Martin 227
de Blasio, Bill 107
Bloom, Dan 7
Bloomberg Michael 108
Borgstrom-Hansson, Carina 34
Burridge, Jenny 147

Casey, Vincent 163
Chandler, Raymond 8
Carson, Kevin 91
Carus, Michael 128
Chu, Prof Steven 129
Clarke, Chris 126
Colau, Ada 250
Crawley, Ian 171–172
Cunningham, Storm
    217, 218

Dafermos, George 251
Davidson, Jane 264–267
Despommier, Dickson 82
Devaney, John 41, 46
Diamandis, Peter 14
Doust, Michael 50

Duggan, Mike 88
Dunsford, Illtud 89

Ehrman, Sigrid 209
El-Gazzar, Rania 220
Elgazzar, Rasha 220
Espinosa, Patricia 107

Fadnavis, Devendra 216
Farha, Leilani 169
Frei, Christoph 106

Gibson, William 7
Gindroz, Bernard 42
Girardet, Herbert 96
Glaeser, Edward 21
Goldman, Michael 176
Gordon-Smith, Henry 85, 87

Hajra, Mayukh 63
Hartenberger, Ursula 233
Harvey, David 201
Heiselberg, Per 192–193
Hidalgo, Anne 107
Hidalgo, Dario 216
Hillman, Mayer 20
Hodgkin, Neil 243–244
Holzer, Dr Peter 194
Holzwarth, Stefanie 204
Howard, Ebenezer 246
Howarth, Catherine, 238
Howe, Sophie 273–274
Hydes, Kevin 185

Index of people

Jones, Glyn 279–280

Kerry, John 106
Kimball, Mary 91
Kluko, Milan 82, 84, 87
Kotler, Steven 14
Kozai, Toyoki 87
Krebs, Lord 108

Lavadinho, Sonia 202
Leeb, Gretel 264, 272, 283
Livingstone, Ken 207
Lorenz, Florian 203–204
Lovelock, James 20
Lubawski, Wojciech 219
Lynch, Matthew 44, 46

MacArthur, Dame Ellen 70, 137
Macron, President Emmanuel 27
Makarigakis, Alexandros 163
Marijewycz, Marco 232
Matan, Annia 202
Matteini, Marco, 30
Marshall, George 8
McNamara, Craig 91
Meikle, Anne 277
Mennicken, Dr Lothar 128
Mepham, David 202
Miller, Jamie 101
Miller, Kirsten 53
Modi, Narendra 215
Montgomery, Charles 177
Moore, Michael 6
Mundy, Ciaran 237

Nabi, Professor Nurun 20
Ng, Jack 86
Nutter, Melanie 216–217, 218

O'Neill, Dan 12–13
Ortiz, Pedro 290
Oung, Kit 142

Paes, Eduardo 106
Palme, Olaf 24
Palmer, Christiana Pas,ca 1, 4
Pearce, Mike 193
Polman, Paul 135
Pulgar-Vidal, Manuel 27

Rapport, David 11
Ramsden, Peter 220
Raworth, Kate 11–13, 233
Rees, William 31
Robertson, George 107
Robertson, Margaret 195
Rodriguez, Oscar 83, 87
Rudlin, David 248
Russell, Steve 136

Sabido, Pascoe 144–145
Sachs, Professor Jeffrey 27
Sadler, Dan 126
Sanders, Bernie 172
Santos, President Juan Manuel 186
Saunders, Lucy 202
Seggers, Peter 75
Seidel, Rebecca Ruth 89
Seth, Nikhil 263
Shideler, Dr John 227
Sippola, Hope 91
Slessor, Catherine 260
Solem, Bård S. 191
Soleri, Paolo 7
Skrobacka, Jadwiga 219
Steer, Andrew 184, 186
Stordalen, Gunhild 69
Sukhdev, Pavan 97

Tang, Audrey 235
Tingvall, Claes 210–211, 212–213

van der Burg, Eric 79

Wackernagel, Mathis 31, 33–34, 35–36
Watson, Sir Robert 19
Watson, Prof Vanessa 176
Weizsaecker, Ernst von 97
Wiedmann, Thomas 34
Wijkman, Anders 97
Williams, Mark 260
Williamson, Theresa 170, 173–175, 179
Wills, Terri 185

Young, Lucy 35
Yunus, Muhammad 21, 233

INDEX OF PLACES

Abu Dhabi 53, 99
Adelaide, Australia 98, 99, 118
Africa 32
Albania 58
Alberta, Canada 41
Ahmedbad, India 215
Amsterdam, Netherlands 70, 79, 113, 117, 203, 221
Anchorage 84
Angola 176
Antalya 69
Argentina 53
Aspen, Colorado 114, 115
Atlanta, Georgia 91, 201
Austin, Texas 107, 108, 117
Australia 144, 146, 160, 229, 237; South-eastern 170–171

Baghdad, Iran 201
Baltimore, USA 41, 82
Bangladesh 19
Bangalore, India 108, 176, 212
Bangkok, Thailand 177, 231
Bangladesh 233
Barcelona, Spain 207, 208–209, 250–252
Basel, Switzerland 99
Beijing, China 84, 150
Belo Horizonte, Brazil 88–89, 118
Berlin, Germany 84, 210
Bern, Switzerland 151
Bhopal, India 215
Birmingham, England 50, 156
Blacktown, Australia 97

Bogotá, Colombia 52, 162
Bolivia 140
Boston, Massachusetts 87, 172
Boulder, Colorado 117
Brasilia, Brazil 165
Bregenz, Austria 189
Brisbane, Australia 97
Bristol, England 41, 99, 237
Brumunddal, Norway 191
Brussels 44, 54, 102
Buenos Aires, Argentina 108, 109
Bulawayo, Zimbabwe 76
Burkina Faso 79
Burlington, Vermont 114, 115, 172
Byron Shire County, Australia 115

Cádiz, Spain 253
Cairo, Egypt 52, 69
Calgary, Alberta 117
California, USA 160, 185, 229
Cambuquira, Brazil 151
Canada 6, 41, 53, 151
Canberra, Australian Capital Territory 117
Caño Martín Peña, Puerto Rico 174–175
Cape Town, South Africa 80, 117, 151–152, 165, 176
Cardiff, Wales 190, 263
Casablanca, Morocco 52
Central America 140
Chandigarh, India, 117
Chennai, India 150, 215
Chicago 50, 82, 107, 108, 172
Chile 140

Index of places

China 21, 41, 46, 70, 84, 136, 142–143, 166, 227
Cleveland, Ohio 236
Cockburn, Australia 118
Coffs Harbour, Australia 115
Colombia 179, 186
Colorado Springs, Colorado 187
Copenhagen 99, 108, 114, 115, 203
Colombo, Sri Lanka 100
Columbia, Maryland 116
Côte d'Ivoire 22
Curitiba, Brazil 200
Cusco, Peru 52, 53
Cyprus 2–4

Darmstadt, Germany 188
Dardesheim, Germany 116
Davis, California 172, 244–246
Delhi, India 208
Democratic Republic of the Congo 22, 176
Detroit 82, 205
Dhaka, Senegal 19–20, 109, 162
Dobbiaco, Italy 116
Dongtan, China 99, 254
Drammen, Norway 121
Dubai 176, 206
Durban, South Africa 162–163

Ecuador 140
Egypt 53
Elephant Park, London, England 258–260
Esklistuna, Sweden 118
Essex, England 248

Feldheim, Germany 116
Finland 41, 60, 139
Flecken, Steyerberg, Germany 116
France 41, 55
Frankfurt 114. 115
Freiberg, Germany 51
Fukushima Prefecture, Japan 115

Genoa 69
Germany 237
Ghana 176
Ghent, Belgium 118
Gothenburg, Sweden 118
Graz, Austria 113
Greensburg, Kansas 114, 115
Grenoble, France 202
Guangdong province, China 166
Guangzhou 50
Gulf of Mexico 18

Haiti 22
Haora, Kolkata 108
Hammarby Sjöstad, Stockholm, Sweden 254
Hamburg, Germany 115, 238
Hangzhou, China 209
Hanoi 108
Harare, Zimbabwe 101, 193
Helsingborg, Sweden 117
Hepburn Shire, Australia 118
Himachal Pradesh, India 63
Ho Chi Minh City, Vietnam 141, 180
Hong Kong 50, 108, 206, 254
Howrah, India 117
Hyderabad, India 150

India 63, 120, 227
Iran 140
Isfahan, Iran 153
Istanbul, Turkey 150, 212
Italy 118–119

Jaipur, India 215
Jakarta, Indonesia 109, 153, 208, 231
Japan 71, 85, 87
Jeju Province, South Korea 113, 115
Jodhpur, India 163
Johannesburg, South Africa 177, 205

Kalundborg, Denmark 160
Kasese, Uganda 116
Kathmandu, Nepal 230
Kaufbeuren, Bavaria 189
Kenya 176
Khulna, Bangladesh 108
Kielce, Poland 219
Kilani, Cyprus 3
Kinshasa 22
Knežice, Czech, Republic 116
Kochi, India 215
Kolkata, India 108, 150
Kristianstad, Sweden 118

Lancaster, California 115
Leeds, England 126–127
Leicester, England 243–244
Letchworth garden city, England 250
Lightmoor garden village, England 248
Lima, Peru 51, 211
Linköping, Sweden 116
London, England 50, 72, 83, 99, 121, 130, 202, 206–207
Los Angeles, California 50, 82, 118, 150, 206, 210

Index of places

Loures, Portugal 117
Ludhiana, India 215
Luzy, France 253

Madrid, Spain 207, 290
Makati, Philippines 218
Makkah, Saudi Arabia 218
Malaysia 141, 229
Mali 22
Malmö, Sweden 80, 115, 118
Manchester, England 50
Manila, Philippines 178
Marrakech, Morocco 204
Marseille, France 58, 165
Maryland, USA 41
Masdar City 99
Medellin, Colombia 52, 79, 185
Mediterranean 69
Melbourne, Australia 177, 202
Mexico City 108, 121, 180, 186, 196, 212, 228
Milan 70, 80
Milton Keynes, England 217
Minna, Nigeria 218
Mississippi River 18
Mongolia 87
Montgomery County, Maryland 172
Montreal, Canada 172
Morocco 53
Mumbai 108
Munich, Germany 115, 117
Munster, Germany 113
Mureck/Steiermark, Austria 116

Nagano Prefecture, Japan 117
Nagpur, India 216
Nairobi, Kenya 90
Namibia 160
Nantes, France 117
Navajo Nation 79
Nepal 54
Netherlands 82, 129
New Buffalo, Michigan 82, 84
New Delhi, India 215
New Jersey, 83
New Mexico 160
New York City 50, 74–75, 87, 107, 109, 114–115, 118, 139, 164, 189, 206, 210, 219–220, 229
New Zealand 196
Nigeria 176
Norway 61, 71
Novo Nordisk, Denmark 161

Oaxaca, Mexico 117
Odenwald District, Germany 116
Ontario, Canada 202, 206
Orlando, Florida, 228
Oslo, Norway 108
Osnabruck, Germany 115, 117
Oxford County, Canada 115
Oxfordshire, England 248

Pakistan 22
Palau 1
Palo Alto, California 115
Paris, France 50, 117, 139, 151, 210
Pays d'Arles 78
Perpignan, France 116
Peru 53, 140
Pfitsch, Italy 119
Philadelphia, USA 156
Philippines 141
Phoenix, Arizona 91
Portland, Oregon 118
Preston, England 236
Puebla Tlaxcala Valley, Mexico 155
Pune, India 215
Punggol Eco-Town, Singapore 254

Qatar 34, 188
Qingdao 108
Québecor, Canada 139

Rennes, France 81
Rhine river, the 164
Rio de Janeiro 25, 106, 109, 170–176, 179, 212
Rochester, Minnesota 115
Russia 41, 142
Rwanda 176

Sacramento, California 82, 91–92, 244
Saerbeck, Germany 116
Saint-Julien-Montdenis, France 116
San Bellino, Italy 119
San Diego, California 115
San Francisco 79, 82, 114, 115, 118, 172–173
San Jose, California 115
San Pedro Magisterio, Bolivia 252
Santa Monica, California 186
Schleswig-Holstein, Germany 116
Seattle, Washington 115
Senegal 22, 162
Seneghe Municaplity, Italy 119
Seoul, South Korea 150, 206

Shanghai, China 50, 85, 99, 108, 176, 186, 209
Shantou, China 108
Shenzen province, China 166
Singapore 20, 84, 86–87, 100, 160, 176, 201, 206, 208, 221, 254
Skellefteå, Sweden 115
Solapur, India 252
Sønderborg, Denmark 115
Spain 226
Stockholm, Sweden 24, 115, 207, 210, 221
Surat, India 215
Suzhou, china 209
Sweden 35, 55–60, 69, 138
Swindon, England 127
Switzerland 36, 60, 122
Sydney, Australia 50, 97, 114, 115, 118
Syracuse, New York State 139

Täby, Sweden 117
Taipei, China 117
Taiwan 87, 220, 234–235
Tanzania 173, 176
Tehran, Iran 150
Thessaloniki, Greece 78, 151
Thisted, Denmark 116
Tirana, Albania 69
Tokyo, Japan 84, 17
Toronto 50, 80, 90–91, 172
Tshwane, South Africa 113

Ulm, Germany 115
United Kingdom 52, 60, 71
Uralla, Australia 115

Valencia, Spain 204
Valletta, Malta 69
Valsui, Romania 80
Vancouver, Canada 51, 54, 107, 115, 172
Vauban, Germany 41, 51, 242
Växjö, Sweden 115
Vernasca, Italy 147
Victoria, Australia 210
Vienna, Austria 116–118, 194, 199, 203, 242–244
Vietnam 34, 141, 164
Virginia, USA 160
Visakhapatnam, India 215

Wales, UK 34, 45, 55, 59, 262–284
Wallonie, Picarde, Belgium 116
Wellington, New Zealand 117
Welwyn Garden City, England 246
Wildpoldsried, Germany 116
Wilhelmsburg, (Hamburg), Germany 116
Wolfhagen, Germany 116
Wuxi Eco-City, China 254

Zimbabwe 22
Zurich, Switzerland 99

INDEX OF TOPICS

Aalborg Charter 42
accreditation 141
affordability of housing 170–174, 246, 252, 259–260
air-conditioning: see cooling
air quality 17, 123, 199, 206, 207–209, 220, 235
agriculture 63, 72, 76, 77, 80, 140, 154, 244; community supported 75, 91; urban 78, 79, 80, 82–88, 90, 139, 163, 196, 244–24, 247
algae farms 85
aluminium, impact of 62, 122
American Chemistry Council 136
American Council for an Energy-Efficient Economy 136
American Society of Landscape Architects 195
anaerobic digestion 71, 105, 119, 128–129, 158–159, 225
animal welfare 72
Apple 139, 144
aquaculture 159
aquaponics 85–86
arcologies 7
artificial meat 88–89
Association of Southeast Asian Nations 231
Australian Petroleum Production & Exploration Association 144
aviation 199, 210, 228

Baugruppen 182, 240–243
biocapacity 11, 13, 19, 31, 287; calculating 31–32, 36, 50–51; *see also* regeneration

biodiversity 1–4, 11, 72, 75, 76, 97, 155–156, 197, 209, 248
biofuels 105, 113, 119, 120, 126–128
biomimicry 101, 137, 193
BioRECO2VER 138
biorefineries 70–71
biosystem services 10
biotechnology 70
blended finance 231–232
Blue Communities Project 151
BMW 135
Bournville Village Trust 248
Bowling for Columbine 6
BP 144
Brazilian Urban Law Laboratory 174
Bristol pound 237
British Standards Institute 41
Brundtland Report 24
BS 8904:2011 Guidance for community sustainable development 41
Building Energy Management Systems 143, 190
buildings 184–198
bus rapid transit 200, 202, 208, 212, 231

C40 Cities 50, 80, 106, 114, 130, 290; C40 Climate Positive Network 258–259
California Farm Academy 91
California Public Utilities Commission 185
Cannery, The 91, 244–246
carbon capture 106, 110, 125, 127–128, 138, 146–147, 191, 248
carbonn® Climate Registry (cCR) 113

carbon footprint 11, 281; calculation 36–37, 50, 59, 60
carbon price 224, 229
car-sharing 204–205, 217, 243
Catalytic Communities 170
cement 145
Center for Land-Based Learning 91, 246
Centre for Low Carbon Futures 128
Centre for Sustainability Accounting 34
Chevron 144
circular economy 44, 51, 70, 79, 98, 102, 135–138, 154, 158, 160, 162, 195, 283, 288
citizen engagement 219–220, 222, 235, 250–252, 253, 263, 268, 286
City Prosperity Initiative 49
City Region Food Systems 81
climate change 8, 11, 15, 16–17, 256; adaptation 106, 108, 109, 165, 169, 227, 231, 256, 288–289
Climate Group 114
closed loop economy: see circular economy
cloud computing 143
Club of Rome 97
CO2 Value Europe consortium 138
coal 108
coastal cities 108–109
combined heat and power (CHP) 106, 109, 110, 120, 138
community land trusts 170, 171–175, 246, 249
community supported agriculture 246
concrete: see cement
Condorcet facility, Paris 139
ConocoPhillips 144
consultation, public 53
consumption, levels and impacts 4, 11, 13, 15, 22, 26, 50, 52, 56, 60, 69, 70, 134–135, 225, 285–286; and carbon footprint 37, 50, 56, 281; and ecological footprint 31, 64, 65, 281
cooling 37, 105, 111, 113, 116, 122, 192–195, 245
contraction and convergence 13–14, 65, 282
Convention on Biological Diversity 1, 2
Corporate Europe Observatory 144
Cotopaxi 140
Council of European Municipalities and Regions 42
Covenant of Mayors 41, 114
cycling 109, 175, 200, 203–204, 209, 211, 217, 246, 252

data 49; for carbon footprinting 37, 57; for material flows 39, 57–58, 62; gaps and limitations 27, 39–40, 61–62, 63; gathering 53–54, 62, 222; on European cities 42; open 51, 137, 216–219, 221
data centres 139, 143
decarbonisation 109, 113, 118, 122, 287
deforestation 56
demand management (of energy) 144
democracy: see citizen engagement
density, of habitation 175, 196, 205
development (unsustainable) 2–4, 15, 20
digitisation 143, 222
diversity 178
distributed (energy) generation 144
district heating 109, 113, 117, 118, 119, 121–122, 255, 256, 259
divestment from fossil fuels 113
doughnut economics 11–14, 66, 282
down-cycling 145
driverless cars: see self-drive cars
drought 151, 154; see also: water scarcity
dystopias 7–8, 19, 98, 215

Earth Overshoot Day 33–34
Eastgate Building 101, 193
EAT Forum 69, 80
eco-cities: see sustainable cities
Ecocity Builders 44, 52
ecological footprint 11, 13, 50, 54, 60, 66, 134–135, 174, 196, 285; components of 30, 60, 61, 64, 66, 69; definition 31–36, 65; of China 21; of food 63, 69, 72–73, 76; of materials 62–63, 283; of Mediterranean cities 63, 70; of Singapore 20; of Switzerland 36, 60; of Wales 45, 59, 267, 280–283; sub-national calculation methods 61, 62
eco-minimalism 190
economics 96, 100, 236
Ecopolis 99
ecosystem accounting 39; see also: externalities, factoring in
Ecosystemic Urbanism 209
electric vehicles 118, 135, 201
Ellen MacArthur Foundation 70, 97, 102
EMBARQ, the Center for Sustainable Transport and Urban Development 211, 216
embodied carbon 157, 161, 191, 197 (note 13)
Emission Database for Global Atmospheric Research 60
energy efficiency 105, 108, 109, 110–111, 120, 123, 139–144, 186, 192, 225, 232
energy intensity 38, 110, 122, 140
Energy Management System 29, 30, 141, 143

energy service companies 141
Energy Star 185
energy storage 105, 113, 118, 125–126, 128–129
entropy 98
European Sustainable Cities and Towns Campaign 41
European Union's Seventh Environment Action Programme 44–45
environmentally extended input-output analysis (EEMRIO) 38, 52, 57–59, 61, 63
EXIOBASE 58–59,
extinction, of species 10, 14, 16, 27
extraction, of materials 26
EU Ecodesign Directive 135
EUROCITIES 81
externalities, factoring in 96, 227
ExxonMobil 144

Fairphone 137
favelas 170, 173–175
fearless cities 252
Ferrock 147
finance: see investment
flooding 108–109, 151, 154–156
Food and Agriculture Organization (FAO) 32, 71, 74, 95
Food Loss and Waste Protocol 72
food supply 16–17, 75–78, 80–82, 89–92, 95, 246; and diet 71–72; impact of 56, 69, 74; poverty 74, 76, 79; waste 70–72, 79, 136
Foresight Future of Cities Project, UK 52
Fraunhofer Institute 114
freight 210
fuel cells 139
FuelsEurope 144
futurology 14

garden cities 100, 246–250
gated communities 181
Genuine Progress Indicator 40–41
ghettoisation 181
Global 100% RE (renewable energy) campaign 114
Global City Indicators Facility 49
Global Commission on the Economy and Climate 224
Global Compact of Mayors 41
Global Footprint Network 31, 39, 60, 69
global hectare 31, 31–32, 33, 54, 65, 267
Global Impact Investing Network 230
Global Initiative for Resource Efficient Cities 54
Global Pact for the Environment 27–28

Global Parliament of Mayors 291
Global Partnership for Sustainable Development Data 49
Global Protocol for Community-Scale Greenhouse Gas Emission Inventories 37
Global Reporting Initiative 102
Global Taskforce of Local and Regional Governments 180
Global Trade Analysis Project 60
globalisation 54–55, 226
Golden Horseshoe Food and Farming Plan 90
Goldman School of Public Policy 205
governance 101, 218–219, 231, 262–283; *see also* policy-making
GPS 216
Grameen Bank 233
green banks 229
green bonds 165, 185, 225–226, 228, 237–238
green mortgages 232–233
Green Revolution 17
green roofs 163, 164, 209
green space 100, 155, 180–181, 209, 247
green walls 85, 163
greenhouses 139
greenhouse gas emissions 8, 14, 16, 64, 106, 109, 113, 126, 142–144, 156, 228; from buildings 111, 122, 145, 184–191; of consumption 62, 50, 70, 72, 73, 74, 279, 281; of materials 63; of production 50, 75, 125, 134, 145; measurement of 37, 43, 45, 55, 59, 130, 279; short-lived 122–123
greywater: see wastewater reuse
gross domestic product (GDP) 26, 41, 64, 140, 142
gross national income (GNI) 19, 20

Habitat III 179
hackathons 221
health 72, 74, 76, 79, 97, 172, 175, 181, 199, 202, 209, 213, 263, 282, 289
heat, reuse of 52, 138–139, 160
heat island effect 122
heat pumps 105, 109, 114, 116, 121, 126, 194
heating 126; *see also* natural gas and heat pumps
hempcrete 147–148
High-Level Expert Group on Sustainable Finance (EU) 237
High Line, The, New York City 219–220
hinterlands 69, 76, 78, 81, 82, 90–91, 100
homelessness 169, 180

housing 26, 179, 180, 240–253
Human Development Index 66
human rights 26, 178
hydrogen fuel 106, 118, 125–126, 210
hydropower 119, 129

IBM 139, 144
Iceland (chain) 75
ICLEI 120, 130, 290
IKEA 71, 144
imagination 8
impact investing 229
incineration 120
indicators 26, 28, 39, 42, 54, 63, 102, 125, 174, 226, 255–260; in industry: 139; of PRINCE project, Sweden 55–57; of the European Environment Agency 44; of REAP 59; of Well-Being of Future Generations Act 268–271, 279–280
industrial cities 142–143
industrial symbiosis 137–138
industry 112, 122–123, 135–145, 155, 160, 166
inequality 19, 14, 22, 24, 58, 69, 211, 234, 282, 291
InfluenceMap 144
informal settlements 26, 52, 151, 169–170, 174–176
input-output modelling: see environmentally extended input-output analysis
intelligent transport systems 108
inspection and enforcement regime 138
Integrated Water Resources Management 151, 153
Intergovernmental Science-Policy Platform on Biodiversity and Ecosystem Services (IPBES) 15, 19
International Energy Agency 106, 109, 110, 139, 145
International Organization for Standardization (ISO) 28
International Panel on Climate Change 11, 16
International Water Association 165, 166
Internet of Things 143, 220–221
investment 109, 139–140, 145, 165, 175, 185, 213, 222, 224–238, 248
iron and steel sector 142–143, 191
ISO 14001: Environmental management 29, 46, 138
ISO 14007: Determining environmental costs and benefits 227
ISO 14008: Monetary valuation of environmental impacts 227
ISO 14030: Green bonds 227

ISO 14040: Life cycle assessment 40
ISO 14046 and 14073: Water footprint 38, 153–154
ISO 14073: Water footprint, further guidance 154
ISO 14097: Assessing investments related to climate change 227
ISO 16075 series: Wastewater reuse 166
ISO 24000 series: Drinking water 166
ISO 37101: Sustainable development in communities 41
ISO 37106: Smart, sustainable communities 43
ISO 37120: Sustainable development of communities: indicators 42–43, 46, 121, 218–219
ISO 37122: Indicators for Smart Cities 43
ISO 37123: Indicators for Resilient Cities 43
ISO 37151: Smart community infrastructures 222
ISO 50001: Energy management 29, 45, 140–142
ISO 9001: Quality management system 29, 46

Koch Industries 144

Land degradation 15–16
land rent scheme 170
land tenure 171, 173, 179, 248–250
LEDs (light-emitting diodes) 84, 88, 111, 251
LEED 38, 185, 189, 195
Leipzig Charter 42
life-cycle assessment 36, 38, 40, 52, 61, 62, 136, 142, 158, 221, 285
living buildings 195
Living Planet Report (WWF) 2, 10–18, 33, 35, 96
lobbying by corporations 144
local currencies 236–237
local multiplier effect 64, 120, 235–236
low-emission zones 206–208

Marks & Spencer 75
material flow accounting 38–39, 50–51
material footprint 26
measurement and verification 142, 222, 271, 286
mega-cities 21–22
Metro-Matrix Method 290
minimum energy efficiency performance standards 141
minimum income scheme 251
mobility: see transportation
motorbikes 212
MRIO: see EEMRIO
municipal bonds 185

National Association of Manufacturing 144
natural gas 106, 110, 111; substitutes for 123–127
nature-based services 165
negative carbon buildings 190
neighbourhood design 180–181, 191–192, 205–206, 217–218, 245, 289
Net Primary Productivity 31–32
Net Present Value Plus (NPV+) tool 66
net zero buildings: see zero carbon buildings
New Buildings Institute 185
New Economics Foundation 64, 235
New Urban Agenda 110, 178, 180
Ngabirano, Amanda 204
Nineteen Eighty-Four 7, 9
nitrogen pollution 10, 17–18, 55, 70, 159
Novacero 140
novel entities 18
nuclear power 110
nutrient recovery 70, 86, 98, 119, 157–158

ocean acidification 10
ocean death 18
'one planet' city criteria 13, 46
oil (fossil) 106
One Planet Council 268, 276
One Planet Development in Wales 11, 60, 196, 247, 267–268
One Wales: One Planet policy 59, 60
overfishing 18
Owens Corning 143
ozone depletion, stratospheric 10, 18

Palau Pledge 1
Paper, cardboard and newsprint, impact of 62, 122
Paris Agreement on climate change 16, 106, 108, 110, 123, 144, 151, 180, 229
Participatory Action Research 53
PAS 2070: Assessing a city's greenhouse gas emissions 37, 50, 66, 130
PAS 8820: Construction materials 147
passive building design 112, 117, 186–188
passive solar energy use 86, 187–188, 245
Passivhaus 188–189, 242, 243, 259
permeable pavements 155, 156, 166
Petropolis 99
Philips 136
phosphorus pollution 10, 17–18, 55, 70, 159
planetary boundaries 10–20, 44, 49, 65, 263, 269; and SDGs 26–27
planning, urban 52, 87–88, 100, 101, 177, 200–202, 205–206, 242, 246
plastic pollution 18, 97, 136

policy-making 53, 98, 101, 106, 107, 137, 141, 156–157, 161–162, 164, 180, 208, 218, 235; *see also* Well-Being of Future Generations Act, Wales
population, human 10, 19–20, 26, 33, 169
post-occupancy evaluation 189, 196
poverty 19, 22, 25, 79, 169, 199, 233
PRINCE project (Sweden) 55–59
privatisation 177–178
procurement, sustainable 65, 74, 75, 135–136, 148, 157, 235–236, 275, 278
property developers 176, 241–245, 246
public ownership 151, 163–164
public-private partnerships 178, 181, 230–231, 248, 258–260
Public Services Boards 272–273
public space 177, 180–181, 204
public transport 109, 175, 200, 207, 208, 216, 231

rainwater collection 86, 139, 153, 155, 158, 161
rapid transit buses: see bus rapid transit
REAP-Petite 60
rebound effect 61
recycling 26, 52, 108, 138, 146; *see also* circular economy
Reference Framework for Sustainable Cities 42
refrigeration 83, 86, 105, 123
regeneration/regenerative cities 96–100, 122, 154–155, 161, 195, 209, 248, 256
regulation 163, 232, 235, 245
renewable energy 104–106, 109, 110, 113, 114–128, 144, 287–288
renewable heat 117
Resources and Energy Analysis Programme (REAP) 59
resource efficiency 45, 110, 137–138, 255; also see circular economy
retrofitting buildings 112, 118, 140, 185, 193, 229
right to the city, the 178–179, 201
Rio On Watch 179
RUAF Foundation 76

safe public spaces 175, 178
Sainsbury's 157
Samsonite 137
sanitation 150
Sankey diagrams 51
sea level rise 17, 108–109, 153
SEEA (UN System of Environmental Economic Accounting) 57
self-drive cars 212–213
sense of place 178

sewage treatment 158–162, 252–253
ShareAction 238
Shell 144
Shinkansen Bullet Train 101
Singapore Centre for Sustainable Asian Cities 254
smart cities 176, 215–222, 251–252
Social Business City Program 234
social equity 11–14, 19, 233–235, 250–251, 263, 274, 289; *see also* inequality, public space, *and* informal settlements
social enterprise 233–235, 249, 250–251
social rent 249
soil loss 17, 95
solar cities 216
solar power 105, 106, 108, 113, 117, 119, 129, 186–189, 195, 245
So.vie.so, Vienna 241–242
sponge cities 166
sprawl 201, 205: *see also* urbanisation and suburbs
SSE 144
Standard Industrial Classification (SIC) System 60
standards, of sustainability 28, 39–47, 51, 135, 142, 290; advantages of 30; choosing 46, 65–66; popularity of 45–46; purpose of 28
Stockholm Environment Institute 34, 59
Stockholm Resilience Centre 10
street vendors 177
suburbs 52, 151, 200, 205–206, 209
superblocks 208–209
supply chain impacts 55–58, 110, 135–136, 146, 148
sustainability 13
sustainable cities 8, 30, 41, 54; compared 254–258
sustainable development (definition) 24–25
Sustainable Development Goals 10, 13, 25–27, 49, 58, 65, 95, 110, 111, 129, 140, 151, 169, 180, 226, 230, 235, 266–268, 275
Sustainable Food Cities Network (UK) 81
sustainable urban drainage systems (SUDS) 156, 158, 166
Sustainable Urban Neighbourhoods Network 181
Swansea, Wales 157
System of Environmental and Economic Accounting, UN 34, 39
System of National Accounts 39, 58

taxes 22, 71, 73, 175, 185, 207, 208, 224–225, 237
textiles, impact of 56, 62, 64
'Think#Eat#Save' tool 71

The City We Need 100
timber skyscrapers 191
Town and Country Planning Association (UK) 100, 246
trade, impact of 55–60
training, need for 141, 185
trees (in cities) 97, 100, 154–155, 209
traffic accidents/fatalities 199, 200, 208, 210–211
traffic congestion 200, 201, 204, 208, 212, 217
traffic management 200, 201, 206, 208–209, 217
Transparency International 260
transportation 69, 72, 118, 175, 199–213, 217–218, 257, 288

Uber 207, 217
UK Green Building Council 260
UNESCO 163
UN-Habitat 19, 49, 99, 178, 204
United Cities and Local Governments 179, 291
Unilever 135, 144
United Nations Committee on World Food Security 80
United Nations Conference on the Human Environment 24
United Nations Economic and Social Council 26
United Nations Environment Programme (UNEP) 24, 44, 51, 52, 71, 122, 165
United Nations Framework Convention on Climate Change (UNFCCC) 107, 279
United Nations Industrial Development Organization 30, 140–141
URBACT Smart Cities network 220
Urban Land Institute 52
urban developments, new 109–110
urbanisation 20–21, 80, 109, 145, 169, 184
urban metabolism 51–54, 98–99
urbanism 176
Urban Mobility Index 201
Urbinsight 53
URBIS 228
utopias 8

ventilation, natural 87
vertical farming 82–84, 86–87
Vienna Wild Garden Housing Project 242–243
Vision Zero 210–211

walking 109, 200, 202–203, 217
waste 70–71, 136

Waste and Resources Action Programme (UK) 71
wastewater 71, 77, 154, 156, 160, 166
watchdog, independent 273–274
water bonds 165
water footprint 13, 55, 56, 154; measurement of 38, 156
Water Framework Directive 154
water pollution 151
water quality 17, 253
water supply 17, 110, 150–167, 287; reuse 86, 157, 160–161, 166; scarcity 150
WaterAid 163
Watersense 157
Water-Wise Cities principles 166–167
Water Technology List 157
Well-Being of Future Generations Act, Wales 45, 60, 262–284
wetlands 15, 150, 154; constructed 159; restoration 100
wind power 105

wisdom of crowds 219–220
World Bank 129, 213
World Benchmarking Alliance 230
World Business Council for Sustainable Development 155
World Council on City Data 44, 49, 54, 218
World Economic Forum 70, 136, 177
World Energy Council 106
World Futures Council 97, 120
World Green Building Council 112, 185, 232
World Resources Institute 130, 184, 208
World Urban Campaign 99
World Water Council 165
World of Three Zeroes, A 20, 234
WWF 2, 10, 27, 34–35, 264, 277

zero carbon 106, 287; buildings 111–112, 184–186
zoning 177,

Eurovision
A Plea for Respect

(Continental Songs and British Attitudes)

Steve Kerr

Copyright © 2021 Steve Kerr
All rights reserved.
ISBN:

Preface

The main purpose of this book is to educate and inform. While the Eurovision Song Contest is held in high esteem in several parts of Europe and beyond, there are still those – often, but not always, in the UK – who, through various media outlets in the last half century or more, have become conditioned to think of the Eurovision Song Contest as an annual event that they view with some degree of negativity, even if a pang of love might sometimes accompany that feeling. Travelling through different parts of Europe and living in a few of its countries over the last fifty years has made me very much aware of the diversity and quality that the continent's music has to offer those outside national borders, which is, more often than not, ignored on this side of the channel. This has often been manifested in varying genres and guises on Europe's biggest stage. Through the decades, the Eurovision Song Contest has come to symbolise an inclusive veritable pan-European unity of sorts: a mega festival of sheer Europeanness and the continent's annual celebration of itself in song. It is now far more than a mere song contest, and dated attitudes towards it must come to an end. In this book, we go on a journey of discovery and uncover the quality found in the contest's multitude of entries as well as its importance in Europe's recent history. The book asks that the contest, as an event, and also its many songs, be respected and indicates why they should be. I have watched the annual event since childhood, in what has become a much-loved family tradition, and this book is also a tribute to this grand institution. At a time when Europe is facing great disharmony through war and political disunity, the contest has the unique power to unite the continent through music – if not governmentally or uniformly then certainly emotionally and melodiously.

Dedicated to my brother, Alistair.

Contents

Chapter 1
Once Upon a Time 1

Chapter 2
Lilli Marlene 13

Chapter 3
The San Remo Connection 19

Chapter 4
Carnations, Fado, and Revolution 76

Chapter 5
A Vous La France 93

Chapter 6
"La, La, La" 136

Chapter 7
Are Rockall And Azerbaijan In Europe? 183

Chapter 8
"Colonels, Culture, Wine, and Sea" 215

Chapter 9
Ein Lied Fur Madrid 237

Chapter 10
British Attitudes 261

Chapter 11
Through the Star Spangled Sky 351

Epilogue 358

Bibliography 359

Eurovision: A Plea for Respect

CHAPTER 1
Once Upon a Time

Once upon a time when the world was young and Europe was only just beginning to leave behind its sick bed and head out, with crutches, on the road to recovery, in those gloomy lean years that followed a devastating war, someone had an idea.

Hollywood had long had a monopoly on the film industry, and American 75s and singers would come to increasingly dominate the European music scene as the 1950s went on. This was the dawn of the Americanisation of Britain and Europe. Chewing gum and Coca-Cola took the continent by storm. In Britain, the very first music Hit Parade had been compiled at the dawn of the decade and was preponderated by the likes of Johnny Ray, Frank Sinatra, Frankie Laine and Patti Page. In the latter half of the decade, singers like Bill Haley and Elvis Presley would make serious inroads into the British top twenty. Often, home-grown British artists would record cover versions of American hits with varying degrees of success. There was grave concern, in disparate parts of Europe, that Europeans were turning their backs on their own culture and local artists. It had been a good fifteen years since Lahle Andersen's *Lilli Marlene* had crossed borders and tanks in war-torn Europe. In 1954, someone had an idea. That someone was Marcel Bezençon, the Director General of Swiss Television and later Director of the European Broadcasting Union. He had the idea of a song contest in which different European countries would compete. The concept had its recent roots in the San Remo Song Festival, held annually in Italy since 1951. Like the Olympic Games, the idea of a song contest between different states had its origins in ancient Greece, where the ancient peoples competed against each other in song. Bezençon thought that such an idea might help bring Europe, which was still recovering from the horrors of war and genocide, closer together. At the same time, it was thought that such a contest might stem the tide of the Americanisation of European culture and cultivate more interest in local European songs and artists. An event such as a *Grand Prix de la Chanson* which would be shown simultaneously in several countries would be an experiment in the pioneering of international live broadcasts. Marcel Bezençon had been born on May Day 1907 into a Europe of gaslights, horses, carts and powerful empires. The creation of the European Broadcasting Union on June 6, 1954 marked further advancements in communication technologies when the first transmission by the network of the Narcissus Festival from Montreux took place. Four million TV viewers in Germany, Belgium, the Netherlands, France, Italy, the UK and Switzerland excitedly watched the live broadcast of the festival's flower-bedecked procession.

This first adventure into a new technical universe was followed by many other shared projects such as the 1956 wedding of Prince Rainier and Grace Kelly. In 1958, news exchanges between EBU states were initiated. In 1961, it again broadcast history in the making when Yuri Gagarin's jubilant return to Moscow was shown live.

The name Eurovision was given to the EBU's network by the journalist George Campey in the London Evening Standard in 1951. The year 1962 saw the launch of Telstar, which linked the Eurovision member countries and the USA. This satellite would cover all historic events of the era. Viewers in Europe were able to watch the funeral of President Kennedy and, in 1969, Eurovision broadcast from the moon in the longest transmission since its creation. World history from coronations to the Olympic Games, from earthquakes to football matches, from the greatest statesmen to the greatest moments of the twentieth and twenty-first centuries – all have been transmitted via Eurovision. The EBU now offers the most extensive global transmission network.

Bezençon presented his idea for a song contest at a meeting of EBU members in Monaco in 1955. His idea came to fruition on May 24, 1956. Little did he know that this one-off experiment would turn into a regular annual event and become the world's most-watched television show. Bezençon retired as director of the EBU in 1971, and

Eurovision: A Plea for Respect

when he died in 1981, Western Europe was a very different place from the one into which he had been born: an affluent society with great technological advances and the Eurovision Song Contest, by then a European tradition that had recently celebrated its silver anniversary.

The Eurovision Song Contest had its more recent origins in the San Remo Festival but its roots in the modern era went back much further. Neapolitan Song was formerly popularised at a song festival which predated San Remo but on which San Remo and, in turn, the Eurovision Song Contest, were based. The song contest was called the Festival of Piedigrotta and was held annually from its inception in 1830 until 1950. The song festival was dedicated to the Madonna of Piedigrotta, a famous church in the Mergellina area of Naples. The winner of the 1830 song contest was *Te Voglio Bene Assaie*, which is often credited to the opera composer Donizetti, although there are no records to verify this. The festival, in which all the songs were sung in Neapolitan, was an important feature of Neapolitan *Canzone* for over a hundred years. RAI re-established another Festival of Neapolitan Songs on Italian radio in the 1950s. It was broadcast for a number of years but then it was also abandoned. In its one hundred and twenty years of existence, the Festival di Piedigrotta produced several popular songs. One of the most famous of the entries was *Funiculì Funiculà* by Italian journalist Peppino Turca and composer Luigi Denza, which was entered into the 1880 song contest. From Piedigrotta, the songs that enraptured the audiences slowly made their way to Rome.

* * *

De Vogels Van Holland – the birds of Holland, the very first song ever performed at a Eurovision Song Contest, was the Dutch entry performed by Jetty Pearl and one of two songs that the Dutch national broadcaster NOS had submitted to the 1956 contest, which was held on May 24 in Lugano, a lakeside resort in the Ticino Italian-speaking region of Switzerland. Jetty Pearl, a Dutch Jew, had been based in London during the Nazi occupation of the Netherlands and had broadcast on free radio to her beleaguered homeland from the BBC in London.

> *"The birds of Holland are so musical*
> *They already learn to twitter in their early youth*
> *The blackbird, the thrush and the nightingale*
> *So they can celebrate springtime in Holland"*

It was appropriate that the opening song at the debut song contest, which heralded a new era in West European relations, was sung by someone whose voice was so much associated with freedom, democracy and resistance to tyranny. The song was composed by Cor Lemaire and Annie M G Schmidt. It was sung in Dutch and was the first entry of the evening conducted by Fernando Paggi in the *Chanson* tradition, which most of the other competing entries also were. In the song, she addressed those listening as "Sir" and "Madam", an example of a much more formal approach and even a distance between singer and audience which has since become one of familiarity. Jetty Pearl was one of two Dutch representatives at the 1956 contests, with the other being Corry Brocken, who sang *Voorgoed Voorbij*. Unlike in all the Eurovision Song Contests since, there was no scoreboard at the contest and the jury's votes for all the songs were never to be revealed. The placing of Eurovision's first-ever song is, thus, unknown.

The EBU rules for 1956 stated that each country should present two songs. Seven countries took part. The original Eurovision Song Contest competitors were France, Belgium, Switzerland, Germany, Italy, Luxembourg and the Netherlands. This meant that a total of fourteen songs competed. Austria, Denmark and the United Kingdom had also agreed to participate but because they missed the deadline for submissions, they were unable to compete at Lugano. The 1956 contest stipulated that each juror from each competing nation should travel to Lugano and that the jury should cast its votes behind closed doors. Some controversy surrounded the voting procedure. This was long before the term "block voting" was invented and concerned the juror from Luxembourg, who was unable to travel to Lugano. The voting rules enabled a jury to vote for any participating country, including their own. In the absence of a juror from Luxembourg, their voting duties were assigned to the juror from Switzerland. The 1956 Eurovision Song Contest was won by Switzerland with the song *Refrain*, sung by Lys Assia. This led to discontent and accusations in some quarters that Switzerland had used its position, as Luxembourg's stand-in jury, to vote itself into first place. There was no strong evidence of this, but the 1956 voting system was never used again. Lys Assia, as the first Eurovision Song Contest winner, occupies a special position in history and was latterly held in high esteem by all who followed the song contest and those who organised and attended the annual event. *Refrain* was the second Swiss song sung by Lys Assia on the night of the song contest. Lys, who had started out as a dancer, had a pan-European hit with the 1950s classic *Oh Mein Papa*, of which she was the original interpreter. The song was a success in English-speaking countries for Eddie Fischer, and an instrumental version of the song topped the UK charts in 1954. Lys Assia's original German version of the song is a far more expressive and lyrically interesting one. Her version of *Oh Mein Papa* features sentences of fractured German grammar with obvious simple grammatical mistakes and brings to life a much-assumed German idea that all circus people were of foreign extraction. Lys Assia's version conveys a sense of tragic-comedy surrounding her "*papa*", a circus clown, which is missing in the English translation. Lys Assia would return to defend her Eurovision title on various occasions and make several attempts at the national finals.

Less than a year after her Eurovision victory, Lys Assia married Johan Kunz, who was ill. This extremely difficult period coincided with the height of her career both at home and internationally. This also meant that she was unable to spend much time with her husband at a crucial period in their relationship. After only nine months of marriage, Lys found herself a widow. She had paid a high cost for winning. The song, *Refrain*, was penned by twenty-five-year-old Emile Gardaz and Geo Voumard. Emile Gardaz was also a bestselling author. Geo

Voumard, on the other hand, was a jazz pianist who founded the Montreux Jazz Festival in 1967. *Refrain*, which Lys Assia sang in French, was of the *Chanson* style, with traces of jazz influence in it. The lyrics express a looking back in regret and sadness at the past loves that the singer experienced in her twenties.

> *"Refrain, Colour of the sky, scent of my twenties*
> *Garden full of sun where I spent my Childhood*
> *I've looked for you everywhere, my love from the distant past*
> *I wait for you on the path where you once held my hand*
> *The days passed and we grew up*
> *Love wounded but time cured us*
> *Lonely and without any springtime*
> *I run through the woods and fields in vain*
> *Do you remember the love we once had, long ago?"*

Although the song came first, due to the secrecy of the voting, the number of points it received was never revealed. In an interview for Dutch TV from her home in Cannes in 2006, Lys Assia lamented the passing of the early Eurovision Song Contests and expressed a preference for the elegance of the contests in the 1950s, with their black bow ties and satin ball gowns, over the glitz and technology of the early twenty-first century contests. For many years, Lys Assia was absent from the Eurovision stage but as the contest grew and matured, interest in its history grew. There was particular interest in Lys Assia, the first winner and, therefore, an icon of the event. Lys returned as an honorary guest to several of the early twenty-first-century contests and even jointly presented the winner's trophy to the winning Norwegian singer in Moscow in 2009. By this time, she was well into her eighties. Only a few years earlier, while on holiday in Denmark, Lys Assia was involved in a fatal car crash in which her second husband was killed. She survived but was seriously injured. Her dedication and professionalism saw her return to show business as highly polished as ever. early 2013, Lys Assia developed severe pneumonia and became

Eurovision: A Plea for Respect

seriously ill but she miraculously recovered and, in her eighty-ninth year, made yet another appearance at the Eurovision Song Contest the following May. Lys Assia stated in an interview that she was proud to be the winner of the first Eurovision Song Contest and described the event as *"sensational"*, citing that reason for its continued success. Not only did Lys Assia enjoy iconic status attributed to being the first winner but she also became an ambassador for the song contest later in life. This is a fact that the EBU should perhaps have recognised with a special award.

"The years pass quickly by like lightening
and upon the roof of my concern pours the rain
Where did the ship that stole my heart leave?
It carried my dreams away from your memory
I wanted you to come back and it to be just as then
When you brought flowers to the
Shutters of my window
and your youthfulness to my home

Refrain, colour of rain, regret of my twenties
Sorrow and sadness at not being a child anymore.
Lonely and far away from you
I walk the roads where you are no longer
I go on my way and cry for those who I once loved in my twenties"

There was still a great demand for sheet music back in 1956, although this was fast subsiding. As a result, the Eurovision entries from Belgium and France were never recorded. While a sound recording of the song contest exists, there is no televised recording still in existence, although a televised clip of Lys Assia singing the reprise of *Refrain* as part of a newsreel has survived. It is one of only two contests for which no televised recording remains. (The other is the 1964 Eurovision Song Contest, which, due to a fire, exists only as a sound recording with stills.)

Three more countries made their debut at the 1957 event, held in Germany. It included those who had missed the submission date in 1956, which brought the number of competing nations to ten. Between the first year of the song contest and the second staging, there had been a huge rise in the number of television sets purchased throughout the continent, making it possible for a large section of those tuning in to *watch* the contest rather than *listen* to a radio transmission. Berkley Smith commented for BBC television. A war veteran, Smith had been appointed British press liaison officer at the Nuremberg trials and was one of the BBC TV commentators for the coronation of Queen Elizabeth in 1953. For the first time, the juries were contacted by phone in their home countries and their votes were made public. The British votes were read from London by the popular DJ David Jacobs, who would commentate at several Eurovision Song Contests for the BBC until the mid-1960s. He was also to be a regular presenter of *Top of the Pops*, the BBC mirror of the 1960s, and would still be presenting his own Sunday evening radio show for the BBC in the 2010s, when he was over ninety.

The voting system which was to prove successful and full of suspense for viewers over several years was introduced. Each jury had ten jurors and each member gave one point to their favourite song. This system of voting provided far more excitement in its sheer unpredictability. A jury was capable of giving its ten points to one song or going to the other extreme of allocating one point to ten countries. Juries were, however, more likely to give less extreme votes, usually allocating ones, twos, fours, and the occasional five, although in 1958 Austria awarded nine points to France. Of course, such a system would prove impractical, if not impossible, nowadays due to the sheer number of participating countries.

Corry Brocken of the Netherlands sang *Net als Toen* for her country. It had been chosen at the Dutch National Song Festival, where it had been one of eight competing songs sung by four singers. The others were Hea Sury, John de Mol, and Marcel Thielemans. *Net Als Toen* was voted the winner by postal vote – a clearly popular choice with far more votes than the runner-up. It was also a manifestation of how popular Eurovision was becoming and how much interest it aroused among the public. At the 1956 Dutch final, the number of postal votes cast was 6,694. In 1957, the top four placings alone received 14,858 votes – a sign of what was to come.

As had happened at the national final, Corry Brockens' *Net Als Toen* led from start to finish at the Eurovision Song Contest. *Net Als Toen*, in fact, set a record. In the system of voting used in 1957, which was used regularly until 1970 and then revived as a one-off in 1974, it was the only song to receive votes from every competing country. Its highest points came from Switzerland, which gave it seven votes. Despite the song's runaway victory at both the Dutch national final and the Eurovision Song Contest, it was not a commercial success and failed to have any impact on the Dutch hit parade or internationally. Onstage at the 1957 song contest, Corry Brocken was accompanied by the accomplished and well-known violinist Sem Nijveen.

The 1957 event, held in the West German city of Frankfurt Am-Main, which had been flattened in World War II, coincided with the impressive rebuilding of the city. The year 1957 witnessed the rejuvenation of the devastated metropolis and the construction of Frankfurt's first high-rise flats. It was an interesting time to live in Frankfurt, one of Germany's leading cities and industrial centres, as well as a European financial stronghold. The population felt a new sense of hope and optimism – a sentiment that welcomed the *"baby"* Eurovision Song Contest as a symbol of a New Europe and a new democratic, prosperous West Germany. It had been widely rumoured that Germany had come in second at the 1956 song contest but there was no evidence to support this. Frankfurt was chosen as the host city because, up to that time, the rules for hosting the event were that each country should stage the contest in turn. As the contest grew, it became more practical for the winning nation to host it in the year following its win. Corry Brocken's victory paved the way for this, and the 1958 contest was held in Hilversum.

The live orchestra at the 1957 Grand Prix played against the backdrop of a picture of Berlin's Branden Burger Tör, a powerful symbol at the time, while the hostess for the evening, the Berlin-born Anaid Iplican, who was German-Armenian, walked down steps that led onto the heavily draped stage. It is of note that the presenter chosen by the German host broadcaster, ARD, was not a full-blooded German but of mixed ancestry. It was an attempt by Germany to cast a new light on itself, far from the nation of Hitler's eugenics and notions of a pure Aryan race. Only the second person to present the song contest, she took over from Lohengrin Filipo, who had acted as host of the 1956 contest in Lugano. The commentator for German radio and television was the Bavarian-born radio host and journalist Wolf Mittler, who had been employed on German Radio's *Germany Calling*, which broadcast propaganda to the UK and America during the Nazi era. He was said to have felt uncomfortable with politics and resigned from his post, which was filled by the British-born Nazi spy Norman Baillie-Stewart. Mittler fell afoul of the Nazi Party and, under suspicion, fled to Italy. While in Italy, he was arrested by the Gestapo but managed to escape to Switzerland, where he lived until the fall of Hitler. He returned to Germany and resumed his career. Mittler commentated for both German and Austrian broadcasting at the first three song contests and was famous for his simultaneous translation of President Kennedy's Cuban Missile Crisis speech of 1962. He spoke flawless English, which he had spoken in childhood with his mother, who was Irish by birth, and this had proved to be a vital tool for the Nazis in his propaganda broadcasts on *Germany Calling*. His voice became familiar to many British listeners in the 1940s, who knew him as Lord Haw Haw. The lyrics of the German entry from 1956, *Im Wartesaal Zum Grossen*, dealt with the refusal by many Germans in the post-war years to deal with the Nazi era and their collaboration with the regime. It was sung by Walter Andreas Shwarz, a Jew who had been deported to a concentration camp with his parents, both of whom perished. Walter's survival was due to one of the Nazi guards at the camp being an old school friend who saved him from a similar fate. The 1957 contest was presented entirely in German. Unlike 1956, when only solo artists were permitted to take part, 1957 brought a change of rules, and duos were allowed to compete. Corry Brocken sang her entry while wearing a long, sleeveless evening dress with long, white evening gloves – an example of the elegance so lamented by Lys Assia. *Net Als Toen* was a ballad in the *Chanson* style. The lengthy violin solo halfway through exhibited a change in genre and showed a jazz-swing influence. It told of a married woman caught up in a mundane, boring marriage and of how she would like her husband to take more of an interest in her, *like he used to, just like then* – "*Net Als Toen*".

France had a notable entry, *La Belle Amour*, sung by Paule Desjardins. It was a typical example of the French *Chanson* of the 1950s. The song placed second, while the beautiful, melancholic ballad *Corda Della Mia Chitarra*, which lasted over five minutes and was sung by Italian singer Nunzio Gallo, was accompanied by a guitarist on stage. Among the competitors was, again, Lys Assia, whose flawless performance received rapturous applause. There was no interval act. Shortly after the notes of the last song had faded, the hostess immediately made contact with the juries. It was the first time a presenter had ever called a jury for its votes. The scoreboard looked extremely different from today's and exhibited only the song titles and not the countries' names. Corry Brocken, who had sung for the Netherlands in 1956, reappeared and sang for her country in 1958 with *Heel de Wereld*. On that occasion, she was not so lucky; she tied for last place and remains the only singer to both win the Eurovision Song Contest and come in last. She enjoyed great popularity and had several hits in the Netherlands in the 1950s and 1960s, many of which were Dutch translations of *Chansons* by Charles Aznavour. Corry was hostess of the 1976 contest in The Hague. She then retired from the music business the same year to study law, in which she subsequently embarked on a career. Some years later, she became a judge.

The earliest Eurovision Song Contests fade further and further into a black-and-white haze. Memories of them are blurred. They slip into history; they and their music, like the Europe that was, are now part of a bygone era. Marc-Antoines Charpenteirs' *Te Deum* sounds throughout Europe and beyond; it is time for the Eurovision Song Contest to start once again.

CHAPTER 2
Lilli Marlene

Lilli Marlene, a truly international hit – a term often used today but the word *international* is so easily earned these days and so frequently given. *Lilli Marlene*, which toiled tirelessly to establish itself as Europe's most popular song of the 1940s, crossing borders, through the blitz, concentration camps, and tanks in war-torn Europe. It had started its life as a poem, *Das lied Eires Junger Soldaten Auf De Wacht* (the song of a young soldier on the watch), by Hans Leip, a teacher from Hamburg who, as a young conscript, composed it in 1915 during World War I. The words were put to music by Norbert Shultze in 1938. The German singer Lahle Andersen first recorded the song as *Das Mädchen Unter Der Laterne* (*The Girl Under the Lantern*) in the spring of 1939. The song, unknown to Lahle at the time, would go on to be the best-known of all the songs associated with World War II. Although Lahle quite liked the song and enjoyed singing it, it was initially not very popular and lay forgotten in the cellars of a Viennese broadcasting station until the Nazis overran and occupied Yugoslavia. They took over Radio Belgrade and, to entertain the troops, brought some records they had found in Vienna. One of the records happened to be *Lilli Marlene*. They broadcast the song with regularity from the radio station and it soon became a firm favourite among the German troops. While the song was rapidly becoming more and more popular, Lahle Andersen, hundreds of miles away in the German capital of Berlin, had no idea that her record was being played in Southern Europe. After a tour of hospitals during which she visited wounded soldiers, she arrived back home in Berlin to find it covered in hundreds of letters from troops, many of them requesting a photo of her along with the lyrics to *Lilli Marlene*. This all came as somewhat of a surprise to the singer. She may have acquired thousands of new fans but Göebels was not happy about her success. The army, which played a major role in Lahle's rise to stardom, was a separate body from the Nazi Party. Göebels had had no say whatsoever in the matter and his pride was slighted. Göebels then gave strict instructions for Lahle to be watched day and night. Nazi spies soon found a piece of news about her life which Göebels found most disturbing: that she had a boyfriend who was a Swiss Jew. The Gestapo arrested her one day in Milan station, where she had attempted to board a train to neutral Switzerland. She was taken back to Berlin, where she was given strict orders to never work with the troops again. If she ignored this, she would be sent to a concentration camp. The Nazis allowed her to briefly return to her flat to collect some personal effects but, while in the flat, a distraught Lahle took a drug overdose. She was found a few hours later. The Propaganda Minister was called. He commanded them to do everything in their power to save *Lilli Marlene*, and Lahle was immediately rushed to hospital. Meanwhile, the BBC broadcast that *Lilli Marlene* had been murdered by the Nazis. News like that would have been a terrible fall from grace for Nazi propaganda. They reacted immediately, vehemently denying this and accusing the BBC of lying. Lahle withdrew to a small East Frisian island in the North Sea, off the coast of Northern Germany, where she kept a low profile, away from the Nazi think tank in Berlin. Although she remained out of favour with the Nazis and was always in their black books, she managed to survive there until the end of the war and Hitler's demise. A mystical transubstantiation had happened: Lahle Andersen and *Lilli Marlene* had become one, something she felt rather proud of. Her song was a true Pan-European hit, one that had far outshone all other European wartime hits and had been written into the history books. The Irish songwriter Jimmy Kennedy soon put pen to paper and wrote the English lyrics to *Lilli Marlene*. He had also composed *The Teddy Bears' Picnic*, his first success, which sold more than four million copies, as well as the wartime hit *We're Going to Hang Out the Washing on the Siegfried Line*. In 1957, he entered the British heats of the Eurovision Song Contest with *I Had Two Dreams*. He entered a second song in 1962, *My Kingdom For A Girl*. Neither, however, was chosen to represent the United Kingdom. Kennedy's best-known composition was the dance song *The Hokey-Cokey*, which has become part of British folk culture and is still traditionally sung at celebrations today. While Kennedy was responsible for the adaptation of *Lilli Marlene* into English, it was Dulwich's Anne Shelton who was the first to sing and record the English version, which she is credited with in the United Kingdom. Marlene Dietrich is credited with a different English language version in the USA. Anne Shelton had a string of hits both during and after the war. In the mid-1950s, she topped the charts with *Lay Down Your Arms*. In 1956, she sang three songs in the Festival of British Popular Songs, meant to choose the British entry at the first Eurovision Song Contest. She did not meet with much success, although one of her songs, *The Heart Of A Child*, came in third. She would again enter the British heats one more time. The lineup of acts against whom she competed included the Johnston Brothers, who topped the British charts in 1955 with *Hernando's Hideaway*, and The Keynotes, both of whom were founded by Johnny Johnston, (AKA Johnny Reine), who also wrote the sequel to *Lilli Marlene: The Wedding* of *Lilli Marlene*. As Allied bombs fell like hailstones on the last weeks of Hitler's Berlin, . On the airwaves, the most popular song was the highly optimistic *Ich will es wird einmal ein wunder*, sung by Zara Leander, a Swede who chose to live in Germany. On her return to her homeland, she was shunned by Swedish society, which saw her as a willing participant in Hitler's propaganda machine. However, she never joined the Nazi party or attended official Nazi functions and was genuinely popular with the ordinary German public who, by 1945, in the big cities, had begun to live a cryptic, rat-like existence. Her pre-end-of-war hit song included the lines, "*Every January is always followed by a May, Every winter by a Spring*". Lahle's career returned after the end of the war. In1959, she manifested her ongoing popularity and competed in the German Eurovision heats but was unsuccessful.Lahle re-entered the German National final for the Eurovision Song Contest in 1961 with the song *Ein mal Sehen Wir uns Wieder*. She competed with thirteen other hopefuls, including Christa Williams, who had sung for Switzerland in 1960. The others in the lineup were Heinz Sagner, Reneé Franke, Rolf Simson, Ernst Lothar, Bobby Franco, Peggy Brown, Deitlef Engel, and Fred Bertelmann. The song, a ballad, is about saying farewell to a lover and making a promise that they will see each other again … next year … maybe. Many of Lahle's most popular songs reflected her life spent near the sea. She had been born in Bremerhaven on the North Sea Coast. Thus, several hits were about longing for her lover at sea, ships leaving the harbour, falling in love with sailors – somewhere in all her songs is the sea. In 1960, she enjoyed great success with the Manos Hadjidakis song *Ein Schiff Will Kommen*, the German version of *Ta Pedia Tou Pireau*, which was featured in the Greek-American film *Never on a Sunday*, sung by Melina Mercouri. Lahle was chosen to sing *Einmal Sehen Wir Uns Wieder* for Germany, which then meant the ARD of West Germany as opposed to the East German broadcaster in the DDR. Europe was still very much divided by the Iron Curtain. The Cold War was in full swing. Germany in particular was at the heart of this, being divided into a communist state and a capitalist state. Berlin was the romantic divided city of espionage, while its western sector was an island in the sea of Eastern communism.

The contest was held at the Palais des Festivals et des Congrés in Cannes on March 18 and was introduced by Jacqueline Joubert, who had also been the speaker at the same venue for the 1959 event. Jacqueline was the mother of British-French TV presenter Antoine de Caunes. At the time of the 1961 Eurovision Song Contest, Lahle Andersen was the most famous singer to sing at the contest due, of course, to *Lilli Marlene*. She was also the oldest contestant ever to compete at the event: aged fifty-five, only a few days away from her fifty-sixth birthday. (She was overtaken by Croatia's Dado Topič in 2007.) During her sojourn in Cannes, she met up with her British wartime contemporary, Vera Lynn. She performed eighth, immediately after the Swedish singer, Lill Babs. The German commentator Wolf Mitter provided the introduction to West German viewers as *Lilli Marlene* made her entry on the stage, which was bedecked in flowers. The song featured a trademark of her work: It commenced with a spoken word singing style. Although the song was sung in German, she did sing the refrain in French. In 1961, the EBU had no specific rule in regards to official languages or singing in another country's language. In fact, there were no language rules at all. During the voting, Germany gave five of its ten votes to Luxembourg's *Nous Les Amoureux*, which won the contest. Luxembourg gave the United Kingdom eight points and Norway gave Denmark eight votes. Each country had a maximum of ten points which they could allocate as they saw fit. This was the biggest number of points given by a jury since 1958, when Denmark gave France nine points. *Einmal Sehen Wir Uns Wieder* managed to pick up only three points and came in thirteenth out of sixteen songs. Lahle received one point each from Denmark, Sweden, and Austria. It was Luxembourg's night, with Jean Claud Pascal taking the Grand Prix. The UK's *Are You Sure* by the Allisons came in second, while Switzerland's Franca di Rienzo and France's Jean-Paul Maurice came in third and fourth, respectively. It was a disastrous result for someone who was not only well-known in Europe but also its *special* star. Maybe Lahle reminded the other European countries too much of the war. At the dawn of what would prove to be a revolutionary decade, the world and its music were beginning to move on ... or were they? The following year, 1962, the American singer Connie Francis recorded the original German version of *Lilli Marlene* and reached number nine in the German hit parade. In 1961, Hollywood made the film *Judgement in Nuremberg*, which starred Spencer Tracey, Burt Lancaster, and Marlene Dietrich, who sang *Lilli Marlene* in the film. By coincidence, Anne Shelton, who had first sung *Lilli Marlene* in English, returned to the British heats in 1961 with the song *I Will Light a Candle*, but was, once more, unsuccessful.

Lahle Andersen died of liver cancer in a hospital in Vienna in August 1972, and a statue of *Lilli Marlene* was erected on the small East Frisian Island of Langeoog where she had once lived. During the Nazi regime, Lahle Andersen/Lilli Marlene had, to the Allied nations, personified the *good* German who had risen above the Nazi value system. *Lilli Marlene* had united Allied and Axis forces, people who had hated each other, through the love of a song. Amid the war, Europe was subconsciously creating the foundations of a new shared culture and collective identity that had not been seen since the days of the Holy Roman Empire. Both *Lilli Marlene* and the Eurovision Song Contest helped unite Europe and contribute towards a collective Pan-European identity both during the conflict and after it. Both have proved that songs can sometimes be more powerful than politicians. The Eurovision Song Contest is proud that *Lilli Marlene,* herself, once graced its stage.

Eurovision: A Plea for Respect

CHAPTER 3
The San Remo Connection

The attractive Ligurian city of San Remo is nestled on the Italian Riviera near the border with France. It has much in common with the cities of Nice and Monte Carlo farther down the coast. A popular holiday destination since the nineteenth century, it boasts a casino and several grand hotels. The Empress Elizabeth of Austria, Tsar Nicholas the Second of Russia, and Alfred Nobel were just some of the seasonal residents attracted to the mild Mediterranean climate. The palm-lined coastal resort is famous for its celebrated showcase of Italian song, the San Remo Music Festival, which has been held every year since 1951.

The Festival Della Canzone Italiana di San Remo, forerunner to the Eurovision Song Contest, was held in the City's Casino until the mid-1970s, when it was relocated to another theatre. It was established to attract investment and rebuild the city's economy after the devastation suffered throughout Italy in the Second World War. It was first held on the last three evenings of January 1951, with only the final evening being broadcast on radio by the Italian state broadcaster RAI. To understand the Festival of Italian Song at San Remo in the 1950s, it is necessary to go back to its birth and the emerging Italian democracy of the time. San Remo was born in the shadow of a right-wing bias that came about as the Cold War was launching a freeze over Europe. It was a different type of war that Europe had never seen before. Italy had just banned left-wing parties from its government as a direct consequence of the Cold War and the fear felt by Europe, whose mercy it was at. This all resulted in a deeply conservative right-wing impact on Italian society which included an extremely conservative government's hold on RAI and the programmes that it broadcast. In 1955, the festival was broadcast for the first time on television. Part of it was shown in five other West European countries via Eurovision. It was important for Italy, having only recently thrown off its fascist garments, to be seen as looking outwards, reaching towards Europe, and not seeming to appear introspective and ultra-nationalist. Several other countries soon began to televise San Remo annually. As its popularity grew, even more countries televised it, turning it into the centre of an international spotlight. However, it did experience some decline at points in the last twenty years.

The song festival has had several formats since its inception in 1951. The first event showcased twenty new Italian songs which were sung by the singers Nilla Pizzi and Achille Togliani, backed by the Duo Fasano, with the winner emerging from the twenty: Nilla Pizzi with the song *Grazie Dei Fiori*. She was to win again the following year with the song *Vola Colomba*, which was bound to become a much-loved Italian classic of the twentieth century.

The rules changed in 1953, when each song had to be performed by two different artists with different arrangements and by two separate orchestras. The format changed once again in 1964, when it was decided that each song would be sung by both an Italian singer and a non-Italian international act who would interpret the song in different styles. This was altered further in 1967, when songs were not obliged to be sung by foreign acts. Double performances were, however, still an integral part of the festival. This lasted until 1972, when each song was sung by only one artist. The 1950s saw among its winners *Grazie Dei Fiori*, *Viale d'Autunno* (Carla Boni), and *Buongiorno Tristezza* (Claudia Villa), all of which had rather haunting melancholic tunes sung beautifully in a manner befitting the music. The composer Mario Buongiorno Ruccione wrote the 1935 fascist battle hymn *Faccetta Nera*. Since the fall of *Il Duce*, he had turned his talents towards composing romantic ballads. He composed the 1955 San Remo winner *Tristezza* and the 1957 winner, *Corde Della Mia Chitarra*. Songs at San Remo in the 1950s often contained lyrics that were nostalgic for the fascist era; others had patriotic tones running through them, such as Nilla Pizzi's *Vola Colomba*. In 1956, it was decided that the winner of this most Italian of song festivals should go on to compete further as

Eurovision: A Plea for Respect

the Italian entry in the Eurovision Song Contest. This annual selection process would last until 1966. After the glory of success at San Remo, the Eurovision Song Contest was to be an often perilous journey through the high waves of a dangerous sea. Only one of the San Remo winners would go on to win the Eurovision Song Contest: Gigliola Cinquetti with *Non ho l'eta* in 1964. Several of the San Remo winners fared very badly and even came in last. It is, however, from this era that one of the most widely recorded and well-known Eurovision songs came. The 1958 winner of San Remo, *Nel blu dipinto di blu*, better known as *Volare*, represented Italy at the Eurovision Song Contest in Hilversum. It was sung at San Remo by Domenico Modugno and Johnny Dorelli and marked Dorelli's debut appearance at San Remo. He was reportedly so nervous that he had problems going on stage. This was rectified when Modugno punched him in an effort to get him on stage. The song was at first rejected by the jury from the festival but was, after some discussion, included in the final twenty songs for competition at San Remo. *Volare* was a monumental composition in the history of Italian music. During the performance, Modugno spread out his arms as if he were going to fly (*Volare*) to illustrate the lyrics – a manifestation of his background as an actor. This was a revolutionary departure from the standard manner of singing for Italian artists who, pre-*Volare*, did not move on stage but stood with their arms covering their chests. His rendering of the song at San Remo is now considered a key moment in Italian music, such was its impact. Nerves very much affected Johnny Dorelli's performance, which had little impact at the festival. Domenico Modugno's version won, beating Nilla Pizzi and Tommy Torrielli's *L'edera*, in second place. Modugno was dispatched to the Eurovision Song Contest amid much excitement and anticipation in the spring of 1958. *Volare* was the first of the entries to be sung on the Hilversum stage on the big night. Among the other competing artists was the Dutch singer Corry Brocken, the 1957 winner who had returned to sing *Heel de Wereld* in an attempt to defend her title. The evening was beset with technical problems, bearing in mind that Eurovision transmissions were still in their infancy. Due to irregularities in the broadcast, *Volare* was not heard in some of the competing countries. To rectify this, Modugno had to sing the song again after the last of the entries was sung. *Volare* came in third. The song, however, became an international hit and reached the number-one spot in the American hot hundred, where it stayed for five weeks. It was to sell two million copies in the USA alone. It was also the most-played song in Italy and worldwide, as reported by the Italian society of authors and publishers. It is estimated that total sales exceeded twenty-two million. Recorded by hundreds of artists in various languages, *Volare* was the only Eurovision song to win a Grammy Award. Written by Domenico Modugno himself, with lyrics by Franco Migliacci, it became the most famous song to come out of San Remo and one of the most successful Eurovision songs of all time. Migliacci's lyrics were inspired by a dream during a wine-induced sleep in which Migliacci saw himself in two paintings by Chagall, painted blue and with the capacity to fly. The song was given two sets of lyrics: one by Mitchel Parish and the other by Dame Gracie Fields, which she sang, adapting the words as she grew older, for many years until she died in 1979 on her beloved Capri. Dean Martin recorded a bilingual Italian/English version of *Volare* in 1958, which reached number two on the UK charts, while Modugno's original version reached number ten. Popular comedian Charlie Drake recorded a version which managed to reach twenty-eight in the UK. A host of Italian artists recorded the song, including Nilla Pizzi, who had come in second place after the song at San Remo. Louis Armstrong, the legendary American jazz musician, recorded a popular instrumental version. As probably the world's most covered pop song, it was recorded by such diverse artists as David Bowie, Barry White, Engelbert Humperdinck, and The Gypsy Kings. It even infiltrated a British Asian film with Sanjeev Bhaskar performing a version in *Anita and Me*, a popular British comedy-drama based on the book by Meera Syal. While all the cover versions are too numerous to mention, one curious aspect of many of them is that the prelude to the main part of the song was left out. This

Eurovision: A Plea for Respect

is by far the most interesting part of the song and conveys a sense of the song being part of a surrealist painting,

> *"I think that a dream like that will never return*
> *I painted my hands and my face blue, then without warning swept*
> *up by the wind and started to fly up into the infinite sky"*

Domenico Modugno's success at the 1958 San Remo Festival and subsequent appearance at the Eurovision Song Contest were to be repeated on several occasions throughout the 1950s and 1960s. In 1959, he won San Remo for the second consecutive year with the song *Piove*. Like *Nel Blu Di Pinto Di Blu*, it was better known by its other title, *Ciao, Ciao Bambina*. It was, as in 1958, performed twice at the festival: once by Domenico Modugno and again by Johnny Dorelli. Once more, Modugno outshone Dorelli and was sent to the 1959 Eurovision Song Contest in Cannes as Italy's representative. His performance received a positive reaction from the audience. He was followed by the entry from Monaco, *Mon Ami Pierrot*, by Jaques Pills. Modugno's song managed to get only sixth place in a contest of eleven countries. It was, however, another international hit for Modugno, although not nearly as successful as *Volare*. Again, many cover versions were made, including one by Hong Kong songstress Kong Ling, who popularised it in Asia. Domenico Modugno enjoyed a certain status in Italy, not merely for *Volare* but also because he was considered the first authentic *Cantautore* – singer-songwriter. The son of a policeman, he was born in Bari but spent the latter part of his childhood in Brindisi. His boyhood dream was to become an actor like those in the films he enjoyed so much on frequent visits to the local cinema. After completing compulsory national service, he started to fulfil his dream by studying drama. Highly talented, Modugno excelled at music and began composing his own songs. This led to one of his earliest successes. His composition *Lazzarella*, sung by a singer from Avellino, Aurelio Fierro, won the 1957 Festival of Neapolitan Song. Modugnos' journey to stardom had started. Elvis Presley recorded a song penned by Modugno and he once again entered the 1960 San Remo Festival. This time he was runner-up with his composition, *Libero*. Claudio Villa, who had won San Remo in 1957, returned to the festival in 1962 with a Modugno composition, *Addio, addio*, coming in first. Together with Domenico Modugno, Claudio Villa was the most successful artist at the San Remo Festival. A tenor from the Trastevere area of Rome, Villa sold a staggering 45 million records in his lifetime. He also won Canzonissima, another song festival staged by RAI, in both 1964 and 1966. *Addio, addio* took him to Luxemburg to compete in the 1962 Eurovision Song Contest. Despite receiving only three votes, he placed ninth; this was a result of several countries' songs receiving very low scores. Four entries each received zero points. Claudio Villa returned to the Eurovision Song Contest in 1967. This was the first time the Italian entry had not been selected at San Remo. His song, *Non Andare Più Lontano*, came in eleventh. His performance was a rousing Latin tempestuous effort with a dramatic ballad representative of his Mediterranean culture and background. With such strong connections to the San Remo Festival, it was an uncanny quirk of fate that he died of a heart attack on the last night of the 1987 festival. His death was announced live from the stage of San Remo during the actual song festival. Domenico Modugno made one more bid for glory at San Remo, this time in 1966, eight years after his success with *Volare*. He was now one of Italy's top male vocalists. The song *Dio Come Ti Amo*, a rather intense, low-keyed, melancholic ballad, swept the board at San Remo, a fourth for a Modugno song. The Eurovision Song Contest of 1966 saw Domenico Modugno's third appearance at the event. There was some disagreement between Modugno and the contest's producers over the stage presentation of his song. Modugno wanted to perform with only a quartet of musicians with him on stage. The organisers disagreed strongly but Modugno was insistent that he sing on stage with the quartet. After a great deal of arguing, the Luxembourgois producers gave in to Modugno's demands. *Dio Come Ti Amo* proved to be one of Italy's biggest disasters at Eurovision and San Remo's biggest slap in the face. Despite a moving, heart-felt performance,

Eurovision: A Plea for Respect

Modugno's song came in last. It was the first Italian entry at a Eurovision Song Contest to do so. What made it worse was that the song received no points at all from any of the European juries. It was the only time in the contest's history that Italy received a "zero". Also, it was the last time that Modugno entered the Eurovision Song Contest and the last time for a good few years that the winning song of San Remo would be sent to a Eurovision Song Contest. The song did have a degree of international success when it was recorded by Gigliola Cinquetti, who had also sung it on stage at San Remo. A film starring the teenaged Cinquetti, also titled *Dio Come Ti Amo*, was released. A production about innocent, young love, its opening scene starts with Cinquetti singing her San Remo and Eurovision success, *Non Ho l'eta*. In 1970, Finnish singer Seija Semola revived the song in Finland, where it became a hit. As one of Italy's top singing stars, Modugno managed to survive the 'nil' points experience in Luxembourg and continued with his successful career in both music and film. He returned to compete at San Remo but never won. His performance at the 1972 Song Festival, ironically, the first year Italy chose to send the San Remo winner to Eurovision since Modugno's 1966 attempt, was a somewhat energetic affair. During the performance of his song *Un Calcio Alla Citta*, he rolled up his sleeves as he took off his jacket, as if preparing to do some manual work. The song climaxed with Modugno running over to sing a hearty chorus with his backing group. Emotions rose in the crescendo as he shouted the final lyrics. The audience gave him an enthusiastic response but it was to be the year of such Italian classics as *Montagne Verde*, sung by the new up-and-coming Sicilian singer Marcella, and the popular bespectacled singer-songwriter Nicola di Bari, whose song *I Giorni dell'Arcobaleno*, a gentle ballad with beautiful lyrics, won San Remo and went on to represent Italy at the Eurovision Contest in Edinburgh. Di Bari, a native of Apulia, on the Adriatic coast in Southern Italy, had, at the start of his career, changed his name from Scommegna to Bari, the capital of the region. He had already won the 1971 San Remo Festival with *La Prima Cosa Bella* and enjoyed great success in South America, where he utilised his fluency in Spanish. *I Giorni dell'Arcobaleno* told the story of a woman who had grown up too quickly when she was a teenager. It tells of how she became romantically involved at a young age, which gave her express passage into adulthood. This, in turn, gave her an elevated mature status among her more immature and innocent peers. The result was her transformation from a girl into a confident young woman of the world. However, the song goes on to say that she really forsook what could have been the best time of her life. The song received ninety-two points in Edinburgh and placed sixth.

San Remo went through more changes in format during the 1970s and 1980s. New rules arrived and departed. New awards were introduced. An established artists section and a young artist category were brought in. The established stars were permitted direct access to the final, while the new up-and-coming singers had to compete in a semi-final. In 1984, the two categories became segregated and two separate contests and two winners were created. In 1998, the three top acts from the newcomers category entered the final, which ended in a victory for newcomer Annalisa Minetti. This result was a controversial one and caused a degree of acrimony. Due to the displeasure it had created, the organisers reintroduced two separate contests for the two categories in 1999.

Except for *Volare* and *Caio, Caio Bambina*, which I have already mentioned, the early Italian Eurovision entries (San Remo winners) produced more than a few internationally popular songs. The 1960 *Romantica*, performed at Eurovision by Renato Rascal, was first heard at San Remo in its two versions. The soft romantic ballad was performed by its composer, Rascal. Its opponent was a powerful rock-and-roll-oriented version by Tony Dallara, classed in Italy as an "*Urlatori*", a screamer – a description of more American-influenced Italian pop of the 1960s. Rascal's version triumphed over Dallara's and he was invited to be the Italian ambassador at Eurovision in London. Rascal, who was somewhat short in stature, with his full height being five-foot-two, had relied heavily on comedy to establish himself in show business. His particular brand of humour was evident in the sixty

Eurovision: A Plea for Respect

films in which he starred. Known as "*Il Piccoletto Nazionale*" (The Tiny Italian), this offspring of opera singer parents, born backstage and baptized in St Peters in Rome, counted *Arrividerci Roma* among his greatest hits. It is one of the three most famous Italian hits of all time. His Saturday night television variety show on RAI in the 1950s was one of the programmes that came about with the initiation of television as Italian home entertainment. In 1977, he starred in the Franco Zefferelli TV film *Jesus of Nazareth*, in which he played the part of the blind man. His song *Romantica* placed eighth in the 1960 Eurovision Song Contest, where thirteen countries competed. Jane Morgan, an American-born singer who found success in France and the UK long before she found fame in her native USA, recorded the song in English and was the singer most associated with it in Britain, where it was a minor hit. The French music star Dalida, herself of Italian heritage, took her French version right to the top of the French hit parade.

The palm trees that lined the seafront at San Remo shimmered in the cool February breeze as *Al di La* took first prize at the San Remo Song Festival of 1961. Its two versions had been sung by Betty Curtis and Luciano Tajoli, with the Betty Curtis interpretation triumphing over Tajoli's. The song was sung by Curtis at the Eurovision in Cannes the following month. At fifth place, it earned a higher placing than the two previous Italian entries. Yet again, an Italian Eurovision entry which had failed to win the Eurovision Song Contest became a huge international hit. The Italian-American singer Connie Francis recorded the song in Italian and also made a bilingual English Italian version. Her Italian version of the song became a massive hit throughout Latin America, while the bilingual recording became a bestseller in the USA, Europe, and beyond. In the 1962 film *Rome Adventure*, the Italian singer Emilio Pericoli sang *Al Di La* in a nightclub scene. The film, a romantic drama, revolves around a young librarian in New England who, disgusted by the prudish and hypocritical values of her hometown after being reprimanded for recommending a book, *Lovers Must Learn*, to a student, leaves for a new life in Rome, where people know the meaning of love. Pericoli's version of the song reached number eight on the Billboard Hot 100 The record was also a hit in the UK and earned Pericoli a gold disc. In 1963, his song *Uno per Tutte* won San Remo, instantly qualifying for the Eurovision Song Contest in London. He came in third. Pericoli had attempted to win San Remo and enter Eurovision the previous year with the song *Quando, Quando, Quando*, which did not win but became one of the best-known Italian hits of the twentieth century.

In the formative years of the Eurovision Song Contest, France and the Netherlands dominated but the big international hits came from Italy. The success of the early Italian Eurovision entries, at that particular period of time when American rock and roll had revolutionised popular music, not only in the USA and the English language world but also across Europe, exhibits the ability of the Italian songwriters of the time to compose material that had great international appeal while being capable of competing with America for pop supremacy. In the aftermath of the American Civil War, Italians were encouraged by the United States government to emigrate to the USA in an attempt to build up the nation via much-needed jobs in general labouring and farm work. Prior to the Civil War, Italian immigrants had been well received in general; however, those who arrived (and they arrived in the thousands) in the latter half of the nineteenth century began to face virulent forms of exploitation, maltreatment, and prejudice from the more established WASP traditions. Anti-Italianism was part of the agenda of the Ku Klux Klan, a white supremacist group that often targeted Italians as foreigners, as opposed to the WASPs, whose superiority they believed in and tried to retain. Anti-Italianism reached a hysterical climax in 1891 when nine Italians in Louisiana were arrested and charged with the murder of the chief of police. They were found not guilty but, despite the verdict, were dragged out of jail by an angry mob and lynched. Other incidents of lynchings occurred in the late nineteenth century, and as late as 1920, two Italians were arrested on trumped-up charges and executed. It is widely believed that their execution had more to do with their Italian immigrant status and anarchist sympathies than any crime of which they had been accused. Italian immigrants in Latin America noticeably faced much less discrimination. Italians in Argentina and Brazil, to which they also emigrated in huge numbers, soon rose to high positions in government with a good many reaching middle-class status in a short period. By the 1950s, Italians had started to make inroads into different aspects of an all-American culture. Joe Di Maggio, the American baseball centre fielder for the New York Yankees, also known as the Yankee Clipper, became an all-American icon and the 'American Guy' personified, though his father had had his property confiscated and was spied upon by American intelligence during World War II. The all-American icon was consolidated by Di Maggio's marriage to the American film goddess Marilyn Monroe. Italian immigrant America was turning into Americanism.

The Italian musical connection in the USA was bolstered by the millions of Americans of Italian extraction who took an interest in Italian music and who had contributed greatly to the concept of an all-American culture. Artists such as Dean Martin, Frank Sinatra, Al Martino, Connie Francis, and Sergio Franchi all popularised Italian Eurovision songs in America. The songs, such as *Volare*, became very much associated with Italian Americans and, in turn, became Americanised. This further embedded Italianism into existing mainstream American culture via the rising pan-American medium of national television and radio, forging a mainstream American culture that was heavily Italian in influence. Name an all-American singer: Frank Sinatra. Name an all-American food: pizza. If asked to provide a typical American surname, many people would give an Italian American name as an answer. American recordings of what were originally Italian songs reached out to an extended, often fragmented group of subcultures; they were an instrument in uniting them and, thus, were a tool in the creation of the USA, a unified nation, and an American national identity manifested in events such as Sergio Franchi's singing of *The Star-Spangled Banner* at Shea Stadium in 1968.

In Uno Dei Tanti 1963, the American singer Ben E. King recorded *I Who Have Nothing*. It was an English version of the 1961 Italian hit song for Joe Sentieri, *Uno Dei Tanti*, composed by Carlo Donida and Mogul. Mogul – Giulio Rapetti – was one of Italian music's most prolific songwriters and composed volumes of hits for major artists. *I Who Have Nothing* went on to be a huge hit for Tom Jones and Shirley Bassey, the latter of whom still performs it in her concerts. In 1963, Cilla Black, a pioneer of British pop and a friend of The Beatles, recorded the English version of the Gino Pauli/Umberto Bindi song *Il Mio Mondo: You're My World*. It was a worldwide hit for her. Around the same time, the Italian teen idol Rita Pavone also charted in the USA and Canada. Some of her Italian hits were recorded in English by other artists. The popular Italian language singer Little Tony, a citizen of San Marino and King of Italian rock and roll, also had chart success in Britain. From 1958-64, all the Italian Eurovision entries, or songs from the Italian prelims (San Remo), except for one, were sizeable hits in either Britain or America. It was certainly, in an international sense, a golden age for Italian pop, which the Eurovision Song Contest was instrumental in creating. This also displays an openness in Britain at that time towards Italian artists and Italian lyrics, which, as the 1960s went on, fast disappeared as the language and centres of pop and rock became firmly established and reinforced. This created a much narrower and more intolerant outlook on what should and should not be listened to. Nevertheless, Italian songs continued to have a great impact on the record-buying public of Continental Europe, and many Italian hits would resound in its borders for years to come.

Sixteen-year-old Gigliola Cinquetti took first place at San Remo in 1964 with the composition *Non ho l'eta (per amarti)*. At the fourteenth Eurovision Song Contest that year at Tivoli in Copenhagen, she had the lead in the voting from start to finish, easily winning the contest with a runaway victory. This marked the consolidation of Italy's place in world pop music and its coronation by the rest of Europe in recognition of its status. Cinquetti faced opposition from other notable ballads such as the Austrian and Monagasque entries. Eastern Europe was there in the form of Yugoslavia, the only communist country to compete as a member of both Eurovision and Intervision. Bulgaria was there in the form of Nora Nova, a Bulgarian by birth who lived in and sang for Germany. Nora's real name, Ahinora Kumanova, came from a Monarchist family that had fled the Balkan state after the imposition of communism. Later in life, she abandoned her music career and returned to Bulgaria, where she became an MP on behalf of the Monarchist party, supporting the return of Tzar Simeon.

Gigliola Cinquetti sang her ballad in a simple, sincere, and direct manner, dressed in a knee-length plain dark dress that was fashionable at the time. Her performance in many ways can be paralleled with that of the 2010 German winner Lena Mayer, also a teenager who dressed in a plain black dress and who delivered her pop song in a very direct manner, free of gimmickry. *Non ho l'eta* went on to top the European charts, and although an English version (*This Is My Prayer*) was made, it was the original Italian version that entered the UK top twenty.

San Remo in 1965 saw competition between such Italian artists as Bobby Solo, Milva, Iva Zanicchi, and Pino Donaggio, while Gigliola Cinquetti returned to sing *Ho Bisogno Di Vederte*. This was during the San Remo era when the rules stated that each entry should be sung in Italian by an Italian artist and that a different version should be sung by a foreign artist. This saw Connie Francis, Kiki Dee, Petula Clark, Gene Pitney, and Udo Jűrgens all singing entries in Italian. The idea had been brought in to judge the quality of the song and to put a strong emphasis on the actual songs – a reminder that it was a song contest and not an artist competition. This concept had been ingrained in the song festival since its inception in 1951.

The winner was the "Italian Elvis", Bobby Solo, with his song *Se piangi, Se ridi*, composed by the highly successful Mogul, whose compositions had featured among the other entries at the 1965 song festival. Twenty-year-old Bobby Solo had participated in San Remo the previous year with his song *Una Lacrima sul viso* but had been disqualified, as he had broken the rules by singing with a backing track, which in 1964 was not permitted at San Remo. However, the song went on to become a classic of Italian pop and a world hit. It also has the honour of being the first record in Italy to be awarded a gold disc. Bobby Solo competed at the 1969 San Remo Song Festival with the winning song *Zingara*, together with Iva Zanicchi, who sang the alternative arrangement. His participation at the 2003 event with Little Tony was his last at San Remo. His song *Se Piangi, Se Ridi*, a gentle Rock n Roll ballad very much in the Presley tradition, which automatically became the Italian 1965 Eurovision entry, came in fifth at the contest, held the following month farther down the Italian coast in the southern city of Naples. The song went to number one in Italy and reached number two in Belgium. At San Remo, another version with a completely different arrangement had been sung separately by the American folk group The New Christy Minstrels. However, 1965 was the year of Pino Donaggios' *Io Che Non Vivo Senzate*, destined to be that year's biggest hit. When recorded in English and other languages by diverse artists, it would be yet another world hit of Italian origins – an Italian evergreen.

After the adulation and triumph of Domenico Modugno at San Remo, followed only a month later by the humiliation and debacle of Italy's worst-ever result at the Eurovision Song Contest of 1966, Italian television chiefs decided to end the automatic right of San Remo-winning songs to represent Italy at the Eurovision Song Contest. Many felt that San Remo was too important an event for a song to win and then to subsequently be rubbished by the international juries at the Eurovision Song Contest. Therefore, the 1967 San Remo winner did not succeed to the Eurovision Contest, and an alternative selection process decided what song should represent Italy. San Remo's reputation and rank in Italian society as "the song festival of song festivals" was, thus, assured.

The Casino of San Remo opened its doors to artists, musicians, journalists, and film crews on the evening of January 21, 1967 for the seventeenth annual festival. This edition of the event would

Eurovision: A Plea for Respect

unwittingly become the centre stage of an event that it had innocently provoked. What was to happen that would make this festival a year to remember, or even a year to forget? A glittering portfolio of foreign artists sang the competing songs, coupled with Italian artists who sang the same songs in Italian. The non-Italian artists included American star Gene Pitney, who had come in second at San Remo twice before. Known as a gentleman and a prolific songwriter who, apart from composing *Hello Mary Lou* and *He's a Rebel* for The Crystals, is best remembered for the Burt Bacharach song *24 Hours from Tulsa*. The shy English singer Marianne Faithfull, who at the time was in a relationship with Mick Jagger and who, in 1967, would descend into drug-induced chaos, would end up destitute, begging on the streets of London, before miraculously curing her habit and making a much welcome return to the music scene. The legendary Sonny and Cher performed the song *Ma piano*. Cher, like many of the other foreign artists, was very nervous at the prospect of singing in Italian. She calmed her fears by writing the lyrics on the palms of her hands, which, visible to millions of TV viewers, she read as she performed alone on the stage while her husband, Sonny Bono, conducted the orchestra. The other foreign singers to compete were the congenial Bobby Goldsboro, who sang *Ragazzo*, Connie Francis, Dionne Warwick, The Hollies, Los Bravos, the American band The Happenings, who had spent months learning the Italian lyrics of their song *Quando Vedro*, and the Irish band The Bachelors, who sang two songs: *Per Vedere Quanté grande il mondo*, a Mogul composition, and *Proposta*, sung in partnership with Giganti. The festival consisted of a who's who of Italian pop with some of the biggest names in Italian music performing: Claudio Villa, Dalida, Iva Zanicchi, Domenico Modugno, Ornella Vanoni, Orietta Berti, Lucio Dalla, Luigi Tenco, and Little Tony. In all, thirty songs and fifty-eight artists were introduced over the three evenings of the festival by presenters Mike Bongiorno and Renata Mauro, who alone had presented the Eurovision Song Contest from Naples in 1964. Luigi Tenco sang his self penned *Ciao Amore Ciao* in partnership with Dalida, who interpreted the song in her own unique style. They had met in Rome sometime before and had soon become lovers. The lyrics of the songs had originally been about a young soldier who fought in the Risorgimiento but had been changed slightly for San Remo. Tenco was a young *cantautore*, a singer-songwriter, the name a conjunction of the Italian *cantore* and *autore*. This usually implied that the songs contained some social critique or that the lyrics were of prime importance. Tenco followed in the tradition of Domenico Modugno. He came from the "Genoa school" in style and nature like Gino Pauli and Sergio Endrigo. A very intense young man who had just announced his engagement to Dalida, Tenco was heavily focused on winning the festival, which meant everything to his career. The sun set over the spectacular scenery of that part of the Riveria; night fell on the first night of the festival as in the Casino the various acts, some nervously, gave all they had. Tenco's song *Ciao Amore Ciao* came in twelfth and was thus eliminated in favour of Orietta Bertis' *Io Tu e Le Rose*. He appealed to the committee, as was permitted. They overruled his appeal but upheld the appeal of *La Rivoluzione*, the song of Gene Pitney and Gianni Pettenati. Luigi Tenco was devastated and returned to his room in the Savoy Hotel. A few hours later his fiancée, Dalida, arrived at the hotel and opened the door of room 219, where Tenco was residing. She was greeted by an awful and unexpected scene. Luigi Tenco lay dead, with a bullet wound in his head and a gun by his side. He had committed suicide. Dalida, overcome with grief, informed hotel staff who, in turn, informed the police. Near his body was a note which he had written before his death. It read,

> "I have loved the Italian audiences but now feel I have lived the last five years of my life in vain. I do this not because I am tired of living but as an act of protest against a public who sends a song like 'Io Tu e Le Rose' and a committee who prefers to see 'La Rivoluzione' in the final instead. I hope this will clarify things to someone."
>
> *Luigi*

Eurovision: A Plea for Respect

Several conspiracy theories abounded in the months and years that followed. Rumours of murder circled the country. The press developed a great interest in the case and were sceptical of the cause of death. As the years went by, growing public disbelief as to the official verdict plus increasing pressure in the media for an enquiry into Tenco's death grew more vociferous, climaxing in the authorities ordering the exhumation of Tenco's body and a further post mortem in 2005. Thirty-eight years after his death, official tests confirmed that Luigi Tenco had died as a result of suicide. The case was closed.

His death had marred the festival and stunned the notes of San Remo into silence. An ice-cold ripple of shock ran up the spine of the Italian music business as it looked at itself in self-examination. The following evening, the Italian public, in a state of shock, tuned into the second night of the San Remo Song Festival. An underlying air of tragedy hung around San Remo for the rest of the festival. The year 1967 ended with a somewhat subdued win for the song *Non Pensare a Me*, sung by the team of Iva Zanicchi and Claudio Villa, although Little Tonys' *Cuore Matto* would prove to be that year's greatest hit. Iva Zanicchi would continue her success and also go on to win San Remo in 1969 with *Zingara* and again for a third time in 1974. Dalida recorded *Ciao Amore Ciao*, which became one of her biggest hits. She performed the song for the first time on an Italian TV show shortly after Tenco's death. Her emotionally charged performance was the best in her career. The audience, deeply moved, gave her lengthy applause. Tenco's death was, however, to be only one of many tragedies in her life. Iva Zanicchi also represented Italy at the 1969 Eurovision Song Contest in Madrid with the big ballad *Due Grosse Lagrime Bianche*, one of the best songs to be presented at the 1969 Grand Prix. It came in a poor thirteenth – a terrible oversight of a much-underrated song and performance in a final in which one of the winners was the ditty *Boom Bang a Bang*. It was an example of how, as is often the case in song contests, the best songs don't always win. Sergio Endrigo's Italian entry in London the previous year, *Marianne*, had come in tenth. It was given English lyrics by Bill Owen and was recorded and became a hit for Cliff

Richard. Bill Owen, an actor, was well known for his portrayal of the legendary character Compo in the BBC's longest-ever running comedy, *Last of the Summer Wine*. The same year, Patty Pravos' top ten Italian hit, a Mogul and Lucio Battisti song called *Il Paradiso*, reached number one in Britain as Amen Corners' English version, *If Paradise Was Half as Nice*. These are two more clear examples of the potential of Italian pop music to enjoy popularity beyond Italy's borders and how, given enough exposure, Italian music could be appreciated by British audiences.

During the 1960s, a good many Italian singers who had first competed at San Remo also found success in films. An example is Domenico Modugno, who featured in many Italian films of the 1960s produced at Rome's Cinecitta, the Italian Hollywood and, in the future, the venue for the 1991 Eurovision Song Contest, the stage of which was meant to resemble a film set. In 1962, Modugno starred in the hit film *Appuntamento a Ischia* alongside another star of San Remo, Mina, who became a superstar and Italy's top female vocalist. One of her greatest hits, *E Se Domani*, came from the 1965 film, *Vagne Stele Dell'Orsa*.

The importance of San Remo and its impact on Italian music during the 1960s must never be underestimated. It had a profound and lasting influence on Italian culture and was also a fertile breeding ground for new talent in the 1960s and successive decades, spawning the likes of Marcella, Iva Zanicchi, Patty Pravo, Andrea Bocelli, and Laura Pausini. While rock and roll merged with more traditional Italian pop, as seen in Little Tony and Rita Pavone, foreign groups like The Beatles and The Rolling Stones, already popular in the USA and Britain, gained a following in Italy. Their works influenced several Italian groups such as I Pooh, formed in Bologna in 1966, who emulated them, singing beat pop songs. Pooh, however, experienced new influences and changed genres, eventually adopting a more Italian influence in their music, which took them on to phenomenal success in Italy. The original vocalist, Ricardo Fogli, won at San Remo in 1982; he also sang *Per Lucia* for Italy at the 1983 Eurovision Song Contest in

Eurovision: A Plea for Respect

Munich and came in eleventh. There were also The Rokes, an English group who, as Italy's top pop group, sang their hits in heavily English-accented Italian. They were unknown in the UK. As the decade passed, there was increasing evidence of the success and growing popularity of the *Cantautori*. The chief exponents were Luigi Tenco, who was the first artist to record with the first Italian recording label, Casa Ricord; Gino Pauli; and Fabrizio De André, who came from what was known as the "Genoa School", heavily influenced by the protest movements in Europe and America and whose work often explored psychological and socio-political themes. Other *Cantautori* such as Modugno remained close to more traditional Italian themes like emotions and romance, although one of Modugno's most famous works, *Amara Terra Mia*, was about rural poverty and emigration. In the song, Modugno sings,

> *"Endless skies and faces like stone, hands hardened yet, without hope*
> *Goodbye, goodbye my love, I am going away*
> *My bitter land, bitter and beautiful"*

Amara Terra Mia was based on an old traditional song from the Abruzzo region in central Italy. Many of the *Cantautori* who appeared at San Remo and who flourished in the 1960s were successors of a rich folk tradition of protest songs which often attacked the exploitation of workers and the hundreds driven from their homes by poverty and hunger in search of a better life. The twentieth-century decline of agriculture in Southern Italy led to a constant stream of migrants from the south to the northern industrialised cities such as Turin and Milan. Others chose to emigrate to more developed countries, taking Neapolitan music with them and planting its seed in the developing cultures of the United States and Argentina. Those who migrated to Northern Italian cities often found no promised land but encountered more exploitation, poverty, discrimination, and racism, in taunts such as "Go back to Africa", often directed at southern migrants.

Fabrizio De André was born in Genoa, part of what was known as the Northern Industrial triangle. He was an admirer of George Brassens, Bob Dylan, and Leonard Cohen, and his songs often centred on those marginalised by society, i.e., gypsies, prostitutes, or the rebellious. The lyrics of his compositions were held in high regard because of the depth they often displayed. De André was instrumental in the valorisation of minority languages in Italy and worked hard to improve the lot of Ligurian as well as Sardinian Gallurese, an Italo-Dalmation Romance language spoken in North Eastern Sardinia, and also Neapolitan. Many of his 1960s works were major hits and enabled his recognition to a wider public. In the 1970s, he formed a relationship with singer Dori Ghezzi, who had appeared at San Remo in 1970 and had, as part of a duo with Wes, an American-Italian act, sung the Italian entry *Era* at the 1975 Eurovision Song Contest, placing third. That same year, the couple set up a home on a farm in Northern Sardinia, a place very close to De André's heart. However, things did not turn out to be as tranquil as they had expected. On a warm August evening in 1979, Fabrizio De André and Dori Ghezzi were kidnapped by Sardinia's notorious "Anonima Sequestri". They were forcibly taken to Supramonte, high in the Sardinian mountains, where they were held hostage for four months. Dori Ghezzi and De André were released a few days before Christmas. It was reported at the time that the couple's family had allegedly paid 500 million Italian lira to secure their release. At the captor's trial which followed, De André showed a fair degree of sympathy for his kidnappers. Dori Ghezzi had appeared at San Remo in 1970 and again in 1973, the same year that Drupi sang *Vado Via*, but was eliminated. His song, however, sold well and was a top twenty hit in the UK. Dori Ghezzi came in second a further time with Wes a year after their Eurovision appearance, in 1976. Ghezzi resumed her San Remo participation after the kidnapping and came in third in 1983 with *Margherita Non Lo Sa*. She competed twice more, in 1987 and 1989, marrying Fabrizio De André in that same year, then retired from the music scene due to bad health in 1990. Ghezzi's Eurovision entry in 1975, *Era*, was a departure from

Eurovision: A Plea for Respect

the norm of Italian Eurovision songs of the time. It was not a traditional Italian pop ballad but, rather, a mixture of balladry with Black American soul. Wes, the other half of the duo, was an African American from North Carolina who had lived in Italy since the 1960s and had collaborated with Ghezzi on several occasions. They became a popular duo nicknamed "Coppia Caffe latte". He was black as night while she was pale with very blonde hair. The juries recognised their song for its unusual quality and rewarded it. Although it came in third, it was by far the superior song of the top three. Fabrizio De Andre harboured anarchist and pacifist tendencies which often surfaced in this work. He was also a man of faith with great spiritual depth who was quick to question and analyse the teachings of the Catholic Church. This spiritual aspect came through in his work, particularly after his experiences as a hostage. It was also evident in his hit song *Preghiera in Gennaio*, a prayer which he wrote in January after the death of his friend. It was dedicated to Luigi Tenco.

While the idea of sending the winning song of San Remo to Eurovision was abandoned after 1966, the winning act at San Remo was still despatched to the Eurovision Grand Prix, albeit with another song, at least for the three years proceeding 1966. In 1970, Adriano Calentano won the San Remo Festival with his song *Chi Non Lavora Non Fa L'amore*. Celentano, once mentioned in an Ian Drury song, and a top male vocalist in Italy, who counted among his many hits the Italian version of Barry McGuire's *Eve of Destruction*, had attempted San Remo before. He was the first artist to win San Remo and not compete in the Eurovision Song Contest since 1956.

The years when San Remo facilitated the selection process for the Italian Eurovision entry were arguably the classic years of Italian pop music, an era that proved itself more than capable of producing some of the best Italian songs in recent history. The period 1956-69 saw the crème de la crème of the Italian music industry compete. The legends of Italian music were all in evidence at San Remo but despite often being eliminated on the first or second evening, legendary stars who suffered hard blows returned to compete again and again. Mina and Milva, the two great divas of Italian pop, both left their mark on San Remo during its golden years. The process of Eurovision selection was re-introduced in 1972. *I Giorni Dell'Arcobaleno* was an excellent song which reached sixth place but perhaps Italian pride rated it higher. The next San Remo winner to represent Italy would not come until 1997.

Gigliola Cinquetti returned to compete at San Remo on several occasions after her 1964 win, repeated at Eurovision. She made a return to the Eurovision Song Contest ten years later, in 1974, with the dreamy romantic Italian ballad *Sì*. She was the last entrant to perform her song on the night and it was a perfect song on which to end the varied musical evening. It was a controversial choice for Eurovision, as it coincided with the long-awaited referendum on divorce in Italy. The divorce issue was both an emotional and divisive topic in Italian society at the time. A law had been passed on December 1, 1970 which, for the first time, allowed divorce in Italy. It was a highly controversial and radical act by the government of the time in a country that, in 1970, was still deeply religious and influenced by the Church. There was strong objection in some quarters. Many heated debates in parliament and Italian society in general ensured legalisation to settle the issue once and for all. A referendum was held in which voters were asked if they wanted to repeal the law passed in 1970. The Christian Democrats and the Fascist Party were very much against divorce and wanted to see it illegal once more. They campaigned for a "yes" vote, meaning the repeal of the law. The Italian Socialist Party and the Italian Communist Party, on the other hand, were in favour of divorce, as was the Radical Party. They campaigned for a "no" vote against the repeal of the existing law. The referendum was held on 12 May 1974.

Despite RAI's selection of the song *Sì*, the song was banned from radio and television, as RAI felt it could be misinterpreted as a form of propaganda and might even influence the public to vote "Sì" – yes – in the divorce referendum. Censors at RAI even made the drastic decision to ban the transmission of the 1974 Eurovision Song Contest in Italy, so Gigliola Cinquetti's performance of *Sì* was not seen by Italian

Eurovision: A Plea for Respect

viewers. The song was, in fact, banned for a whole month on RAI even though the referendum was in May and the song contest had been held over one month before. The lack of airplay and the ban on transmission sealed the fate of the song in Italy, where it failed to enter the Italian top forty. *Si* did very well at the Eurovision Song Contest and came in second with eighteen points, six behind Sweden. Wearing a blue dress and accompanied by four backing singers, Cinquetti sang of her love for a man whom she pursues. She described her exhilaration when, in the end, they finally start their lives together and she says wholeheartedly and sincerely, "Si". Although the song made little impact in Italy, Cinquetti recorded it in several languages. Unusually, the country where it became the biggest hit was Britain, where it had been a Radio 1 record of the week and had been given plenty of airplay. The record peaked at eight on the UK chart with the English version, *Go*, as the A side and the Italian *Si* as the B-side. It was Cinquetti's second top twenty hit in the UK. The song was also a hit in Germany and Belgium, while in Finland, Lea Levan shot up the charts with the Finnish version, *Niin*, the other side of which was a Finnish version of Cher's *Dark Lady*. It reached number five and stayed on the Finnish chart for over a year. Unlike Cinquetti's 1964 winner, *Si* did not become a huge Pan-European hit. The referendum ended in a victory for those in favour of divorce; therefore, the "No" (to change an existing law) won. While governments came and went with great regularity in Italy, a spate of high-profile abductions and murders of those in authority dogged the country in the 1970s. Terrorism was rife in the form of the Red Brigades and other groups. Discontent with government policies and politicians became a part of Italian life. The Radical Party had been a stronghold of Liberalism in Italian politics since its creation in 1955 but it was not until the mid-1970s that the party was voted into parliament. Among those who became increasingly drawn to its form of progressive left-wing politics was the *Cantautoro* Domenico Modugno. His interest grew in supporting them beyond the ballot box and he became a party member. In 1976, Modugno made yet another appearance at San Remo, this time as a special guest performer. Modugno's musical abilities began to take second stage to his radical politics. He stood as a Radical Party candidate and was elected to the Italian parliament in 1987. Modugno, who bore the same name as a town in his native Puglia, fought on behalf of many social issues. One of his most passionate causes in later years was the improvement of psychiatric patients' conditions in a Sicilian hospital. It was while he was in Sicily in the hot summer of 1994 that Domenico Modugno died suddenly of a heart attack. His legacy in terms of music and social progress was a great gift to the Italian state. His contribution to both San Remo and the Eurovision Song Contest was groundbreaking. He had given the stages of both song festivals international relevance. Modugno posthumously received both the San Remo special award and the award for creativity for *Volare*. In Copenhagen in 2005, at the celebration of the Eurovision Song Contest's fiftieth anniversary, *Volare* was voted the second most popular Eurovision song of all time. The actual golden anniversary of the song *Volare* was commemorated in Italy by the release of a special stamp.

    Domenico Modugno's sojourn in politics was matched by another San Remo winner and Eurovision contestant, Iva Zanicchi, who was elected to the European Parliament under the flag of Silvio Berlusconis' "Forza Italia" party. She was joined in Strasbourg by other Eurovision participants turned Euro MPs – Dana and Nana Mouskouri among them. San Remo participant and fellow 1960s singing star Rita Povone was also elected to the Italian parliament as an MP.

    The Italian music scene of the 1970s was dominated by the likes of Lucio Battisti and Claudio Baglioni, who had hits such as *Piccollo Grand Amore*. Both were no strangers to San Remo, while *Cantautori* began to include even more "protest" in their songs. Rafaella Carra, a popular singer and dancer, as well as show and song exponent, already with a chain of films to her credit, had had a hit in the UK in 1977 with the English version of her popular Italian success *A Far l'amore Comincia Tu*. Many years later, Carra would go on to be a highly

Eurovision: A Plea for Respect

successful TV host. Music was only one of Carra's outlets. She was the variety show personified and an Italian icon of the 1970s and 1980s. She cannot be compared to Italian rock artists of note who made an impact on the Italian music scene such as Le Orme and Area! Those who do so completely miss the point when it comes to Raffaella Carra. She was sheer showbiz, an enigma when it came to the Italian music scene in the 1970s and 1980s. Her music was often accompanied by elaborate choreography. Her brilliance as a show woman was all too obvious. In 1983, she entered San Remo but was not successful with the song *Soli Sulla Luna*. For many years, she worked in Spanish broadcasting and in 2011 was invited by RAI to announce the Italian jury results at the Eurovision Song Contest. It was a significant role and quite an honour, one that reflected her popularity and celebrity, as she was chosen to be the first Italian announcer to do this in fourteen years after Italy's long absence. Despite being an international star with a particular foot in Latin America, (she had lived in Argentina at the time of the colonel's brutal dictatorship), Raffaella remained only a one-hit wonder in the UK. The Italian offerings of the 1970s at the Eurovision Song Contest, now free of San Remo, were a broad range of Italo-pop in the form of the popular husband and wife duo Al Bano and Romina Power, daughter of Tyrone Power, who were all set to storm the European charts in the 1980s with hits like *Felicita*, which was rejected at San Remo but proving to be exceedingly popular almost everywhere. Their 1976 Eurovision entry was sung partly in English and partly in Italian and was the first Italian entry to not be sung totally in Italian. Ballades in the form of Gianni Morandi's *Ochi di Ragazza* and Massimo Ranieri's more traditional Italian ballades *Chi Sara Con Te?* and the dramatic 1971 effort which he had sung two years previously in Dublin, *L'Amore E Un Attimo*, all scored respectable placings, although none of them were in the top four.

Numerous female vocalists enjoyed repeated success in the Italian hit parade during the 1970s. Rafaella Carra, Marcella, Patt Pravo, and Nada were just drops from a never-ending fountain.

Steve Kerr

In the latter half of the 1970s, one of Italy's most prominent female vocalists, who had received many accolades, was Mia Martini. A native of Calabria in the South West of Italy, she was the sister-in-law of the tennis player Bjorn Borg. She was much associated with the song *Piccolo Uomo*, which was a major hit in 1972. Other best-sellers, such as *Credo* and *Donna Sola*, helped establish her as a leading force in Italian music of the 1970s. She appeared regularly at the San Remo festival. In 1977, RAI invited her to sing for Italy at the Eurovision Song Contest at Wembley. The song that was selected for her to sing was a dramatic ballad entitled *Libera*, a song that celebrated the liberation of women in its lyrics. Martini recorded the film for the song contest previews in the stalls of an empty opera house. In this recording, the song is still interpreted as a ballad but bearing in mind that these were the 1970s, it was thought to bring it more in line with international tastes of the time. This changed the song from a ballad to an up-tempo disco/dance number, which completely changed the character of the song. Martini recorded the disco version in Spanish, French, English, and Italian and recorded the original ballad version in Italian. The song lost a great deal of its integrity in the transition. Also, Martini's vocal abilities were more evident in the original ballad version. The disco gave the song a blandness and did no justice to Martini's rendering. It was just another 1970s disco song. The song was performed after the Spanish entry *Ensename a Cantar* by Mikki, a rather silly, repetitive ditty that was not even catchy. Despite the very strong vocal standing in positive contrast to Spain's entry, *Libera* came in a poor thirteenth in a field of eighteen. Through years of perfecting her stage presentation and the vintage of a more mature voice scarred by the pains of time, Mia Martini's performance at the Malmo Eurovision Song Contest of 1992 proved to be a classy, highly professional affair. Her song was *Rapsodia*. Mia was, by this time, a much-celebrated artist in Italy. The song was a somewhat emotive ballad tinged with a feeling of pathos and despair. It tells of two former lovers who have left behind everything dear to them in search of their lost youth. Desperately reaching for an eternal youth of sorts, they

Eurovision: A Plea for Respect

question themselves and take stock of their lives. Who are they really? They feel "ashamed" yet "sincere" and admit, as they sit and drink in a bar, that they can't seem to get their lives together. Martini sings of their situation as "This immense and hopeless tragedy." As the old lovers depart, they once again go their separate ways, and as the song ends, she sings that they become songs of once upon a time. A classic in the Eurovision hall of golden ballads, *Rapsodia* came in fourth. Mia Martini's artistry moved into immortality in May 1995 when she suffered a heart attack brought on by a cocaine overdose. Martini's strong association with the San Remo Song Festival was commemorated in the special critic's award for best artist, which she had won on three occasions. It was renamed the Mia Martini Award and became an integral official award presented annually at the song festival.

* * *

Italian pop music had developed from the popular Canzone Napolitana, which had existed for a few hundred years before twentieth-century compositions such as *O Sole Mio* and *Santa Lucia* became totally associated with the genre. A feature of this genre is that it exhibits a simple harmony containing both a narrative and a refrain often in contrasting moods. Most of its earlier exponents were solo male vocalists who graduated as a separate genre around 1830. Neapolitan songs (Canzone Napolitana) earned a degree of familiarity outside Italy due mostly to mass Italian immigration in the forty years spanning the last decades of the nineteenth century and the first decades of the twentieth century. Enrico Caruso was one of the earliest exponents of the genre. Canzone Napolitana were always sung in Neapolitan. In more recent times, a popular interpreter of the Canzone has been Massimo Ranieri from Santa Lucia in Naples, who sang at the Eurovision Song Contests of 1971 and 1973. He became a popular teen idol in 1969 when he sang *O Sole Mio* on live Italian TV to a chorus of screaming girls in the audience. Like opera and folk music, "Canzone Napolitana" formed an iconic angle in Italian social-cultural history and in its national identity. Some of the world's earliest musical recordings were made in Italy. Before Italy's reunification, a separate and distinct musical tradition had developed which was particularly significant in the development of sacred music such as Gregorian chant and the Renaissance choral works of the great Palestrina. There had been increasing foreign influence in Italian music since the late 1900s but with the rise of Mussolini and the Fascist cause in the 1920s, there came about "Autarchia". This led to the arts taking on a more introspective angle, encouraged by the official government policy of cultural isolation. During the Fascist era, there was a distinct veering away from music outside these boundaries, such as jazz, and a moving towards more traditional and pure Italian music such as the Neapolitan songs and opera, an affirmation of national culture. Due to the Fascist regime's isolationist policies, a hunger for more international music developed and was fulfilled by the arrival of hundreds of American GIs at the liberation of Italy. Italian music took on a mixture of foreign, mostly American, styles which led to the birth of truly Italian pop. One of the first works to exhibit this was the song *Ciao To Diro*, sung by Adriano Celentano at San Remo in 1957. Italian pop was also highly influenced by Latin American music such as Rumba and the Bossanova. The Musica Leggera – the more traditional Italian-based mainstream popular music – gave way to different Italian music. During the 1960s and 1970s, Italian pop music changed its direction and new music evolved, incorporating elements of rock and pop into mainstream popular music. Growing industrialisation, as Italy slowly moved away from an agricultural culture, further increased the weakening of traditional Italian popular music in society. Other sources, such as immigration from Africa, fuelled the diversification of Italian music. There was a counter-movement in favour of the Leggero, most prominent in Toto Cotugnos' 1983 international and iconic Italian hit *L'Italiano*, which was a more traditional style of Italian popular music. In the late 1970s, Italian popular music, due to new developments in technology, often reflected experimental electronic music, led by Franco Battiato. Music in Italy forms a strong

Eurovision: A Plea for Respect

component of national identity, though it has taken its form from the diverse cultures that have ruled or been influential in the country. Opera, an Italian contribution to world culture, is a tradition that grew before there was any notion of an Italian identity or even any longing for a unified Italian peninsula. Unlike in other European cultures, opera became embedded in Italian popular culture and popular music. It could even be classified as a genre of popular Italian music often witnessed by elderly Italians who sang along to well-known arias such as *O Mio Babbino Caro*. Opera spread throughout the world, and the popularity of Neapolitan songs internationally was due to the vast number of southern immigrants who left their homeland. The international popularity of Italian pop music was carried into a wider world by the platforms of San Remo and the Eurovision Song Contest.

In 1970, Sandie Shaw, who had recorded several of her hits in Italian, sang *Che Effetto Mi Fa* at San Remo. She repeated her performance at the festival in 1990 with the song *Deep Joy*, sung in Italian, at the same event as *Sono Felice* by Milva. In the 1960s and 1970s, many British artists who were popular in Italy recorded their Italian releases in Italian. At the 1971 San Remo Festival, the English group Mungo Jerry sang *Santo Antonio*. Meanwhile, at the same festival, the Scottish group Middle of the Road featuring Sally Carr sang *Lo Schiaffo*. Middle of the Road was at one point based in Italy. Their UK number-one single *Chirpy, Chirpy Cheep Cheep* is considered the first pop song to be classed as Euro-pop. Although they counted several UK hits to their credit, they were far more successful on the continent than they ever were in Britain. As history evolves and time moves on, it is rather sad that most British singers who enjoy success in Italy now would never think of recording their work in Italian. The 1971 San Remo Festival was won by Nicola Di Bari but the runner-up, *Che Sara*, by José Feliciano and Ricchi é Poveri, became one of the most famous and popular songs from San Remo. The Spanish version was a worldwide hit for Feliciano. Ricchi é Poveri, who sang *Questo Amore* for Italy and came in twelfth at the 1978 Eurovision Song Contest, was one of the many Italian acts to score great chart success all over the continent in the 1980s with their Italo-pop hits *Mamma Maria* and *Sara Perché ti amo*. Their performance at Eurovision was preceded by the Norwegian entry *Mil Etter Mil*, sung by Jan Teigen, who through the years became a much-celebrated contestant. His song was the first, since the new system of voting was introduced in 1975, to score zero points – an "incident" that he capitalised on and drew strength from. Italy went on to send the group Matia Bazar to sing *Raggio de Luna* at the 1969 contest, held for the first time outside Europe, in Jerusalem. Antonella Ruggiero, the lead vocalist, had a particularly notable vocal range and the ability to sing in four octaves, which gave the group a somewhat distinctive sound. Matia Bazar was one of Italy's leading groups in the 1970s and 1980s, as winners of San Remo in 1978 – a win which they repeated in 1983 and 1985. Their Eurovision appearances gave them some limited success across the continent. The group's 1970s combination of progressive rock and pop gave way in the 1980s to more Italian pop-based material. The songs often had intriguing melodies which blended well with Antonella's unusual voice. Antonella, who embarked on a solo career, competed in several more appearances at San Remo and came in second in 1996, 2003, 2005, and 2007. Later in her career, Antonella developed an interest in preserving Alpine music. The 1970s flew by in a haze and before we could say "1980", it already was. As with the Eurovision Song Contest, technology was slowly changing the face of San Remo. The rather small oblong stage of the Casino theatre was, in 1972, a fairly simple affair decked in parts with colourful flowers to add a touch of beauty to the proceedings. In 1976, the San Remo Festival moved away from the Casino to another venue. Italian TV did not have the facilities to broadcast in colour, so brightly coloured flowers and lights all came over in shades of grey. RAI made its first colour transmissions, inside Italy, in 1977, and the quality of stage design and production gradually changed. As the advancement of technology became more rapid, the productions grew more lavish. In a fast-changing Europe in the fields of technology, demographics, economy, and politics, and the move in Italian music further towards mainstream international pop, an ingredient somehow disappeared at San Remo. It had lost something of the quintessential Italianness that had been so evident in the 1950s and 1960s. In Italy, by 1980, San Remo had far outgrown its child, the Eurovision Song Contest, in terms of ratings and public interest. The disconnection of the two festivals had not benefited the Eurovision Song Contest and had led to a slow loss of interest among the Italian public by the 1980s, although it must be stated that most of the Italian representatives at Eurovision continued to be established artists, even household names, who had won or competed at San Remo. At the first Eurovision Song Contest of the new decade, an anniversary edition, the Italian song was sung by Allan Sorrenti. It was a rather run-of-the-mill Italo-pop ballad but one which was totally in keeping and very much representative of the contemporary Italian pop of the time. Sorrenti, a Neapolitan singer, was fluent in Italian and English. He was born to a Welsh mother and spent a large part of his younger years in Aberystwyth. His early career in the first few years of the 1970s, including his first album, was of the progressive rock and experimental genres. By the mid-1970s Sorrenti had abandoned this in favour of a more dance-orientated repertoire which gave rise to major European hits. *Non So Che Darei*, although not the strongest entry at the 1980 song contest, did, after the winner *What's Another Year*, prove to be the biggest success commercially and charted in various European countries. Alan Sorrenti took part in the 2006 O'Scia Festival on the Sicilian island of Lampedusa, where Domenico Modugno passed away and which is far nearer Tunisia than it is to mainland Italy. In 1981, for the first time in its history, Italy did not compete in the Eurovision Song Contest. This was, however, a milestone year in the relationship between Italy and the contest, which for the next three decades would become more sporadic. In 1982, Italy chose, yet again, not to enter a song. Italy further withdrew in 1986. The 1980s in Italy saw the rise of Italo-disco, which had evolved from Italo-pop and early forms of electronic dance music. The main exponents of the genre usually sang in English, and its music was often characterised by the inclusion of drum machines and synthesizers. There is evidence that Italo-disco was first put together in the recording studios of disco-stomping Germany. It was certainly very closely related to Euro-disco and echoed the music of Giorgio Moroder, a top Italian producer. The genre produced a multitude of pan-European hits and scored a degree of success in the UK, bearing hits such as *Tarzan Boy* by Baltimora, *Dolce Vita* by Ryan Paris, *Call Me* by Spagna, and *Boys, Boys* by Sabrina. Italian composer/singer Raf's *Self Control* was an English cover version by Laura Branigan, who would be an important link in the delivery of Italian music to the British and American public. Italo-disco highly influenced top British electronics acts of the 1980s, in particular Erasure and Pet Shop Boys, whose records soared up the British top 30. Italian pop was also once again making great inroads into the European hit parades as well as providing the airwaves with popular turntable numbers. Eurovision and San Remo singers Al Bano and Romina Power, Ricchi e Poveri, Toto Cotugno, Ricardo Fogli, Uberto Tozzi, Raf, and Eros Ramozzoti all enjoyed great pan-European success even in the Soviet Union, where Al Bano and Romina Power could do no wrong. With the return of Italy to the Eurovision stage in 1983, with Ricardo Foglis' *Per Lucia*, it committed itself to winning the contest by entering two major acts. Al Bano and Romina Power represented Italy for a second time in 1985 and, as in 1976, came in seventh with their song *Magic oh Magic*, which, like their 1976 attempt, was sung partly in English, though only the title. Like all the records from their era, it proved to be very popular. This catchy Italo-pop song followed in the footsteps of the 1984 entry, a very different type of song: Franco Battiatos' composition *I Treni Di Tozeur*, which he sang as a duet alongside the powerful vocals of Alice. Franco Battiato, a multi-talented Sicilian singer-songwriter who also adjusted his talent to film and painting, had been one of Italy's most accomplished progressive rock musicians of the 1970s. He collaborated on several psychedelic-progressive rock works with leading rock bands. As the 1970s went on, he showed an increasing interest in the development of experimental electronic work. So important and influential are his elementary works from his highly creative period in

Eurovision: A Plea for Respect

the early 1970s that they are much cherished and sought after by rock aficionados across the world. The electronic experimental phase was soon replaced by works that would more closely fit into Minimalism, which had its roots in the New York "downtown scene" of the 1960s and had initially been called the "New York hypnotic school". Some of his other material developed into derivations of Musique Concréte, whose music did not fit into the recognised norms of melody and rhythm. It often relied on synthesizers and more natural sounds from landscapes or open-air situations. The basis of Musique Concréte was originally credited to Pierre Schaeffer in the 1940s. In 1977, Battiato won the Stockhausen Awards for contemporary music. After a move towards more Italian pop, he initiated a songwriting partnership with the musician Gusto Pio which resulted in a chain of new works, some of which touched on Middle Eastern melodies that were extremely popular in Italy and opened up his talent to a broader audience. His new wave of more pop-orientated material was often characterised by the quality of his lyrics, with spiritual and philosophical themes set to electro-pop rhythm. His success lay in his ability to produce pop that had an intellectual current running through it. In 1989, he recorded the song *Veni l'autunnu'*, which displayed an interest in and a desire to promote regional and minority languages, being sung in Sicilian and Arabic. Sicily was once ruled by both Arabs and Normans – a melting pot whose descendants bear substantial DNA of both peoples. His 1992 anthem *Povera Patria* was originally written as a criticism of Italy's cultural ignorance and lack of honesty. However, it took on a different direction when it became a hymn of a populace tired of the violence of the Mafia in the wake of the Mafia killings of two judges. Battiato is only one example of the many artists of exceptional calibre who have appeared at Eurovision. In the Yorkshire Post, journalist Sarah Freeman wrote, in May 2005, "The Competition has never required much deep thought. Instead it's about pointing and laughing at the outlandish costumes, the off key notes and the often questionable choreography of our European neighbours". Meanwhile, a CNN online report prior to the 2013 Eurovision Song Contest described the contest as being full of "dubious talent"; this was, of course, written by a CNN reporter based in London. The Yorkshire Post article displays what outright rubbish is often written about the contest in the British press. The article displays not only an extreme ignorance of the contest but an even more appalling ignorance of continental music and artists. Ivan Raykoff, an arts professor in New York, explains that British audiences tend to view Eurovision as an entertaining farce rife with continental quirkiness. Sarah Freeman no doubt thinks that Franco Battiato is a type of pasta on offer at the Italian restaurant at Armytage Bridge. Battiato's Eurovision entry relied heavily on Alice's strong vocals. Alice Visconti won San Remo in 1981 with the song *Per Elisa*, written by Giusto Pio, Alice herself, and Franco Battiato. It was a dramatic, passionate rock song about destructive emotions such as jealousy, anger and the divisive actions that result from them – betrayal and revenge. *Per Elisa* was based on Ludwig Van Beethoven's *Für Elise*. Alice sang the song partly as falsetto but at the same time utilised the full strength of her distinct contralto voice. *Per Elisa* was a very radical departure for San Remo in 1981 and became the first true rock song to win. Alice, who forsook the traditional evening dress in favour of tight jeans and a leather jacket for her San Remo performance, was warmly received not only by the audience and juries at San Remo but by the country in general, and her star status as a top Italian vocalist was consolidated. *Per Elisa* established Alice as a star across Europe, and she achieved hit status in many West European countries. By the mid-1980s, Alice had become a major recording star in West Germany. She also gathered a considerable following and great commercial success and acclaim in Scandinavia and Benelux, where her popularity by 1985 had exceeded that in Italy. The song *I Treni Di Tozeur*, the trains from Tozeur, a rather unorthodox composition, concerns the train that ran in the first half of the twentieth century through the Atlas Mountains in Tunisia down to the Algerian border to the oasis town of Tozeur in the Tunisian Sahara desert. The railway was built by the King of Tunisia to transport him in luxurious comfort from his palace in Metlaoui in the north of the country to his winter palace in Tozeur.

Eurovision: A Plea for Respect

The railway line, built only for the king's personal use, had cost a fortune; not only had it dug deep into the Tunisian treasury but its construction had involved an enormous loss of life. The train itself was a gift from France to the King of Tunisia and was made and decorated in the finest-quality materials. It was nicknamed the "Red Lizard" and contained a stunningly beautiful interior of velvet chairs and artefacts of mahogany and brass. When Tunisia became a republic in 1957, the train was taken by the state and lay unused and decaying for years. However, after thirty years, it was restored and the line from Metlaoui to Tozeur was opened once again but this time for public use. As *Per Elisa* was based on a work of Beethoven, there was also, once again, a link to another great classical piece. The song featured a quote from Mozart's opera *The Magical Flute* – "Doch Wir Wollen dir Ihn zeigen und du wirst", sung in German by three mezzo-soprano opera singers. The line in the opera is, by tradition, usually sung by three young boys and is little more than a second long. At the Eurovision Song Contest, the three opera singers were each dressed in a colour of the Italian flag, while the recording of the song featured music provided by the orchestra of La Scala Milan. The song was performed by Alice and Battiato on stage in Luxembourg, while Guisto Pio conducted the orchestra. Alice wore a white trench coat-style dress and black gloves reminiscent of a character in a 1950s neo-realism film made in Rome's Cinecitta. Franco Battiato wore a plain white shirt and black suit, a deliberate attempt perhaps to not draw attention from the importance of the song. On the left, on a raised platform, stood the three opera singers, while on the right, on an opposing platform, a single musician played along with a synthesiser. The song received several top marks from the various juries. It was thought that the combination of a good song and Alice's popularity throughout Europe would help it secure a victory. The song, disappointingly, did not win but came in fifth and turned out to be that year's best selling song from the contest. It topped the Italian charts, where it ended up in the top twenty best sellers of 1984. A curious irony was that Sweden did not vote for Italy but *I Treni Di Tozeur* went on to do very well on the Swedish charts. Perhaps here we can see what a difference the televote would have made if it had been in operation under the 1984 rules. It proved to be, by far, the most popular Italian entry inside Italy itself for many years and was also to be heard in the 1985 Italian comedy film *La Messa e Finite*, which was a big box office hit.

American-Italian films of the 1950s and 1960s often centred around relationships with Italian servicemen and local Italian girls, such as the widely successful *Buona Serra Mrs Campbell* starring Gina Lollobrigida, Phil Silvers, and Telly Savalas. Like many Italian counterparts of its time, it was filmed in Rome's Cinecitta. The film is a comedy on which the musical and box office smash film *Mamma Mia*, which revolves around Eurovision winners ABBA's hits, is based. Sophia Lorens' 1960 *It Started In Naples*, which also starred Clark Gable, was another successful comedy filmed on location in Rome, Naples, and Capri. Both films are set against a background of poverty and women entrenched in an Italy of post-war deprivation who are rescued by a handsome prince in the form of a comfortably well-off American. In *It Started In Naples*, Sophia Loren sings the tongue-in-cheek song *Tu Vuò Fa l'Americano* – *You Want to Be American*, composed by Renato Carosone, who also wrote the popular *E La Barca Turnó Sola*, a parody of the Gino Latilla song from the 1954 San Remo Festival. Other films such as *Two Women*, based on Alberto Moravias' *La Ciocciara*, also starring Sophia Loren, dealt with the suffering and hardships of ordinary people in the aftermath of war in Italy. In its infancy, San Remo had inhabited an Italy devastated by war. In the 1950s and 1960s, it was still an impoverished Southern Mediterranean cousin of the affluent north whose southern cities, like Naples, were rife with cholera and deprivation. As both San Remo and Eurovision grew towards middle age, they moved in a very different Italy, a much more affluent society whose wealth rivalled that of the liberal democracies of Northern Europe. The Italian pop that had evolved as a result of American musical influence in the 1950s, witnessed in the films *Buona Serra Mrs Campbell* and *It Started In Naples*, produced an ocean of stars, a great number of whom have been revealed in this

Eurovision: A Plea for Respect

chapter, who appeared at San Remo or Eurovision and, in many cases, both. Italian pop was, in its early days of evolution, quite unique compared to other pop music. It had an inkling towards sentiment and was often a ballad in style. Its exponents were often crooners who interpreted a distinct fusion of Italian folk rhythms and modern pop forms found in British and American music or, as often as not, rhythms found in Latin American pop. *Schlager* and *Chanson* were also a profound influence on the genre. Italian pop reflected a national appreciation of beauty and emotionality. It was often dramatic, as manifested in San Remo competitor Patty Pravos' *Tutt 'Al Piu*. Her 1973 hit *Pazza Idea* was much influenced by sacred music. It was number one for nine weeks in Italy and was the second bestselling single of 1973, surpassed only by Elton Johns' *Crocodile Rock*. Italian pop always stayed faithful to its romantic heritage and was the dominant musical trend in Italy from the 1950s through the 1990s. Unlike 1980s Italo-Disco, whose lyrics were often penned in English, Italian pop never strayed from the national language of the peninsula. Its structure was composed of romantic ballads with rhythms that were melodious, often with a background of more traditional arrangements. The music was often highly orchestrated, as in Gigliola Cinquetti's *Non ho l'eta* and *Alle Porte Del Sole*, Gianni Morandi's *In Ginochia di Te*, or Iva Zanicchis' 1969 Eurovision entry, *Due Grosse Lagrime Bianche*. Its style has influenced many ballad singers and songwriters all over the continent, with prime examples being Lilli Ivanova of Bulgaria and the Welsh singer Shirley Bassey, one of whose greatest hits was *Never, Never, Never*, an English cover version of Minas' great ballad *Grande, Grande, Grande*.

One song that seemed to be on everyone's lips wherever one went in Europe circa 1977 was *Te Amo*, the Umberto Tozzi hit which he had also recorded for the Spanish market as *Ti Amo*, sung by shop assistants in Lima, factory workers in Havana, and teenagers in Seville. It was the Italian world hit of the decade. Dalida recorded a French version in 1977 which enjoyed considerable success. Umberto Tozzi had initially joined the group Off Sound when he was just sixteen and had composed for the duo Wes and Dori Ghezzi. This was his first real success and catapulted him to further recognition with a series of successful albums and hit songs in the years to come. Tozzi's most famous song in the English-speaking world was his 1979 Italian language hit *Gloria*, a huge hit for him in Italy and Europe but a chart-topper for Laura Branigan in the USA and Britain. Brannigan's version went platinum and hung around in the American hot hundred for thirty-six weeks. Tozzis' original recording of *Ti Amo* was number one on the Italian charts for an astounding seven months. Brannigan's *Gloria* was nominated for a Grammy in 1982, further bringing the name of Umberto Tozzi to the attention of the English-speaking world and the USA in particular. While *Ti Amo* turned into the second *Volare* of the Italian music business, Tozzi also basked in the limelight of San Remo, which he won in 1978 with *Si Puo Dare di Piu*. He sang it together with Gianni Morandi (Eurovision 1970) and Enrico Ruggeri (Eurovision 1993).

About the same time, Raf, born Raffaele Riefoli, was just another Italian living in London – a city whose Clerkenwell area once boasted a lively and thriving Italian community where Caruzo and Mazzotti both lived. Raf's first musical venture was a plunge into the new wave scene of the 1980s, which in 1984 produced his first single, *Self Control*, a popular European as well as Italian hit which was simultaneously recorded by the American vocalist, popular in the 1980s, Laura Branigan, who once again popularised an Italian hit in the USA and Britain and had considerable success worldwide with her own version, which shot to number one in a variety of countries. Raf continued to record his singles in English but after a few less successful recordings, he decided to record his work exclusively in Italian, which initiated a chain of best-selling singles and albums. His albums Metamorfosi (2008) and Numero (2011) were both highly acclaimed and secured places in the Italian top ten albums charts almost thirty years after his first hit. In 1987, Umberto Tozzi and Raf teamed up to sing as a duo at the Eurovision Song Contest in Brussels with their composition *Gente di Mare*. It is a song about the lives of the people of the sea that, through the eyes of city dwellers, appear to be idyllic. It speaks of how those who have their origins in sea-faring communities are deeply nostalgic for the sea but after visiting their roots for a day are faced with the harshness of their reality as the nostalgia goes and they are again desperate to leave.

> *"We prisoners of the city*
> *always live for today*
> *and for yesterday*
> *Pinned down by reality*
> *And the people of the sea are free to go"*

The song proved to be extremely popular on the night of the Eurovision Song Contest. *Gente di Mare* was very much in the running during the voting, at the end of which it had scored 103 points, which situated it in third place, after Germany and Ireland. The Israeli song *Shir Habatlanim* – in the same vein as Fred Astair and Judy Garland's *We're A Couple Of Swells* – by the comedy duo of Datner and Kushnir came in eighth. A fun song, it had caused outrage in the Israeli "Knesset", prompting the Minister of Culture to threaten to resign if it went to Brussels. *Gente Di Mare*, which was a blues-influenced Italian ballad, did not reach the Italian top ten but it managed to enter the respective top tens of Sweden, Belgium, Austria, and Switzerland and was also a hit in Germany and the Netherlands – a reflection of Raf and Umberto Tozzi's popularity throughout Europe.

Italian songs at the Eurovision Song Contest have not been exclusively the prerogative of Italy. When the language restrictions were loosened in 1999, it produced a willingness by Romania to sing its entries in Italian and manifested that most Romanians easily understand Italian and Spanish. There is also a veneration of Italian culture in Romania and a feeling that Italy is its cultural motherland. The Romanian Eurovision entries of 2006, 2007, and 2008 were all at least partly sung in Italian, while its 2012 entry was partly sung in Spanish. Romania's best placing to date was third at both the 2005 and

2010 contests, when it had chosen to sing its entry in English. Mihai Trăistariu's *Tornero*, however, came in fourth in Athens in 2006 and was seen in the lead-up as a potential winner. *Tornero* was the country's highest placing song in Italian and was the biggest Romanian hit song to originate from the Eurovision Song Contest. Its chorus was sung with great gusto not only by the falsetto Trăistariu but by a great section of the audience at the Eurovision event and was to have chart success in twenty-two European countries. The Greek singers Anna Vissi, who had sung at the same contest, and Tamta, also had success with the Greek version of *Tornero*. Italian was also heard in the 2007 Latvian entry, a pop opera attempt called *Questo Notte*, sung by a group who called themselves Bonapartio IV, a new band made up of six Latvian tenors. One, Roberto Meloni, was an Italian-born Latvian immigrant. The group was formed by the Swedish composer Kjell Jennstig, who had originally composed the song in English with the title *Tonight*, which he presented to the Latvian national selection for Eurovision. The rules stated that the Latvian representatives had to be Latvian citizens, so Jennstig set about forming the group. The language of the song was changed to Italian and the title *Tonight* was dropped in favour of *Questo Notte*. There were high hopes for the song in Latvia but despite qualifying for the semi-final, it was to come in only thirteenth at the Grand Final in Helsinki. The small republic of San Marino made its debut at the Eurovision Song Contest in 2008. An Italian-speaking nation of only 32,000, where Romagnol is also spoken, the Most Serene Republic of San Marino, to give it its full title, has been independent since 301 AD and is completely enclosed by Italy. The small state's broadcasting company made a statement in late 2007 confirming its first participation in a Eurovision Song Contest. The entry was selected internally by a committee from a choice of songs submitted to the broadcaster. The selection committee was headed by San Marino's biggest star, Little Tony. The fifty songs and singers that were considered included numerous Italian artists as well as songs that had been rejected from the Moldovan and Azerbaijani national selections. In the end, the pop/rock band Miodio was chosen to

Eurovision: A Plea for Respect

represent San Marino with their song *Complice*, sung in Italian. The group was a mixture of both Italian and San Marinese members. Their debut was to be a disappointing final. *Complice* managed only nineteenth in the semi-final. Several artists expressed an interest in singing for San Marino at the 2009 contest, including San Marinese Italian King of Rock and Roll, Little Tony. Miodio, meanwhile, went on to compete at the Golden Stag Song Festival in Romania as well as the Mediterranean Song Festival in Montenegro. They also competed twice at San Remo. Due to financial problems, San Marino declined to enter a song in 2009. In 2010, the country had already chosen the Italian duo Paulo & Chiara to represent it but the national broadcaster was let down by much-needed sponsorship and funding and again had to withdraw. San Marino made a welcome comeback in 2011, represented by the Italian singer Senit, an ethnic Eritrean, whose song was performed in English, as was the 2012 entry sung by the San Marinese Valentina Monetta. The somewhat controversial *Facebook Uh, oh, oh*, which had to be changed to *The Social Network Song* due to the lyrics breaching the rules, was a composition of veteran German Eurovision songwriter Ralph Siegel. It was sung in English. Little Tony's desire to sing for his homeland at Eurovision was never realized. After suffering a heart attack in 2006, Little Tony survived only to die of lung cancer in 2011. Valentina Monetta was again chosen to sing for San Remo in 2013, returning this time with an Italian song, again penned by Ralph Siegal, *Crisalde*. Like all the San Marinese songs that preceded it, *Crisalde* failed to qualify for the semi-final. It had been predicted to win by many and was a bookie's favourite. This prospect made 2013 an even more disappointing year for San Marino. Several rumours abounded on the internet early in 2013 that the song *All In Your Head* by Lys Assia, the 1956 winner, which had been submitted but rejected for the 2013 Swiss national selection, would represent San Marino but this was confirmed by San Marino TV officials as nothing but rumour. Later in 2013, the national broadcaster announced that Valentina Monetta would represent her native country for a third consecutive year in 2014, making her parallel only to winners Lys Assia,

Corry Brocken, and Udo Jűrgens, who all sang at three consecutive Eurovision Song Contests. Monetta's lack of success did not deter her from representing her native land yet again and she appeared for a fourth time at the 2017 Eurovision event but once again failed to reach the Grand Final. San Marino, one of Europe's smallest countries, was, prior to its 2008 Eurovision debut, an obscure mini-state that many ordinary Europeans not too high up in the geography factor would have had trouble naming in a list of European sovereign states. After its first four years of competition in the event, its recognition as a European sovereign state is now widespread throughout Europe. Its place as a European nation is firmly embedded in the psyche of greater Europe. The potential for tourism and its contribution to the national economy is enormously beneficial to a small country like San Marino. Another asset that has come with Eurovision participation is that San Marino has shown the world that it has a thriving music scene and that music is not the exclusive prerogative of the larger, politically potent nations. A Eurovision win would also focus global attention on the small republic, which could bring international investment to the country.

The Swiss canyon of Ticino enjoys a Florida-type status in Switzerland. Its beautiful scenery makes it not only popular for holidays but also an area where people chose to retire to. Its Italian spontaneity and southern zest for the dolce vita are much sought after by the German and French-Swiss who swell Ticino's population in the summer. Despite Italian being one of Switzerland's official languages, Switzerland has chosen only nine Italian songs to represent it at the Eurovision Song Contest, from the period 1956-2013, out of a possible fifty-two. Lys Assia's 1958 *Giorgio* made references to Italian culture food, and place names but was, despite its title, sung in German. Anita Traversi, a popular Italian-Swiss singer of the 1950s, 1960s, and 1970s, participated in the Swiss national final seven times from 1956 to 1976 but was selected for the Eurovision Song Contest only twice, in 1960 with her *Cielo e Terra* and in 1964 with *I Miei Pensieri*, which came in last with zero points. Anita Traversi, along with Barbara Berta's *Dentro*

Eurovision: A Plea for Respect

di me in 1997, were the only two Swiss-Italian artists to sing for Switzerland at the Eurovision Song Contest from 1956-2000. Of the seven other artists to sing in Italian for Switzerland, none were Swiss-Italians. Peter, Sue and Marc, who raced up the scoreboard at the 1981 Eurovision Song Contest with *Io Senza Te* and ended up in fourth place, were Swiss-German. Gianni Mascolo, who sang *Guardando Il Sole* for Switzerland in 1968, was an Italian citizen. Mariella Farré, although of Italian descent, was, in fact, a Swiss-German from Thurgau. Her 1983 Eurovision entry *Io Cosi Non Ci Sto,* an Italian pop ballad that finished in fifteenth place, was sung in Italian but she returned to the contest again in 1985 in a duet with Pino Gasparini with the song *Piano, Piano,* which was sung in German but had an Italian title. Farré had also competed in the Swiss national final the same year as a solo artist with the German language entry *Oh, Mein Pierrot.* Sandra Studer, the popular Swiss-German television personality, competed for Switzerland in 1991 under her former stage name, Sandra Simo, with the Italian song *Canzone Per Te.* She enjoyed more success than most of the other Swiss-Italian entries and came in fifth. In 2000, the Swiss entry was once again in Italian but sung by a Swiss-German singer. Jane Bogaert sang *La Vita e Cosi.* She was at the time in a relationship with Al Bano, who had separated from Romina Power in 1999. At the contest, Al Bano acted as her backing singer in the rousing ballad, which came in only twentieth. From 2000 to 2013, Switzerland was represented by only one Italian song at Eurovision. The Ticino-born Paulo Meneguzzi won the 1996 Viña del Mar Song Festival in Chile and slowly built up his career in Italy. At the 2001 San Remo Song Festival, Meneguzzi came in seventh in the New Artists Section. He then went on to compete at the 2005 and 2007 festivals, which led to him becoming one of the most popular Italian rock artists of the 2000s. In 2008, he represented Switzerland at the Eurovision Song Contest with the soft rock ballad *Sto Pregando.* The rules of San Remo state that to compete in the Italian section of the festival, an artist must be an Italian citizen. Gianni Mascolo represented Switzerland but also competed at San Remo, as he was an Italian citizen. Meneguzzi, although Swiss by birth, held an Italian passport and was thus eligible to compete in both song festivals. *Sto Pregando*, a relatively good song and a strong contender for the Eurovision Grand Prix, did not, surprisingly, live up to the high expectations and did not even qualify for the semi-final – a heavy blow to Swiss expectations and to European fans of Italian music and classy contemporary ballads

Lys Assia's *Giorgio* had been a German *Schlager* in genre although in many respects it was a pseudo-Canzone in its melody and rhythm, similar to Neapolitan rhythms. Its lyrics of aspects of stereotypical Italian life with a quasi Italian classical music at its foundation. Italian place names such as Lago Maggiore also feature in the song. It was to be Lys Assias third attempt at Eurovision and was almost another winner. It was beaten only by the Netherlands and pushed the Great Italian song *Volare* into third place. Lys Assia had a solid Eurovision degree to her credit, having sung two songs for Switzerland at the 1956 contest and, in the same year, sung in the German national final. She had gone on to defend her title at the 1957 Eurovision Song Contest, singing for her homeland. At the 1956 Swiss national final, she had sung five compositions, one of which, *Addio, Bella Napoli*, had an Italian title. The others in French or German reflected the multi-lingual nature of the Swiss republic. Lys Assia returned to the Swiss national final of 2012 when she was already well into her eighties. Aliki Vouougklaki, the National Star of Greece, popularised the Greek version of *Giorgio*.

As always when it came to Eurovision, the San Remo connection was never far away. Not only did Swiss-Italian artists aspire to sing at San Remo but top Italian singers who had conquered San Remo also eyed singing for Switzerland at Eurovision as a platform for more international recognition. In 1969, the much-celebrated Mina competed at the Swiss national final with *Dai, Dai, Domain* and *Non Crederei*. Ten years later, in 1979, Rita Pavone also competed in the prelims with *Dieci Cuori*. Neither of them, however, was selected for the Eurovision Song Contest despite their superstar status.

Eurovision: A Plea for Respect

Anita Traversi's songs at the Swiss national final often reflected the warmth and temperament of Switzerland's Latin side, with several of her songs reflecting life in the Canyon of Ticino, her birthplace, i.e., *Bandella Ticinese* (1956) and *Malcantonesina* (1960). Ticina was, in fact, a popular lyrical topic at the Swiss national prelims, which can be seen in the Georges Pilloud attempt in 1964, *Amore in Ticino*, and the 1981 national final when the Italian language *San Gottardo*, referring to the famous mountain pass, came in second to *Io Senza te*, which, complete with Italian lyrics, Mongolian flute, and folksy, crystal-clear vocals, won the passport to Dublin.

As the twentieth century approached its final decade, Italian music was riding high all over Europe and, indeed, worldwide. The rock singer Zucchero made his first appearance at San Remo in 1982 with two of his own songs. In 1990, he had an international hit with his most renowned work, *Senza Una Donna*, with which he duetted with the popular English singer Paul Young, who sang the English lyrics. Zuccheros' own solo Italian version was a successful single both at home and abroad. Toto Cotugnos' worldwide hit from 1983, *L'Italiano*, was a landmark in Italian pop music in that it rejected the influences of American rock, British pop, and Latin American Rumba and looked inwards to its native Italy for inspiration. It is of note because it was the first successful pop record for some years to return to a more traditional Italian sound. The lyrics of the song reflected its traditional theme and rhythm. "Let me sing with my guitar in hand," says Cotugno. He makes references to Italian cuisine ("Good morning Italy with your Spaghetti al dente") and the veteran politician President Pertinni ("A partisan for President"). He mentions the other little everyday aspects of culture in Italy: "With the car radio in the right hand and a canary over the window". Also, he references the Americanisation of Italy ("With too much America on the posters), Italy's deeply embedded spiritual traditions ("Good morning Maria, with eyes full of melancholy" and "Good morning God"), and the role of women in Italian life ("With more women and always less nuns"). He ends the song by saying that he is proud to be an Italian – a true

Italian. As for the song, it was an affirmation of the real Italian. Toto Cotugno participated for the first time at a San Remo Festival in 1976. Over the following years, he would return to compete no less than thirteen times. *L'Italiano*, written by Cotugno and Cristiano Minellono, was presented at the 1983 San Remo Song Festival. Cotugno had originally written it to be performed by Adriano Celentano but he declined. The song qualified for the final round but managed only fifth place in the voting. Throughout the 1980s, other Cotugno songs such as *Innamorata* quickly became huge hits both in Italy and abroad. *L'Italiano's* popularity was revived in 2006 after Cotugno performed it at a charity event celebrating the Italian football team's triumphant victory at that year's World Cup. Toto Cotugno was chosen to represent Italy at the 1991 Eurovision Song Contest in Zagreb. Although Cotugno had won the 1980 San Remo Song Festival, the victory had occurred in the period when the San Remo winners were not sent on to Eurovision.

The 1990 Eurovision Song Contest is of great historical merit in the history of post-modern Europe. It was a Pan-European celebration of the end of the Cold War and aspirations for a new united Europe without frontiers, which was reflected in the lyrics of many of the songs. There was an optimistic feeling of togetherness as Europe headed closer to 1992. Cotugnos' entry, *Insieme 1992*, was unashamedly upfront in support of European integration and unity: "We're more and more free", "And you a woman without borders, with you under the same flags", "Our stars, one single flag we're stronger and stronger", and "This is an Italian song for you together unite, unite Europe". He sang with great conviction in a white shirt and suit, backed by the group Ashes and Blood, which had sung *Dan Ljubezni* for Yugoslavia in 1975. An emotional Toto Cotugno walked back onto the Eurovision stage to receive the trophy and a cheering auditorium. His 149 points were clearly displayed on the scoreboard with Liam Reilly of Ireland, with yet another European theme, *Somewhere in Europe*, coming in a joint second with France at 132 points. The contest was also of major significance in that Yugoslavia, the only

Eurovision: A Plea for Respect

communist state to compete in the Eurovision Song Contest, had won in 1989, which was the year of the fall of the Berlin Wall and the collapse of communism in Europe. The Zagreb contest, held in Croatia because Yugoslavia's 1989 winner, Riva, came from Croatia, was ironically the last year of the original Socialist Federal Republic of Yugoslavia's existence. The country collapsed in 1991 and descended into a bitter war. The last Yugoslav Eurovision entry was in 1992, when it was by then only the rump state comprising Montenegro and Serbia. It would be banned from the Eurovision Song Contest until 2004, when it would return under the name "Serbia and Montenegro". The other former Yugoslav states all entered their own respective songs, which included, in 1993, the first entry by an independent Croatia: *Don't Ever Cry*, sung by Put, which was a tribute to the suffering of the Croatian people during the recent war, when 15,000 Croatians were killed or simply disappeared, with a further 300,000 displaced. Immediately before singing the reprise of the 1990s winning song, Cotugno gave a speech in which he dedicated his song to all the Italian singers before him who had taken part in the Eurovision Song Contest but had missed out on winning first place. He ended by calling for a united Europe, a call that he repeated three times. He could not have expressed his belief in the song more. In retrospect, it seems sad that the hopeful optimism and aspirations so evident at the 1990 song contest should have degenerated only a few years later into the first war on European soil since 1945, with unending bickering and disagreement among the various member states of the European Union, the abuse of its human right charter, and the most severe economic crisis to hit the continent since the depression. All of this helped to shatter the expectations manifested at the 1990 Eurovision Song Contest and the sentiments expressed so well in Toto Cotugno's winning song.

The following year, Italy hosted the contest for the first time in twenty-seven years. The audience at Rome's Cinecitta included several well-known faces from the Italian world of show business, including Gina Lollobrigida. Italy's two Eurovision winners, Gigliola Cinquetti and Toto Cotugno, acted as hosts and even gave a rendition of their winning songs while contestants sang excerpts from famous Italian songs in between the various entries.

Since the 1980s, RAI had started to broadcast the Eurovision Song Contest as a delayed transmission, often in the wee hours of Sunday morning, unlike San Remo, which was still broadcast live. An apathy towards Eurovision in Italy set in, partly due to the times of RAI's broadcast and their eagerness to promote the song festival at San Remo as the premier song contest. When Rome hosted the 1991 Eurovision Song Contest, only three million Italians tuned into the delayed transmission, while only a few months before, an amazing fifteen million had viewed the San Remo Festival. Italy continued to submit songs for Eurovision, usually performed by big stars, most of whom had competed at San Remo. RAI sent representatives to the Eurovision Song Contest from 1991-93. From its first appearance in 1956 until 1993, Italy had been placed in the top five fourteen times. This was not such a bad record, yet, in that period of thirty-two years, it had, despite its great musical heritage and the place of music in Italian society, produced only two winners – twenty-six years apart. Citing a diminished interest in the contest, Italy did not compete from 1994-96. There was an underlying sense that they had given their all with consistently good songs and top artists and it still wasn't enough. Toto Cotugnos' *Insieme 92* failed to make much of an impression on the Italian charts and did not enter the Italian top forty, although it was a hit throughout Europe. Well-known commercial Eurovision songs which were often massive hits across Europe, such as *Puppet on a String*, which Sandie recorded in Italian as *La Danza Delle Notte*, *Save Your Kisses for Me*, *Making Your Mind Up*, and even *Waterloo*, were not very big hits in Italy, as none of them entered the top ten. Jalisse, a popular duo riding high on a wave of success in Italy, won the 1997 San Remo Festival with their song *Fiume di Parole*. In the same year, Italy made a return to the Eurovision Song Contest, and it was decided that *Fiume di Parole* should be sent to Dublin as the Italian entry. For the first time since 1972, both the winning artist and the winning song of San Remo

Eurovision: A Plea for Respect

were to represent Italy at the Eurovision Song Contest. *Fiume di Parole* was a very strong Italian rock ballad which owed much to the excellent vocal refrain. It was a veritable slice of contemporary Italian balladry par excellence and certainly by far one of the better entries. It came in fourth but was the best entry in the top five placings (though it did deserve a higher placing). If it could not win, then it certainly should have been runner-up. An excellent example of contemporary rock – *Canzone*? A San Remo winner? Only fourth place? The Italians were slighted. Italy did not appear on a Eurovision stage for another fourteen years.

A generation of Italians had grown up, had been born, or had grown older without the Euro Festivale, as it was known in Italy, with the younger generation familiar only with San Remo. Jalisse submitted a song in a later attempt this time to represent San Marino at the 2008 Eurovision Song Contest but it failed to impress the selection committee. A few years later, RAI UNO produced a biographical film shown shortly after the San Remo Festival, entitled "Volare, La Vera Storia Di Domenico Mudugno", which starred Beppe Fiorello as the Italian folk legend, who even appeared in the reproduction of the 1958 San Remo Festival wearing the actual suit that Modugno himself had worn at the 1958 event. The Eurovision Song Contest without Italy drew parallels to a banquet table that had a place set for a guest of honour but the guest's chair was empty. Italy, one of the original seven countries to compete in Lugano in 1956, was, after all, a very special guest. Did not the European Union originate from the Treaty of Rome and Italy, a founding member? In 2008, Rafaella Carra, the darling of Italian TV, hosted three Eurovision-orientated shows for the Spanish National Network, TVE. In the same year, there were calls by Toto Cotugno for Italy to return to the Eurovision Song Contest. His voice was joined by that of leading musician Vince Tempera, who voiced a degree of sadness at Italy's continued absence and the severe decline of the Eurovision Song Contest in Italian society. Rafaella Carra also appeared to be active in the campaign for Italy's return, and Russian superstar Dima Bilan was one of the guests on Rafaella Carras' hit show on RAI UNO, *Carramba Che Fortuna*, as part of a Eurovision special which also featured other recent Eurovision contestants. The EBU made a determined effort to bring Italy back into the Eurovision fold where she so rightly deserved to be – and, more than that, where she firmly belonged. In the years following 2009, the EBU concentrated on a determined mission to persuade Italy to return. It paid off and, in 2010, Italy announced that it would compete at the 2011 Eurovision Song Contest in Dusseldorf. In addition to this announcement, the EBU issued a statement in which it said that Italy would also return with an automatic entry to the Grand Final and, as one of the EBU's greatest financial contributors, would join the big four – the UK, France, Spain, and Germany – to become the big five. In the early months of 2011, rumours circulated around Europe. They hinted that RAI was to cut up the contest. It had stated in press releases that it had returned to Eurovision "with one condition to be able to adapt it to our needs and fit into our formula". It was thought that RAI would not show the postcards of the songs but would return to RAI studios with panelists discussing the songs, a popular form of programme in Italy. There was criticism in certain quarters by those who said that RAI was lucky enough to be given big five status and that the EBU was already bending over backwards to accommodate it. Its return, however, was warmly welcomed in most areas and Europe was happy to have the long-lamented missing link in the chain. It looked forward to some more quality music from Italy, which it had never failed to deliver before.

The much anticipated Italian entry accumulated even more interest when RAI announced that its Eurovision selection would make a return to San Remo for inspiration and talent. The winter waves in the harbour of San Remo ebbed and flowed as history returned to repeat itself in its mild sea breeze. At the 2011 San Remo Festival in mid-February, thirty-year-old jazz singer Raphael Gualazzi won the newcomer section of the festival with his jazz/blues composition *Foglia d'amore*. His self-penned number also received the Mia Martini Award, presented by the critics for best newcomer, and the Sala Radio TV

Eurovision: A Plea for Respect

award. RAI announced that both Gualazzi and *Fogllia d'amore* would represent Italy in Dusseldorf. The song was an amazing success. It was expected that most of the song would be sung in English as *The Madness Of Love*, or at least part of it would be. At the Eurovision Grand Final, however, Gualazzi sang most of it in Italian, with only a few verses in English. The song owes much of its inspiration to American music popularised in the 1920s. *Fogllia d'amore*, unlike most previous Italian entries, was a stride piano-style work which did not rely on influences from Leggero or traditional Neapolitan music. The song reflected the profound musical influence on its composer of such American artists as Duke Ellington and Oscar Peterson. It raced up the scoreboard, ending up as the runner-up with 189 points, beaten only by Azerbaijan's first-ever win. Jazz had proved it could be a well-received genre at the Eurovision Song Contest.

The tall palm trees of San Remo smiled broadly and awaited the arrival of its annual privileged position, that of the musical capital of Italy.

As the artists and songwriters gathered at the Ariston Theatre, home to the song festival since it was moved from the Casino in 1976, the established acts and newcomers both eyed the lucrative first place with a mixture of emotions. Among the established acts in competition for 2012 were the veteran San Remo participants Loredana Berte and Lucio Dalla as well as former Eurovision Song Contestant Matia Bazar. The Italians once more showed their great national passion for the song festival and watched it at home in large numbers presented by, among others, Italy's 1970 Eurovision representative, Gianni Morandi. Nina Zilli had come in first in the established artist section on the second night with her song *Per Sempre*, a big Italian ballad, but it sank to seventh place at the final, which was won by Emmas' *Non e L'inferno*. A special jury voted on which of the finalists would represent Italy at the Eurovision Song Contest in Baku. The winners of the 2011 song contest, the duo Ell and Nikki, announced to the viewers that Nina Zilli and her song *Per Sempre*, would represent Italy at Baku. This was changed at a later date and it was decided that *Per Sempre* would be replaced by the more up-tempo contemporary pop song *L'Amore E Femmina*, which came in ninth in Baku. Zilli had sung the song half in English, the language in which it had originally been written, but it was *Per Sempre* that proved to be the major success, reaching fifth in the Italian top ten. RAI's live broadcast of the Eurovision Song Contest was beginning to bear fruit, and viewing figures in Italy rose steadily. The seeds of interest were again growing. Italy once more manifested how committed it was to the Eurovision Song Contest in 2013 by announcing that the Italian entry would be chosen on the final evening of the San Remo Song Festival.

The Italian passion for San Remo is part of Italy's collective identity. It unites the once peninsula of city-states and bickering kingdoms and produces in it sentiments of Pan-Italian nationalism. As Hollywood films and Italian songs helped to forge an American national identity, so did San Remo give the children of Garibaldi a deeper sense of Italianism, a post-war alternative to the darker aspects of Mussolini's ultra-nationalism and the fascist years and a link to that factor deeply embedded in Italians – music. Maria Serafovič's *Molitva*, a Eurovision winner, also became an anthem of national solidarity among Serbs, sung at moments of Serbian victory in football and other areas of international competition. A Serb win at Wimbledon gave rise to a chorus of *Molitva*, while in Tel Aviv, Israeli independence day street celebrations echoed with the community singing of Dana International's *Diva*. In the same way that those peoples feel a direct national connection to the evergreen Eurovision songs and express nationalist emotion through the singing of them, so is the heartfelt emotion of being an Italian connected to the high Italian celebration of San Remo. It's an annual ritual and celebration like a holy sacrament – a sacrifice of the mass that must be offered for some sort of Italian redemption necessary for the salvation of the music of Italy. Like the songs *Molitva* and *Diva*, the festival of San Remo expressed a wider national sentiment, one of being an Italian but also of being accepted internationally and recognising that identity as part of a much wider mass European choir.

Eurovision: A Plea for Respect

The passing of Nila Pizzi in 2011 marked the completion of an end of an era in Italian music that was already long gone. As the waves in the harbour at San Remo gently whisper in the darkness of a summer's evening, the song festival long passed, the lyrics and tunes of over half a century echo through its elegant streets and piazzas. Its participants have aged and fallen to stardust while the newly born of other decades are the new stars. Its stories of jubilation, joy, disappointment, anticipation, and tragedy move to and fro with the breeze, together with the tall palm trees that line the shore. Dalida recorded several songs from San Remo and Eurovision in French, which, due to her interpretations, gained popularity in France, where she lived most of her life, and beyond. She was crowned Miss Egypt 1954 but never competed at the Miss World Pagent in London due to the Suez crisis. Shortly after, Dalida arrived in Rome from Egypt one Christmas Eve to work as an actress in Italian films but it was her voice that launched her into an incredible singing career. Although Dalida never competed in a Eurovision Song Contest, she was much associated with it, as she was with San Remo, where, as a participant, she had been a potential Italian Eurovision entrant. In her career, which spanned four decades, she became one of the greatest European pop singers and sold more than 150 million records in her lifetime. Her 1974 Pan-European hit *Gigi L'Amoroso*, a more traditional Italian song, was particularly successful. It tells the story of a Neapolitan singer who goes off to Hollywood to seek his fortune there. References are made to *Volare* in the song. Many years later, Gigi returns to his village, Americanised. It is interesting that in her vastly successful career, she declined all efforts to establish herself in the USA. Her life was beset with tragedy, of which the death of Luigi Tenco was only one example. Dalida's ex-husband, with whom she remained on amicable terms, shot himself in 1970. Shortly after Tenco's death, she had a brief affair with an eighteen-year-old Italian student and became pregnant by him. A subsequent abortion left her unable to conceive any more children. Mike Brant, a Cypriot-born Israeli singer and close friend of Dalida, was one of the most popular Francophone singers of the 1970s. On the same day as the release of a new album in 1975, he jumped to his death from the window of a Paris hotel. A former long-time boyfriend also killed himself in 1983, leaving Dalida severely depressed and desperately alone. One May evening in 1987, Dalida returned to her home in the Montmartre area of Paris to find her beloved pet bulldog dead on the doorstep. On that very same evening of May 2, 1987, Dalida, unable to cope anymore, took a fatal drug overdose and entered immortality as a tragic icon of French music. She is commemorated by a statue in Paris, one of only three women to have a statue erected to their memory in France. Dalida had a major impact on French society. A poll by the Encyclopaedia Universalis placed Dalida second only to General de Gaulle as the public figure who has most influenced France.

As the lights of San Remo reflect onto the calm sea on warm summer nights, the strains of *Grazie Dei Fiori* and *Vola Colomba* echo in the night sky and perhaps the spirit of Nila Pizzi walks along the promenade and past the old Casino where she won the first San Remo Festival. Does the sound of *Ciao Amore Ciao* resound from the old Casino on balmy serene nights? And in a romantic nocturnal scene framed by the palm trees on the shore, do Luigi Tenco and Dalida walk happily hand in hand at last?

Securing a high-profile artist to sing for Italy at the 2013 Eurovision Song Contest was a major priority on RAI's agenda. It also sought a major Italian star who wanted to advance their career by cracking the European market. An extra effort was put forth, with intense interest and backing from major record companies. Among the established acts from San Remo rumoured to be RAI's Eurovision act was the extremely popular Marco Mengoni and the winner of the fifth Italian *X Factor* and rapidly rising star Francesca Michielin. Mengoni had also risen to stardom by competing in *The X Factor*, with a gold disc awarded for his first single, and was runner-up at the 2010 San Remo Festival.

As the 2013 San Remo Festival drew to an end, Marco Mengoni, with his song *L'essentiale*, was announced as the winner, a popular choice. RAI, capitalising on the winner's popularity, selected it as the

Eurovision: A Plea for Respect

Italian Eurovision entry. The outright winner of the established artists section at San Remo was once more to represent Italy. In the weeks preceding the Malmo contest, Mengoni, born on Christmas day 1988, was the golden boy of Italian pop with a staggering ten million views of his entry on YouTube. On the fourth night of San Remo, he had performed Luigi Tenco's *Ciao Amore Ciao*, adapted to his own particular soulful voice, which had rock overtones. At San Remo, he had competed with two songs: the winner and also *Bellisima*. Luca Benedetti of Italy's national evening paper "Corriere Della Serra" gave Mengoni's voice a most unusual description, calling it a "captivating Miaow". Mengoni's musical style was far removed from the leggero or Neapolitan canzone, which were both an integral part of Italian pop music from the 1960s to the 1990s. This shift in genre is reflected in both Guallazi and Zilli's Eurovision efforts, which were more in tune with international contemporary tastes – a sign of the times and the ever-changing, ever-evolving nature of the ephemeral world of Italian pop that Modugnos' *Volare* had so much impact on and was so informative on regarding which direction Italian popular music was ultimately to travel. It was a different world from Gigi's singing, which was so much enjoyed by the people of his village. The one that looked onto the bay of Naples in Dalida's popular song *Gigi L'Amoroso*, in which he is scolded for his new American ways and told by her that all they have in America is rock and the twist and know nothing of real music. Well, the leaves of the seasons are ever-changing. Yet, *L'essenziale*, a soft rock ballad unlike *L'Amore E Femmina* and *Foglia d'Amore*, did retain certain limited features of Italian balladry in its makeup. Mengoni describes his own music as British Black. Growing up in an international pop music culture in the wake of talented artists like Michael Jackson and David Bowie, who impacted his works, as well as in the shadow of homegrown Italian rock and pop, which was a major contributing element to his career, the Italian glam-rock singer Renato Zorro had a 1970s grease painted, androgynous look and spangled costumes which never ceased to fascinate a generation of Italians. He was often compared to Britain's Marc Bolan. As Mengoni walked on stage at the fifty-seventh Eurovision Song Contest wearing a shiny blue suit and tie, his album featuring *L'essenziale* was number one in Italy, the third to consecutively do so. Mengoni's song came in seventh in the voting – a respectable placing for Italy in a European event that it had been absent from for far too long.

The following years saw Italy rocket up the Eurovision scoreboard on several occasions. The rousing sound of the pop opera group Il Volo's, *Grande Amore*, the last song to be performed at the 2015 song contest, brought the house down and earned Italy third place. Italy once again returned to the top five, to the astonishment of the bookies, in 2018 with Ermal Meta and Fabrizio Moro's *Non Mi Avevo Fatto Niente*, a song that had been penned shortly after, and inspired by, the terrible bombing of the Ariana Grande concert at the Manchester Arena in 2017. It dealt with terrorism in Europe and the suffering endured as well as the pain caused in countries torn apart by conflicts carried out in reaction. The following year, the Milanese singer Mahmood, of mixed Sardinian and Egyptian heritage, exhibited a different and changing Italian culture with his song *Soldi*, sung partly in Arabic. Mahmood, despite his background, does not speak Arabic but is fluent in Sardinian. However, he describes his music genre as Moroccan pop. *Soldi* went to number one in Greece, Israel, Lithuania, and his native Italy. It was also a top ten hit in several other countries and holds a record-breaking number of streams on Spotify. It was placed second to the winner from The Netherlands at the Grand Final but was heavily criticised by the then Italian right-wing deputy Prime Minister Matteo Salvini because of Mahmood's Arab connections. The pandemic of 2020 saw the cancellation of what should have been the sixty-fifth Eurovision Song Contest. The San Remo winner of that year, *Fai Rumore*, by the vocalist Diodato, which went to number one in Italy a week after its win, had been chosen to represent Italy but due to the extraordinary circumstances never actually competed. The EBU broadcast a substitute for the event on the evening when the song contest had been meant to take place, entitled *Europe Shine a Light*, a poignant and reflective emission that captured the mood of a Europe

Eurovision: A Plea for Respect

deep in the throes of Covid-19. Diodato sang his entry alone in an empty Verona Arena, a very moving scene, from a nation where so many had died. Italy's Eurovision entries were doing well internationally, as was the song contest itself, and RAI confirmed its participation at the next contest in Rotterdam after the EBU and Dutch broadcaster agreed on a strategy to allow the return and revival of the world's greatest music event under special conditions in 2021. It was to be the glam/alternative rock group Maneskin, Danish for moonlight, who won the 2021 San Remo Contest with their rock song *Zitti e Buoni*, translated as "shut up and sit down." This once more displayed another side of Italian music that had external influences and was yet again far removed from *The Leggera*. It was decided that it should go to the 2021 Eurovision Song Contest in Rotterdam. The group already had six platinum discs and one gold disc to their credit before they flew to the still Covid-conscious Netherlands to perform. As the voting proceeded in the Ahoy Arena, a former coronavirus hospital, Italy placed fourth in the jury vote but soared further up the scoreboard to first place after receiving a huge number of points from the televote. Spectacularly, Italy won the contest. It was Italy's third win, thirty-one years since Toto Cotugno's triumph, and it had, once more, come from San Remo. The San Remo Song Festival and the Eurovision Song Contest, are so entwined as to be blood-related, their histories forever bound together.

Questo capitolo è dedicato ai miei amici Siciliani. Mille grazie per il cibo, il vino e soprattutto le canzoni.

CHAPTER 4
Carnations, Fado, and Revolution

Portugal is a country on the extreme west coast of Europe and the Iberian peninsula that was once the centre of a great empire. Its pleasant climate and lifestyle have seduced tourists from other parts of Europe for many years and continue to do so, yet not so long ago, it was a very different place. The Estado Novo was the name of the conservative authoritarian regime set up in Portugal in 1933. Its establishment followed the right-wing military *coup* of 1926 which overthrew the so-called First Republic and its democratic constitution. The father of the Estado Novo (new state) and its chief architect was António de Oliveira Salazar, who was a devout Catholic and traditionalist. He believed that forces and trends such as economic modernisation and liberalism had to be curbed so that the country's Catholic and rural values would be maintained. His hardline, dictatorial Estado Novo differed from such contemporaries as Hitler and Mussolini in Germany and Italy, where mass state violence and genocide took place. Salazar believed in the use of a more moderate form of violence to keep him in power. Many dissidents were, nevertheless, murdered, most of them by the secret police, known as PIDE. Those who weren't murdered were incarcerated for years in Santiago on the Cape Verde Islands, off the coast of West Africa. The regime enforced nationalism on the educational system which glorified Portugal and its territories in Africa and Asia (the Ultramar). The population was fed a diet of ultra-nationalistic and ultra-Catholic values which were imposed on them. The press, television, and radio faced harsh state censorship, while the three robust values of God, Fatherland, and Family formed an eleventh commandment which was the mother of the Estado Novo. The regime embraced and pursued an economic model called corporatism. This type of economic policy was implemented to defend the oligarch and promote an extreme form of

Eurovision: A Plea for Respect

capitalism as the state economic system came under the watchful eye of the regime. For a large part of its domination of Portuguese life, the regime was preoccupied with clinging onto its overseas colonies in the wake of British, French, and Dutch decolonisation. While most of Continental Europe was ravaged by World War II, Portugal saw huge economic growth due to exports to nations devastated by the war. This continued for most of the 1950s. Economic stability came to Portugal. This was reflected in the many bridges and primary schools that were built. Only the elite, however, were able to pursue any form of further or higher education for, like many things in Portugal under Salazar, it was closely supervised. This might sound strange to people who have grown up in Western democracies but Salazar's ideology believed that too much education for the masses was harmful and could lead to the destruction of the ultra-traditional and ultra-Catholic values of Portugal. It was available only to those with close ties to the regime. The only political party was the National Union, which was really a broad coalition of fascists, nationalists, monarchists, and extreme capitalists. The Boy Scouts were replaced by the Mocidade Portuguesa, and the Legiá Nacional was established. This was a militia for youths and both of these encouraged a military way of life.

While most of Western Europe experienced a liberalisation of society and an economic boom in the 1960s, things in Portugal were very different. The economy faced massive problems. Some politicians supported the introduction of more liberal reforms similar to those that had been introduced by General Franco in Spain, where they had met with a degree of success. These reforms were rejected in Portugal, as the regime feared their results would lead to its destabilisation and open the back door to the communists and socialists. The 1960s also saw a clampdown on student activities or, more precisely, democratic ideas that were becoming popular in the universities. The government closed the National Secretariat of Portuguese Students. In 1962, students demonstrated in the thousands in protest of the clampdown. These protests were backed by the Communist Party. A mega demonstration that took place in Lisbon in May of that year was violently put down by the regime. This led to a strike which was to be the first chime of the bell that would herald the regime's demise. Portugal became more heavily involved in suppressing the independence struggles in its African colonies, which led to many thousands of young conscripts being killed in action. Additionally, Portugal was one of the poorest countries in Europe, plus the African wars saw millions of citizens emigrate to other European countries as well as North and South America to escape poverty and conscription. It was in this environment that Portugal made its debut at the Eurovision Song Contest of 1964 in Copenhagen. António Calvário sang *Oração* and received zero points, as did fellow contestants Sabahudin Kurt (Yugoslavia), Anita Traversi (Switzerland), and Nora Nova (Germany), while Italy's *Non ho l'età* by Gigliola Cinquetti won, with the UK's Matt Munro winning second and Monaco's Romuald winning third. Shortly after Anita Traversi had performed the Swiss song *I miei pensieri*, a protester who had posed as a stagehand walked onto the set carrying a placard with the words '*BOYCOTT FRANCO AND SALAZAR*'. The protestor managed to appear in front of the cameras for just a few moments and make his point. Salazar's rule came to an end in 1968, when he suffered a stroke, but it was by no means the end of the dictatorship he had created. The new leader of the Estado Novo was Marcelo Caetano, one of Salazar's closest advisers through some of the toughest hours of the regime. Although Portugal had managed to keep a tight grip on the colonies, due mostly to the elite paratroopers and special army units, Soviet aid to guerrillas in Angola, Mozambique, and Guinea enabled them to cripple the Portuguese army. Thousands of soldiers were killed. The colonial conflicts were deeply unpopular in the country and could be compared with the US intervention in Vietnam. António de Oliveira Salazar died in 1970. Caetano embarked on a process of democratisation within the framework of the dictatorship, which included the position of Chief of the Mocedade Portuguesa. Caetano's reforms included the introduction of pensions to workers, who had been unable to pay social security contributions. He also authorised the construction of an important oil-processing centre. He reformed the secret police and renamed them. Opposition parties were permitted to stand at the 1969 elections under controlled conditions. This brought a token change to the national assembly. The president of the republic had much more control over the government, which differed considerably from Salazar's regime. Caetano was, therefore, unable to take the democratisation too far. The president was a hard-line right-wing traditionalist. At one point, Caetano thought he could do more if he stood for president. This, however, never took place. Caetano was up against a hard-line ultra-conservative of the old school but when he made concessions, the opposition groups looked upon them as insignificant drops in the ocean. Opposition to the regime began to grow. Twentieth-century Europe was changing culturally, socially, and economically at a phenomenal rate. London was at the heart of the swinging 1960s and a permissive society was well under construction. The student protests of 1968 led to the resignation of President de Gaulle. The black flag of anarchy flew over Notre Dame in Paris. Alexander Dubček's Prague Spring of 1968 gave a human face to communism, while Pope John XXIII's Second Vatican Council reformed and shed new light on the Catholic Church. Laypeople took part in the Mass. They read from the scriptures. By the late 1960s, many women in Spanish cities had stopped wearing the mantilla to Sunday Mass. The Estado Novo, bogged down by traditionalism and conservatism, could not keep up with the intellectual changes of the New Europe.

One popular singer and actress to emerge during the dictatorship was Lisbon-born Simone de Oliveira. Simone had originally started her career in school. Born in Salazar's Portugal in the 1930s, she started to achieve popularity in the latter half of the 1950s. Her successes in the theatre included *A Tragédia Da Rua Das Flores*, *Maldita Cocaina*, and *What Happened to Madalena Iglésias* – a homage to popular signer Madalena Iglésias, who represented Portugal at the 1966 Eurovision Song Contest. One of her greatest theatrical successes, however, was the much-acclaimed *Passa Por Mim No Rossio*. Her film career was no less successful. She appeared in the Portuguese film *Canção da Suadade* in 1964 and then in *Operação* D*iamante* in 1967. After a break of nearly ten years, Simone starred in *Cântico Final* in 1976 and *A Estrangeira* in 1983. She also acted in popular soaps such as *Roseira Brava*, *Vidas de Sal*, and *Senhora Das águas* and was a member of the new faces-type talent show *Chuva de Estrelas*, which showcased new singers and musicians on Portuguese TV. Simone was crowned the 1964 Queen of Portuguese Radio and won the 1965 Portuguese Song Festival, which led to her appearance at the Eurovision Song Contest with the song *Sol de Inverno (Winter Sun)*, a classy, melancholic ballad about lost love: *"I live in longing my love my life has lost its meaning. I feel like the winter sun shining without warmth in a winter sky"*.

Simone made a dignified and composed entry onto the Eurovision stage in Naples, wearing a sparkling evening dress. She gave a relaxed and emotive performance in front of the original star-spangled Eurovision logo that decorated the stage of the RAI Sala di Concerto. The song failed to arouse much interest in the juries and, disappointingly, ended up in joint thirteenth position along with Kirsti Sparboe of Norway, both of whom received only one point. The contest was a victory for Luxembourg's France Gall, while Conchita Bautista (Spain), Ulla Wiesner (Germany), Lize Marke (Belgium), and Viktor Klimenko, a Cossack who represented Finland, were all awarded no points. Simone was awarded a ticket to the Eurovision Song Contest of 1969 after she came in first in RTP's Portuguese Song Festival. At the contest in Madrid, Simone sang *Desfolhada Portuguesa*. It was a very different type of song from her 1965 entry, more up-tempo and dramatic, with innovative lyrics which included several hidden political overtones, written by José Carlos Ary dos Santos. The music had been written by Nuno Naraeth Fernandes and was unmistakably Portuguese in sound. Her performance was the penultimate of the evening, as she performed directly before the Finnish entry. The Portuguese delegation was certainly the most elegant on stage at the Teatro Real that night. Simone wore a formal evening dress and was backed by two female singers and a male musician who accompanied her on the Portuguese guitar. The backing singers glided onto the stage as Simone took up her position in front of the Salvador Dali-designed stage. Her performance was superior to her 1965 effort. She used her dramatic skills to express and interpret the deep and meaningful lyrics. Her appearance was dignified, as if she was Portugal personified. Unfortunately, *Desfolhada Portuguesa* fared little better than *Sol de Inverno*. Rather unfairly, it chalked up only four points, placing fifteenth. It was one of the song contest's greats that, sadly, got away. The song, however, became incredibly popular and went on to be a huge hit in Portugal. Simone also recorded the song in French, Spanish, and Italian. Her representing Portugal in 1969 saw her career at its highest peak. Although *Desfolhada Portuguesa* was a massive success and became a Portuguese classic of the 1960s, Simone did make several other records that enjoyed great popularity, among them *Maria Solidão* and *Deixa lá*. Shortly after the 1969 contest, Simone became ill, as a result of which she lost her voice for some time. This almost ruined her career. Gradually, however, her voice returned and was more potent than ever, leading to a comeback in 1973. She returned to song contests in 1980 when she represented Portugal at the OTI Festival of Latin Countries. In 1987, Simone was diagnosed with cancer but conquered the disease and survived to become a legend in Portuguese show business. In 2006, at the age of sixty-eight, she starred in the leading role of TVI's new soap, *Womans Angel*. Recordings by Simone de Oliveira and her 1950s and 1960s contemporaries were bought by a Portuguese public whose annual income was only $160, bearing more of a relation to life in Africa than in Europe in the 1960s. At that time, the per capita yearly income in Turkey was $219, while in the United States, it was $1,453. A large percentage of the population was illiterate. The well-off were hugely affluent, which contrasted sharply with the poverty of the majority. Portugal was more or less locked into endless poverty. Salazar saw Portugal's backward situation as an advantage in maintaining stability. Due to Salazar's rigid economic policies, there was no national capital available to build up local industries. European and American interest in investing in Portugal was not at all well received by the regime. The country's economy relied on the export and re-export of commodities. A large percentage of the Portuguese population was exploited in the name of profit and suffered profound material inequality. When Simone De Oliveira represented Portugal at Eurovision in 1969, she represented a European country where only 18% of the population was eligible to vote. The early part of the 1970s saw public anxiety dominated by Portugal's overseas colonies and independence wars. There was deep discontent over the colonial wars in both civil and military sectors of society. The military felt frustrated by the regime's professional suffocation.

The song *Tourada* ('*Bullfight*') was chosen by Portuguese TV to represent the country at the 1973 Eurovision Song Contest in Luxembourg. It was written by José Carlos Ary dos Santos and Fernando Tordo. The song was performed by Tordo himself. It was full of metaphors that severely criticised the Estado Novo. The regime had started to fall apart at the seams. As Fernando Tordo took to the stage of the Nouvelle Theatre dressed in a grey flared suit, blue shirt, and grey tie, complete with 1970s feather cut and backed by an all-male trio, he gave a cocky performance against the backdrop of the RTL orchestra and a shield of the golden lion of Luxembourg. He approached the microphone and confidently went into the opening verse of *Tourado*. *"It doesn't matter, sun or shade, cabins or barriers, we fight the beasts shoulder to shoulder. Nobody deceives us. We fight hand in hand. Only waiting can hurt us"* was a bold statement! Fernando Tordo heralded the end of the regime. The time was ripe for revolution. Another very popular singer around that time was Paulo de Carvalho. He had been a drummer/singer with the group The Sheiks, popular in the mid-1960s. It was not until de Carvalho left the group and went solo in 1968 that he shot to fame, becoming extremely popular. His music is reputed to be the most interesting and innovative dance music to be produced in Portugal during the late 1960s and 1970s. He combined R&B with funk and ethnic African music. De Carvalho had several hits in Portugal and competed at the 1970 RTP Festival of Song. Portugal did not compete at the 1970 Eurovision Song Contest

Eurovision: A Plea for Respect

in Amsterdam. Only twelve countries chose to do so, due mostly to discord over the four-way tie in 1969. De Carvalho went on to win the RTP Festival in 1974 with the song *E Depois do Adeus*, composed by José Calvário and José Nisa. It then went on to the Eurovision Song Contest in Brighton. This was considered a vintage year by many, or at least a very memorable one, as it saw ABBA's *Waterloo* triumph over the seventeen other songs, including the one performed by Italian singer Gigliola Cinquetti, who had won the contest ten years before. She returned with her song *Si*. Originally, eighteen nations were scheduled to perform on the big night but Dani of France withdrew as a sign of national mourning for President Pompidou, who had died a few days before and whose funeral took place on the morning of the contest. Contemporary bands Poogy from Israel and Korni from Yugoslavia brought interesting and different sounds to the contest, which had not really seen their like before. *E Depois do Adeus* was a big ballad with a very romantic feel. Paulo de Carvalho, dressed in a dark blue three-piece suit with a dark blue tie to match, a contrast to Fernado Tordo's light grey, ascended the Brighton stage as the penultimate act. The song was well suited to his voice and he gave a powerful interpretation with a dramatic ending. Sandwiched between Piera Martell's *Mein Ruf Nach Dir*, a powerful and emotive ballad from Switzerland, and Gigliola Cinquetti's *Si*, also a ballad, *E Depois Do Adeus* did not stand out enough to prove popular with the juries. He was awarded only three points and came in joint last along with the Swiss song that had preceded him. It was a big disappointment after Tordo's 1973 effort, which had placed tenth – one of the few times Portugal had ever reached the top ten after all the votes were counted. However, Paulo do Carvalho and his song would go down in Portuguese and Eurovision history.

Tordo's 1973 placing may have gone down well with Portuguese national pride but the dictatorship was not at all pleased with RTPs choice of *Tourado*. Shortly after its selection, its composer, Carlos Ary dos Santos, was arrested and sent to prison. He was, however, released before Tordo performed it in Luxembourg. The 1974 song contest, introduced at the Dome by Katie Boyle, was at an era in its history still dominated by ballads. ABBA's success was yet to be experienced. ABBA is considered the best thing ever to happen at Eurovision. Their many hits and million-selling albums are now legendary, yet at the same time, ABBA's hold on the Eurovision Song Contest in the years proceeding their 1974 victory brought gimmickry, bubblegum pop, and a musical *"competitionitis"* unseen at the contest before 1974. We also see the start of an erosion of particular national styles of music and the emergence of the dominance of Anglo-American-influenced Europop clones, from Ankara to Albufeira and Lillehammer to Limassol. It was to become a feature of the contest until the end of the 1980s. The British music weekly *The Disc* announced, *"The old Eurovision Song Contest is dead, long live the new Contest"*. One of the *World Review* books of 1974 described ABBA's opposition in the contest as *'boring and old fashioned'*, yet was this a reflection of the ignorance and intolerance of any other type of European pop music that did not come out of the Anglo-American stable in an ever decreasing acceptance of the boundaries of *listenable* pop music? Artists like Carvalho were merely *irrelevant* to British music critics. The days of the dominant continental ballads, which had given us twelve of the Eurovision winners from 1956-74, were numbered. The French ballad, by far the most successful, was to triumph only twice after 1974: in 1977 with Marie Myriam's *L'oiseau et l'enfant*, and again in 1983 with Corinne Hermès *Si La Vie Est Cadeau*, the last of Luxembourg's five winners. Paulo do Carvalho may not have aroused much interest in the international juries on that April evening in Brighton but his song would go down in Portuguese history and hold a special place in the Eurovision hall of fame. During the long years of the dictatorship in Portugal, the regime encouraged interest in and promoted *Fado* music. *Fado* is regarded as the national music of Portugal. Its name comes from the Latin for fate. As a genre, it is almost without exception always melancholic. Its origins are obscure and varied. One theory suggests that it came from Africa in a dance form, another that its sorrowful rhythm was influenced by homesick sailors. Others suggest Arabic roots and even that it originated from the Brazilian music hall. Death, loneliness, love, and betrayal are all themes that figure strongly in *Fado*. A traditional *Fado* performance features only one singer and a Portuguese guitar. Purists say *true Fado* can be performed only by a woman (who would dress in black with a shawl); the *fadista* must be able to deliver a sense of '*saudade*' (nostalgia, longing). Emotion is expressed through gestures and facial expressions. The audience of a good *fadista* is often moved to tears. An authentic *fadista* would never move her/his body during a performance. It is the perfect music to feel sad to, created, as it would seem, for this purpose. This made *Fado* very popular with the Portuguese diaspora who had been forced by economic hardship to seek a living in America and Northern Europe. In *Fado*, the emigrants found sentiments that reflected their homesickness and '*saudade*'. *Fado* first found popularity in the early half of nineteenth-century Portugal. Its chief exponent was Maria Severa, famed for her scandals as well as her singing. The *Fado* of Maria's era was the music of Lisbon's poor and became very much associated with the working class. The *Fado* of Lisbon and the *Fado* of Coimbra became two different styles of the genre. The *Fado* of Lisbon had always been the earthy music of the streets and reflected the pain of the lives of the underprivileged, while Coimbra *Fado* (Coimbra being a university town) has always been considered more refined and sophisticated, favoured by the academic classes. The Estado Novo forced all fadistas to perform in special *Fado* houses. It is this period that saw its golden age and from which emerged its greatest-ever exponent, Amália Rodriquez, who communicated '*saudade*' at home and internationally from the 1940s until her death in 1999. For many, Amália Rodriguez *was Fado*. Other top *Fado* singers to emerge around that era were Alfredo Marceneiro and Lucilia do Carmo. The latter owned her own *Fado* house in Lisbon, the *Adega De Lucilia*, which was known for many years as The Faia. *O Faia* was considered the best *Fado* house in Lisbon and was a highly popular venue from the 1950s-70s. Lucilia herself was considered one of the most authentic *Fado* artists of the twentieth century. Her son Carlos do Carmo,* a lawyer by profession, gave up his career in law to perform at his mother's club. His many hits have included *Gaivota, Lisbon Girl,* and *Moca*. He recorded a lot of material written by Frederico de Brito, Ary Dos Santos, and Paulo de Carvalho. Carlos do Carmo was one of the few male *fadistas* to receive international acclaim and have a career beyond his Iberian homeland. He had a distinct style and often blended *Fado* with Bossanova and Sinatra-type ballads. He was highly influential and respected and was noted for the high quality of the works he performed. As of the time of this writing, Carlos do Carmo remains one of Portugal's most popular *fadistas*. Sadly, due to the dictatorship's promotion of extreme nationalism and its association with *Fado*, the genre became tainted with its connections to the Estado Novo and by 1974 had begun to lose popularity.

A couple of months into the new year of 1974 saw Gaetano replace the disobedient General Spinola, who had been at odds with the regime over the direction of their policy in Africa. Cracks began to appear within the highest ranks of the state. In 1973, Otelo Saraiva de Carvalho, an army officer, had formed a secret movement of army elite called the Movement of Armed Forces (MFA), made up of left-wing officers who had been disillusioned by the government's attitude towards the colonial wars. This organisation saw that the time was ripe for a revolution. This significant event in Portuguese and European history took place on April 24 1974, only a few weeks after Paulo de Carvalho had sung *E Depois Do Adeus* in Brighton (and come in last!). The revolution started with two secret green lights. RTP first played De Carvalho's *E Depois Do Adeus*. Then, shortly after midday, the radio broadcast the revolutionary *Grandola, Vila Morena* by Zeca Alfonso. At that point, the MFA began its takeover of the country's nucleus and commenced with the revolution. The rebel captains insisted that civilians should stay indoors but thousands of ordinary people ignored their plea and ran into the streets, cheering and chanting support for the MFA. By 6.30 p.m., the Estado Novo, Europe's oldest dictatorship, had collapsed. As democracy was born in Lisbon, its many supporters thronged the streets with red car*nations*. Some soldiers placed the flowers in their guns, and the historic event became known as the Carnation Revolution. The carnation became a symbol of the non-violent nature of the *coup d'etat* and April 25 is commemorated by the national holiday known as Freedom Day. Caetano and President Thomas went into exile in Brazil and the country embarked on a difficult path to democratisation, marked by a bitter feud between right and left. Many political exiles returned to their homeland and were met with adulation from their supporters; this led up to the first democratic election, which took place on the first anniversary of the Carnation Revolution. So, Paulo Carvalho took up his niche in Portuguese history; the Eurovision Song Contest and its connection with the revolution were not yet over.

One of the army officers involved in the overthrow of the regime was a certain Duarte Mendez, who was also a singer. As he had actively taken part in the revolution, it was fitting that he was selected to represent Portugal at the first Eurovision Song Contest since the revolution. Both the music and lyrics were by José Luis Tinoco. The song began with the lines,

> *"From those who died without knowing why*
> *From those who persevered in silence and in cold*
> *From strength born in fear and rage liberated in the early morning*
> *The banks of my river are made"*

It was entitled *Madrugada (Dawn)*. Its lyrics were certainly very representative of Portugal at that time. Mendez had been persuaded by the song contest organisers in Stockholm to not perform in full military uniform, which he had wanted to do. He was backed by the Swedish trio The Dolls and finished his performance with the lines, *"There can never be enough songs like this."* The moustached Mendez wore, of course, the by-then emblem of Portugal, the red carnation, with his very 1970s wide lapels and with his shirt collar out of his jacket. Other memorable lines that would have touched the hearts of the Portuguese were: *"I'm singing about people who have just discovered themselves"* and

"*Arm in arm and flower in each gun*", both of which vividly described the feelings and sights of April 25.

The 1975 contest was warmly introduced by Swedish TV personality Karen Falck. Nineteen countries competed in the Grand Prix. German blues singer Joy Fleming gave a lively performance but faced her Waterloo by ending up in seventeenth position. The UK had "*resurrected the Shadows from the dead*" as one popular daily had put it, to sing for them. After ABBA's very 1970s-sounding win the year before, and the tidal wave of euphoria and success that had followed, BBC viewers had expected a singer or group more representative of current pop. In the end, the Shadows were given a very mediocre song and flew off to Sweden disgusted at the lack of support given to them by the British media. They came in second to the Netherlands' Teach-in with *Ding, Dinge, Dong*, which at the time...seemed catchy. Apart from *Madrugada*, the most interesting song was the first-time entry from Turkey, *Seninle Bir Dakika*, sung by Semiha Yanki, who gave one of the best performances ever seen at a Eurovision Song Contest. Her song, a melancholic and emotive ballad, was obviously of higher quality than many of the other entries. The Turkish singer fell to her knees as she interpreted the lyrics but it was not to be rated by the juries and was awarded only three points. The song came in last. The young singer was heartbroken and left the hall in tears. Despite the Portuguese singer having participated in a left-wing revolution, the political left in Sweden chose to give the annual contestants' arrival in the Swedish capital a hostile welcome. Demonstrators thronged the streets, protesting against commercial music and, in particular, what they saw as its *arch-perpetrator*, the Eurovision Song Contest. This topic was much covered and discussed by Radio Sweden's sometimes rather cynical International Service. In future years, Swedish attitudes towards the event were to be very much turned. Under the watchful eye of veteran Eurovision scrutineer Clifford Brown, who in his tenth year saw that no voting irregularities took place, all nineteen nations cast their opinions on the continent's new songs. Europe showed it was not really interested in what happened at dawn on April 25, at least not enough for Portugal to be highly placed. Duarte Mendez was awarded sixteen points and placed sixteenth in the slot he had performed in, with only Germany, Norway, and Turkey receiving fewer points.

Due to the old regime's associations with *Fado*, one of the first reforms the new democratic government wanted to carry out was to ban *Fado*. This, however, was overruled and Carols Do Carmo won the RTP Song Festival the following year in 1976 with the *Fado Uma Flor de Verde Pinho* (*A Green Pine Flower*), written by José Niza and Manuel Alegre. This surely must be one of the most beautiful songs in the history of the song contest but, sadly, was completely ignored by the Fiftieth Anniversary event in Copenhagen in 2005, where the public voted on their favourite Eurovision songs of all time. The song must be heard several times to appreciate its beauty and, therefore, did not fare well at the Congresgebouw in The Hague. The contest was won by the UK's Brotherhood of Man with the ditty *Save All Your Kisses for Me*, which as a song was one of the weakest winners but, nevertheless, proved amazingly popular Euro-continent wide. Catherine Ferry of France came in second with the bouncy, up-tempo *Un, Deux, Trois*, and Monaco's *La Musique et moi*, also an up-tempo 1970s dance number, sung by Marie Christy, came in third. Up-tempo was in! The lyrics of the UK winner ("*Save your kisses for me, save all your kisses for me bye, bye baby bye, bye, Don't cry honey don't cry*") contrasted sharply with the lyrics of Manuel Alegre:

> "*There is no boat or wheat, there is no clover, no words to write this song, loving you is a poem that I don't dare write.*
> *There is a river without water and I'm without heart*"

Carlos scored twenty-four points and placed twelfth. The song's composer, Manuel Alegre, would later take part in Portuguese history again, this time in the role of a presidential candidate. Alegre had lived in exile for many years while the dictatorship was in power. An active socialist, he was elected to Parliament in 1974 and remained an MP until 2002. To his credit, Alegre was also a prize-winning author who had also penned songs for Portugal's Queen of *Fado*, Amalia Rodriguez. His time in Parliament included a spell as deputy speaker. Alegre was not to be so successful in his bid for the presidency and came in as the runner-up, with only 20% of the vote, to Cavaco Silva. His opponents in the 2006 presidential election included Mario Soares, Jeronimo de Sousa, Fracisco Louça, and Garcia Pereira. The result for Alegre, a highly respected and popular politician, since his election in 1974 was a disappointing one but the democratic process of Portugal in 2006 seemed a far cry from the days of Salazar and his Estado Novo. The baby born of the notes of *E Depois Do Adeus* had grown up and was a thirty-two-year-old adult.

A group of Portuguese vocalists formed to compete for their country at the 1977 song contest in Wembley. Os Amigos included, in their lineup, both Fernando Tordo and Paulo de Calvalho and sang yet another song of the revolution, *Portugal Nao Coracao*, whose prelude was, "*My brother fought and died for Portugal*". Needless to say, they illustrated the song's theme by wearing red carnations on stage. Portugal continued to participate in the song contest but its entries were never as interesting as they had been in the 1960s and 1970s. In 1978, it dropped the politico-revolution theme in favour of an Abbaesque effort, *Dai-Li-Dou*, sung by Gemini, a repetitive clone of Sweden's best export which managed yet another seventeenth place. Europop would be the order of the day for many of the Portuguese entries after that. Manula Bravos' *Sobe, Sobe, Balai Sobe* in 1979 and 1980's *Un Grande, Grande Amor* by Jose Syd managed top ten placings, and a gentle ballad of some note, *Selencio e tanta gente*, by Maria Guinot, ended the 1984 contest with a bit of class but just failed to achieve a top ten placing, coming in eleventh. Nevada's pleasing ethnic effort of 1987, *Neste Barco a Vela*, made little impact on the juries. Some top ten placings in the early 1990s climaxed in the Lucia Moniz interpretation of *O Meu Caraçou Naó Tem Cor*, which placed sixth at the 1996 contest in Oslo, the highest-ever-placing Portuguese song up until then. This was followed by a '*nil point*' in Dublin the following year. Europop was again replaced by ballads which were

Eurovision: A Plea for Respect

replaced, in turn, by more Europop, which was then followed by the *Senhora Do Mar* in 2008. This was the cause of much speculation. It was a *Fado*-orientated ballad acted out on stage by singer Vania Fernandes. It qualified for the final from the semi-final but disappointingly did not reach the top ten. *Fado* again happily raised its gracious head in 2012 with Filipa Sousa but it would seem that *Fado* did not travel well and had never gotten very far in the contest despite some truly beautiful and inspirational songs.

Portugal's story is eternally entwined with the Eurovision Song Contest. Portugal, however, was not the only country to have, as its entry, a revolutionary call to arms. In 2005, as host nation, Ukraine, in the wake of its successful Orange revolution, in which its 2004 winner Ruslana had played a crucial role, chose its revolutionary anthem, *Razom Nas Bahato*, but gave it much less overtly political lyrics and entered it as the home entry, sung by Greenjolly. A gritty and confrontational rap song, it was, in many ways, reminiscent of the protest songs of the liberation movements in Chile and Nicaragua. In 2011, Portugal entered a song entitled *A luta e Alegria*, which translates into *Struggle Is a Joy*. It was a somewhat peculiar yet original entry performed by a satirical comedy group, Homines Da Luta, famous for parodying well-known songs from the years of the Carnation Revolution. Each member of the band took on the role of a character who was a caricature of singers from that time of great change in the 1970s. Europe did not comprehend the satirical element and the very Portuguese humor. The band did not succeed to the Grand Final.

Portugal also failed to qualify for the final in 2004, 2005, and 2006. After Finland's long-overdue victory in Athens, Portugal was, for many years, the most unsuccessful country in the contest. This was rectified in 2017 when Salvador Sobral's mellow ballad *AmarPelos Dois*, composed by his sister, triumphed in Kyiv and beat Bulgaria's Kristan Kostov into second place. Following his win, Sobral gave a controversial speech on stage, in which he stated, "*I think this might be a victory for music that actually means something. Music is feeling. Music is not fireworks*". This angered some of the other contestants who had relied heavily on special effects. Despite its win, I believe that Portugal should receive a special award for tenacity, for being such a good sport, and for entering some of the most meaningful lyrics in the contest's history.

Lucillia Domingues, maid at the Viscount de Richmond's home in Grosvenor Square, sang me the Portuguese entries of the 1960s and 1970s as she cooked me dinner back in 1988. This chapter is dedicated to her. I wonder where she is now.

obrigado e esperança onde quer que esteja está bem

* Carlos do Carmo died in January 2021. His death was marked by a day of national mourning throughout Portugal.

CHAPTER 5
A Vous La France

Katie Boyle's charm captivated viewers all over Europe as she introduced the 1968 Eurovision Song Contest from the Royal Albert Hall. For the third time, the contest was being held in London. On both previous occasions (1960 and 1963), it had been hosted by Italian-born Miss Boyle, whose pedigree was Russian/Italian nobility with a smattering of Australian blood as well. The contest bounced onto TV screens with a confidence and freshness that only the London of the Swinging Sixties could offer. Technology was changing and it was the very first time that a contest had been broadcast in colour. The UK had hoped for a two-on-the-trot win, having triumphed the previous year in Vienna, and had asked the popular Cilla Black, pioneer of British pop and one-time associate of The Beatles, to do the honours. Cilla turned down the invitation to sing for the UK, and Cliff Richard, a frequent tenant in the British top ten since 1958, took up the offer. Massiel of Spain pipped Cliff at the post and robbed the UK of a second victory, as is well documented in the history books. In third place was France with the song *La Source*, sung by the petite blonde Isabelle Aubret. The song had been composed by Daniel Faure with lyrics by Henri Dijon and Guy Bonnet. Isabelle Aubret was born Therese Conquerelle into a poor working-class family, one of eleven children in the northern industrial town of Lille in 1938. As a teenager, she won the French National Gymnastics Championship of 1952. Ten years later, in 1962, Isabelle won the Eurovision Song Contest in Luxembourg with the ballad *Un Premier Amour*, composed by Claud-Henri Vic and Roland Stephane Valande. Her 1968 entry dealt with an unusual and even disturbing topic for a song of its time: rape and murder. It tells the story of a young girl:

"*She was blonde; she was gentle*

She liked to rest in the wood
Lying down on the moss
Listening to the birds singing"

As the girl walks into town one day, she passes through the woods, where she views her attackers: three men staring at her.

"Three men like wolves. She a lamb with
tender flesh and they with a ravenous appetite"

The song goes on to say that as they picked up her dead body from among the leaves and rocks, a pure spring gushed out. The refrain ends the song:

"The spring sings in the middle of the wood and I wonder if I should believe this legend of a girl that they found there"

It received a total of twenty points on the big night and Isabelle went on to record a German version of the song, which became *Such mich* dort wo *die sonne scheint*. The song had originally featured in the French National heats of 1967 along with:

Do You Wanna Kiss Me?, L'acordeon de Broadway
Ce soir ils vont s'aimer, Mireille Mathieu
Il doit faire beau la bas, Noelle Cordier
Les amoureux, Vicky Leandros
Il est mort le soleil, Nicoletta

Noelle Cordier was selected to sing for France that year, while Vicky Leandros would go on to sing *L'amour est bleu* for Luxembourg in the same contest. Mireille Mathieu had already competed with the same song in the 1966 French selection. ORTF1 decided that *La Source* would best represent France in London. The decision was made internally. When Katie Boyle started to contact the European juries,

Eurovision: A Plea for Respect

La Source soon soared into the lead with far more points than either the UK or Spain. After the sixth jury had voted, only three other countries voted for France, and between them, they awarded only three points. The progress to victory was halted abruptly when none of the final six juries gave France any points. Isabelle Aubret failed to be the first person to win the song contest twice. She had competed in the 1961 French national heats with *Le gars de n'importe ou* and come in second. She tried again two years after *La Source*, when she duetted with Daniel Beratta, and again placed second with *Olivier, Olivia* in 1970. A 1976 effort, *Je te connais deja*, was less successful, ending up sixth out of seven songs. Her last attempt, the patriotic *France, France*, came in third at the 1983 French national final. Isabelle had worked in a mill prior to singing and was a dedicated supporter of the Communist Party. She formed a close friendship with the songwriter and singer Jean Ferrat, also a communist and political activist, whose works celebrated both love and the workers' struggle. His great masterpiece, *La Montagne*, which she recorded in 1964, was also recorded and regularly performed by Isabelle Aubret. It dealt with the sometimes-complicated relationship between city and countryside in France. The song became a million seller for Ferrat at the height of the *Ye, Ye* era in French music, when the scene was dominated by Claud Francois, Sheila, Johnny Halliday, and Sylvia Vartan. Ferrat once famously said, "I wonder what will happen to a civilization that can put a man on the moon but has forgotten how to make soup". Another of Ferrat's greatest and most controversial works was his 1963 *Nuit et brouillard* (*Night Fog*), which dealt with the Petain regime's willingness to round up Jews for the Nazis in Vichy France. Jean Ferrat's father had been one of them; he was to die in Auschwitz while the young Ferrat was sheltered by communists. Isabelle Aubret and Ferrat collaborated on several projects. Aubret recorded many of Ferrat's works. Both were often subject to a lack of airplay on French national radio, controlled by the government, because of their political beliefs. Ferrat was often banned from radio or TV in the 1960s and 1970s. Although a communist, Ferrat was nevertheless often critical of the Soviet Union and in particular of the Soviet invasion of Czechoslovakia. Throughout his career, he never sang or gave any concerts in the USSR, unlike Aubret, who made several appearances in the Soviet Union and behind the Iron Curtain.

In early 1968, the French Communist Party and the French Socialist Party united to form a political force intent on the formation of a left-wing government and to oust the ruling Gaullist Party in the elections. In March, far-left groups, which included some prominent poets and musicians as well as 150 students, occupied a building in Paris University. They held a meeting and discussed the topic of class discrimination in France along with political bureaucracy. The police were called and after the occupants had stated their wishes, they left the building peacefully. Many months of conflict involving students ensued, however, and the university was closed down on May 2, 1968 with the expulsion of some of its students. The next day, the Sorbonne's students met in Paris to protest. The police invaded the Sorbonne, which led to a demonstration by student and teachers' unions on May 6. As they marched in protest towards the Sorbonne, the police charged towards them, wielding batons. The protestors retaliated and the police had to retreat but returned later with tear gas. There were many arrests amid violent scenes. Other demonstrations and riots unfurled over the coming days on the Left Bank. Hundreds were injured. The brutal heavy-handed actions of the police gave rise to a wave of sympathy for the students. Workers' unions called for a one-day general strike, while many of France's more popular and well-known singers voiced their support for the students. On May 13, over a million workers were on strike, representing a staggering 22% of the workforce. These strikers were not organised by the unions but were a manifestation of more radical views that demanded the overthrow of General de Gaulle and his right-wing government. The riots even displayed an anti-union stance. As events escalated, President de Gaulle fled the country. Many predicted civil war or even another revolution. On May 30, de Gaulle stated his refusal to resign but announced a general election. This was followed by a huge Gaullist rally through the centre of Paris in support of the president. The Communist Party agreed to the election, and the revolutionary fever that had swept through the country slowly ebbed. The elections, ironically, gave the Gaullists a resounding victory but at the same time, the country had sent a strong message to the president, whom they regarded as too authoritarian and old-fashioned to govern. Jean Ferrat and Isabelle Aubret watched the riots and consequent national crisis unfold, both profoundly moved by this exceptional episode in French history. Shortly afterwards, Isabelle Aubret recorded and performed the Ferrat composition *Ma France*, a moving portrayal of the aftermath of the 1968 riots. Gaullist state censorship banned anything that might have been seen as a threat or an enemy of the state. de Gaulle in particular saw the potential of television in forming public opinion. Were not protest songs a potential focus for political and social discontent? Between 1952 and 1964, the left-wing singer-composer Georges Brassons suffered greatly from this, with over half his songs banned. Ferrat's material often suffered the same fate and he was prohibited from singing *Ma France* on TV or radio for many years. Aubret recorded and had successes with a number of Ferrat compositions, notably, *La Montagne*, *Potemkins*, *Ma France*. Others, such as *Eh, L'amour*, as well as *Deux Enfants au Soleil* and *C'est Beau La Vie* were recorded by both Aubret and Zizi Jeanmaire. After Jean Ferrat's death, a grieving Aubret gave a very personal and highly emotional interview for French media. In 1956, Andre Claveau recorded the song *Les Yeux d'Elsa*, which was a poem by Aragon set to music by Jean Ferrat. Claveau went on to win the 1958 Eurovision Song Contest for France with the song *Dors Mon Amour*. The poet Louis Aragon, a lifelong communist and a founder of Surrealism (Aragon had also been involved in Dadaism), provided Aubret with more songs and recordings of his works set to music. Among other singers to record his poems were Nicole Rieu, who represented France at the 1975 Song Contest and placed fourth with *Et Bonjour a Toi l'artist*, written by Pierre Delanoë, who had also composed for Piaf, and Alain Barriere, who sang *Elle était si jolie* for France at the 1963 song contest. Others who interpreted the works of Aragon included Georges Brassons, Leo Ferré, and Helene Martin.

Aubret suffered severe injuries in a car accident, the circumstances of which were rather vague. This resulted in fifteen operations. As a consequence, she was unable to walk for several years. Before her hospitalisation, she had been asked to play the lead role in the film of the musical by Michel Legrande, *The Umbrellas* of *Strasbourg*. Because she was unable to fulfil the role, the part was instead given to Catherine Deneuve. The leading role had been offered to the First Lady of French *Ye, Ye*, the Bulgarian-born, Sylvie Vartan, but her agent had turned it down on her behalf without consulting her. Vartan was devastated. The film went on to become one of the greatest box office successes and a classic of French cinema. Aubret also recorded several songs by Jaques Brel, including the exquisite *Amsterdam* and *Rosa*, but it is her interpretation of his composition *Fanette* that is most associated with her. It was to become one of her most popular recordings. Jaques Brel had, in fact, competed in a song contest himself, though not Eurovision. He had taken part in the 1954 Knokke Song Contest in Belgium but placed only twenty-seventh out of twenty-eight songs. In an act of generosity he, Brel, bequeathed the rights of *Fanette* to Aubret so that she could survive financially during the long convalescence that followed her terrible car accident. The Communist Party, which Aubret so ardently supported, saw the students in the 1968 riots as anarchists and, although they supported them, it was with much reluctance, and they ended up siding with the government on the issue. It is perhaps, even despite that fact, fitting that a singer with such radical anti-establishment views as Aubret should have sung for France in such a historic and monumental year as 1968.

Ten years later, in 1978, the Eurovision Song Contest was held in Paris due to France's 1977 victory, *L'oiseau et L'enfent*, sung by the Congolese-born Marie Myriam, who was of Portuguese extraction. As the cameras scanned the audience at the Palais de Congress on the big night, they focused for a few moments on French music legend Serge

Eurovision: A Plea for Respect

Gainsbourg and his partner Jane Birken. They were known in the UK for their 1969 chart-topper *Je'taime. Moi Non Plus*, which Gainsbourg described as "the ultimate love song". Its sexually explicit lyrics led the Vatican to condemn it as immoral. Several European countries, including the UK, banned it, which, of course, surrounded it with great controversy. Ultimately, probably due to this, it charted in many European countries. The song had originally been recorded by Serge Gainsbourg and Brigitte Bardot, with whom he'd had an affair, but she had backed out of the project and been replaced by the English actress/singer Jane Birken, with whom Gainsbourg would also have a relationship until they separated in 1980. Contrary to popular belief, they were never married.

Gainsbourg had several different involvements and associations with the Eurovision Song Contest. In 1967, he composed the Monegasque entry *Boum Badaboum*, sung by Minouche Barrelli, the daughter of the jazz musician Aime Barelli and singer Lucienne Delyle. Minouche, a Parisian, later applied for Monegasque citizenship and, thus, inherited that very rare item: a Monegasque passport. She died in Monaco in 2002. Critics of the contest only too eager to malign it will cite the song title as typical saccharine, Eurotrash lyrics in the *Boom-bang-a diggy doo – ding dong* genre but a closer inspection reveals that the lyrics are not about happy, carefree love that the genre so often celebrates but about nuclear war and the bomb. Minouche begs to live a full life before it is ended by a bomb: *Boum-badaboum*, the sound of an exploding bomb. Serge Gainsbourg, regarded internationally as one of the world's most influential musicians, was born Lucien Ginsberg in Paris to Jewish/Ukrainian immigrants. His mother had fled to France after the Russian Revolution of 1917. Gainsbourg was profoundly affected by the Nazi occupation of France and, in particular, the compulsory yellow stars worn by Jews. His family managed to escape occupied Paris for Vichy France using false identity papers but their situation in Limoges, their adopted town, remained precarious. After the war was over, Gainsbourg taught music at a school set up for the orphans of deported murdered Jews. He then changed his name from

Lucien to Serge and his surname to Gainsbourg after the artist Gainsborough, whose paintings he admired, then took up work playing piano in bars and clubs. His early songs were very much influenced by the *Chanson* and, in particular, the music of Boris Vian. He then experimented with jazz and, in the 1960s, with pop, in particular the *Ye, Ye* sound that was coming out of France. He went on to introduce funk and reggae into his compositions in the 1970s and toyed with electronica in the 1980s. Many of his songs had morbid themes, others had sexual twists, while still others were often scandalous and extremely provocative. He pushed the boundaries of acceptance to the extreme. His works illustrated the sexual revolution of the 1960s and the changes in French and Western society at that time. In 1965, he wrote one of his biggest international hits, the song, *Poupée de cire, Poupée de son*. This work was written for the sixteen-year-old France Gall and was the Luxembourg entry for the 1965 Eurovision Song Contest, held in the Southern Italian city of Naples. It went on to win the contest with a total of thirty-two points. Kathy Kirby of the United Kingdom came in second with twenty-six points. Third was Guy Mardel of France and fourth was Austria's Udo Jürgens. An English version entitled *A Lonely Singing Doll* was recorded by Twinkle and released in the UK, although it would not enter the chart. Twinkle had already had a hit with the teenage tragedy song *Terry*, which had led to a public outcry and was banned by the BBC. The recording featured Jimmy Page as a session musician. Twinkle had also been a classmate of Camilla, Queen Consort. Although the English version failed to chart, Gall's original was a massive hit all over Europe and beyond. Having peaked at number two in France, it went on to top the charts in Quebec and Norway and reached number three in Germany and Luxembourg. It also climbed to the top ten in Flanders, Holland, Finland, and Wallonia as well as Japan and Singapore. Gall recorded it in German, Italian, and Japanese although there were other Japanese versions by Mieko Hirota and Saori Minami. The song was covered by a myriad of artists across the globe in local languages, from the Danish version by Gitte Haenning and the Spanish version by

Eurovision: A Plea for Respect

Karina to *Bup Be Khong Tinh Yeu*, the Vietnamese recording by Hgoc Lan. Korean and Arabic versions were also recorded.

Poupée de Cire was the first song that was not a ballad to win the Eurovision Song Contest. On the surface, the song may seem a catchy contemporary *Ye, Ye* genre composition but a closer look at its lyrics reveals something far more complex. Nuances of lyrics, double meanings, and puns make it difficult for non-native French speakers to fully understand the lyrics. They are lost in direct translation. For a number of reasons, the song was groundbreaking and daring, which current attitudes and trends toward Eurovision in Britain, and particularly the British media, seem happy to stubbornly disregard.

In the song, the singer sees herself as a wax doll, a *poupée de cire*, a sawdust doll, a *poupée de son*, and a fashion doll, a *poupée de chiffon*. Her heart is engraved on her songs and she sees life through the bright rose-tinted glasses of her songs. Her records are like mirrors where anyone can view her. Through her recordings, she has been smashed into a thousand shards of voice and scattered everywhere so that she is everywhere at once. The song's listeners are the rag dolls, the *poupée de chiffon*, who laugh and dance to the music and allow themselves to be seduced for any and no reason. What good is it, the singer asks, to sing about love when she still knows nothing of boys? She is nothing but a wax doll, a sawdust doll under the sun of her blonde hair. Is the doll, then, not France Gall herself? Someday, however, she, the wax doll, the sawdust doll, will be able to live her songs without fear of boys. *Poupée de son* can also be translated as a song doll. France Gall is the song doll through which Gainsbourg speaks. *Poupée* is written for a teenybop singer to sing about herself. The sun of blonde hair, surely Gall herself, singing songs written by adults with adult themes which she, the teen singer, does not totally comprehend. Again, the question is asked: Who is the *poupée de cire, poupée de son*? Again, the answer is Gall herself. We see Gall as naïve, manipulated, and exploited and Gainsbourg as the controlling adult. Serge Gainsbourg knew she was too young to fully comprehend this even while Gall herself sang a song about the dynamics. Gainsbourg, who wrote *Poupée*, exploits the subject of the song. It is at this point in particular that the song is at its most groundbreaking and daring. As Gall grew older and more mature and aware of what was going on around her, the adults who manipulated her could be seen more clearly. Only then did she see herself as the subject of the song, an unpleasant realisation that made her most uncomfortable. This ultimately led to France Gall's disassociation with *Poupée* and her refusal to perform it in later years.

Throughout the 1960s, France Gall had a chain of hit records. Her songs were as often as not written by top French songwriters of that era, who included Pierre Cour, Eddy Marnay, André Popp, and Vline Buggy, all of whom had written Eurovision entries in the 1960s and 1970s. André Popp and Pierre Cour co-wrote *Tom Pillibi*, sung by Jacqueline Boyer, which won the Grand Prix in 1960, as well as *Le Chant de Mallory* (France, 1965), sung by Rachel and *L'amour Est bleu* (Luxembourg, 1967). In later years, André Popp would work with Roger Whittaker, a partnership that would produce such hits as *Durham Town* (*Mon Pays Bleu*), *The Last Farewell* (*Le Dernier Adieu*), and *I Don't Believe In If* (*Aprés La Guerre*). Eddy Marnay, from Algiers, wrote the lyrics for Frida Bocarra's outstanding winning French entry of 1969, *Un Jour, Un Enfant*. He composed a great number of songs including for Celine Dion, who named her son after him. Pierre Delanoe had written Piaf's *La Goualante du Paure Jean*, Joe Dassan's huge 1970s hit *L'ete Indien*, and Michel Sardues' *Les Vieux Maries* and *Le France*. He also composed France's 1958 winner, *Dors Mon Amour* for André Claveau. Other Eurovision compositions included Nana Mouskouri's 1963 effort and Sophie & Magly's 1980 effort, both for Luxembourg. His most famous and enduring work has to be *Et Maintenent* (*What Now My Love*), with lyrics by Gilbert Becaud and covered by a vast array of international artists. Gall's partnership with Gainsbourg had begun when he was invited to write some material for her in the 1960s. He had, by that time, already written for the legendary Juliette Greco. The result was Gall's 1964 summer hit *Laisse Tomber Les Filles*, which was followed at the end of the year by *Sacre Charlemagne*, a children's song written by her father, which sold two million. Her follow up to *Poupée de Cire*, also a Gainsbourg composition, was *Les Sucettes*, full of risqué explicit double entendres. Gall was filmed for TV shows acting out the sexual references in the lyrics, which at that time the teenaged Gall was too naïve to understand. The song became a massive hit but caused a furor in France. So shocked was France Gall when she found out the song's true meaning that she was unable to face the public and led the life of a recluse for some months. The scandal damaged her career and, despite the great success of *Les Sucettes*, this marked the death knell of what had been the golden partnership of Gall/Gainsbourg. The young France Gall felt bitter and betrayed by those around her. Despite suffering a major dent, France Gall's career made a steady drive upwards, although a 1967 release, *La Petite*, about a young girl who becomes the object of desire by a friend of her father's, also aroused a lot of controversy. All of the aforementioned thus cast *Nefertiti* by Gainsbourg, which she also released that year, into the shadows. Her chart success became leaner as the swinging 1960s turned into the 1970s. In 1972 she recorded two of Gainsbourg's works, *Frankenstein* and *Les Petits Ballons*, which failed to enter the hit parade. It was the last time that Gall would record a Gainsbourg song. Her subsequent pairing up with Michel Berger saw rejuvenation in her career, and Berger's song *La Declaration D'Amour* in 1974 was the start of a *Belle Epoque* of hits for Gall. Romance also blossomed between them and they were married in 1976. Gall's career went from strength to strength, and in 1988 she had an international hit with Bergers' *Ella, Elle L'a*, a tribute to Ella Fitzgerald. The successful partnership both on and off the stage lasted until Berger's sudden death from a heart attack in 1992. Tragedy struck again in 1997 when Gall's daughter died of cystic fibrosis, after which France Gall withdrew from the limelight. She performed very rarely after that until her death from cancer in 2018. She is buried amongst the *Greats* of France in Montmartre Cemetery. When Gall and Gainsbourg's professional relationship hit the rocks after *Les Sucettes*, and their last collaboration failed to achieve any lasting success, Gainsbourg had centred on the career of his partner

Jane Birken, with whom he sang on several recordings. Three years after his last collaboration with France Gall, Gainsbourg released the LP *Rock Around The Bunker*, a retro rock composition completely preoccupied with the Nazis, at whose hands Gainsbourg, as a child, had suffered. Another Eurovision contestant who suffered as a child from the Nazi occupation of Europe was the Dutch singer David Alexandre Winter, who sang *Je Suis Tomb du Ciel* for Luxembourg in 1970. As a Jew, he grew up in a Nazi concentration camp.

Gainsbourg later explored the world of reggae and produced two reggae albums. His *Aux Armes Et Cetera* was a 1978 reggae version of the French National Anthem, which made him extremely unpopular in some right-wing circles. By the 1980s, his work had become more eccentric and even more controversial. His fondness for alcohol began to take its toll and he was admitted to hospital for liver surgery. He began to appear in public less and less. In 1990 he again, and for the final time, wrote a song for the Eurovision Song Contest, held that year in Zagreb in the dying years of the old Yugoslavia. The song was entitled *White and Black Blues*, which, despite its title, was sung in French. Again, the song was not without controversy. Gainsbourg had originally titled the song *Black, Lolita Blues*. The French representative that year, the West Indian singer Joelle Ursull, however, refused to perform the song, as she did not approve of the lyrical content. Joelle Ursull, from the island of Guadeloupe, was a former beauty queen and a former Miss Guadeloupe. She had also been a member of the band Zouk Machine. *White and Black Blues* is all about the issues of living with skin colour, how other people perceive you, and its associations for a black woman living in French society. The song begins with a desire to overcome racial prejudice:

> *"When someone talks to me about skin colour*
> *I have the blues which runs down my spine*
> *I feel as if I'm in a tale of Edgar Allan Poe"*
> She repeats the refrain:
> *"We the blacks*

Eurovision: A Plea for Respect

We're a few million and a dime a dozen."

It is a reflection that life for blacks in French society could be a lot better. She makes reference to Africa, which reflects the deep, emotional ties that she still has with that continent:

"Do you hear the beat of the Tom, Tom?
They go through your heart and pierce your soul"
In the second verse, she sings:
"Africa, my love. I have you in my skin"

On stage in Zagreb, Joelle Ursull was accompanied by an accordion, a synthesiser, and steel drums as well as two dancers who did little but interfere with the strong performance and distract viewers from the song's meaning. The recording charted in France and reached number two. It remained on the charts for twenty-six weeks. It was also a top ten hit in Austria and reached number eighteen on the Swedish charts. At the contest, it was awarded a total of 132 points by international juries and came joint second, tying with Liam Reilley's *Somewhere in Europe*, which, again, had a "coming together of Europe" at its core. It was a contest in which a large portion of the national entries focused on the democratisation of Central and Eastern Europe, which had only just recently occurred. Many countries had sent songs whose lyrics manifested the jubilation felt on the fall of the Berlin Wall, that powerful symbol of a divided continent. We once again see the historical and political progress of Europe expressed in the songs performed at the song contest. The 1990 event was held at a time when there was great hope and expectation throughout Europe for its forthcoming closer political union and the much anticipated European Union of 1992. A sentiment of strong Pan-Europeanism filled the air that particular night, and this was its own celebration party. This fervour was blatantly displayed in the juries choice of winner: Italy's Toto Cotugno and *Insieme 1992*, an anthem that unashamedly called for Europe to unite, while *White and Black Blues* was in itself a musicological pioneer. Along with the Spanish entry, *Bandido*, by Azucar Moreno, it initiated a new type of song at the song contest, which was a mixture of ethnic genres and popular dance rhythms to be found in contemporary pop/dance hits. This genre would gradually take hold of the contest and dominate it in decades to come. Ethnopop numbers such as Sertab Erner's Turkish winner in 2003 or Eleni Paparizou's Greek winning entry of 2005, which fused *Nisiotika* with pop, were prime examples. Other notable examples were the Cypriot and Greek entries of 2012, which fused *Tsifteteli* with modern dance rhythms. All of them would also go on to do very well internationally except, of course, in Britain, where they were more or less boycotted. In the case of Azucar Moreno's *Bandido*, the ethnic element was flamenco, while Joelle's *White and Black Blues* relied heavily on Calypso influences. Azucar Moreno, the first act of the evening in Zagreb, started with backing track problems; they had to walk off stage and restart singing their entry, which they did with cutting-edge adrenalin after one of the most nerve-racking and embarrassing moments in the contest's history. It, however, proved to be a lucky omen, and although their flamenco-pop number did not enter the top five, the sisters from Badajoz went on to be one of the most successful Spanish acts ever, selling over three million recordings on both sides of the Atlantic.

Serge Gainsbourg's success with *White and Black Blues* was to be one of his last. He died of a heart attack on March 2, 1991. Thousands lined the streets of Paris as his funeral cortege passed on its way to the Jewish area of Montparnasse Cemetery, where he was buried. His contribution to the arts in France had been immense. He was not only a singer and songwriter but had also acted in films such as the 1960 Italian production *Revolt of Slaves*. He also appeared with Jane Birken in *Les Chemins* de *Katmandou* and directed the films *Equateur*, *Charlotte Forever*, and *Je T'aime ... Moi Non Plus*. He was an author and wrote the novel *Evguénie Sokolov*. Among his works recorded by artists in the UK were a 1994 version of *Je T'aime ... Moi Non Plus* by the high priest of punk, Malcolm McLaren, and a recording of *Poupée de Cire, Poupée de Son* by the Scottish indie-pop band Belle &

Eurovision: A Plea for Respect

Sebastian. The British-French singing star Petula Clark, whose career had excelled in France with the aid of Gainsbourg's compositions, recorded a tribute to him, *Le Chanson de Gainsbourg*, in 2003. In 2010, the film of his life, *Gainsbourg*, was released in French cinemas. His progressive music had covered many different fields through a mountain of genres. On his death, President Mitterrand, in a tribute, said, "He elevated the song to an art".

* * *

A common genre in the initial twenty years of the contest's history was the *Chanson*. It was a constant feature of the entries that represented France, Monaco, and Luxembourg and also the French language songs that, on occasion, came from Switzerland and Belgium. It proved to be a successful and popular ingredient and provided the early days with several winners and high placings. Isabelle Aubret, Serge Gainsbourg, Pierre Delanoe, Frida Boccarra, Jean Valée, and Alain Barrier all came from the *Chanson* stable. The *Chanson* originated from the epic poems that were set to music. The *Chanson de Geste*, as they were called, were usually set to simple monophonic melodies, often about great heroes. *The Song of Roland* is probably the most famous. Although the early *Chansons* survive in literature, very little of the music has survived. In the twelfth and thirteenth centuries, the *Chanson Courtoise* developed. An adaptation to Old French from *Occitan Canso*, as its title implies, they were the songs of courtly love written by men who were usually under love's spell. Those *Chansons* were often sung by troubadours. One of the joint winners of the 1969 Eurovision Song Contest was the Netherlands' Lennie Kuhr with the song *De Troubadour*, a rather folksy affair of considerable merit, sung as a tribute to one of those troubadours of old. The lyrics describe who the troubadour sang to:

The nobility...
"For Knights in their hall
He sang in strong, tough language, a long and bloody story
The troubadour."

And the servants...
"But also workmen in the barn
Heard his song of adventure
Heard by the night's kitchen fire."

It tells of his demise...
"Like this he sang his whole life long
His very own song, himself singing
But death as always, still passes on its way
The troubadour. The troubadour"

Lennie still performs the song today but to a much more contemporary folk arrangement, accompanied by more authentic medieval instruments than her original 1969 winning performance in Madrid. Later types of *Chanson Courtoise* became polyphonic rather than monophonic as the genre broadened and became more elaborate. For the first time, they featured refrains. The importance of the Crusades featured strongly in the *Chansons de Croisade*. The *Chanson* slowly developed into Burgundian *Chanson* in the late Middle Ages or Renaissance and was a polyphonic French song. In the fourteenth century, *Chanson* was either a ballad, *Rondeau*, or Virelai, some of which were set to a variety of popular poetry of the time and became known as Burgundian *Chanson* due to the area from which it had originated: Burgundy. Among the most important composers of that era (1420-1470) were Dufay, Gilles, and Binchois. Their songs were of a simple, uncomplicated nature and were usually, but not always, sung in three voices.

The Parisian *Chanson* was a later development of the genre from the fifteenth and early sixteenth centuries. Personalities much associated with it included Josquin Purez, whose work, unlike previous works of the genre, was not confined to fixed forms and bore some similarities to liturgical music. Its music was often written to portray imagery and many of its most notable composers were, in fact, highly influenced by Italian madrigals. Much earlier works were variations on

Chansons. These, in turn, came to be known as *Canzone*, an ancestor of the Sonata. The *Chanson* walked and developed slowly up the path of time until it had transformed into the lute and keyboard accompanied works of the French composers Boesset, Gaultier, and Lambert who, among others, enjoyed great popularity in seventeenth-century France. Genres such as *Chansons de Loire* and *Air du Cour* flourished at the time and were genuinely popular. These genres were, however, overtaken by opera in eighteenth-century France as the most widespread form of singing. *Chanson* did experience a revival in popularity in the nineteenth century, often highly sophisticated and with a great deal of German influence. The highly popular *Melodie* of the latter part of the nineteenth century must not be confused with *Chanson*, which it was not. By the 1880s, a new type of *Chanson* had begun to take root in France; this was to be a genre that first saw light via the cafés and cabaret venues of the Bohemian, Montmartre area of Paris, the *Chanson Realiste*. This particular style was very much associated with the literary realism of the time, to which Emile Zola belonged, and was influenced by the poorer sections of Paris society. Although this was a genre generally performed by female vocalists, the father of the genre was, in fact, a man, Aristide Bruant, whose songs were groundbreaking in that they were the first *Chansons* whose lyrics were full of slang and the street language of the working class. The popularity of *Chanson Realiste* and its place in French society spanned from the 1880s until the 1940s. Its chief exponents, Edith Piaf and Trehel, were both young women from somewhat dysfunctional French families. The genre's female singers put great emphasis on their emotive facial expressions, with heavily made-up, whiter-than-white faces and bright red lipstick. They almost always wore black dresses on stage. The songs dealt with the underworld of French society, orphans, poverty, motherhood, and the role of the Mother. The genre was in stark contrast to the post-World War II *Picturesque Chanson*, which was full of romantic love, idyllic Parisian streets, and the stillness of the Seine by night with accordion music as its backdrop. Piaf, although strongly identified with *"realiste"*, was known to prefer the love and general optimism found in the *"picturesque"*. The early Eurovision Song Contest of the mid to late 1950s saw an abundance of *Chanson* entries. The contest title in French was *Le Grand Prix Eurovision de la Chanson*, which was also used as its title in non-French-speaking countries such as Germany and Austria, inferring that the contest itself was a contest of the *Chanson* genre. The title was abandoned in the latter years of the twentieth and early twenty-first centuries, when the genre had all but disappeared from the contest and had been overtaken largely by pop. It was, thus, substituted by its English title, reflecting the pop genre. Austria's 1966 winner, *Merci Cherie*, penned and sung by Udo Jürgens, owed much to the *Chanson* genre and was even classed as such when sung in French on a Belinda Carlisle CD of classic French *Chanson*. Among the popular *Chanson* singers who appeared at the first Eurovision Song Contest was Michel Arnaud, who had recorded many songs penned by Gainsbourg, Jaques Brel, and Eddy Marnay. Michel sang two songs at the 1956 contest, both on behalf of Luxembourg: *Ne Crois Pas* and *Les Amants de Minuit*, both from the post-war stable of the *Chanson* genre. While *Les Amants de Minuit* is all about midnight lovers who keep secret all that has passed between them from midnight until dawn, *Ne Crois Pas* is an up-tempo number in which she tells her lover that although he is good-looking at the moment, that is meaningless and is all due to his youth. Just like everyone else, he will gain weight, lose his hair, and also probably his teeth. A more melancholic introspective 1956 entry closer perhaps to a pessimistic *Chanson Realiste* was the entry from Belgium, *Messieurs Les Noyes de la Seine*, sung by Fud Leclerk, who had been the pianist for Juliette Greco. The song is about a deeply unhappy man who is trapped in a loveless marriage, unable to get out.

> *"The girl I love never loved me*
> *The Friends I had have all left me*
> *I'm travelling around the world to find springtime*
> *But since the earth is round, it might take a long time"*

Eurovision: A Plea for Respect

The chorus of the *Chanson* reveals how he envies those men who have drowned themselves in the Seine and how he wishes he, too, were dead.

"The drowned men of the river Seine
Open the Watery gates for me
I'm tired of watching the strange appearance
Of the betrayed children of Puteaux"

Fud Leclerk also represented Luxembourg in 1958 with *Ma Petit Chatle* and came in fifth. He placed sixth when he sang *Mon amour pour toi* for the Grand Duchy again in 1960. He was back to singing for Luxembourg for the third time two years later in 1962 with *Ton Nom*, which came in sixteenth. On that occasion, he came in last and enjoys a place of honour in the history of the song contest, of note, because he was the first singer ever to be awarded the infamous *"nul points"*. None of the juries had voted for his song.

For many British enthusiasts of *Chanson*, the Eurovision Song Contest was its prime outlet and a major source of interest. While Piaf and other French songs did receive some airplay and enter the top twenty, they were not considered mainstream enough to be allocated equal airplay status with homegrown singers. They were often considered slightly highbrow and usually followed by a Francophile specialist minority. In the 1950s and 1960s, we see a great number of *Chanson* followers whose interest in the genre was born from watching the contest from year to year. The dominant genre of *Chanson*/ballads was so established at Eurovision that when France Gall first performed *Poupée de Cire, Poupée de Son* at the rehearsals in Naples, she was reportedly booed by a section of the assembled for no other reason than that the *Ye, Ye* genre which the song came from differed so much from the other genres of the contest, i.e., *Chanson*. For the followers of *Chanson*, and European pop in general, the BBC had monthly Saturday evening radio emissions that linked up to Europe and its music. *Pop Over Europe,* presented by Katie Boyle (and in later years by Marina Von Senger) made musical stopovers at various continental radio stations where local presenters provided information and airplay of their national hit parade and top local singers, often playing their country's number-one record. These included links to capitals behind the Iron Curtain which provided BBC listeners with knowledge of the music scene in Poland, Hungary, and Yugoslavia, which in those far-off days of the Cold War was all very mysterious. *European Pop Jury*, on the other hand, was a monthly mini Eurovision Song Contest with links to various European broadcasting stations whose DJs played a popular local record for which audiences in the broadcasting centres voted. It was a much more lively and informal affair than its sister programme. In the same time slot, there would be other "*specials*" in the same vein: highlights of the Sopot Song Festival, the Knokke Song Contest, and the Slovene Song Festival, or one-off reports on music in a certain European country. The great interpreter of *Chanson*, Barbara, was once featured on such a programme dedicated to the French music scene. These emissions were often presented by Colin Berry, who for many years read out the results of the UK jury at Eurovision.

The *Chanson*, which had an almost unbroken line of popular acclaim in France for centuries, with perhaps a short period of rivalry from opera, suddenly found itself in conflict with new musical genres. The American-led rock and roll revolution, closely followed by the British pop invasion of the early and mid-1960s, posed a threat to the more traditional *Chanson*. It is around that time that we see the emergence of a new French genre, very much influenced by pop and Rock n'Roll but a genre that, despite this, was still very much a product of France: *Ye, Ye*. This had its origins on the French radio show *Salut Les Copain*, hosted by Daniel Filepacchi and aired in 1959. It featured young teenage singers and one who was the special guest or "*Sweetheart of the Week*". *Salut Les Capain* was taken from a 1957 song title by Gilbert Becaud and Pierre Delanoe, both of whom had scant respect for the new *Ye, Ye* which the radio emission showcased. The genre was almost entirely a European phenomenon performed generally by young female singers, albeit in France the top *Ye, Ye* star was, in fact,

Eurovision: A Plea for Respect

Johnny Halliday who married Ye, Ye's glamorous queen, Sylvie Vartan, amid a media circus at Loconville, to become the golden couple of France. Sylvie had escaped her native Bulgaria with her Hungarian-Armenian parents and had been brought up in France. Johnny Halliday had family connections with America. His French version of *Let's Twist Again* topped nearly every chart in Europe but did not make any impression on the UK chart. The genre took root in France and then spread at a later date to Spain, Italy, and Portugal. Maria de los Angeles Santamaria, who called herself "Massiel" became Spain's top Ye, Ye star, superseded some years later by Karina, who sang *Un Mundo Nuevo* for Spain at the Eurovision Song Contest of 1971 in Dublin. Italy's top Ye, Ye singers were the incredibly popular Mina and Rita Pavone. Its young exponents were often sensual and sultry, yet almost naïve, still with an aura of innocence around them. Their songs often reflected this, such as Françoise Hardy's *Tous Les Garçons Et Les Filles*. Mina, the first Italian girl to sing Rock 'n' Roll, also embodied these characteristics but Mina's innocent young image was shattered after a massive scandal resulting from an affair with a much older married actor. The shy Françoise Hardy recorded *Comment Te Dire Adieu*, an American hit, also covered by Vera Lynn in the late 1960s. Its original English lyrics were replaced by French lyrics written by Serge Gainsbourg. Despite being very much part of the Ye, Ye era, Hardy, much more than any of the genre's other singers, retained a quintessentially "*Chansonesque*" quality in her delivery. A good many decades since the heyday of Ye, Ye, it is Françoise Hardy and France Gall who are the most highly regarded of these 1960s artists, possibly due to their most original and most French material, unlike many other Ye, Ye stars who sang French versions of Anglophone songs such as Sylvie Vartan's *Twist and Shout*. Françoise Hardy was elevated to iconic status in French culture, enjoying a similar position as Juliette Greco. In 1963, Hardy, Queen of the Mods, sang her self-penned composition, *L'Amour S'en Va*, at the eighth Eurovision Song Contest in London and competed with such names as Nana Mouskouri, Esther Ofarim, and Alain Barriere. Her song came in joint fifth in what has gone down as one of the most controversial contests, certainly in terms of voting. During the juries' delivery of points, there was a technical problem on the line with Oslo. Katie Boyle could not hear the jury but the audience in the hall could. When all the other juries' votes were in, Switzerland was in the lead and Esther Ofarim appeared to be the winner. Katie Boyle returned to the Norwegian jury, who read out their votes. On the second hearing, those votes differed from the first set, which the audience had heard but Katie Boyle, due to a technical fault, had not. On the second reading, the Norwegian jury had altered their votes, which they allocated to their neighbour Denmark. This put Denmark in the lead, relegating Switzerland's Esther Ofarim, with her Francophone song, *T'en Va Pas*, into second place and awarding Denmark's *Dansevise* the Grand Prix. Hardy's *L'Amour S'en Va* went on to win the prestigious *Grand Prix due Disque* 1963; it is still considered to be Hardy's theme tune. By coincidence, the composer Pierre Delanoe, whose song *Salut* le *Capain*, after which the popular radio show that launched *Ye, Ye*, was named, wrote the Luxembourg entry that year: *A Force de Prier*, performed by Nana Mouskouri.

Ye, Ye continued to take over French music in terms of popularity and *Chanson* began to lose ground. Some more traditional *Chanson* singers gave in to the new genre and withdrew from the music scene completely. The great Belgian composer Jaques Brel continued to compose within the *Chanson* genre, a decision that did not affect his popularity. The singer Alain Barriere stubbornly refused to pander to *Ye, Ye* and proudly continued in the *Chanson* genre. Alain sang *Elle Etait Si Jolie* for France at the 1963 contest in London. He placed joint fifth along with Monaco. The song became his biggest-selling single and Barriere went on to have a chain of top twenty hits in Francophone countries, which turned him into a French mega-star. His 1975 *Tu T'en Vas*, a duet with 1967 French Eurovision entrant Noelle Cordier, also provided him with a best seller. Many French *Chanson* interpreters *did* survive the *Ye, Ye* onslaught, nor was it killed off as a genre by *Ye, Ye*. Instead, what we see is a crossing over, a fusion of *Chanson* and *Ye, Ye* pop which becomes more evident as the 1960s move on. A prime

Eurovision: A Plea for Respect

example of this is the 1966 collaboration of Gainsbourg and Michel Arnaud, *Les Papillons Noirs*, which coincided with Arnaud's son Dominique Walter singing for France that year at the song contest, ten years after his mother had competed in the very first contest. Another credible example of the fusion of *Chanson* and *Ye, Ye* pop is, again, Arnaud's 1967 hit *Ne Dire Rein*, a far cry from her 1957 *Zon, Zon, Zon*, which displayed a much more traditional form of *Chanson*. This fusion, a crossing over of genres, becomes very evident at the song contest, which, as a descendant of medieval *Chanson* genres through its Francophone *Chanson* entries, proves its distinguished pedigree and value as a notable example of the development of European vocal and popular music. We also see, around this time, in the song contest and in French popular music in general, a new phenomenon in the merging of *Chanson* and *Ye, Ye*: the birth of modern French pop.

During the 1960s and 1970s, popular music in France was also an important factor in socio-cultural modernisation; music began to be seen more and more as an important economic and cultural component in answer to the many social changes and increasingly elevated position of television and radio. Music grew in importance as a necessary expression of popular culture in a widening and expanding media. It provided answers to questions asked on the understanding of culture and also politics preceding the 1968 riots as well as in their aftermath. As the media slowly moved away from the centralised control and censorship of de Gaulle's state, which had curbed the music of Ferrat, Brassens and, to an extent, Isabelle Aubret, it began to develop and cultivate its own freedom in a milieu free from government domination. Political activism began to concern itself with the rights of minority languages such as Breton and Catalan and the cultures of regions such as Corsica, Basque, Brittany, and Provence, including traditional popular music. At the 1963 Eurovision Song Contest, Francophone songs dominated the event, and although the winning song was sung in Danish and third place went to Italy's Emilio Pericolli, all four French songs ended up in the top ten, with three in the top five. Luxembourg's French language winner of 1961, *Nous Les*

Amoureaux, followed Jacqueline Boyer's 1960 victory for France and preceded Aubret's 1962 French victory for *Un Premier Amour*. Since 1956, five of the winning songs had been in French. France, incapable of hosting the 1963 contest due to mounting costs and having already hosted the event in Cannes in 1959 and 1961, handed over the organisation and hosting to another EBU member. As had happened in 1960, the BBC graciously stepped in and offered to hold the contest in London, despite never having won it to date. The French winner on that occasion in 1960, eighteen-year-old Jaquline Boyer, was the daughter of another popular French singer, Jaques Pills who had sung for Monaco in 1959 and come in last. She was also the step-daughter of Edith Piaf, whom her father had married in 1952. He was Piaf's first husband and Marlene Deitrich had been their bridesmaid. Boyer's mother was also a top French chanteuse of that era: Lucienne Boyer, whose hit songs included the classic French ballad *Parlez Moi D'Amour*, which Sam played on the piano in Rick's Café as Ingrid Bergman made her entrance scene in the film *Casablanca*. Pills was Jewish and Jaqueline had been born in Nazi-occupied Paris in 1941. This was a harrowing time for them but miraculously they survived both the war and the Holocaust unscathed. Pills continued his career and died aged 64 in Paris in 1970. He holds the dubious honor of being the first ever Eurovision contestant to die. The protagonist of the 1960 winning song, Tom Pillibi, owns two castles, one in Scotland and the other in Montenegro, as well as some boats. Young women of royal blood want to be his girlfriend. Jaqueline tells us that Tom is perfect; the only thing about him is that he tells lots of lies and you can't believe a word he says. An English version with somewhat different lyrics was recorded by Julie Andrews. The 1962 contest not only saw France win yet again, but the runner-up and third-place finishes went to French language songs (Monaco and Switzerland). Other genres of the non-*Chanson*/Francophone variety, such as *Schlager* or operetta, fared badly. Austria's *Nur In Der Wiener Luft*, an operetta-type song sung by opera singer Eleonore Schwarz, came in last; she shared her zero points with Belgium, Spain, and the Netherlands. The German *Schlager* had the

Eurovision: A Plea for Respect

last laugh, however, as their *Zwei Kleine Italiener* by Connie Froboess became that year's big international hit and earned a gold disc even though it managed only sixth place. Sweden's 1963 representative, the highly acclaimed jazz vocalist, Monica Zetterlund, could manage only zero points with her mystical jazz ballad *En Gang i Stockholm*. However, she still managed to smile for photographers. Alain Barrier's *Elle etait Si Jolie* was only one of a number of French records to take the charts by storm in Chile, where *Chanson* and *Ye, Ye* were gathering a great deal of interest. The song had already been a great success when recorded in Serbo-Croat by Yugoslav *Chanson* singer Dragan Stojnić; many years later it was covered by German and Croatian punk bands.

The song contest and Francophone music had great cultural and political significance. The Eurovision successes and expansion of French music outside of French-speaking nations became a visible and audible manifestation of France attempting to hold onto her *"cultural exceptionalism"* in the approaching storm of English and American cultural imperialism and world dominance which gathered more and more momentum in the final decades of the twentieth century.

In the UK, although French language music had never met with consistent mainstream success, being often confined to TV shows such as *International Cabaret* and *The Nana Mouskouri Show* or the radio's *Pop Over Europe*, French records met with sporadic chart success. Petula Clark, an English-born singer of Welsh origins who migrated to France, where she married a Frenchman and became a big part of the *Ye, Ye* movement, had a chain of hits in Britain in the 1960s, a couple of which, including *Chariot*, were sung completely in French. Petula, a child star in the 1940s, had also competed in early BBC Songs for Europe in the late 1950s.

Mirielle Mathieus' *La Dernière Valse* reached the top thirty. Françoise Hardy's *Tous Les Garcons et Les Filles* reached thirty-six in 1962 and *Et Même* reached thirty-one in 1964, while her *All Over The World*, sung in English, was her biggest hit, peaking at sixteen in the UK top 30. Belgium's Singing Nun hit the top ten with *Dominque* in 1963 and was featured on the first edition of *Top of the Pops*. Donavan's

1968 top ten hit *Jennifer Juniper* featured a verse in French. A year later, Serge Gainsbourg's infamously banned *Je'tame ... Moi Non Plus* reached number one. To most in the UK, he is known only for this song and, irritatingly, as a one-hit-wonder. Séverine's original French recording of her 1971 Eurovision winner for Monaco, *Un Banc, Un Arbre, Une Rue*, reached the UK top ten, sending her English version into obscurity. Vicky Leandros reached number two with an English adaptation of her 1972 winner *Apres Toi*, and a few years after, Marie Myriam's *L'Oiseu et L'Enfant* became a minor hit. Anne Marie David's 1973 *Tu Te Reconnaîtras* was recorded in English as *Wonderful Dream* and reached the top twenty. Charles Aznavour's English version of *Elle* topped the charts as *She* but, again, an English adaptation replaced the original. *She* was also the theme song of a Friday night drama series televised in the mid-1970s. Plastic Bertram, the Belgian King of Punk, plunged onto the charts loudly with *Ca Plan Pour Moi*, as did Desireless with *Voyage, Voyage*, which, along with Vanessa Paradis' *Joe Le Taxi*, enjoyed chart success in the 1980s. Most were classed as one-hit wonders or novelties such as Gilbert Becaud's *A Little Love And Understanding*, regarded as fit only for programmes such as *Euro Trash*. Why, considering the hundreds of high-quality French recordings, have so few had any success on the British singles chart? There is always the reply from radio stations that French language music is not popular in Britain. A better answer would be that it is not popular due to a lack of airplay, which explains why the public knows very little about it: because it is simply never heard. In the 1980s, former eastern bloc countries began to broadcast a mixture of French, Italian, and English language recordings that were popular in Europe. By 2013, it had become very rare in countries like Lithuania to hear French or Italian contemporary hits on the radio, as the airwaves had become dominated by English language music, usually from Britain or America. These days, in the UK itself, the chances of a mainstream French language recording making the chart is even more remote, although some contemporary French groups have, in the last few years, had hits, but in English.

Eurovision: A Plea for Respect

In the 1990s, French music again revealed its political and cultural significance as it fought against Anglo-Americanisation in culture and in popular music in particular. It was for the same reasons that the Eurovision Song Contest had come into being. Set up to stem the Americanisation of European culture and to encourage and focus on interest in homegrown musical acts, it was no less relevant in the 1990s than it had been in the 1950s, in fact, perhaps more so. Increasing time allocated to the presence of Anglophone music on the airways was seen as a worrying trend by sections of the government and French society. To counteract this problem, laws were implemented enforcing minimum quotas of French music broadcast in France. As far back as the 1970s, competition for the airwaves was fierce. ORTF, the state-run radio and television, had imposed restrictions on airplay. Airtime given to foreign records could not exceed that accorded to French artists. No record could be played more than once per week. This made stations like Radio Luxembourg and Radio Monte-Carlo, which had no restrictions, very popular. Due to limited airplay on the national airwaves, exposure for artists was difficult. This made touring very important for promotion. The Olympia in Paris was France's greatest music venue and always a true testing ground for French artists. It was an important theatre in terms of promotion and exposure. In the 1970s, an aspect of all this was the *Tour de France de la Chanson*, a mobile music contest in which singers appeared in different towns each evening and whose audiences voted for the favourite. In 1973, the American music magazine *Billboard*, famous for its weekly chart, the Billboard Hot 100, published a special feature on French music. Almost half of the various Francophone artists featured had appeared at the Eurovision Song Contest or had some association with it. An excerpt from the July 1973 edition of *Billboard* tries to explain Francophone music in relation to the international market:

"However a glance at the International hit parades shows that the *Chanson* can rarely count on an international career, the exceptions being Eurovision Song Contest successes like Séverine's '*Un Banc, Un Arbre, Une Rue*' and Anne Marie

David's *'Tu te reconnaîtras'* which had enjoyed success in Holland, Belgium, Luxembourg, Switzerland, Norway, Germany and the UK. The language barrier still persists when it comes to French as can be seen by the fact that Cliff Richard's 1973 Eurovision Song Contest [entry] *'Power to All Our Friends'* has enjoyed a much wider international success than the winning song of Anne Marie David."

Despite being the King of *Ye, Ye* in the 1960s, the legendary Johnny Halliday remained a powerful force on the French music scene of the 1970s. He was one of the record company Philips' greatest sellers along with the likes of Serge Lama and Mirrielle Mathieau. Serge Lama was selected to represent France at the 1971 Eurovision Song Contest in Dublin with the song *Un Jardin Sur La Terre*, whose music was composed by Alice Dona. She wrote many classic French songs for artists like Dalida, Joe Dassin, and Sylvie Vartan. The song was selected internally. Other artists who had made it to the national selection finals were Marcel Aumont, Guy Bonnet, Jean Ferrat (the radical communist *chanteur* of *Chanson* whose songs were widely recorded by many artists), and Séverine, who, after missing out on singing for France, was asked to sing for Monaco and won. Her song in the French final was *Viens*, which became the B-side of her winning *Un Banc, Un Arbre, Une Rue*. By the time the Eurovision Song Contest had celebrated its silver anniversary in 1980, France had won the contest on no fewer than five occasions and French language songs no more than ten. France had come in second twice (*La Belle Amour* in 1957 and *Un, Deux, Trois* in 1976), and third on six occasions (*Qui, Qui, Qui* in 1959, *N'avou Jamais* in 1965, *Il Doit Faire Beau* la Bas in 1967, *La Source* in 1968, *Il y Aura Tojours Des Violins* in 1978, and *Je Suis Enfent Soleil*, Anne-Marie David's 1979 effort. By the twenty-fifth song contest, France had been in the top five at eighteen of them. Their entries had constantly been of high quality and often by not only well-established artists but by highly acclaimed and respected acts such as Serge Lama, who had also written Isabelle Aubret's runner-up at the 1970 French national heats and had adapted Greek compositions by

Eurovision: A Plea for Respect

Manos Hadjdakis to French for Nana Mouskouri. Although Francophone songs had won the song contest three years in a row, i.e., Monaco (1971) and Luxembourg (1972 and 1973), none of them were a victory for France. Serge Lama could manage only tenth place and the French entries of 1972 and 1973 were to end up even further down the scoreboard, earning eleventh and fifteenth places, respectively.

Edith Piaf died of cancer in 1963 and left *Chanson* mourning its iconic exponent. A few years later, a young singer by the name of Mireille Mathieu turned up on the airwaves. She was from Avignon and spoke with a strong Provençal accent. Her strong voice and passionate delivery of French *Chanson* soon earned her the title of the new Edith Piaf. Some critics disagreed and said that Mireille was only another pop singer. Within a few years, two other singers appeared on the French music scene who, the same critics said, were far more worthy of the title. They were Betty Mars and Georgette Le Maire. The latter appeared at a gala evening broadcast live on BBC Radio at the Royal Albert Hall, when in January 1973 Britain officially entered the Common Market. It was a tribute to Europe that also starred Louis Neefs, who sang his 1969 Belgium entry, the Flemish *Jennifer Jennings*, and Luxembourg's 1971 entrant Monique Melson, one of the very few Letzebourgish singers who had sung for their birthplace. Betty Mars, on the other hand, had enjoyed great success with *Monsieur L'etranger* and *Mon Café Russe*, both very Piafesque *Chansons* of the original mould. In 1972, Mars took part in the French national selection. She competed against Anne-Marie David's *Un Peu Romantique*, Martine Clemenceau's *Con Un Monde de Couleurs*, and Noelle Cordier's *Un Monde Neuf*. Betty Mars' song *Come-Comedie* was considered the best and was elected to the Eurovision Song Contest in Edinburgh. The song was an archetypical French *Chanson* and recreated the *Chansons* of the Paris music hall. The writer, Frederic Botton, had written various songs for Juliette Greco. *Come Comedie* was an optimistic song, although with just a tinge of sadness in its notes. Betty Mars also recorded a German version, *Komödiant der Lieve*. On stage at Edinburgh's Usher Hall, Betty Mars gave her performance dressed in a stocking-thick white net gown with wide, puffy sleeves, as if destined for a show of classic *Chansons* at the Moulin Rouge. She cut a most gracious and stunningly beautiful picture. She interpreted the song accompanied by arm movements and gestures. There was most definitely a similarity to Piaf in her voice and presentation. The song, unfortunately, was not really what the juries were searching for; it harkened back perhaps *too* much to another era of *Chanson*. Betty Mars continued her successful career with more hit recordings like *Le Voyageur* and *Un Matin Comme Les Autres*, which secured her popularity. By the late 1980s, her career had started to wane. In 1989, beset by emotional and financial problems, Betty Mars threw herself out of a window and tragically died three weeks later. She is interred in Paris at the New Puteaux Cemetery, ironically, the same resting place mentioned in Fud le Clerk's *Chanson* about the suicides of the Seine, which he sang at the very first Eurovision Song Contest.

While the 1970s saw Philip's artists as some of the biggest names on the French music scene, Barclays, in an effort to increase their artists' impact on the international scene, saw many singers making multi-language versions of releases. Martine Clemenceau made English, German, and Italian versions of her 1973 French Eurovision entry *Sans Toi*. Patrick Juvet recorded an English version of *Lundi au Soleil*, a song originally written for Claude Françoise, and was also planning a Japanese recording of *La Musica* at the same time. Martine Clemenceau had sung three songs at ORTF's Eurovision national final: *Sans Toi*, the winner, *Je Suis Venue de Loin*, (fifth), and *L'Arlequin* (sixth). Meanwhile, Anne Marie Godart, who had sung for Monaco the previous year, performed two songs: *Pas Un Mot, Pas Une Harme* (third) and *Lui, Il ne Saura Jamais* (fourth). Jean-Pierre Savelli sang only one song, *Qui Je'taime*, which was the runner-up. Despite success in the French final, Martine Clémenceau could manage only fifteenth at the Eurovision final in Luxembourg in April of 1973. Her song was a ballad, which started gently with piano accompaniment and built to a crescendo. It was performed in the same slot as the 1972 winner, *Aprés Toi*, and France had high hopes. Martine gave a good performance, dressed in a long purple dress with backing singers, but it was immediately followed by the first-ever Israeli entry, Ilanit's *Ey Sham*, which was the final song of the evening. It was in a similar vain and genre to the French entry but was a far more powerful ballad with a much more dramatic chorus performed by a much more experienced singer. It would seem the juries overlooked Clémenceau's *Sans toi* in favour of the Israeli entry, which came in fourth. If Israel had not entered that year and *Sans toi* had been the last song to be performed, it may have received more points. Martine Clémenceau, however, did go on to win the 1973 Tokio Song Festival (Yamaha) with her song *A Jour L'Amour*, although it was not until the 1980s that she really came into her own. Her 1981 hit *Solitaire* was covered by Laura Branigan and peaked at number seven in the US hot hundred. She dueted with Claud François on their hit *Quelque fois*. In the 1990s, she composed several songs for Herbert Leonard and was awarded the Rene Jeanne Prize. In the 1960s and 1970s, *Chanson* and exponents of French music stopped being seen as exclusively French and became part of a melange of Francophone music that was the repertoire of both singers from France and the Francophone countries of Europe and Canada. Likewise, Patrick Juvet, who enjoyed success internationally, was not looked upon as only a Swiss artist, nor was his compatriot, Henri Des, who was a big star in Belgium. Both had represented Switzerland at Eurovision, in 1970 and 1973. Like many singers who performed and recorded in French, they were looked upon as Francophone artists. Although France had fared badly in the Eurovision Song Contest of the early 1970s, Francophone music and many of the Francophone Eurovision singers were doing well in both Europe and Canada. The *Billboard* of July 1973 stated, "In the field of popular repertoire CBS now has a really solid roster of local talent and has seen consistent chart success achieved by Mike Brant, Joe Dassan, Gerard Le Norman (Eurovision France 1990) and most recently Anne Marie David following her Eurovision triumph."

Although British and American pop had made significant inroads into the French charts in the 1960s, the situation had changed and French artists were going from strength to strength. The article from *Billboard* 1973 went on to say "Altogether the French talent scene has rarely looked brighter and in consequence it has never been more difficult for a foreign act to make it into the French top ten".

As *Chanson* receded at the Eurovision Song Contest, in favour of other genres, the French entries began to reflect other musical faces of its by-now multicultural society and ethnic minorities. By 1990, one of the largest of these ethnic groups was the North Africans, a leftover of France's colonial rule. It had become commonplace to hear Arab and Berber music blaring out of Algerian and Tunisian cafés in Marseilles. One of the most popular French-Tunisian singers who blended *Chanson* with more traditional North African genres was Amina Annabi, born in Carthage. One of her earliest successes was *Sheherezade*, a rap song with Arab rhythms. Her debut album, *Yalil*, was released in the USA and peaked at number five on the Billboard World Music Chart. Her numerous hits and popularity in France led to her winning *Le Prix Piaf* in 1991 as best female vocalist. The same year, she was chosen by internal selection to represent France at the 1991 Eurovision Song Contest in Rome with the song *Le Dernier Qui a Parté*, which she co-wrote with the young Senegalese musician Wasis Diop, a Lebou who had gone to France to study engineering in Paris, where he joined an African band. In 1979, he left the band to embark on a solo career and had a degree of success with Marie-France Anglade and the jazz saxophonist Yasvaki Shimizu. The film *Hyenes* featured work from Diop's first album as its soundtrack. Wasis Diop is currently one of Senegal's top musicians, writing most of his work in French. *Le Dernier Qui a Parté* was, at the time, a highly original song for Eurovision. Its melange of *Chanson* and North African rhythms with accordion background music impressed the international juries and came in first along with a very poppy number from Sweden's Carola. Because 1969 had ended in a tie, with four countries sharing the Grand Prix, new rules had been introduced by the EBU in the event of another. Amina's moment of victory was short-lived and a tie-break resulted in Sweden's Carola being proclaimed the winner, with a much

Eurovision: A Plea for Respect

inferior song. The tie-break rules were again changed, and if the current rules had been in place in 1991, Amina would have won, as would have France's Frida Boccarra in the 1969 four-way tie. Frida Boccarra was from Morocco, and therefore North African like Amina. Loreen, the 2012 winner for Sweden with *Euphoria*, was an ethnic Berber whose parents were Moroccan. Morocco, a member of the EBU and, therefore, entitled to enter the Eurovision Song Contest, entered only once, in 1980. They were represented by Samira Ben Said's *Message D'Amour*, which came in last. This result infuriated the Moroccans and led to their long absence from the contest.King Hassan himself was reported to have called the result an insult and vehemently stated that Morrocco would never again enter. Samira's career exploded soon after and she became an Arab superstar, beloved from Casablanca to Cairo and beyond. She also gathered a following of world music fans in the USA and Great Britain, where she received several World Music Awards in the 2000s. Amina Annabi, after the success of *La Dernier Qui a Parté*, also turned her talent to acting. Among her many film roles was the lead part in 1993's *La Nuit Sacreé*, in which she starred with the Spanish actor Miguel Bosé. In 1994, she recorded an album with Malcolm Maclaren which also featured Catherine Deneuve and Françoise Hardy. It is at this point that we must take note of participants in the Eurovision Song Contest and artists who are considered to be "world music" being often one and the same despite Eurovision being classed by the media, or the British media at least, as inferior in merit to so-called "world music". A closer study reveals that several artists revered on the "world music" scene have also been Eurovision artists. Another example of this is the exponent of Yemeni Jewish music, Ofra Haza, who sang for Israel at the 1983 contest in Munich, marking the first time Israel had competed at a German-hosted Eurovision. Ofra's song stated, "*We live, we, the people of Israel live*". Its lyrics told of the survival of the Jewish people through centuries of persecution which had culminated in the Holocaust. By coincidence, the first three Israeli winners, i.e., Ishar Cohen, Gali Atari, and Dana International, were of Yemeni Jewish backgrounds, although some of Dana International's antecedents also originated from Transylvania. Israel would, nine years later, enter a song that made reference to the establishment of the State of Israel and the centuries-long wanderings of the Jewish people. The song *Kan* by the duo Datz, translated as *Here*, employed biblical quotations that reinforced the belief that Israel was the Jews' God-given land. It is interesting to ponder the fact that many of France's musical greats were, in fact, of Jewish, Mahgreb, Armenian, or East European stock, with Charles Aznavour, Edith Piaf, Michel Le Grand, Regine, Serge Gainsbourg, Sylvie Vartan, and Barbara among them. Many Jewish artists have competed at Eurovision through the decades, including the French winner Frida Boccarra, born in Casablanca in 1940 into a pied-noir Jewish family of Italian heritage that came from Tunisia. Her version of *Somewhere* (*Un Pays Pour Nous*) from the film *West Side Story* was its composer, Leonard Bernstein's, favourite recording of the song.

At the 1992 song contest in Malmö, we again see France represented by a singer from an ethnic minority and the song, again, with a very ethnic rhythm. Kali was chosen by French television to sing *Monté La Rivié* and placed eighth. It was the first time in the history of the song contest that France had not sung entirely in French, with *Monté La Rivié* being sung in both Antillean Creole and French. The song dealt with life, for which the river (*La Rivié*) was a metaphor. The theme of searching ran through the lyrics, i.e., searching for peace and love and the hope that one day you will find the source of the river. Kali, alias Jean Marc Monnerville, was both performer and composer of the song, which was much inspired by the traditional music of Haiti. Kali, from Fort de France, the capital of Martinique, adopted Rasta-style dreadlocks when he was a member of 6éme Continent in the late 1970s. They concentrated on a range of traditional Caribbean music and reggae. Pressure from the record company on what type of music they produced led Kali to leave the group and go solo. His appearance at the 1992 Eurovision Song Contest is of great significance. Although his banjo-playing delivery of *Monté La Rivie* was not enough to convince the juries, his performance led to a great increase in his popularity in France and helped boost sales of his album that followed, a song from which was named Best Song of the Year in 1994. Kali regularly angered the political right and some French nationalists with his often powerful lyrics. He appeared at the centenary celebrations in 1996 marking the first anti-colonial uprising in Zimbabwe. He continues his fight against globalisation through the lyrics of his songs and often gives concerts in schools in his native Martinique, where he educates the young on their musical roots.

These ethnic rhythm language entries of the early 1990s were a reflection of the profound changes that were happening in French culture and society in general. Since the end of World War II, immigration to France had been on the rise, mostly from the French colony of Algeria and other Maghreb countries such as Morocco and Tunisia. During the Algerian War of Independence, the immigrants who supported separation from France found themselves increasingly unpopular. After Algeria gained independence, thousands of *Pieds Noirs*, Algerians of European heritage, fled to France. They were often resented by the French, many of whom looked upon them as oppressors who had brought everything on themselves. The *Pieds Noirs* soon found themselves in limbo, neither French nor Algerian. They were joined by the *Harkis*, Muslim Algerians who favoured French rule and had fought against the independence movement. Deeply resented at home, they settled in France. Maghrebi immigration was followed by immigrant workers from the French sub-Saharan colonies as well as immigration from Sicily, Portugal, Greece, and Turkey. Asylum was granted to citizens of authoritarian regimes in Eastern Europe and South America. Additional immigration from Pakistan meant that by 1975 numbers had increased from 350,000 to 700,000 and continued to rise. The boom in immigrant numbers also saw a surge in the popularity of anti-immigrant parties that linked immigration to crime. At the time that *White and Black Blues* was performed at the Eurovision Song Contest in 1990, the far-right National Front had recently won 11.7% of the vote for the European elections. The entries by Amina and Kali in 1991 and 1992 represented France at a time when the Le

Pen-led National Front (FN) was at its most radical and most xenophobic. At the 1993 legislative elections, the FN vote soared to 12.7%, although due to the electoral system at the time they did not win any seats. This result faded into insignificance when in 1995 Jean Marie Le Pen polled 15% of the vote in the presidential election. In the French entries of the early 1990s, we see a fusion of French and ethnic immigrant cultures. It is a combination that states that ethnic culture and music in particular had moved out of the shadows, away from the fringes and therefore able to take part in the Eurovision Song Contest.

Canada is not a member of the EBU and cannot compete in the Eurovision Song Contest. It is, however, an associate member of the EBU and can therefore broadcast the contest, which it has done in the past. Jeanne Manson, a singer from Canada's all-powerful neighbour, the USA, had been popular in the Francophone countries in the 1970s and had represented Luxembourg at the 1979 contest in Jerusalem with the much underrated *J'ai déjà vu ca dans tes yeux*. The Quebequais singer Celine Dion, already with a degree of success in the Francophone countries, represented Switzerland at the 1988 contest in Dublin and won the Grand Prix with *Ne Partez Pas Sans Moi*, pushing the UK's Scott Fitzgerald into second place. Celine, from Charlemagne in Quebec, had come from a poor but very happy musical family. After her arrival into the world in March 1968, her parents had her baptised as Celine after a 1966 song by Hughes Aufrey. Aufrey covered many Dylan songs in French and represented France with *Des Que la Printemps Revient* at the 1964 Eurovision Contest in Copenhagen. Dion's success in Quebec and Europe went largely unnoticed in the UK and little interest was shown in her by the media after the Eurovision win. Although several English language albums had been released, it was really through her success in the States, and in particular with the theme song to Disney's *Beauty and the Beast*, that Dion began to be noticed in Britain. Her Eurovision victory had come in the late 1980s, a time when interest in the contest in the UK had slumped to an all-time low. Established British musical stars would not touch it.

Eurovision: A Plea for Respect

The national finals were full of unknowns and songs of little merit. Its prestige and glow as an event in the UK needed a much overdue polish. In this climate, Dion's success at Eurovision did not precede her, and music critics and chart followers were unaware that she was the victor of the 1988 Eurovision Song Contest. Therefore, biased criticism and prejudices did not bar her from airplay on Radio One and other pop channels. If she had become a household name in Britain by winning Eurovision, as the 1987 winner Johnny Logan had done, then, due to the negativity directed towards her, it would have been much harder for her to have made it in the UK. Negativity and bias against Eurovision artists from the British media and sections of the music industry carried over into the public domain and informed the public perception of the contest and its artists. A prime example of this is Johnny Logan, a thrice Eurovision winner (twice as an artist and once as a songwriter) whose career was strangled in Britain by negativity and lack of airplay, never really taking off. That perception affected how the artist's records were received and treated, a perception which Celine Dion managed to avoid. By the time large sections of both media and public realised that she had won Eurovision, she already had a strong firm grasp on the music industry in the UK and there was no going back. Celine, along with ABBA, became one of the world's best-selling artists by the close of the twentieth century and a truly international star. Two years previous to Dion's Swiss victory, a fellow compatriot, Sherisse Laurence, although not a French Canadian (she was, in fact, from Stirling in Manitoba) sang a ballad in French, *L'Amour de Ma Vie*, for Luxembourg. Like so many songs in the contest at that time, it was very much influenced by syntho pop. First to perform on the stage that evening, Cherisse came in third after all the results were announced. Both the winner (Belgium) and runner-up (Switzerland) were also sung in French. Luxembourg was again represented by a Canadian citizen of Belgian-Sicilian background, Lara Fabian, in 1988 with the song *Croire*. She placed fourth. Lara had spent the first five years of her childhood in the Sicilian city of Catania. As a lyric soprano, she had a wide vocal range and performed regularly in French, English, and Italian. Constant touring of Quebec contributed to the huge success of her 1994 breakthrough album *Carpe Diem*, which turned gold, spawned three hit singles, and led her to win the 1993 and 1994 Felix Awards for Francophone Artist with the biggest success in Quebec. Disney invited her to be the voice of Esmeralda in the film *The Hunchback* of *Notre Dame*. She also sang *Que dieu aide les exclus*. The Paris Match voted her *"Revelation of the Year"* in 1998. She again won a Felix Award in 2000 for *"Artist with Most Recognition Outside Quebec"*. With a catalogue of international hits to her credit, Fabian has sold more than eighteen million records worldwide. One of her most interesting musical ventures was the concept album *Mademoiselle Zhivago*, released in Russia and Ukraine, which featured a selection of songs in diverse languages by the Russian composer Igor Krutoy. The songs all interact with each other through a mysterious woman who crosses different lives at different periods in history. This album helped consolidate her status as a superstar in both Russia and Ukraine. French Canadians joyously opened bottles of Beaujolais and shook the foundations of old Quebec City in celebration of Celine Dion's 1988 win and Fabian's 1988 fourth-place finish. They had reason to celebrate, for not only did the Canadians all achieve top four placings but they also went on to be the biggest international stars to come out of the contest in the second half of its fifty years of existence. With the demise of the *Chanson* and the dominant hand of the Francophone songs loosened, *Ne Partez Pas Sans Moi* was the last Francophone winner for well over twenty years and *Se, la Vie est Un Cadeau"* was the last *Chanson* genre winner in 1983.

The Francophone countries' lot in the contest was further put into peril when the EBU relaxed language rules in 1999 and allowed countries to sing in a language other than their official one. It further pushed the idea of *the language of popular music* as *English*, an issue that the song contest had originally been set up to stem. As the years passed, French also began to lose much of its status, even as the language in which the Eurovision Song Contest was itself to be presented to its multitude of world viewers, which further consolidated English as the

Eurovision: A Plea for Respect

world's international language. English had become the usurper of Esperanto as the language that the world, at different times, thought it would one day communicate in. In 2001, France called upon yet another French Canadian to represent them: a singer from New Brunswick, Natasha St Piere. The song *Je N'ai Que Mon Âime* was a power ballad and, as a reflection of changing times, was sung partly in English on the big night, marking the very first time a French entry in the entire history of the song contest, up to then, had featured English lyrics. While the EBU relaxed the language rules from 1973-77 for countries that found singing in their official language a disadvantage, i.e., The Netherlands and Scandinavia, it re-enforced the rule of official language usage again after 1977, which lasted until 1999. At the 1999 contest in Jerusalem, by far the majority of the twenty-three competing nations, fourteen in all, chose to sing in English or at least partly in English, while only eight – Lithuania, Spain, Croatia, Poland, Cyprus, Portugal, Turkey, and, of course, France – chose to perform in their national language. There was strong opposition to having part of Natasha St Pierre's 2001 entry sung in English. The opposition seemed to stem from a fear of French losing its status as a prominent language on the world stage and France itself losing its elevated status in world politics. The arguments and controversy were to be even more vehement in 2008 when electronica musician Sébastien Tellier sang for France. His entry was largely in English. Over the years, France had firmly established itself in the history of the culture of the song contest as one of the countries that simply never entered a song sung in English. The others were Spain, Serbia, Andorra, and Portugal. Criticism and opposition to the language of the 2008 entry were, from some quarters, quite scathing. One social network comment deplored the neglect of French, citing, "The loss of linguistic prestige as directly associated with the loss of identity as well as France's cultural and political advantage as leader of the Francophone World". Sébastien Tellier's entry caused so much debate that it was even discussed in the French parliament. One government MP stated, "Fellow citizens don't understand why France is giving up defending its language in front of millions of television viewers around the world". His view was echoed by many of his compatriots who advocated the use of French at the song contest as symbolic of France itself, its culture, and its special position in Francophonie. One comment on a social network stated, "A song in English to represent France! What kind of image are we trying to create of ourselves? An image of a conquered, colonized country which has renounced its identity? Our language is not just ours, but it is found on all five continents! Is the principal country of the Francophone world sending the signal that it is abandoning its language? How shameful!" Something more than national pride was at stake. As one of the EBU big five, France, as a major financial contributor to the EBU, is entitled to immediate qualification at the final, unlike lesser contributors who must compete in the semi-finals. Before the 2008 final, Sébastien Tellier, due to the controversy, added more French lyrics to the song but came in only eighteenth out of twenty-five countries on the big night. Natasha St Pierre's *Je n'ai Que Mon Âime* fared much better and came in fourth, a contrast to the previous year's representative, the Moroccan singer Sofia Mestari, whose song came in second last. Natasha's entry reached number two in France and Wallonia. Her singles and albums met with considerable success in France, Wallonia, Switzerland, and Eastern Europe but were much less successful in Quebec. *Je n'ai Que Mon Âime* was by far the best song at the 2001 song contest and would have been a much more deserving winner by far than the song *Come on Everybody*, which gave Estonia its first victory. Performed by Tanel Padar and Dave Benton, it proved to be an unpopular winner with limited international success. Benton, who originally came from the island of Aruba, was the first black person to win the contest and, at fifty, was, to date, the oldest-ever winner.

Sébastien Tellier's appearance at the 2008 contest proved to be a turning point in French entries and established French artists' approach to the contest. Tellier, a well-known and highly respected multi-instrumentalist whose compositions featured guitar, piano, synthesizer, keyboards, and percussion, crossed a diverse range of genres: electronica, R&B, French pop, ambient, and new wave. British music critics had given him rave reviews in the press. His album *L'incroyable Vérité* was a pop work that included lo-fi electronica and, among other points of interest, bizarre cabaret tunes. His follow-up album was a collaboration with the Nigerian drummer Tony Allan and featured the string-led *La Ritournello*, which was remixed in Britain by Metronomy. Sébastien Tellier's participation caused great interest not only with the French public but also with the French music industry, which resulted in a renewal of interest in the contest by big established artists. As a result, the Eurovision Song Contest became a great advertising campaign in the French media. This paved the way for one of France's true contemporary mega-stars, Patricia Kaas, to sing at the 2009 contest in Moscow. Kaas, a native of Lorraine, had a deep, husky, throaty quality to her voice, or a smoky voice, as some who compare her to Marlene Dietrich and Hildegard Knef would say. It displayed a degree of grit which bore similarities to Edith Piaf as well. As a teenager, Kaas took great pains to develop her career as a singer. She was initially rejected by record companies that said the music industry did not need another Mirielle Matheiu, who was her idol and by whom she was much influenced. Eventually, at the age of nineteen, she was noticed and sponsored by the actor Gerard Depardieu. He produced her first single, which was written by his wife, Elisabeth. It turned out to be a flop but was the beginning of *"the happening"* era in the life of Patricia Kaas. She again collaborated with Elisabeth on the song *Kennedy Rose*, a tribute to the head of the Kennedy dynasty and mother of President John F. Kennedy. It became a minor hit in France. Since then, Kaas has had a number of highly acclaimed hit albums and singles that have proved popular not only in France but internationally. Patricia Kaas was not only a well-known established artist when she entered the Eurovision Song Contest in 2009 but also one of the most highly respected and critically acclaimed artists to grace the Eurovision stage in recent years. At the Echo Awards in Germany, she was nominated for Best International Female Vocalist along with Cher, Whitney Houston, and Madonna. One of her English language recordings was a cover of the James Brown classic *It's a Mans World*, which featured on a mainly German language album. Another track in English from the same album was accompanied by the British rock musician Chris Rea. One of Kaas's numerous laurels was her achievement of being the first Western singer to perform in Hanoi since the Vietnam War. She also toured Cambodia and gave a benefit concert at Chernobyl. In 1999, Kaas came in third in a rather unusual poll, *Marianne*, i.e., a poll for the national symbol of France. Meanwhile, in April 2001 she gave a concert to mark the occasion of the accession of Prince Henri as Grand Duke of Luxembourg. The song *Et s'il fallait le faire* was chosen by Kaas fans who took part in an international online poll in which they chose their favourite track from her *Kabaret* album to represent France at the Eurovision Song Contest in 2009. It was a jazz song heavily informed by *Chanson*. Perhaps we should remind ourselves, while referring to both jazz and *Chanson*, that jazz was itself the offspring of *Chanson* and African music. Kaas exuded a polished and professional profile on stage. Her presence had the aura of a great star as her clear voice cut like a diamond knife into the expectations of the hall. She was received with great applause by the appreciative audience. The song placed eighth, scoring a total of 107 points in the final on May 16. It brought France back into the top ten after some recent very poor results, giving it its highest placing in several years.

In 1994, Kaas was offered the main role in the film *Falling in Love Again*, about another of her idols, whose style she was highly influenced by: Marlene Dietrich. However, the project never advanced further than the basic stages and was later dropped due to financial difficulties. Latterly, Kaas has worked on the show *Kaas Chante Piaf*, which commemorated the fiftieth anniversary of the death of Edith Piaf. *Kaas Chante Piaf* was also released as an album.

La France was to have a few more lean years at the song contest after Kaase's appearance and suffer some very low placings but *Chanson* was to make a dramatic comeback and take the Eurovision stage by storm in Rotterdam in May of 2021. The *chaunteuse* Barbara Pravi represented France with the song *Voila*. It was welcomed with open

Eurovision: A Plea for Respect

arms by lovers of the *Chanson* and the old classic Francophone entries that they so much lamented. As stated before in this book, like numerous persons of note in French music, Pravi came from a very cosmopolitan background. On her father's side, she was of Serbian and *Pied Noir* Jewish Algerian extraction, while on her maternal side she was Polish/Jewish and Iranian. Her grandfather was the acclaimed Iranian artist Hossein Zenderoudi, a pioneer of the Saqqa-Khanea movement, a genre of neo-traditional art found in Iran. The song received much attention and critical acclaim before the contest and was hotly tipped as a possible winner. Pravi gave a passionate rendering of her self-penned song which received an equally passionate response from the audience at the Grand Final. Throughout the jury voting, both France and Switzerland vied for the top position on the scoreboard. The Swiss entry, a ballad, *Tout L'universe*, sung by Gijon's Tears, was the first Swiss entry to be sung in French for over a decade. The scenario evoked reminisces of years gone by when the French ballads battled for first place. Due to the huge number of votes cast for Italy in the televote, France did not win but came in second, achieving, by far, the highest French placing at a Eurovision Song Contest since Amina tied for first place, then subsequently placed second, with Sweden in 1991. The direct and powerful lyrics of *Voila* portray the situation of an oppressed woman who overcomes her insecurities and lack of confidence to find strength and realise her dream to write songs and sing. She pleads, "Accept me for what I am".

The Eurovision Song Contest is only crass, cheesy, wall-to-wall bad taste. Where, I ask, is there evidence of this in this chapter? On the contrary, several songs and acts of exceptionally high standards have graced the Eurovision Song Contest carrying the tricolour of France and the banners of Francophone Europe.

Steve Kerr

CHAPTER 6
"La, La, La"

The Spanish Civil War was one of the most significant events in twentieth-century European history. It was a brutal conflict that raged from July 1936 to April 1939 between the Republicans and the Nationalists, with international involvement. The war began after a declaration by a group of generals in the armed forces against the elected government of the Second Spanish Republic. The rebel *coup* was given strength by support from several conservative right-wing organisations such as the Carlists, Alfonsists, CEDA, and Fascist Falange. All were to fight under the Nationalist flag led by Francisco Franco. E.G. (Euzako Gudarostea -the Basque Army), communists, Catan leftists, P. A. R.T., Galician Nationalists, anarchists, and other socialist groups fought under the Republican flag. Unlike the Nationalists, they were a much broader platform with less to unify them. Both sides attracted a significant number of foreign volunteers. The Spanish Civil War was different from other European wars of the century due to the political division and passion it aroused, a passion that extended far beyond Spanish borders. Mexico and the Soviet Union supported the Republican cause, while Nazi Germany, Italy, and Portugal gave support to the Nationalists. Volunteers came from as far afield as China and Romania. The Catholic Church came out in favour of the Nationalists. During the war, atrocities were committed on both sides and thousands of civilians were killed for their political views and religious convictions. After the end of the war, those who had sided with the defeated Republicans were executed, imprisoned, and persecuted by the victorious Nationalists. The years that followed saw the exodus of thousands of Republican sympathisers. Many fled for their lives to neighbouring countries. Those who sought asylum in the south of France soon found themselves under Vichy rule. Most of them were then rounded up and deported to Nazi concentration camps. Meanwhile, Franco established himself as the "Caudillo" of a fascist dictatorship and all Conservative and right-wing groups merged to form the new regime.

In the years of the Second Republic, Catalonia, Basque, and Galicia were granted autonomy from Madrid and, for the first time, the Spanish state recognised Galician, Basque, and Catalan as official languages. Franco's Nationalist government revoked the autonomy and set out to establish a strong ultra-Nationalist centralised state. Castellano (Spanish) dominated Spanish life and took precedence over all the other languages of Spain. All minority languages and cultures in the country were repressed in favour of the dominant Spanish culture and Castillian language. Franco, himself Galician, held a particular dislike of Basque and Catalonia for their support of the Republican cause during the Civil War. Franco called the Basque country "traitor provinces" and the Nationalists' bombing of the city of Guernica was one of the most condemned acts of the war. Shortly before the end of the conflict, Franco had invaded Catalonia.

From the time of their triumph in 1939, the Francoist Nationalist regime pursued an aggressive campaign of Nationalism across the whole of Spain. British, American, French, and other foreign films were permitted only if they were dubbed, while films of the Spanish cinema were produced only in Castillian. Obviously, under the dictatorship, all leftist names, such as Lenin, were banned but the ban also extended to regional names such as the popular Catalan name Jordi, after St George, Patron of Catalonia, or Joan, the Catalan for Juan. Spanish versions of Catholic and classical names were the only ones the regime permitted. Names that were banned were often forcibly replaced in official records. One of the regime's harshest policies was to demote the minority languages to the level of dialects – in other words, speeches that were not highly developed and were not real languages. Therefore, Catalan and Galician were considered debased forms of Castillian, as was Asturian. Basque, which was not a Latin-based Romance language and came from a completely separate language group, could not be construed as such but was nevertheless despised and looked down upon as a language of country bumpkins not fit for any form of educated discussion and too coarse for high society. In Catalonia, the regime tried to eradicate any vestige of Catalanity and attempted to Spanish-ize the country by enforcing the adoption of Castillian culture and language. Public notices displayed Francoist linguistic propaganda: "IF YOU ARE SPANISH SPEAK SPANISH" and "SPEAK THE LANGUAGE OF THE EMPIRE". Catalan speakers who conversed in their native tongue in public were often rebuked with remarks such as "Speak in the Christian tongue" or "Stop barking!". Some public signs were deliberately designed to offend: "PROHIBITED TO SPIT AND SPEAK CATALAN" was just one of them. The despised language was prohibited from public life and so retreated to the home, where its development was severely stunted. Franco's policies, often subtle and concealed, on the topic of Catalan did great damage to the language.

Galician had emerged as a literary Romance language in medieval times and was at its golden age in the twelfth and thirteenth centuries. Due, however, to hegemony from the culture of powerful Castilla, Galician fell from literacy and court life, and there followed what was known as the dark centuries, which ended with a renaissance in Galician literature in the nineteenth century. Like Catalan and Basque, its use was suppressed under the dictatorship of Generalisimo Franco.

Spain first entered the Eurovision Song Contest in 1961 when Sevillian singer Conchita Bautista sang *Estando Contigo* in Cannes. Like the Greek Colonels, Franco was keen on improving Nationalist Spain's image abroad. It was also important for him to woo the Americans. Conchita Bautista wore a traditional Andalusian costume onstage but Spain's first effort was not to be and it ended up in ninth place out of sixteen entries. The Eurovision Song Contest was of the utmost importance. A win would catapult Spain into the news all over Europe: an ideal road to positive propaganda.

Due to Spain's isolation from the rest of Europe, it suffered severe economic difficulties, and a trade and military alliance with the

USA came into being. This was mostly in reaction to the Cold War and Spain's strategic position in the wake of East-West tension. This alliance also led to Spain's admission to the United Nations. Franco, in return for American help, had made several promises to polish the sharpness of the regime. In the 1960s, as part of this promise, Franco loosened his grip on Spanish culture and the state's strict views on music. On Spanish radio and television, foreign artists were overlooked in favour of the Castillian language and Spanish singers. In 1963, Karina appeared on the scene as a *Ye-Ye* singer. *Ye-Ye*, which had come from France, developed a distinct sound in Spain very different from its French roots. It borrowed heavily from flamenco and often featured more traditional Spanish instruments such as castanets and guitars. The film star Concha Velasco hit the Spanish charts with the song *La Chica Yé-Yé* in 1965 and firmly established *Ye-Ye* on the Spanish scene. Other singers who would interpret *Ye-Ye* with a distinctive Spanish sound, such as Massiel, soon followed. In the latter part of the 1960s, groups such as Los Mustang, Los Brincos, and the duo Dinamico acquired a great following among the youth of the day. They emulated foreign bands such as The Beatles.

As part of Franco's loosening grip on cultural control, initiated to please the Americans, the Catalan, Basque, and Galician languages, the public use of which had been expressly repudiated for most of the dictatorship, slowly crept into certain aspects of public life. This was particularly evident in music. In 1961, the Basque singer Mikel Labeguerie, much inspired by the French George Brassens, released material that inspired a whole new wave of Basque song. The group Ez Dok Amairu, with the vocalist Mikel Laboa, came together in 1965 as a group of folk singer-songwriters who concentrated on the renewal of interest in Basque folk music. One of Laboa's most famous compositions was the song *Txoria txori*, which went on to be a Basque separatist anthem.

Castillian language folk singers such as Maria Jimenez and Perret (Eurovision, Spain, 1974) were less vulnerable to censorship due to most of their work being performed and recorded in Spanish. Raphael, who sang twice for Spain at the Eurovision Song Contest, in 1966 with *Yo Say Aquél* and again in 1967 with the tempestuous *Hablemos del Amor*, was an exponent of Latin/*Chanson*/pop and rose to become one of the greatest Spanish music stars of the 1960s. Meanwhile, Marisol, already mentioned in the chapter about the Greek Colonels, enjoyed a chain of hits as one of the most successful *Ye-Ye* singers. As for Gypsy music, Franco was very happy for them to express their sentiments in the traditional soulful flamenco (it was, after all, a great tourist attraction) as long as they didn't start singing about their poverty and how bad their lot was under the regime. In the 1970s, several bands emerged in Andalucia that were exponents of a new genre, flamenco rock. The leading group, Triana, which took its name from the Gypsy Quarter of Seville, merged flamenco with progressive rock. In the latter part of their career in the 1970s, they would turn to writing music with more politically inspired lyrics.

In Catalonia, a new genre was "created" in the 1950s, although it was more of a movement than a mere genre. It was called "Nova Cançó", New Song, and strived to establish the Catalan language in music, rejecting the policies of the Fascist regime. Its inspiration often came from contemporary British or American music of the period. After the thawing of the Fascist policy on Catalonia, and in Spain in general, new cultural developments sprang up in Catalonia. The first recordings of Catalan music were made. The Serrano Brothers made an LP of international hits sung in Catalan, while José Guardiola made an LP on a similar theme. Guardiola had sung for Spain at the 1963 Eurovision Song Contest in London and had placed twelfth. Known as the "Spanish Crooner" he was famous for his versions of foreign songs such as *Mack the Knife* and *Sixteen Tons*.

After the first recordings of Nova Cançó appeared in the early 1960s, several singers and songwriters followed the trend. At the 1963 Festival of Mediterranean Music, a Catalan artist, Salome (Maria Rosa Marco), together with the Valencian singer Raimon won first prize with the composition *Se'n va anar*. Salome would later go on to win the 1969 Eurovision Song Contest (along with three other countries)

Eurovision: A Plea for Respect

with the up-tempo *Vivo Cantando*. Raimon later became more involved in the composition of protest songs. After giving a concert at the University of Madrid in 1968, Raimon became the catalyst for anti-government student riots. Maria Del Mar Bonnet was another interpreter of Nova Cançó who enjoyed popularity. She had collaborated professionally with a young Catalan singer called Joan Manel Serrat (Juan Manuel Serrat). Both had, at one point, been pioneers of Nova Cançó in the band Els Setze Jutges. Serrat secured a recording contract in 1965 with the newly established Catalan company Edigsa and gave his first live performance at the Palace de la Musica Catalan in 1967. This helped to establish him as the most important exponent of Nova Cançó. His popularity led to his election to sing for Spain at the 1968 Eurovision Song Contest. It was highly beneficial for the reputation of Franco's regime abroad if a Catalan singer was seen singing for Spain. Serrat was to sing a composition by the duo Dinamico entitled *La, La, La* and was flown to London to perform it. Serrat, however, insisted on singing the song in his national Catalan instead of the official language of Spain, Castillian. The regime insisted that he sing the song in Castillian and would not agree to a Spanish entry sung in Catalan. Franco had relaxed the language rules slightly but not as much as that! Serrat defiantly insisted that he sing *La, La, La* in his native tongue and blatantly refused to sing the Castillian version. A furious Franco recalled Serrat back to Spain and the popular Massiel, who was at the time in Mexico, was commanded by the Caudillo to London to sing *La, La, La* in Spanish. Massiel's version differed slightly from Serrat's original recording, which was sung in a more laid-back, mellow style. Massiel's version was a bouncy, more poppy rendition with a much stronger vocal in the chorus. She went on to win the contest, beating the UK by one point. An exuberant Massiel, dressed in an archetypical 1960s white mini dress, was presented the Grand Prix by a boa-clad Sandie Shaw. Serrat's actions did not go unpunished and his music was banned from the radio while supporters of the regime spat at him in the street. This, however, gave him more potency and credibility and he became the greatest Catalan artist of the twentieth century. A popular Spanish cover of *La, La, La*, which was a gentle, suave version of the song and one of the superior reproductions of the winner, was, in fact recorded, by the legendary Portuguese singer Amalia Rodrigues. Franco arranged lavish national celebrations on Massiel's return from London, which included jubilant street parties throughout Spain. Some forty years later, a documentary made for Spanish television uncovered evidence that not only was Franco keen that Spain should be represented at the annual contest to give the impression that his Spain was equal to the European democracies and that life was as normal there as anywhere else in Western Europe, but he was extremely impatient for a Spanish victory and was determined to claim Eurovision glory for his country, which would give his regime a perfect stamp of approval. In fact, "El Caudillo" went to the extent of sending executives from Spanish TV to other European broadcasting organisations to purchase favour in the months leading up to the 1968 event. The TV executives toured the length and breadth of "Eurovision" Europe, purchasing TV programmes that Spanish TV had no intention of broadcasting. Those purchases and contracts with the other European countries were then changed into goodwill votes by the indebted TV companies at the Eurovision Song Contest. Critics in some quarters have disagreed with this, stating that Massiel won due to her television appearances on other continental stations prior to the contest, while critics hostile to the contest have pointed out that it is ludicrous to imagine Franco going to such lengths to win a "silly song contest". However, this is precisely where they have their wires mixed up. The contest is not a silly song contest. It is even more than a song contest. It is a pan-European event that was extremely prestigious in 1968 and still holds prestige and is afforded a high degree of respect in many parts of Europe.

The state-run TVE decided that they would pull out all the stops for the 1971 contest. A big, lavish national final was planned which featured the crème de la crème of Spanish pop, a "who's who" of the music industry. The year 1971 also saw a new system of voting introduced to the song contest to replace the long-established phone delivery. Instead, two jurors (one male and one female, one below and one above twenty-five) were dispatched from each competing country to the host city. Each juror had to award a minimum of one point and a maximum of five points to every song. This eliminated the "nil points" scenario and created an extremely boring and tedious voting procedure which did nothing to enhance the song contest. This system also proved to be unpopular with viewers and was dropped in 1974. In 1975, a new system, which has since proved to be exceedingly popular, was introduced and is still in place today. The 1971 Spanish entry *En un Mundo Nuevo*, performed by Spain's original *Ye-Ye* vocalist, Karina, at the contest in Dublin came in second to Monaco's Séverine. It was a ballad with a slow start which gathered pace and reached a feel-good, rousing crescendo. The lyrics said that at the end of the road there would be a new, happy world full of love and peace. It was, in fact, a thinly veiled critique of the by-then ageing Franco regime that would surely soon end and the better, free, democratic order that would hopefully replace it. After the contest, certain jurors complained they had been accosted and told that they had to vote for Spain. The 1971 Contest was also a memorable date in fashion history as it coincided with the height of the Hot Pants craze which had taken Europe by storm and which was very much in evidence on the stage of the Gaiety Theatre, Dublin on that memorable Eurovision night

Massiel, who knows how to deliver a song, while at a concert in Andalucia in 1983 defended *La, La, La* in response to critics who had latterly condemned the song as a "Canción de leche". "On the contrary," stated Massiel, "it is a song dedicated to the mother, a song dedicated to the earth."

Franco took full advantage to exhibit a positive angle of his Nationalist Spain to the rest of Europe when Spain hosted the 1969 contest from the Teatro Real in Madrid. The TV cameras focused on Carmen Pollo de Franco, wife of the "Generalissimo", who was part of the audience. Austria boycotted the event due to its opposition to the fascist regime.

One singer destined to become an international star, who had come to prominence by the end of the 1960s, was the former Real Madrid football player Julio Iglesias. His own composition about a former girlfriend, *Gwendolyn*, was chosen by Spanish TV to represent Spain at the 1970 contest in Amsterdam. Iglesias did not win but the exposure helped him acquire interest from outside Spain. His 1972 recording *Un Canto a Galicia*, secured him an international following and was a hit in many other European countries. The song was sung in Galician, again, an example of Galician re-entering public life and popular music in response to Franco's relaxing of the language restrictions. The song dealt with the homesickness of a Galician immigrant exiled from his homeland, Galicia. In response to poverty during the Franco era, many Galicians left their birthplace and went to live in the big industrialised cities such as Barcelona, Bilbao, and Madrid. Others emigrated to various countries in Latin America.

As the Franco era drew to an end, Spain had notched up a favourable record at the Eurovision Song Contest. From 1968-75, the monarchy without a king managed to come in the top five on five occasions. The Basque folk-pop group Mocedades (Youths) enjoyed great success with their 1973 runner-up *Eres Tú*, which charted all over Europe and Latin America, where it became one of that year's best sellers. This was the year that General Franco and Carmen Pollo de Franco celebrated their golden wedding anniversary, and a third win would have met with great approval, but maybe that was too much to expect. It was, however, a close finish and if not a win then a close nearly. The Castillian lyrics were adapted into English as *Touch the Wind*, which was released in the USA with the original *Eres Tú* as the B-side. However, it was the Castillian version which was preferred by DJs for airplay. *Eres Tú* climbed the Billboard Hot 100 and entered the top ten, making it the only song sung entirely in Spanish to reach the American top ten. Eydie Gorme recorded the English adaptation and had a minor hit in the States.

General Franco died in November 1975 after ruling Spain with an iron hand for thirty-six years. On his death, Richard Nixon declared that "Franco was a loyal friend and ally of the United States". King Juan Carlos succeeded Francisco Franco as head of state, and the difficult and unpredictable transition from dictatorship to democracy started. Catalonia, Basque, and Galicia were given autonomy and their respective languages were elevated in status. Joan Manuel Serrat was once more to be heard on the radio.

Felipe González was elected as Spain's first socialist prime minister in 1983. Shortly after his election, the new government expressed its desire for Spain to be represented by a true singer of the people, one of the new socialist administration's first acts after coming to power. Right-wing critics said the only thing they had done was to choose a Gypsy singing on the streets of Seville as the Spanish entrant for the forthcoming Eurovision Song Contest in Munich. Certainly a true singer of the people and a member of an oppressed minority, the singer was Remedios Amaya and the song chosen was a heavily ethnic flamenco affair, *Quién Maneja mi barca*, which she performed barefoot. The Gypsies, whose culture was strongly identified with Andalucian culture, were often harassed and sometimes persecuted under the dictatorship but more often than not they were simply ignored as an irrelevance. It was also a symbolic gesture to send a Gypsy singer to Munich, which had witnessed so much Nazi euphoria in the 1930s and 1940s and where, under Hitler, the Gypsies had been classed as an inferior race and relentlessly pursued and murdered. The song did not appeal to the juries. It was at a time in the song contest's history when "ethnic" music did not guarantee many votes. Remedios went down in history as having the honour of receiving the dreaded nil points, which she shared with Çetin Alp of Turkey and the song *Opera*. Questions were asked. Someone was to blame. Who? Remedios Amaya, in response, said it was no one's fault and that the result had just been bad luck. The Spanish press, however, made much of the bad result. One paper blamed the result on sending an "ignorant peasant" to Munich – a derogatory remark aimed at her Gypsy roots. Others called for Spain's withdrawal from the contest altogether. There was, in particular, a lot of criticism from the right. Some viewers complained to TVE that it would never have happened under Franco. Several newspapers were supportive of Remedios and her performance, blaming commercial pop, the music industry, and the song contest. Some saw her as a protagonist in a political drama, an innocent at the centre of a sneaky socialist plan. Spain's two-years-on-the-trot victories of 1968/69, which had been followed by public celebrations, had been a propaganda gift d'or for Franco and now the wicked socialists were trying to undo all that Franco had strived and achieved for Spain at the Eurovision Song Contest. They either were too young or had very short memories. Spain had been on the receiving end of nil points in 1962 and again in 1965. Viewers associations called for the persons responsible for selecting *Quién Maneja mi barca?* to be sacked. Shortly after the song contest, Remedios appeared in a flamenco film and then, disgusted by the music business and flamenco rock, faded back into obscurity somewhere in Seville.

In 2007 a Gypsy, Maria Šerifović, won the Eurovision Song Contest for Serbia with her song *Molitva*, a heart-wrenching ballad. She was the first Gypsy to do so. In 2012, the Gypsy singer Sofia represented her country, Bulgaria, with the song *Love Unlimited*. It did not qualify for the semi-finals. The semi-finals were introduced in 2004, when the list of participants grew too long to perform on only one night. Sofia became the first singer ever to use any Romany at the contest, even though the song contained only a short sentence in the language. The following year, 2013, saw Bulgaria's neighbour, North Macedonia, or the former Yugoslav Republic of Macedonia as its official title was at the contest, enter a song that was 50% in Romany and 50% in Macedonian. It was performed by the queen of Macedonian Gypsy music, Esma Redzepova, a legend in her homeland, and Vlatko Lazanoski. In an interview in the lead-up to the 2013 event, Esma emphasised the historic connotations their entry had in linguistic terms. Despite many of the entries being sung in English at recent contests, the Eurovision Song Contest has a good track record when it comes to minority languages. Marianne Mendt of Austria's 1971 effort, *Muzik*, was sung in Viennese. Ireland's 1972 entry, *Ceol an*

Eurovision: A Plea for Respect

Ghra, sung by Sandie Jones, was the only time they sent a song in Irish, even though several songs in Irish throughout the years made it to the national final. The ballad *Ceol an Ghra* (*The Music of Love*) was performed third on the night between the French and Spanish entries. Its lyrics contained references to "Tir na nôg", a mystical place that exists only in Irish mythology. Legend says that whoever visits this place is given the gift of eternal youth. It is often said that "Tir na nôg" is a reference to Ireland itself. In the preview film shown throughout the eighteen competing countries a week prior to the event, Sandie Jones was seen wearing hot pants, which sent waves of disapproval through what, in 1972, was still deeply conservative and Catholic Ireland. At the actual song contest in Edinburgh, Jones wore a shimmering emerald green evening dress, symbolic of Ireland. As she performed, an unemployed musician threw chemicals as a protest at the commercialisation of the music industry. No one was hurt and the young musician was arrested. Over thirty-five years later, RTE made a documentary about the song and the events that had led up to its being selected at the National Song Contest. *Ceol an Ghra* was also voted the most popular Irish Eurovision entry ever in a poll in Ireland even though the song managed to come in only fifteenth at the song contest itself. Finnish entries have been sung in Swedish by members of the minority ethnic Swedish population and one of the Finnish entries was sung in the Turku dialect. Staying in that area of the world but heading even farther north, the Norwegian entry of 1980 featured some Sami, the language of the Lapps. It was meant as a protest against a proposed government plan to build a power station in Lapland. Some Norwegians opposed to this went on a hunger strike. The building of the station went ahead regardless. The song featured some very typical Sami-style folk singing. Lapland was also to feature at the song contest in 1977 when Finland's Monica Aspelund sang the anthemic *Lapponia*. Travelling farther south to milder climates, some of Antonis Pappas' lyrics for the 1995 Greek entry *Pia Prosefhi* in Dublin were in Ancient Greek. While it would be almost unthinkable for the BBC to enter a song in anything except English, even in Gaelic or Welsh, the French entries of the early 1990s which had been performed by ethnic minority singers and even sung in Creole went on, in the middle half of the decade, to feature minority languages from the regions of European France itself. Patrick Fiori's *Mama Corsica* (1993) was the first time the French entry had been sung in Corsican and it was not to be the last. The language returned to the Eurovision stage in Dusseldorf in 2011 when the pop opera singer Armaury Vassili sang *Sognu* in Corsican for France. If we consider Creole as a regional French language, given that Martinique and Guadaloupe are departments of France (as once was Algeria), then the 1997 contest saw the French entry performed in a third regional language. In fact, it was sung entirely in Breton, the language of Brittany. The representatives, however, were made up of singers not only from Brittany but also from other Celtic countries. The group L'Héritage des Celtes was headed by Dan Ar Braz. Dan had previously sung with Alan Stivell and together they had performed a repertoire of Breton, Scottish, and Irish traditional folk music. Dan Ar Braz had been born Daniel le Bras but changed his name as an expression of his Breton culture, identifying himself more with Brittany than with France. With Alan Stivell and his musicians, they had concentrated on such instruments as the harp, bagpipes, and Irish flute. Dan fused electric guitar with these Celtic instruments to create a different and exciting sound. In the 1970s, Dan left Stivell to form his own group, Mor, as it was considered much more middle-of-the-road (MOR) than the music he had concentrated on with Stivell. The group was not successful and Dan moved on to other projects, an example of which was an LP made up entirely of Irish jigs. In the mid 1970s, he joined the contemporary folk group Fairport Convention. After some years, he returned to Brittany, inspired by his time with Fairport, to record more Celtic-based albums. In 1990, he recorded an album in English, which included works by Donovan and Richard Thompson, both British folk legends. Dan, by this time, had begun to make inroads into the American music scene, where his talent for matching traditional folk music with more contemporary styles earned him a loyal following. The group L'Héritage des Celtes had

Eurovision: A Plea for Respect

evolved from a group of vocalists and musicians whom he had collaborated with at the Festival de Cornouaille in Quimper, where he had utilised his musical connections in the Celtic world and presented a much-acclaimed repertoire that led them to enjoy a high degree of popularity throughout France.

By the 1990s, Celtic folk-pop had become highly influential in the Eurovision Song Contest. Secret Garden, which enjoyed success in the American New Age chart and later in the British Classic FM chart, won the 1995 Eurovision Contest with the song *Nocturne* for Norway. Unlike most winners, it was primarily an instrumental with a limited vocal refrain. Instead of a vocalist at its centre, the song featured a violinist who, it was later revealed, was Irish. Eimear Quinn gave Ireland its fourth victory in six years and its seventh win in the contest with the Celtic folk number *The Voice*, which was Ireland's 1996 entry.King Charles III recorded a show in 2021 in which he revealed his ten favourite songs and *The Voice* was amongst them . The up-tempo syntho pop that had been a feature of many of the 1980s contests was now almost totally gone from the musical agenda. The 1995 contest was noticeably void of dance-pop /up-tempo numbers and was dominated by ballads and folk-pop such as the German entry. The United Kingdom nevertheless sent a rap song, *Love City Groove*, which enjoyed considerable chart success in Britain. The days of "boom bang a ding dong diggy loo", which counted for only a small percentage of Eurovision songs, were long gone. The British media had based its opinion of the entire display of Eurovision entries on those songs, thus creating an idea in the British psyche that all the contest songs were of this genre and therefore unworthy of any positive critical acclaim. How far from the reality that was ended up being exhibited on the eve of the 1995 event, when one Irish newspaper even complained about the lack of up-tempo happy tunes and how morose and melancholic all the entries were. The shift in styles from *Chanson* and Europop to Celtic folk-pop at the contest, plus Dan Ar Braz and L'Héritage des Celtes riding on the crest of a wave in terms of popularity, led them to be selected to represent France in Dublin. The song, *Diwanit Bugale*, was sung in Breton. L'Héritage des Celtes was a collective group of singers and musicians that crossed the spectrum of Celtdom. The members were Donal Lunny from Ireland, Karen Matheson from Scotland, Elaine Morgan from Wales, and Brittany's Alan Stivell, who, as an expert in Breton harp, was already well known in Scotland, Wales, and Ireland. His Celtic music was also very much influenced by his childhood in Paris and especially by the Algerian and Moroccan music that surrounded him as he grew up. Karen Matheson was a former member of Capercaillie, whose mother came from the Hebridean island of Barra. One of Scotland's top folk signers, Karen almost always performed in Gaelic. Elaine Morgan, on the other hand, performed most of her work in Welsh and formed the group Rose Among Thorns. Donal Lunny was an artist who has been at the forefront of traditional folk in Ireland for nearly forty years and had popularised many old Irish folk tunes. The musician Bagad Kemper played the bagpipes in the group. Hats off to France for exhibiting some of the culture that its diverse regions had to offer. La France, douze points! One of the problems foreign language music has in the UK is that there is an extreme intolerance of anything not sung in English. This applies not only to European languages of the continental variety but also to other domestic languages native to the islands. There is a history of languages being suppressed and neglected, with Manx, Norn, and Cornish being tragic cases of linguicide, now all extinct as everyday spoken languages. Meanwhile, Welsh and Gaelic are now both spoken as everyday languages only in limited circles despite Welsh and Gaelic having their own TV channels in Wales and Scotland. For many, Gaelic is only an occasional language, a language kept for occasions, music, poetry, and broadcasting. Could we see our Eurovision entry chosen in the annual "Mod", the festival of Gaelic song? Or Bonnie Tyler and Marina singing a future Eurovision entry in Welsh? Could a well-known British songwriter such as Gwenno resurrect Manx or Cornish and write a UK entry in them? At this current stage, most probably not. These languages are rarely given airplay on mainstream channels and are considered confined to certain geographic and cultural borders.

Eurovision: A Plea for Respect

Although they receive funding for their preservation and both languages appear on British passports, they are not seen as an integral part of contemporary British popular culture. Exposure to the public at large is limited to off-peak specialist folk shows on Radio 2. Meanwhile, Anglo-American pop dominates the airwaves. While the vocalist Gwenno was nominated for a Mercury Award, this is perhaps the reason people like Karen Matheson and Elaine Morgan turn to countries like France for Eurovision. Yet, in nationwide indifference there is a sense that the UK as a whole misses out on a lot of talent whose exposure to the wider nation, outside of their geographic and cultural boundaries, would contribute to a richer culture and a revival of more ancient cultural aspects throughout Great Britain to which they are closely related. The Hebridean Gaelic singer Ishbel Macaskill, who died as a result of an accident at home a few years ago, was an example of how so many had missed out on her sensitive and warm interpretation of Gaelic Scots music. Almost unknown beyond the borders of Gaeldom in the United Kingdom, she won the Mod after many years of performing to limited audiences. Her voice of exceptional quality puts her in a parallel with Edith Piaf and Amalia Rodriguez – recognition for which she never received. In 1957, the popular Gaelic singer Calum Kennedy sang a ballad in his native tongue at an international music festival in Moscow at the height of the Cold War. He won first place and was awarded a medal by Kruschev. Meanwhile, in 1981, Mary Sandeman, another popular Gaelic songstress, donned a Kimono, changed her name to Anika, and, in disguise, topped the British charts with the disco number *Japanese Boy*. While Gaelic and Welsh were perceived, at least recently, as politically correct in the positive sense of the word, Scots was not and fell into a pitiful state, ridiculed by the dominant culture as a language of the "peasantry" and the ignorant even though it was also the language of poetry and music, once spoken by several social classes. Its place in society fell dramatically after television became more widespread. In the 1950s and 1960s, singers who sang in Scots often enjoyed the same status as pop singers, which is what, in reality, they were, rather similar to how the "Nedersalndse talliche" singers and hit songs fit into the more international pop in the Netherlands hit parade. Northumbrian and Scots should not be dismissed as inferior dialects of Standard English, as they are dialects only in the sense that you might as well say Dutch is a dialect of English.

The annual Hogmanay/New Year extravaganza televised by Scottish Television/BBC Scotland in recent years has featured a variety of homegrown Anglo-American rock-influenced groups plus a selection of contemporary Scottish folk music – a far cry from the old traditional folk-pop songs of the White Heather Club and its clones that were so much a permanent fixture of Scottish Hogmanay celebrations. It has become an in-vogue form of entertainment to show old footage of these shows from the 1950s, 1960s, and 1970s complete with tartan, dorric Scots songs, accordions, eightsome reels, etc. The contemporary Scottish presenters of the recent Hogmanay shows laugh and poke fun at the camp and kitsch of it all, at the same time perhaps trying to erase part of their own past and cultural information that they desperately want, for some reason, to discard.

Languages such as Scots and Sardinian have both been rigidly oppressed through education systems. Neapolitan is often looked upon, even by its own speakers, as the uneducated, inferior cousin of standard Italian, yet how can this be when it is clearly a language of literature and music? The use of Sardinian in everyday life was severely restricted under the fascist regime. In fact, in a period of 100 years, this living language has severely declined. Robert Burns, who sometimes composed in Scots-English and at other times in Scots, shows similarities to the Greek poet Simonides, who would at times write in Doric and at other times in Ionian. Doric was the language of Sparta, which was considered second to the Attic, high-prestige Koine Greek by which it was ultimately subsumed. As for the popular Scots language singers, it would not have been out of place for one of them to have represented the United Kingdom at the Eurovision Song Contest in the 1950s or early 1960s, but since then there has been a consolidation such that the mid-Atlantic English of Anglo-American rock pop is the

Eurovision: A Plea for Respect

only acceptable form of expression. The Dundonian dialect of Scots, which is very similar to Dutch/Flemish, has much in common with its cousins in Belgium and the Netherlands in both pronunciation and vocabulary. The Netherlands won the contest in 1957 with Corry Brokken's *Net als toen* and again in 1959 with Teddy Scholten's *Een beetje*. Ten years later, Lenny Kuhr's *De troubadour* came in first. It was to be the last Dutch language winner to date, as Duncan Laurence's composition *Arcade* and Teach-In's *Ding dinge dong* were both in English. The contest's critics choose to site the former as a typical Eurovision song, even though the majority of songs at the 1975 Grand Prix were ballads. The irritatingly catchy *Ding dinge dong* is now infamous. The Netherlands' entries over the last decade have consistently been in English except for 2010, when a very traditional type of Dutch song was sent to Oslo and performed entirely in Dutch. On home ground at the 2021 contest, Jeangu Macrooy sang *Birth Of A New Age* partly in English but also in Sranan Tongo, a lingua franca of Suriname. The fall from grace of the Dutch language entries, and certainly their lack of success in the winner's lane, as well as the language's failure to reach out to the international music scene, is an example of how the image of popular music and singing in Dutch are now so removed from each other. In the 1950s and even 1960s, this was much less the case. The Americanisation of Europe had started and was even well underway, but the juries at the song contest could still see *Net als toen* and *De troubadour* for what they were, that is to say, good songs. The hegemony in international popular music culture has led to an erosion of homegrown, home-language music that, outside of its national borders, is not tolerated. In several West European countries, homegrown contemporary singers often record most of their output in English. This is particularly so of Sweden, Denmark, Netherlands, and, to a lesser extent, Germany. Apart from what they think will give them a wider appeal and an international audience, it is also considered trendy to sing in English and slightly old-hat to record in local languages. This relegating national and official languages to secondary status and alienating or abandoning of minority languages and dialects only succeeds in even further condemning them to the road to extinction. If we omit the fact that English is now the world's international language and go back in time to the late 1950s and early 1960s, then we could easily say that a United Kingdom entry sung in the Dundonian dialect might have achieved similar aims in that it would have been understood in Flanders and the Netherlands, or if the entry had been in Dorric Scots, then we could add Scandinavia and Germany to that list. Take the dialect of Carlisle in Cumbria, which reflects many words and pronunciations of languages found in contemporary Sweden. Geordie, the direct descendant of Northumbrian English, was usurped by Oxford English as the standard form and the language of politics because the speakers of Oxford English held political power. The Yorkshire dialect and the Lancashire dialect are both rich in the linguistic traditions of Norse, Danish, and Anglo Saxon, now all so sadly dying, so often maligned in the media and very much associated with "lower class" habits. All are socially inferior relatives of the standard or at least media BBC English. In the last sixty years, there has been an increasing breakdown of the class system and continuing upward social mobility. In a society that has, at its core, a population that has advanced socially in a reasonably short period, there is an over-readiness to throw off their former garments, to which they apply negative tags, and be seen in the clothing of the elite. Therefore, ignorance attaches negative concepts to local dialects, accents, and even minority languages rather than see them for what they really are and try to preserve them. On the other hand, the BBC media-speak that has evolved from standard English over the years is speech that social advancement associates with the elite, the educated, the beautiful, and the trendy, which the up-and-coming try to emulate. These social attitudes have done irrevocable damage to languages in the British Isles. As we have seen in other European countries, such as Germany and Italy, regional variations are looked upon as different but not exactly wrong and, to a degree, enter the ranks of public life and are common in music. Elvis Presley's *Wooden Heart* was not sung in high German, as is often thought by non-German speakers, but in

Eurovision: A Plea for Respect

Schwabisch, the Swabian dialect. Although Southern Italians often complain that Northern Italians consider themselves superior because they speak in standard Italian, there is still, at least to some degree, a healthy respect and an everyday usage with a fair degree of pride for the regional dialects and minority languages of the former Italian kingdoms and city-states. Forty-five years ago, on the Isles of Scilly, which then were very remote and visited by few with the exception of Harold Wilson, who loved the islands, the population which had once spoken the extinct Cornish language, closely related to Welsh and Breton, then spoke with a rich and heavy Cornish accent. A return visit in 2013 saw this beautiful way of speaking almost completely gone, with most of the younger generation speaking variations of media English. While banning minority languages, dialects, and accents would have given them potency and made them popular, they faced a far worse fate, that of ridicule which, once it set in, sealed their fate when even the speakers themselves in the regions and countries of origin became ashamed to be associated with them. At best, the languages and dialects are neglected and their former speakers now indifferent. In the 1970s, the West Country group the Wurzels had a few hits singing in West Country accents (not dialects). The popular Goodies, who had a series of comedy shows in the 1970s, also had a hit with *Black Pudding Bertha*, a song with Yorkshire accents and dialect words thrown in. All these recordings fared very well in record sales and would even have made popular Eurovision entries. Both these groups had highly comical aspects to their works and all the aforementioned songs were comedy songs. Here, we see Yorkshire and West Country accents in music that are acceptable to the media and music critics but only because of the comedy element of the songs. This endorses public perceptions and the musical establishment's view that dialects and accents only merit ridicule. If a haunting ballad on the theme of unrequited love or a protest song about social problems in the North were sung completely in Yorkshire or Lancashire dialect, it would not be accepted by the pop establishment as credible and would be perceived as comical. The Fife duo The Proclaimers met with a fair degree of success nationwide, albeit with their songs *A Thousand Miles* and *Letter from America*, which they sang in Fife accents but with few traces of Scots language. A main point in the linguistic area of the vocals is the heavy emphasis on the "R's", which would seem to be a reaction to the current fashion among younger Scots to intentionally not pronounce it or to roll the "R", as is often the case in Irish pronunciation.

In rediscovering the traditional ethnic languages, dialects, and accents in the United Kingdom, we rediscover our continental roots from the Norman/French, Norse, Danish, Celts, Saxons, and others who came to these shores and assimilated with the Picts and other Ancient Britons to form the culture we have today. Ever changing with every new wave of immigrant, the United Kingdom has a vast wealth of minority languages, dialects, and accents and, thus, a great heritage of musical tradition within its shores. All this gives us some shared history with other nations that compete in the Eurovision Song Contest, just as Disney films helped Europeans rediscover their folk tales, which migration had taken to the Americas, or black gospel music, born of traditional African rhythms and Free Presbyterian Scots Gaelic psalms in the Americas, the popularisation of which by Mahalia Jackson and the like led to a rediscovery of part of Scottish culture, appreciated in Scotland and throughout Europe. In London, the Cockney dialect and accent once familiar in Whitechapel High Street and Hackney is now disappearing due to changing demographics. It is gradually being replaced by what is known as MLE, multi-cultural London English, which developed through immigration to London but is spoken now by all ethnic backgrounds. Lena Meyer, the 2010 Eurovision winner for Germany, topped almost every chart in Europe with her winning song, *Satellite*. Her rendition of the song displayed an obvious attempt at a Cockney accent, which didn't quite work but gave the song a certain "je ne sais quoi". British newspapers were quick to point out similarities to the very pronounced Cockney vocals of the London singer Lily Allen. Would Lena's song have sounded better sung in German? Would she have won if she had sung in German? In a

Eurovision: A Plea for Respect

recent YouTube survey, most people believed a German version would have sounded better, and a majority of those who commented believed that all songs in the contest should be sung in their own national languages. Yet, despite the dominance of English in recent contests, as is seen, minority languages and dialects have taken to the stage on several occasions. Is this an indication that the song contest is more a reflection of a true, more inclusive Europe which organisations such as the European Union fail to reflect?

In the Forest of Dean in South West England, a former royal hunting forest and a source of timber, limestone, coal, and charcoal, due to the isolation of its communities, one can still find more than a trace of Anglo-Saxon in its distinct dialect, some of which reflects similarities to German, such as bist (are), which in Modern German is also bist.

It would be interesting to see how home entries developed if, as has often been suggested of late, Scotland, England, and Wales entered separate songs as they do football teams in the World Cup. Could a future Scottish entry see a song with music penned by Annie Lennox and sung by Da Fustra in Shetlandic? Would Donny Munro perform a folk ballad in Gaelic? Would Bonnie Tyler really ever sing a gritty power ballad in Welsh for Wales? Or Katherine Jenkins team up with Mal Pope for an interesting dramatic, highly orchestrated ballad sung in their countries' language? Or would they go down the same road as Ireland and Malta and simply send an English language pop song? For many years, Irish songwriters were encouraged by RTE to compose Irish lyrics for the National Song Contest. There was even an incentive of more prize money for the composers if the winner was in Irish. Malta, that tiny Mediterranean island popular with British tourists, submitted its first two Eurovision songs in Maltese: *Maria Maltija* (Joe Grech, 1971) and *L'imhabba* (Helen and Joseph, 1972). Both came in last. Malta then had an absence of two years while it contemplated why it had done so badly. Malta returned in 1975 with the Carnival ditty *Singing this Song*. This time, Maltese had been swapped for its other official language, a leftover of British rule, English. The song did not come in last but still ended up in a very poor placing in Stockholm. The two Maltese language entries had come in last not because they were sung in Maltese but because they were rather poor songs, as was their 1975 effort, which was the worst of them all. Malta felt slighted and, sure its songs would never place, did not compete in 1976. In fact, they did not return to the song contest until Rome in 1991, an absence of sixteen years.

In the 1960s and 1970s, Luxembourg was the most successful country in the contest. Europeans over the age of fifty-five will automatically identify the Grand Duchy with Eurovision. One of its keys to success was that it employed songwriters and artists popular in other European countries, sometimes Francophone and sometimes German schlager stars, or artists who ticked both boxes, such as Vicky Leandros. This reflected the country's geographic position and culture, which is a fusion of Romance and Germanic Europe. The entries were always sung in French, which, all ingredients put together, created a winning combination. Luxembourg is a trilingual country, with German, French, and Luxembourgish being the three official languages. Luxembourgish is a High German language and an example of Moselle Franconian. The French singer Patricia Kaas spoke Lorraine Franconian as a child. Her father was a miner from Lorraine and French, while her mother was German from Saar. Patricia grew up in Stiring Wendel between Forbach and Saarbrucken, on the French side of the border. It is an area whose culture, like Luxembourg's, is a fusion of Germanic and Romance. Moselle Franconian, spoken in Luxembourg, is similar to the language spoken by Germans in Transylvania. Although all its winners and high placings were Francophone, Luxembourg's own language, Luxembourgish, did feature in its entries, if only twice, in 1960 when Camillo Felgen sang *So laang we's du do bast* and again over thirty years later when Marion Welter sang *Sou Fräi* in 1992. Luxembourg did badly in the 1993 contest and was relegated. As a result, it did not compete in the 1994 contest, the first time since 1959, and it has never yet returned. Like Luxembourg, Switzerland has several official languages, four: German,

Eurovision: A Plea for Respect

French, Italian, and Romansh. For years, their entries reverberated around French, Italian, and German. They followed no set rota. The language the entry was sung in just depended on which was the best song in the national heats, unlike the Belgian entries, which alternated between French and Flemish each year. As the most musical tongue, the Francophone entries usually fared much better than their Flemish counterparts. This, of course, often ended in bitter Flemish/Walloon arguments. When Flemish songs did badly, the Walloons blamed the result on the "ugly sound of the Flemish language". Lys Assia, who won the first-ever contest for Switzerland in 1956 with the song *Refrain*, sang it in French at the contest, although she was a German speaker. It should be noted, however, that the other song she had sung in 1956 (each country had two songs), *Das alte Karussell*, was in German. Lys returned in 1957 and 1958 to defend her title with *Giorgio* and *L'enfant que j'étais*, the former sun in Italian and the latter in French. The country did enter a song once in the Romansh language, the least spoken of the official languages. In 1989, the group Furbaz sang *Viver senza tei* on home territory in Lausanne. It was the first and only time to date that the Swiss have sung in Romansh. Unfortunately, it was not the best debut for the Latin language at Europe's most-watched television programme. A rather mediocre song, it failed to stir much interest and placed thirteenth. The group had attempted to win a ticket to the contest in 1987 when it came in third at the national selection with another Romansh song, *Da cumpignia*. It tried unsuccessfully again the next year with *Sentiments*, coming in second to Celine Dion. In 2012, Austrian television brought their Eurovision selection right up to date with the national final "Austria Rocks the Contest", which featured an array of contemporary artists and songs, many of which kept true to the show's title. A panel of celebrities commented on the songs. This included Marianne Mendt, who sang in Viennese in 1971. The winner chosen to represent Austria was a group by the name of Trackshittaz, with its *Woki mit deim Popo*. Sadly, despite an energetic performance in Baku, it did not qualify for the final. The song was, however, notable; like the 1971 entry, it was not sung in High German but in a Bavarian dialect spoken in Upper Austria, Mühlviertelish. It was not the only minority language to be heard at the 2012 contest. The Russian entries had, in the years following the country's 1993 debut at the contest, been in Russian, often very Russian styles of pop, which some Western countries found hard to relate to. They included songs by Russian mega-stars Philipp Kirkorov and his ex-wife Alla Pugacheva, one of the world's best-selling artists who enjoys iconic status in the countries of the former Soviet Union. She is a regular in the Russian tabloids, and her success is due to her mezzo-soprano voice and display of genuine emotions. Her 1997 appearance at the Eurovision Song Contest, where she sang *Primadonna*, although perfectly acted out on stage, managed only fifteenth place. This was perhaps due to the importance of the lyrics in the song, which most of the judges did not understand. Russia turned to English for its entries in the 2000s, with major artists such as the controversial girl duo Tatu, which had enjoyed a fair amount of chart success in Britain. Success came hand in hand with the change over to English and even climaxed in Dima Bilan winning the contest in **Belgrade** in 2008 with *Believe*, which he had performed in English. It turned to a Ukrainian singer singing in Ukrainian for its 2009 entry in Moscow, at what has been termed the Peking Olympics of the Eurovision Song Contest. A group of singing grandmothers, Buranovski Baboushki, from the village of Buranova in Udmurtia, between the Yurals and the Volga, participated in the 2010 Russian national selection with the catchy *Dlinnaja-Dlinnaja Beresta I Kak Sdelat Iz Nee Aishon*, which, translated, meant something like "Very long birch bark and how to turn it into a turban". They finished third but returned to compete in the 2012 national final. Among the competition were strong contenders such as top Russian pop singers and 2008 Eurovision winner Dima Bilan and Yulia Vulkova (ex-Tatu). In the end, the Baboushki won with their number *Party for Everybody* and pushed Dima into second place. They set off to Baku, where, dressed in traditional Udmurt costumes, they sang their song in Udmurt, albeit with an English title. Udmurt is a language very close to Finnish and Estonian a1n6d0 is spoken in the

Eurovision: A Plea for Respect

Russian Republic of Udmurtia. They finished as runners-up on the big night. This was the highest-placed-ever minority language song. The Baboushki, who hoped to build a church in their village with the song's royalties, fulfilled their wish. Money from the same source also helped to pay for other much-needed developments in their village. The last church in Buranovo was torn down by Stalin in 1939.

Võro, another language closely related to Finnish and Estonian, made its voice heard at the 2004 Eurovision Song Contest in Istanbul. The language spoken in South Eastern Estonia is classed as endangered and there is a real fear among native speakers of it dying out. The song, entitled *Tii (Road)*, was performed for Estonia by the girl group Neiokoso – a language threatened with extinction highlighted on the Eurovision stage.

As the artists retired backstage and braced themselves for the onslaught of the voting at the 1964 contest in Naples, the audience was entertained by Mario Del Monaco, who sang a selection of Neapolitan songs. The Neapolitan dialect differs considerably from standard Italian. Twenty-seven years later, when Italy next hosted the contest following Toto Cutugno's first place finish in 1990, held at "Cinecitta" in Rome, Neapolitan had become the language of the Italian entry. A romantic ballad in which the singer compares his love's beauty with that of the ocean, *Comme è ddoce 'o mare* was a Maracchi/Artegiani composition sung by Peppino di Capri in his native tongue. It was only the second time that an Italian entry had not at least partly been sung in Italian and the first time it had been in a minority language.

As a linguist, I was always fascinated by different languages, which led me to an interest in other cultures. I also had a great love of music. All this led to an interest in and fascination with the Eurovision Song Contest. The contest was, in my opinion, a far more interesting place when each country sang in its national (or minority) language. More often than not, it was the only time some of those languages were ever heard on British radio or television. In 2006, Monaco entered a song which was sung partly in Taihitian – more a reflection of changing demographics in Southern France than in the tiny principality in that area of the world. Jean Gabilou, a Taihitian, had sung for France in 1981. Meanwhile, in 2011 Norway was represented by a black singer, Stella Mwangi, who sang some of her song, *Haba, Haba*, in Swahili. This failed to qualify for the final but there was to be another, more chilling tale about this song, far away from the Eurovision Song Contest. Angry and dissatisfied with the promotion of multi-culturalism and immigration to Norway from Asia and Africa, the right-wing extremist Andreas Behring Breivik had made an entry in his diary shortly after the selection of Stella Mwangi and *Haba, Haba* as the Norwegian entry, which was also a chart-topper in Norway. Breivik wrote about his displeasure that a black woman singing in Swahili should represent Norway. It was one of his many grievances regarding what he saw as the erosion of Norwegian identity and culture. He would later go on to bomb government and civilian targets in Oslo. His terrorist agenda culminated in the massacre of young delegates at the Workers Youth lead – events that would shake Norway to the core.

Twelve countries competed at the 2011 Liet International Song Contest, held in the Italian city of Udine, home to the Friulian minority language. The Liet is a "Eurovision Song Contest" held annually for the minority languages of Europe. It originates from a small local event organised in the Netherlands and, in 2011, saw songs competing in a variety of languages, among which were Romansh, Gaelic, Friuli, Asturian, Burgenland-Croatian, Sami, and Fries. Many of the artists who have competed at the Liet have turned to increasingly more creative work in an attempt to preserve their language, such as singing rap in Plaat Deutsch (Low German). A previous winner went on to achieve national fame in France after he came first with a rendition of a song in Corsican, while a Finnish Sami (Lapp) band that won in 2008 has successfully toured Europe and former Frisian winners have gone to number one in the Benelux Countries. The 2011 event ended in yet another victory for the Frisian language, with Janna Eijer's power ballad *1 Klap (One Breath)*. In a statement to the press shortly after her win, Janna described the plight of her language.

Eurovision: A Plea for Respect

She spoke about a major weapon in the death of minority languages and dialects, which is to link them to social inferiority and ignorance. The Coffee Shock Company and Austrian band singing in Burgenland-Croatian were the 2011 runners-up with a Blakan-influenced mix of rock and reggae. Their song *Gusta mi se je znicila* may have been very imaginative and creative but it is exactly the sort of thing that certain British TV viewers and DJs would find hilarious simply because it's sung in another language. Even rap or soul sung in national languages such as Danish and Finnish is enough to send them into hysteria and ridicule, judging it as bad. It is exactly at this point that we recognise British attitudes that come out in Eurovison: an intolerance of foreign language music fit only for Euro-trash. Is not the twinning of the two words "Euro" and "trash" unique and native to this side of the Channel? The Armada? Napoleon? Trafalgar? Who exactly ruled the waves? There are an estimated fifty million minority language speakers within Europe's borders. According to UNESCO, out of all the world's languages, a total of 2,500 face the danger of extinction. The same organisation also claims that over the last three generations, 200 of them have died out. Some languages are in such a state of abandonment and disrepair that they have to beg to be given the status of minority language. All this despite a European Charter promoting and stimulating Europe's endangered languages, which came into being in 1998, ratified by twenty-four European nations. While many journalists have written about the Liet contest, there is a tendency for them to see it as some antidote to the *"cheesy"* English language of the Eurovision Song Contest, without seemingly doing much research into minority languages featured in past Eurovision Song Contests. It is also important to the revival or the keeping alive of minority languages that they venture further than their little enclaves and that their music is sung by artists of a different culture. This was the case with the French Eurovision 2012 entry, which, although sung in Corsican, was performed by a singer from Normandy. As the singer Alfredo Gonzalez, a 2009 finalist at Leit from Asturias in northern Spain, said in conversation about the Asturian language, "At the moment it's in serious risk of dying. So the fact we can come here to sing in our language and be recognised outside our borders is very important." In England, a country whose culture is often blurred and double-identified with Britishness, the re-discovery and revival of dialects such as Lancashire, Yorkshire, and Geordie, as well as the regional variations in accent, would help them rediscover their own sense of *"Englishness"* and national culture as opposed to Britishness, which does not revolve around only the media language of television and radio. This may also start a wider rediscovery of more traditional English music, which Steeleye Span manifested so well in the 1970s. In 2013, Denmark won the Eurovision Song Contest for the third time with the song *Only Teardrops*, sung by Emmelia de Forest. Prior to her victory, she had toured folk venues for five years and was a protégé of the Scottish folk singer Fraser Niell. Denmark's high regard for folk music had led the Danes to confidently enter a rising star of folk music into the song contest. After Emmelia's win, Fraser Neill had the following to say: "It's hard to imagine this happening in Britain where a peculiar disregard for folk still lingers. Mumford & Sons for example is arguably the biggest band in the world right now. The world thinks of them as English and Folky and does not have a problem with that but the English still do". For minority languages to survive, they must be seen as relevant to the young, and to do that, minority languages must be made to look trendy. This is an aspect that Eurovision song organisers in the UK must also look into more deeply. In Great Britain, this is a *"Great"* stumbling block. Not only must the heavy independent pop radio stations, due to their insecurity and vulnerability to the ephemeral world of pop, follow the strict and narrow line of the word *"IN"* to its extreme but their DJs must also be seen to have attitude and converse in the streetwise-orientated speech of some subcultures. This makes the linking or pairing of dialects and minority languages with *"trendy"* all but impossible. Recordings in dialects (in their pure form), such as Yorkshire and Geordie, or a minority language which is not currently PC, such as Doric Scots, would not be tolerated by such stations, all of which boycotted the recent mainstream and contemporary pop of Eurovision winners *Satellite* by Lena Mayer and *Euphoria* by Loreen. These were extremely popular and charted all over Europe in a big way. *Euphoria*, despite jumping onto the British charts at number three, was more or less banned from Radio One, the BBC pop channel, and, with the exception of *The Graham Norton Show*, was hardly heard on Radio 2, either. This created more or less a form of open musical apartheid.

At the 2011 Liet, organisers even went so far as to ban English from the song festival. It would seem that Liet has turned into what the Eurovision Song Contest originally set out to do: promote local European culture and language and inspire an interest in European singers. National and official languages, of course, have their place in Europe and its culture, just as minority languages do. It would seem to many that the Liet has, in some European quarters, become the alternative Eurovision Song Contest or the song contest for better songs, but this is not how it should be seen, nor should it be seen as a fringe non-mainstream event. An amalgamation of Liet, its aims and objectives, into the Eurovision Song Contest would give minority languages a central stage to promote themselves. It would also add diversity and make the contest more interesting and reflect a wider Europe, which it already has done up to a point. Eurovision has, since its early years, been a platform from which small countries that do not wield as much political clout as the more powerful nations can compete on equal terms and even, as has often been seen, defeat them. Malta, Monaco, Ireland, Luxembourg, Andorra, San Marino, Iceland, Norway, and Finland are just a few. While we have seen how successful Luxembourg and Ireland have been, tiny Monaco also triumphed at Eurovision.

Finland's 1968 representative, Kristina Hautala, and her song *Kun Kello Kay*, was the first Finnish entry to be chosen by popular vote. Kristina had competed in the national final against Ank, Johnny, Irina Milen, Aarno Raninen, and Inga Sulin to come out top. She was very much a star of the Swing Sixties in Finland and one of her biggest hits was *Rakkautta Vain*, a cover of The Beatles' *All You Need Is Love*. She flew to London on a tidal wave of positive publicity and backing from the Finnish media. A prospective first-time Finnish winner? Finland to win Eurovision? At the Royal Albert Hall, Kristina, an ethnic Finn from Stockholm, came in last with only one point. She was broken-hearted and the country, its media included, was flabbergasted. Subsequent low placings in the years that followed created an inferiority complex regarding Finland and the Eurovision Song Contest, out of which rose a conspiracy theory and a resignation that Finland, whether they sang in Finnish or English, and no matter what type of song they entered, would never win the Eurovision Song Contest. As Finland is a country on the periphery of Europe, with a relatively small population, Finns believed that they did not have the power or influence to win the contest. In the same month, the Finnish newspaper quoted the following: "The idea of Finland as a small country has had an important influence in Finnish Nationalistic thought and politics since the nineteenth century. In the Cold War era Finnish security politics emphasised that, in order to remain independent, Finland had to recognise its smallness and take this 'fact' into account in its relations to the great World powers" (*Missä on Scromi*). In 2001, something happened to change their mindset. Estonia, a small Baltic state that was once part of the mighty Soviet Union, won the contest, followed by another small Baltic state, Latvia. These two events boosted Finland's confidence, giving it renewed hope in the Eurovision Song Contest and a determination to do what its neighbours across the Gulf of Finland had done: win! This paid off and the anonymous monster-clad group Lordi, which arrived at the contest at Athens Olympic Stadium amid a whirlwind of media publicity, with accusations of Satanism thrown at them and Greek businessmen threatening to sue, won the contest with *Hard Rock Hallelujah*, which, as a heavy metal rock song, was sung in English. Lordi already had a following in the UK, where they had received awards from the music industry. The song charted in the UK but due to a total lack of airplay, it was only a minor hit and did not climb any higher. Lordi returned jubilantly to an overjoyed Finland. Turkey was once, not so long ago,

Eurovision: A Plea for Respect

a great world power and was at the helm of the Ottoman Empire. The Turkish language was a commonly heard tongue in many Eastern European countries and there still exist large communities of ethnic Turks in some Balkan countries. Turkish influence on the world stage has much waned, although, strategically, it is still of great importance given its position between East and West. Its language, however, is no longer a world or international one, even though it can often be heard on the streets of London, Amsterdam, and Hamburg due to the huge number of Turks forced, by poverty, to emigrate to Western Europe. Turkish entries at the Eurovision Song Contest were always sung in Turkish. Turkey, an active member of the EBU, like Israel, is therefore allowed to take part in the Eurovision Song Contest. However, unlike Israel, and what is often overlooked in Western Europe, is that it is also a European country. The province of East Thrace lies in Europe. The Turkish language has always gone hand in hand with Turkish identity, and to many especially more nationalistic Turks, it is an obligatory part of Turkish culture and an object of great pride. Turkey competed regularly and thus became an integral part of the Eurovision Song Contest. For a variety of reasons, Turkey was absent from the 1976, 1977, 1979, 1994, and 2013 contests. In 1976 and 1977, the country still felt slighted after coming last with its first-ever entry in 1975. In 1979, the contest was held in Israel. Turkey, under pressure from other Muslim states, decided to stay away. Maria Rita Epik, a Turkish singer of Italian extraction, and her group, 21 Peron, had been chosen at a national final to sing *Seviyorum* and were bitterly disappointed when they were forced to withdraw. Turkey faced relegation under the old process in 1994, and in 2013 there was some mystery as to why they did not compete. There were rumours that Islamist political groups had objected. In 2012, the contest was held in Baku, Azerbaijan, a Muslim country, and interestingly, Turkey was represented by a Jewish singer, Can Bonomo, with *Love Me Back*, a very Jewish-sounding song. Can, who was considered alternative and not at all mainstream, later performed at his own concert in New York. Up until 2003, Turkey had finished last on three occasions and had received "nil points" in both 1983 and 1987. Its performance at the song contest in general had been very poor and apart from only a handful of occasions it had usually finished nearer the bottom of the scoreboard. In the 1990s, Turkish coastal resorts had started to become a magnet for British holidaymakers searching for the sun at affordable prices. Its image abroad much improved from the general Western impression of the country back in the 1970s and 1980s, which had been formed by the film *Midnight Express*. The film portrayed Turkey in a very negative light. The popular million-selling artist Sertab Erener was invited to sing for Turkey in 2003. Sertab had an international career and had appeared on television and at concerts throughout Europe. Her records had sold ten million copies and she had recently hit the charts with her composition *Kiss, Kiss*, which was covered in the UK by the Australian Holly Valance. Well established in her homeland by 2003, she had already attempted twice before to sing at the Eurovision Song Contest, in 1989 and again in 1990 when she entered the Turkish national finals. Sertab accepted the offer to sing for Turkey in Riga but the acceptance did not come without conditions. Sertab stated that she would sing for Turkey only if TRT would allow her to perform in English. Turkish television agreed to the conditions and Sertab thus became Turkey's 2003 Eurovision singer. This condition, however, ignited a political fire in Turkey. The person in charge of Turkey's influential language academy, the Turkish Language Society, made a vehement attack in extremely condemnatory words on the decision to not sing in Turkish. He went on to stress that doing so would "lead to a new alienation in our languages and every area of life". Sertab was insistent on the English language issue. She believed that a Turkish singer needed to sing in English to reach out to a world market. Her decision led to an awkward confrontation in the Turkish Parliament when a government MP reproached the Minister of State (who was responsible for the running of the TRT) and demanded to know why he had approved the singing of the Turkish entry in English. Sertab got what she wanted and performed her song on the Eurovision stage in English. *Everyway That I Can* was also released in Turkish and became

Eurovision: A Plea for Respect

a big chart success in Turkey. The song was a very ethnic Turkish affair. While sung in English, it gave out a European-friendly, but still within the bounds of Turkish, cultural vibe. The song, self-penned by Sertab, has much in common with *Makam*, modes found in Turkish classical music as well as pop and folk. However, there the similarities end. The melody features a repetition of phrases in the lower part of the octave which gradually expands the range upwards. Another Turkish feature in the song is its "*Arabesk*" element, which is particularly prominent in the string section, where instruments play short figures and are interpolated with the actual sung phrases and the melody. The music also features what is called a *Darbuka*, an instrument found in the musical traditions of various areas of the Middle East. Its sound can be heard on recordings from Egypt's Om Keltum to the *Tsiftelei* pop hits by Greece's Katy Garbi. It is a goblet-type drum much associated with belly dancing and Gypsy music and is a particular feature of most *Arabesk*. Sertab may have insisted on singing in English but the Turkish version sounds miles better. The Turkish language seems to blend naturally with the *Arabesk*, *Makams*, and *Darbakas*, while the English lyrics do not seem to fit into the music. This is often the case with oriental-based songs at the Eurovision Song Contest that are interpreted in English. Some good examples are the 2005 Albanian entry "*Tomorrow I Go*, by Ledina Celo, and the 2008 Armenian entry, Sirusho's *Qele Qele*, both of which were performed in English. It is rather like the Byzantium liturgy at the Cathedral of *Agia Sofia* in London's Bayswater, which on occasion sung in English rather than Koine Greek. The English of the liturgy coupled with the Byzantine chants/hymnody just does not sound right. Sertab's insistence that the song should be performed in English, however, did pay off. It would even seem that, despite the angry disputes that the entry provoked in Turkey, she was proven right, as the song gave Turkey that very first, much overdue, victory. In Riga, an ecstatic Sertab performed the reprise of her winning song, while in Ankara, Turkish politicians danced at what they expected would now be the EU's approval to join the European Union. Turkey had waited in the wings for years while

Johnny Come Latelys had pushed in front of them and become full European Union members. "Why?" was the question many Turks asked. The English version of Sertab's *Every Way That I Can* seems to give more than a veiled fleeting reference to this, which can clearly be seen in the lyrics. Mathew Gumpert's essay on "Auto-Orientalism at Eurovision 2003" describes the complex nature between Turkey and the European Union as characterised in Sertab's winning song:

> "That performance may be summarised as a series of contradictory demands, simultaneously erotic and political in nature. To dominate or be dominated, to possess or be possessed, to love or be loved, to seduce or be seduced – demands that need to be made in the first place because lover and beloved, like Turkey and the West, are *"separated"* what *"separates"* them are a number of *"facts"* familiar to anyone with even a cursory knowledge or erotic poetry or political history, distance, delay and difference."

The lyrics certainly seem to reflect a metaphorical, much deeper significance than they initially suggest. *"I feel you moving on a different course, making the way for a different coast"*; that is the relationship that Turkey has with the West and the European Union in particular. Turkey is ever enthusiastic at prospective European Union membership but Europe would seem to have other ideas, a different agenda. Turkey, the lover, sees the object of her affections, the European Union, flirt with other admirers, that is to say, the other candidates for membership whose hand the European Union accepts in marriage while Turkey, rejected and alone, is left out in the cold. *"Tell me what you see in other girls all around"*: Turkey's sheer frustration at the rejection of having to play second fiddle, what seems to be an endless game of waiting. It asks, "Why?"

"Come on closer and tell me what you don't find here". This lets Europe know what it is missing. *"I wanna show you all again what it would be like if you just let go and let me love you"*: If Europe would only change its approach to Turkey and allow Turkey to embrace it, what Turkey has desired for so long. The *Makams*, *Arabasks*, and *Darbakas*,

Eurovision: A Plea for Respect

Turkey and the English lyrics, the wider world and the European Union. The music and the lyrics, two different people having a conversation: *"Every Way That I Can"* is Turkish enough for the Turks to recognise it as a Turkish song and Western enough for Western Europe to recognise it as an oriental song and still appreciate it as a pop song. It is not too heavily oriental, such as, for example, the popular songs by the Lebanese songstress and Pan-Arab star Fehrouz, or even the oriental *Gozün Aydin*, a hit for Turkish diva Aija Pekkan in the 1970s and a classic in Turkish pop, still heavy in its orientalness. Sertab's winning song struck just the right balance between East and West and bore some similarities in this respect to the ethnic pop of the *Tsefteteli* or *Indianithika* Greek hits which often start life as heavy in the oriental aspects of their (deeply Middle Eastern) music in Egypt. Often, cover versions are made by Turkish language singers who take them into the hit parade there. These same songs are very often Westernised/Hellenesized and become very popular, frequently topping the Greek charts. An example of this is Katie Garbi's *Xechasmena, Xechasmena*, with her other hit, *Pesto me ena fili*, following in the same vein. On the occasion of her appearance at the 1994 Eurovision Song Contest, Katie Garbi represented Greece with *Hellada Hora Tou Fotos*, a *tsefteteli* song. Many aspects of older forms of *Laiko*, Greek popular music, are heavily influenced by Turkish and Egyptian music, as was Serbian pop "turbo folk". This was also true after Sertab's win and performance in English at the contest. Sertab had not only proved that she was right but also insisted that all Turkish entries from then on should be sung in English. From 1975 to 2003, Turkey had ended up in the top five at the Eurovision Song Contest only once, in 1999 when Sebnem Parker sang *Dinle*, which came in third. Sertab's victory heralded the start of a new era in Turkey's relationship with the song contest. The following year in Istanbul, their entry *For Real* by the group Athena came in fourth. This was a most interesting oriental/ska/punk number. In 2007, Sibel Tüzun came in fourth again with *Shake It Up Sekerim*. A year later, in Belgrade, Mor Ve Otesi came in seventh with *Deli*, while *Dum Tek Tem* by Hadise, an ethnic Turk, who won the Belgian *Pop Idol*, came in fourth again. In 2010, it was almost a second victory by MaNga with their *We Could Be the Same*, which came in second, while in 2012 Can Bonoma managed a seventh-place finish in Baku with *Love Me back*. This regular run of success made it all the more shocking when Turkey decided not to participate in 2013, nor for several years that followed. The placings were, after all, a far cry from the Turkish efforts of the 1970s, 1980s, and 1990s.

Following Turkey's 2003 win, and Sertab's perseverance on the language issue, there was something of a twist in the tale because although Sertab had given her song international appeal, she had also set a precedent in the song contest for a move towards more ethnic music not only from Turkey but from various other European nations. The year 2004 saw strong entries with ethno/folk-pop roots. These included Ukraine's Ruslana with *Wild Dances*, which drew on Carpathian folk music. Jelko Josomovič from Serbia and Montenegro came in second with the folk-pop *Lane Moje*, with the singers dressed in full traditional Serb costume. The song featured traditional Serb folk instruments, the *Kaval*, violin, and Turkish *Saz*. The following year, in Kyiv, Turkey's Gulseren again sang a very oriental Turkish song. *Rimi, Rimi, Ley* was unashamedly oriental and ethnic, far more so than Sertab's winner. The song did, however, go against Sertab's wishes and was sung in Turkish, though it managed only a thirteenth placing. It would be correct to say that, due to its sheer orientalness and the colour of its staging, it would not have looked out of place in Bollywood. Greece sent the Swedi-Greek Elena Paparizou with the ethno-pop *My Number One*, sung in English. It was a mixture of pop and *"nisiotica"* (island music) with traditional Greek dance routines accompanied by the Lyra (a Cretan string instrument). Unlike *Everyway That I Can*, the English lyrics fitted like poetic gloves over the music.

"You're my lover, undercover, you're my secret passion and I have no other"

It also reflected the deeply Greek Orthodox Christian culture of the singer's and songwriter's roots

"Your my passion, my belief, my crucification"

Eurovision: A Plea for Respect

The combination was perfect and Greece won the contest for the very first time since its first entry on a Eurovision stage, sung by Marinella back in 1974. Paparizou's song was a convincing win, thirty-eight points ahead of the runner-up, Malta, with four points from Turkey. The Albanian, Hungarian, Cypriot, Serbia-Montenegran, and Macedonian entries also featured a more ethnic-orientated pop. Even the United Kingdom's entry that year, *Touch my Fire*, sung by Javine, had borrowed much from Middle Eastern genres. Another example of this was the debut Moldovan song *Boonika Bate Toba*, which was a clever and innovative blend of punk, hip-hop, and Romanian folk coupled with comical lyrics, a trademark of its performers Zdob și Zdub, who returned to sing for Moldova in Dusseldorf in 2011. The trend in ethnic pop had set in back in 1990 when both the French and Spanish entries featured this genre. There had always been some flirtations with ethnic music in the contest before that. Greece's protest song of 1976 was a prime example. Others included Spain's 1983 entry, the Turkish entry of 1985, *Di, Di, Di* by MFO, and their 1988 effort *Sufi*. Avi Toledano's *Hora* for Israel in 1982 was yet another example. The Israeli entry of 1974, *Netati La Khayayay*, by the group Poogy, was an earlier work that displayed elements of Hebraic music tradition. The lyrics were rich with surrealist allegories. A member of the group, Danny Sanderson, many years later confirmed that they had been a shrouded criticism of Prime Minister Golda Meir's politics and that they had also expressed a desire for a Palestinian state to co-exist with the state of Israel. The Israeli national selection of 1978 included several Israeli folk-pop compositions although they all lost out to Izhar Cohen's *Ab a ne bi*, a typical 1970s disco-funk number. It was not until post-2003, that is to say post-Sertab, that ethnic-pop and ethnic-dance began to dominate the contest.

There were, of course, many answers from different quarters as to why, after so many years in the darkness of the right-hand side of the scoreboard, Turkey had succeeded and won. One theory was that it was due to their stance on the invasion and war in Iraq. Turkey had voted against the use of their country as a staging area in the American-led war. This parliamentary vote had come as a massive shock to the country's ruling politicians and was a particular slap in the face to the USA. Fears, however, lurked in the Turkish government of the rise of Kurdish nationalism within its borders, where there was a substantial Kurdish minority. Would a more democratic Iraq see the establishment of independent Kurdistan and, with that, the desire for unification with their brothers in the Kurdish lands in Turkey? Throughout Europe, there was general disapproval of the war, which led to the question: Was the victory at the 2003 Eurovision Song Contest an endorsement of Turkey's decision to say no to America from an anti-war Europe? Was it just a coincidence that the USA's biggest ally in the war, the United Kingdom, not only came in last but received no votes at all? The British duo Jemeni, students from Paul McCartney's music college, sang out of key. This was due to technical difficulties with the backing track. There were accusations of sabotage. During the performance, their dressing room was trashed but then nothing was ever really proved. Other theories for Turkey's victory included the televoting factor that had recently been brought in by the EBU to replace the national jury panels. For the first time, the EBU also approved voting by SMS. While the televote ensured a genuine popular vote, one problem had arisen since it was implemented in 1997: that of diaspora voting. This seemed to be particularly prominent in the high points that Germany had allocated to Turkey since the changeover to televoting. This was blamed on the huge number of Turkish nationals and Germans of Turkish ethnicity who resided in Germany. This contrasted sharply with the pre televoting years. Germany had awarded Turkey nothing, completely ignoring its entries from 1993-96, while it had given Turkey high points in 1998 and 1999 when other countries either did not vote for Turkey or gave it low points. It was assumed that the high points awarded after the introduction of televoting were due in particular to the large ethnic Turkish population residing in Germany who had voted for the Turkish songs. The state-run TRT had even told Turkish viewers abroad watching the contest via satellite TV to phone in and vote for their homeland. Another theory, which is probably the most credible, is that Turkey won due to the changing face of Europe after the fall of communism and the break up of Yugoslavia. More countries had begun to compete in the contest as independent states, many of which shared a similar socio-cultural background to Turkey and, thus, had shared musical tastes. At the 1980 Eurovision Song Contest held in The Hague, Morocco and Turkey both entered songs. In the voting, they received few votes from the juries. Turkey ended up in fifteenth place and Morocco eighteenth, coming in second last only to Finland. The only country to vote for Morocco was Italy, which awarded seven points. Italy also awarded its eight points to the Turkish song. Turkey had not voted for Morocco but Morocco, on the other hand, had given its twelve votes to Turkey – an example indicating that, despite the distance, both Turkey and Morocco shared a cultural heritage and musical tastes. The Turkish entry, *Petrol*, was a very Anatolian oriental number, bulging with Arabesks and Makhams which, due to the political map of Europe at that time, as well as the overwhelming number of Western European nations competing, went completely over the heads of most of the national juries. The singer Ajda Pekkan's songs often reflected shades of feminism and the lot of women in Turkey, and she was a leading figure in the Turkish feminist movement. If she had returned to the contest and sung *Petrol* in the first two decades of the twenty-first century, it would have been a completely different scenario and her song may even have won, given how different the Europe on the scoreboard is now compared to the "Eurovision" Europe of 1980. The same could also be said of Morocco, which has never appeared on a Eurovision stage since then. It must also be said that Pan-European viewers are now very familiar with ethno-pop as a genre, and it has even proved exceedingly popular, no longer strange to Western ears. This is also a contributing factor to why ethnic pop has fared so well in recent contests. Although, surprisingly, at the 2012 contest, favourites Cyprus, with *La, La Love*, and Greece, with *Aphrodisiac*, both ethnic-dance songs, expected to take the contest by storm but did very badly. Sertab had insisted that all Turkish songs should be sung in English post-2003, but several still continued the tradition of singing in Turkish. The 2006 entry *Superstar* was sung entirely in Turkish on the big night except for one sentence being sung in English. It was decided at one point that the entry was to be sung in English but then the usual passions were aroused on the language issue.

In the tsunami of Anglo-American rock/pop culture, it would seem that even official and national languages are threatened by its force, as are the more ethnic styles of music once found in abundance on the European continent. Sertab's victory and the subsequent success of *"ethnic"* music at the song contest are manifestations of a desire to safeguard and hold onto more traditional aspects of national culture in an ever-increasing homogeneous Europe faced with globalisation. Countries that, a few years ago, competed in English are now sending entries sung in their national (or minority) languages. An example of this is Estonia, whose entries in Estonian have, in the early twenty-first century, done very well e.g., *Koula* (2012). Denmark, The Netherlands, and Iceland, among others, also regained confidence in their native languages. There is also a new confidence in entering ethnic-folk-pop. Countries that had done badly in the contest now have a new self-confidence; they are not afraid of entering songs that, in the 1970s and 1980s, would have been thought of as too ethnic. Turkey had received *"Nul Points"* in 1983 with Çetin Alp's *Opera* and in 1987 with Seyal Tanner's *I Melody*. The contest in the 1980s still lived very much in the shadow of ABBA and *Waterloo*. There had been several Turkish attempts to clone ABBA at the contest; most, though not all, have done badly. The Turkish efforts of 1983 and 1987 were attempts to do Western pop a la ABBA, which the Turks were particularly bad at. This had been, perhaps, an understandable reaction to how the juries rubbished the excellent Turkish debut *Seninle bir Dakika* in 1975. It is a rather profound, melancholic, very Turkish ballad with beautiful lyrics. There was also the dismal and very disappointing result of Aija Pekkan's 1980 attempt with *Petrol*, which was Turkey being very much itself. Turkey felt it was not allowed to be itself at the contest, so it attempted to do Western pop, which it took on rather self-consciously. Seyal Tanner, who sang *I Melody* in 1987, was the maverick of the Turkish music scene. She was also an accomplished film star who, at one point, was married to a member of Los Bravos. Looked upon as zany and over the top, she was very much her own boss and never danced to the tune of others, especially when it came to what music to record. A cross between Malcolm Maclaren, Zandra Rhodes, and Cilla Black, her style and particular fashion set the trends in the Turkish world of mode for decades. At times outrageous, the unique Seyal Tanner, like Serge Gainsbourg, frequently shocked. She would often rock the more conservative elements of Turkish society and, just like Gainsbourg, she recorded a wide range of genres. It was Tanner who was instrumental in bringing the "musical" to the theatre in Istanbul and Ankara, which, until then, was completely alien to Turkey. While her Eurovision appearance catapulted her into the "*Nul Points*" hall of fame, it had done her no harm at all. It is, at the same time, a great pity that Europe, in general, did not get to know or respect this interesting and highly original artist a lot better.

The 1983 entry *Opera* was the first of the Turkish songs to receive "O" at the song contest. As a result, its singer, Çetin Alp, became an object of vilification by the Turkish press, which blamed him outright for the disaster, accusing him of bringing shame to the Turkish nation. Çetin's career, due to this, quickly declined and, despite a degree of success at the Balkan Song Festival, his star faded from the Turkish pop scene, though his humiliation was resurrected by the press every spring in the lead-up to the Eurovision Song Contest. On the eve of the 2004 event, hosted by Turkey in Istanbul, Çetin was invited for an interview on Turkish TV, which he accepted. In the process of the interview, he was given a harsh grilling and, once again, became the scapegoat for scarred national pride, which, of course, had been redeemed by Sertab the year before. Three days later, Çetin Alp, who suffered from a heart condition, tragically died.

In 2011, Greece chose a song in the tradition of heavy *Laiko*. *Demotika* (folk) and "*light*" songs dominated the Greek music scene in the earlier years of the twentieth century. The "*light*" or "*Elafro*

Tragoudia" had much more in common with the international popular music of its era than the very traditional *Dimotika*, which often featured clarinets. This changed with the arrival of the *Rebetiko* with the multitude of refugees from Asia Minor, after the sacking of Smyrni, in the 1920s and 1930s, which had, to say the least, an unprecedented influence on twentieth-century Greek music and evolved into what is now known as *Laiko*, popular Greek music. The golden era of *Laiko* was the 1960s and 1970s, which saw the careers of such singers as Marinella, Stelios Kazantzidis, Tolis Voskopolos, Stratos Dionysiou, and Rita Zachellariou blossom. It also featured heavily in the heyday of Greek cinema, with songs by film stars Aliki Vougouklaki, Dimitris Papamichaël, and Mary Chronopoulou becoming classic hits of the golden age. This form of *Laiko* was known as *"light" Laiko* and was often of a light-hearted, optimistic nature. In the 1980s, however, *Laiko* transformed into what is often referred to as *"contemporary Laiko"* with its various blurred and often vague branches such as *"Laiko pop"*, *"Heavy Laiko"*, and *"Pure Laiko"*. Singers who clung to the less Western-influenced traditions of the genre included Angela Dimitrou, who competed in the 1986 Greek National Eurovision Final, Pascalis Terzis, Vasillis Karras, and Natassa Theodoridou. The singer Eleftheria Arvanitaki, on the other hand, sang a more folksy *Laiko*. From contemporary *Laiko* arose *"Modern Laiko"*, which was a meeting between the Eastern Mediterranean and the West in Greek popular music. The Greece of today is a Greece that still clings to its traditional culture but that, due to emigration to America and Germany, mass tourism, European Union membership, and exposure to Anglo-American rock, has changed gradually over the last few decades. Modern *Laiko* is, in itself, a reflection of modern Greece, much influenced by Western rock and pop, which is often an integral component of Modern *Laiko*. Its main exponents include the megastars Anna Vissi and Despina Vandi. The 2011 Greek national final was broadcast, due to financial restrictions, from the very modest set of ERT studios in Athens and presented by Lena Aroni. It was a rather dour affair with no live audience, completely devoid of any atmosphere. In the years following Greece's 2005 victory, it had been represented by mega-stars Sakis Rouvas and Anna Vissi, who topped the American charts and performed at sell-out concerts at London's Royal Albert Hall. Both of them managed top ten placings but were kept out of the top five. There had been accusations at the 2006 final that Anna Vissi, who was chosen to sing all the finalist songs, had deliberately sung the songs flat and out of key, all except for her own composition, *Everything*, which she sang with great passion. *Everything*, a dramatic ballad, sung in English, was chosen as the Greek entry. Although a chief exponent of Modern *Laika* and one of the first Greek singers to bring rock into *Laika*, Anna Vissi had chosen to sing a song that was more in tune with international pop. Top established singers such as the Greek/Lebanese Sarbel, with his *Tsefteteli* dance song *Maria*, and Alkeos with the very traditional *Hopa*, another lively dance number, which had more than a shade of *"Dimotiko"* in it, also followed them in the parade of Greek entries of the 2000s, not to mention Greek-American pop star Kalomira's *My Secret Combination*, sung in English with a rich ingredient of *Tsefteteli* and a classic example of the popularity of ethnic pop/dance music at the contest in that decade. It came in third. The song charted in the UK, but was given no airplay, as is so often the case, and so as a result did not climb much further up the chart. With the star-studded parade of superstars and popular established acts, Greek TV decided to try something new and give new up-and-coming contemporary acts a chance. It was thought that the publicity brought on by the song contest would propel their careers and benefit their record sales. A pop idol event was thought to be the answer but this was cancelled due to the lack of time left to organise it. Instead, ERT held a national final with six acts that had appeared in recent episodes of *Greek Idol* and *The X Factor*. It was also hoped that this would instil an interest in the national final in a younger generation. It was a national selection that was to be almost cursed. From the beginning, there was a barrage of discontent from various corners, i.e., artists, songwriters, and managers, on a variety of subjects. There were accusations of political bias and lawsuits were filed. To cap it all off, there was much criticism of the low-key, basic production of the show itself as well as much media displeasure at Lena Aronis' presentation. The show had the lowest viewing figures of any previous national final in that decade. It was bad news for ERT. What unexpectedly did come out of this was one of the most interesting Greek entries in some time: the song *Watch My Dance*, which started with an English hip-hop introduction sung by Stereo Mike, who attended both Leeds Met and Westminster University and had worked with several British hip-hop bands such as Kalashnekoff and Skinnyman. The song then broke into a heavy *Laiko* sung entirely in Greek by Loukas Giorkas, a Cypriot from Larnaca who had won the first season of *The X Factor*. At the semi-final of the 2011 song contest in Dusseldorf, the Cypriot entry sung by Christos Mylordos, *San Aggelos S'agapisa*, was also a more traditional *Laiko* song and their first entry to be sung in Greek since 2008. However, despite a good delivery by Mylordos, it failed to qualify for the final, whereas Loukas Giorkas of Greece did. Mylordos' poor eighteenth place in the semi-final was due to the stage full of bizarre antics being acted out to illustrate the song, while Mylordos and his backing singers wore heavy black tops with heavily chained black jeans. They swayed back and forth, from side to side, as the song reached its climax. It was a reasonable song that did not do well due to far too much action on stage and too much moving background scenery. Giorkas, on the other hand, wore a black suit, white shirt, and black tie and incorporated a *sirtaki* into his performance. It was very much a feature of Classic Greek 1960s cinema. *Watch My Dance* was a great song for Greeks but was it a song that Europe would vote for in the final? Greece's efforts at ethnic pop/dance had nearly always ended fairly successfully, but this was different. Would it end up in the top ten or would it be a gamble that would end up with zero? Would it be another Remedios Amaya, too confined to its home culture to appeal to ears outside its borders? Although the song did not win the contest, it came in a very respectable seventh. However, what is of note is that after the rundown of votes was given in the following week, it was revealed that *Watch My Dance*

Eurovision: A Plea for Respect

had come third in the international televote, which with the combined jury vote had placed it seventh on the scoreboard. Not only had the song done much better than was thought but it also showed that a very ethnic Greek song could be popular with a wider European audience. Was this, then, not a green light for record companies to do some serious marketing of ethnic Greek music to a wider world? Had the world not eagerly listened to Nana Mouskouri's renderings, in Greek, of the songs of Manos Hadjidakis and others in the 1960s, 1970s, and 1980s? Then could it not also happen again and open Greek *Laiko* up to a much wider audience? Political power that the United States and Britain wielded, along with affluence, had enabled these countries to afford advanced musical technology. It was not a case of people in Russia and China being tone deaf, as is more or less often implied by some music critics in the West. Through the Eurovision Song Contest, there is a cry from smaller music scenes, national languages, small countries, minority groups, and ethnic rhythms that they are there and are alive and want to survive, stand up, and be counted! Let us return to Spain where this chapter began. The days of General Franco are now long gone and more than a generation away. The lives, loves, and antics of the *Caudillos* grandchildren and great-grandchildren are still frequently found in the pages of glossy magazines such as "*Hola*! and *Semana*, still avidly read by many Spaniards. The country is now an established, stable, liberal democracy. Catalonia now displays a confident and public national identity, which was very evident when it hosted the Olympic Games in Barcelona in the early 1990s. Joan Manel Serrat's brave attempt to sing *La, La, La* in Catalan failed; now Catalan is seen and heard all over Catalonia, although it would not be until 2005, thirty-seven years later, that Catalan would be heard for the first time on a Eurovision stage, and the song would not come from Spain but from the tiny principality of Andorra, a Catalan speaking country tucked away in the Pyrenees. Ironically, the song *La Mirada Interior* was sung by a Dutch woman, Marian Van de Val, who lived in Andorra.

Remedios Amaya, the Gypsy singer who, with her ethnic flamenco renderings, managed "*nil points*" along with Turkey's Cetin Alp in Munich in 1983, retreated from the limelight, where she remained for several years. However, unlike Turkey's tragic Cetin Alp, who, according to Turkish media reports, lived as a broken, lonely recluse after his 1983 failure, Remedios returned to the music scene, this time with a bang! She went on to become Spain's top flamenco singer.

Eurovision: A Plea for Respect

CHAPTER 7
Are Rockall And Azerbaijan In Europe?

The tiny, uninhabited island of Rockall in the Atlantic is remote, inhospitable, and dangerous. It is one of Europe's farthest-flung outposts, situated as it is on its extreme periphery. In the same area, not too far away, lie the islands of Ireland and Iceland, Europe's most westerly nation-states. The high waves and blustery wild winds off Scotland's North West Coast gaze over towards Rockall, another world. Between its briny, twilight shores and the Scottish mainland lies the isolated island of St Kilda, whose population was evacuated en mass by the government in 1930 and never returned. Now the island is home to ruined villages and puffins as seagulls fly and waves crash over another desolate and remote North Atlantic island, that of Rona, which is even more remote than St Kilda and the farthest-flung island of the British Isles ever to have been continuously inhabited. It had a hardy population until 1685 when rats from a shipwreck swam ashore and ate all the islanders' food supplies, which led to the starving to death of all the inhabitants. Bluebirds gather over England's south coast. England is cut off from France and the rest of Continental Europe by "*La Manche*" – the English Channel. Culture geography and location are so very much related, as Britain's attitude towards Eurovision is so very much informed by its history and geography. Ireland, whose once impoverished states' biggest exports were its people and its faith, first competed at a Eurovision Song Contest nearly a full ten years after it had started. Butch Moore sang *Walking The Streets In The Rain* in Naples in 1965 and came in fourth. Moore came from the show band era in Ireland, which spanned the 1950s, 1960s, and 1970s. His elevation to star status enabled his Capitol show band to ascend to the pinnacle of Irish show business. The Irish show band culture was centred largely around ballrooms and the parish halls of Ireland. The bands usually toured the country and their vocalists sang mostly cover versions of other people's hits. The show band culture and Irish rural life formed the backdrop for the successful TV play *The Ballroom of Romance*. The most famous bands in the 1960s, the height of show band culture, were The Miami and The Royal, who both played regularly in packed halls. Butch Moore's ballad was to become a hallmark of Irish Eurovision entries. The Irish ballad is always a reliable bet.

Dana's 1970 winner *All Kinds of Everything*, written by Derry Lyndsey and Jacky Smith, was a folk-pop number with music adapted from a traditional folk song found in the fishing villages of Aberdeenshire and Banff. It gave Ireland its first win at one of the most difficult and bloody times in Irish history. As the *"troubles"* escalated in the North, Dana, a teenager from the Bogside in Derry, an area that saw some of the worst violence of that era, somehow managed to transcend the politics of the time with her song, the lyrics of which were full of beautiful things:

Snowdrops and daffodils
butterflies and bees
sailboats and fishermen
things of the sea
wishing-wells
wedding bells

She returned to her native city carried triumphantly on the shoulders of a British soldier through crowds of cheering locals from both sides of the sectarian divide. She would again bring both factions together several years later on her wedding day as thousands watched her make her way to St Eugene's Cathedral. Dana had done what so many politicians had attempted to do and had failed miserably at. In middle age, Dana, as Rosemary Scallon, went on to become an MP in the European Parliament and stood twice as a candidate in the Irish presidential elections. The year 1970 proved that Irish culture as an export could have some clout. Second place at the song contest in

Eurovision: A Plea for Respect

Amsterdam went to *Knock, Knock, Who's There?* by the UK's Mary Hopkins, a cultural victory for Ireland over its traditional colonial oppressor. Dana, however, like some mystical figure from an ancient Celtic saga, managed to detach herself from all that and rise above colonialism, pseudo-religion, sectarianism, and violence. The Republic of Ireland, alongside the United Kingdom, joined the European Economic Community in January 1973. While the UK basked in the glory of Eurovision in the late 1960s and early 1970s with their top artists, songwriters, and high placings, they were courting the EEC and learned that the musical and political lingua-franca of European aspirations walked hand in hand. This was a major lesson that Ireland was also to learn. A relatively small country that had been independent for just fifty years, Ireland was no industrial giant like Britain or Germany. Culture was high on its agenda of exports. As the years went by and Ireland became the golden child of the EEC, the Irish settled down to be, unlike their cousins across the Irish Sea, dedicated Europeans. Its cultural reputation abroad did not rely only on Eurovision but on a host of musical acts that waved their Irish passports through international barriers, from Boyzone, Westlife, and the Boomtown Rats to U2 and Rory Gallagher. In the years that followed Dana's victory, Ireland, as a whole, did reasonably well at the song contest. Their songs were nearly always respectably placed in a hall of fame that saw Angela Farrell, The Swarbriggs, Tina Renaulds, and Red Hurley take over the 1970s baton from the likes of Sean Dunphy, Dickie Rock, and Pat McGuigan, who had carried it back in the 1960s. In 1969, Muriel Day, who came from Newtonards in County Down, was the first singer from Northern Ireland to sing for the republic. She appeared in the film *Billy Liar*, in which she mimed the song *Twisterella*, stepping in at the last minute for a pregnant singer who was unable to perform. Her *Nine Times out of Ten* became a Northern soul classic. The year 1971 saw Anglo-Irish relations plummet to a low and a further escalation of violence in the North, which had also spread to England. The BBC, moved by Dana, who had come from Northern Ireland, and her role as a pacifying force between the Nationalist and

Unionist groups, asked Clodagh Rodgers, a singer popular in the UK, a Catholic born in Northern Ireland, to sing for them at the 1971 contest in Dublin. She accepted. It was seen by the British as an act of appeasement but the idea backfired terribly. Clodagh arrived in Dublin, in the early spring of 1971, to a cold and hostile reception, seen by a section of Irish society as promoting their colonial oppressor. Clodagh, some years later, revealed that she had received death threats from the IRA before the 1971 song contest. Ireland's 1966 entrant, Dickie Rock, destined to be a big star in Ireland, was at one point part of the Miami show band. The band's lineup changed over the years and Rock left the group. Late one summer evening in 1975, while the band was travelling back to Dublin from a performance in County Down, their minibus was stopped by the UVF, a loyalist paramilitary group. The gunman shot three of the Miami show band dead and wounded two others. The killings caused revulsion throughout Ireland. Frank McNally, the Irish Times diarist, summed up the massacre as "An incident that encapsulated all the madness of the time." A month later, the IRA bombed a Belfast pub, killing four civilians and a UVF member. Then, two days later, they shot dead a Portadown DJ, allegedly in retaliation. A senseless vicious cycle of violence and evil had engulfed the country.

The massacre had a detrimental effect on the show band culture in general. Many bands came to fear playing in Northern Ireland and, as a result, it was often excluded from tours. The demise of the show band culture started with the advent of disco and by 1984 the Irish public had all but lost interest in it.

In 2010, stamps were issued in Ireland to commemorate the golden age of the Irish show bands. One of them included a photo of Dickie Rock when he was a part of Miami.

The tenth anniversary of Dana's win saw Ireland celebrate its second victory, Johnny Logan's *What's Another Year*. This, however, coincided with a difficult patch for Ireland economically. Although it hosted the 1981 Eurovision from Dublin and competed at the 1982 contest in Harrogate, RTE was in a bad financial state and, as a result,

Eurovision: A Plea for Respect

did not take part in the 1983 Eurovision Song Contest in Munich. Johnny Logan returned to the Eurovision Song Contest like a Merlin sprinkling magic stardust in 1987 and won yet again with *Hold Me Now*, from the Irish tradition of ballads. It is worth noting that although the 1980s was a notoriously soul-destroying decade for the contest's self-esteem in the UK, *What's Another Year* was a number-one hit, as was the following year's winner Bucks Fizz's *Making Your Mind Up* and the 1982 winner Nicole's, *A Little Peace*. *Hold Me Now* was also a major UK hit; although missing out on the number-one slot, it peaked at two. Overcome with emotion, Logan was unable to reach some of the notes as he sang the winning reprise to *Hold Me Now*. His 1980 winner, *What's Another Year?*, penned by Sean Healy, has a distinctive saxophone introduction and tells the story of a man coming to terms with the death of his wife, his companion. The question is asked: *"What's another year for someone who's getting used to living alone?"* His second winner, *Hold Me Now*, which he composed himself, was a typical power ballad found at the Eurovision Song Contest. Logan continued to write songs for the contest, in particular, *Terminal 3* for Linda Martin in 1984 and again for Linda Martin who took the Grand Prix in 1992 with *Why Me?*. The late Balkan superstar, Macedonian singer Toše Proeski, also recorded a version of *Hold Me Now*. Ireland continued to enjoy phenomenal success at the Eurovision Song Contest, winning four times between 1992 and 1996. Ireland loved Eurovision and Eurovision loved Ireland. In the latter half of the 1990s, Ireland's economy started to boom and the term "Celtic Tiger" was the nickname applied to its runaway economic phenomenon. Emigrants who, some years before, had bid farewell to the Emerald Isle and flown to Boston or London, following in the paths of a generation back in the early 1950s, returned home. Brian Singleton wrote in the Irish Times, *"Its economic restoration and reputation drove it to international prominence alongside the Celticisation of music, theatre and dance. Ireland as a successful national economy had no further direct need to invest in a European contest in which all but one of the entrants ultimately fail."*

Steve Kerr

During the boom years of the Celtic Tiger, Ireland's financial fairness became a place not where the population emigrated from but where Middle Eastern asylum seekers and East European immigrants wanted to emigrate to. Soon, Ireland had become a far more multi-cultural society than it ever had been. The Irish televote at the Eurovision Song Contest awarded high votes to Lithuania and Poland. Ireland had strong ties with Poland; the first elected Irish Nationalist MP had been a Polish countess. As the years moved on, Ireland began to lose a grip on the song contest. Its ballads, which had been the staple diet of its Eurovision glory, no longer appealed to a wider Europe that had moved on to other types of music. The votes gradually got fewer and fewer. In 2001, Gary O'Shaugnessy came in twenty-fifth; thus, Ireland was relegated from the 2002 event. Estonia, Latvia, Ukraine, Turkey, Greece, and Serbia, all from the eastern parts of Europe, replaced Ireland as the winners. Ireland was no longer the true spiritual home of Eurovision. Marty Whelan and Pat Kenny, the long-time RTE commentators who took the contest seriously, were usurped by Wogans' BBC commentary, which was also widely available in Ireland. This was an example of the former colonial oppressor still enforcing its cultural value system on the once colonised, and with an Irish accent as well. Wogan proved popular and Whelan had to compete against him for ratings. The Irish perspective on how the contest should be perceived competed with the English perspective. A well-known alternative Irish comedian, when talking about Ireland's role in Eurovision, said of the event, "*It's shite*". He sounded convincingly hip, punctuating his sentences with the right expletives in the right places but unaware that he had taken on the cultural mantle of the former colonial power and that he was an organ of the scapegoat. After failing to qualify for the semi-final in 2005, Ireland invited one of its biggest stars, Brian Kennedy, to sing for them in 2006. He had recently sung at the footballer George Best's funeral. His song *Every Song Is a Cry for Love*, once again a Celtic ballad, placed in the top ten in Athens. The 2007 entry was a very ethnic-sounding folk song, *They Can't Stop The Spring*, which was reportedly about East European immigration to

Eurovision: A Plea for Respect

Ireland. It featured the line *"and Europe's all one stage"* and came in last. Ireland's national final of 2008 was a very different affair from previous years and was won by the popular puppet Dustin the Turkey, who sang a parody of Ireland at the Eurovision Song Contest. One of the panel at the national final was Rosemary Scallon, formerly the 1970 winner Dana, who was not at all pleased at the selection and called it *"a foul decision*!" She later told reporters that if countries did not want to take the contest seriously, they should not compete. Dustin's song *Irelande, Douze points* was really a rather clever, entertaining song with references to Johnny Logan and Terry Wogan as well as the mention of the many competing nations. Dustin was extremely funny. He was accompanied by two boa-clad female vocalists who sang the chorus soulfully. If they had sung it alone, I am sure it would have sounded much better and might have gone on to the final. However, not only did Dustin not qualify for the final but the Irish entry was booed off the stage in Moscow. Ireland's Eurovision garlands lay withered and in tatters. In the same year, Ireland's economy began to shake nervously. Pure pop girl bands and a former winner were drafted to sing for the country in 2009 and 2010 but only one scraped into the final, where it came twenty-third. The United Kingdom and Ireland buried their old animosities and allied themselves in the song contest. Both were glad of each other's annual exchange of votes. An example of bloc voting? In the last decade, Britain and Ireland have usually given each other a reasonable number of votes. The UK was glad of any points that would keep it from last place. RTE continued with the national selection process on the popular and often controversial *Late, Late Show*. Once the patch of Gay Byrne, a veteran Irish broadcaster, it had replaced the former Irish selection, *The National Song Contest*. In 2011, the formula remained the same but RTE invited leading music industry promoters to find the songs and search for suitable singers to interpret them. The result was the terrible twins Jedward – John and Edward Grimes – already well known in the UK from their *X Factor* days. Their Eurovision entry was a highly polished pop offering in keeping with the time, entitled *Lipstick*. During Eurovision week, the twins and their song were a welcome national relief from bailouts, unemployment figures, and austerity measures, which otherwise dominated the news. The twins, with their zany haircuts and impish looks, somersaulted across the Eurovision stage all the way to eighth place. Their song became a huge hit in Ireland and Europe, boosting their careers and leaving the chrysalis of the Irish ballad to become the youthful, fresh, upbeat pop butterfly. The reliable Irish ballad had been abandoned for the heart of international youth culture.

According to Brian Singleton, Ireland's 2011 entry was the first to be a truly commercial success and one that was reported to have earned a small profit for RTE. Now that Ireland's national reputation had declined economically, it had reverted to its former size – a size that makes it possible once again to use Eurovision for gentle political purposes in terms of redefining a country's image.

The seagulls fly over the harbour at Kinsale on the Irish Sea. On the flight of the seagull, it is a long, long way from Ireland, on Europe's western periphery, to the land of Azerbaijan on Europe's most eastern frontier. "Is Azerbaijan in Europe?" is the question on many people's lips. The answer to that seems to be both yes and no. It is part of the Caucasus area, located at the crossroads of Eastern Europe and Western Asia. Some reliable sources class it as Europe, while other, equally reliable sources class it as Asia. To the north, it borders Russia; to the west, it shares a border with Georgia and Armenia; and to the south, Iran. With this in mind, is there such an entity as a pure, clear-cut defined Europe in cultural or geographical terms? It is bounded on the east by the Caspian Sea. The Palestinian academic Edward Said wrote, "One can talk about a European tradition in the sense of an identifiable set of experiences, of states, of nations, of legacies, which have the stamp of Europe upon them. But, at the same time this must not be divorced from the world beyond Europe. In the Algerian context there is a good phrase for this: 'Complementary enemies'. There is also a complementarily between Europe and its others and that is the interesting challenge for Europe, not to purge it of all its outer affiliations and connections in order to try to turn it into some pure new thing."

The EBU, European Broadcasting Union, the governing body of Eurovision, was formed at a time when the former imperial states of Europe still had protectorates, or governed in a colonial sense, in much of the lands on the periphery of North Africa and Asia. Their national broadcasting stations were very much related to the national broadcasting companies inside the European continent. As a result, they, too, became active members of the EBU, making the term "Eurovision" a bit more flexible and inclusive than what the definition of Europe had normally been. The mighty Soviet Union, the bastion of communist ideology and the antonym of the USA, showed a more than keen interest in participating in the Eurovision Song Contest. In 1987, at the time of unprecedented political reform, a motion was forwarded by the Minister of Education suggesting that the country could benefit from entering the Eurovision Song Contest. The idea behind it was concerned with the USSR's until-then cool relationship with the capitalist West. It was thought that a win would have an impact on the West. A proposal that Valery Leontyev sing the first-ever USSR entry at the Eurovision Song Contest was, however, rejected. Gorbachev did not agree with the Education Minister's suggestion and believed that such a move would be too radical for the Soviet Union to take.

Valery Leontyev was a popular singer from the Komi Republic in the Russian Federation whose long career peaked in the 1980s. Described in Time magazine as being something between Mick Jagger and Mikhail Baryshnikov, he appeared at the very first Soviet Union rock/pop festival staged in Yerevan (Armenia) in 1981.

Ten former Soviet Republics would compete at the Eurovision Song Contest as independent countries: Russia, Estonia, Lithuania, Latvia, Ukraine, Belarus, Moldova, Armenia, Georgia, and Azerbaijan. Five of them would win the contest in the years 1994-2022.

In the 1970s, government policy in several states drew up a scorched earth policy for Eastern Europe's rock and pop scene, particularly so in Czechoslovakia, which was still living in the shadows of the Prague Spring and the Russian invasion.

Songs and groups with Anglo-American lyrics and names were forced to change them, while long hair and exotic Western fashion were banned. Karel Gott was a Czech singer who was also popular in German-speaking countries and who sang for Austria at the 1968 Eurovision Song Contest. A Czech singer residing in Czechoslovakia, representing a country west of the Iron Curtain in the spring of 1968, was not only unusual, it was also highly symbolic of the Prague Spring and Dubček's reforms. In 1964, Gott recorded a cover version of a Beatles song, which was followed by many other hits. In 1970, he recorded his own composition, *Hey, Councilman*. It was a light, catchy ditty about someone who was in love petitioning the local town council. The Communist Party leader Husák said it promoted "irresponsible illusion" among the youth. The record was banned from the radio. Got,t however, continued to perform it. He also fell afoul of the authorities with his song *I May As Well Flip A Coin*, about a lover lamenting the fickleness of his actions. The song was banned. The communist authorities deemed it insulting to Czechoslovakian currency. Gott came under heavy scrutiny from the regime. He found it hard to secure bookings. His songs faced strict censorship, their lyrics carefully examined. At one point, he spent hours trying to persuade the authorities not to cut his hair. Long hair, he tried to explain, was important for his image in the West, but they refused him the right to perform with his hair touching his ears. Geography, as well as politics and ideology, played a major part in the authority's attitude towards pop and rock music in the communist states of Romania and Bulgaria in the 1960s. The Black Sea and its miles of beaches provided both countries with several popular holiday resorts. The governments upgraded the facilities and advertised Romania and Bulgaria to the West as cheap holiday destinations. Millions of Western Europeans flocked to the Black Sea resorts as an alternative to the crowded beaches of Spain. Due to the number of foreign tourists, Bulgaria and Romania began to import Western products such as Coca-Cola. The tourist

Eurovision: A Plea for Respect

trade also brought with it denim-clad youths and other vestiges of Western decadence, often leading to an exchange with local youth. Pop groups were permitted to work in the resorts as cabaret in the numerous coastal hotels. Through the Western-style pop at these resorts, Romania and Bulgaria began to absorb a more Western-influenced pop culture. Romanian popular music at the time often reverberated around what the authorities dictated people should listen to, songs like *Glory to Our First President* and *The Party, Ceaușescu Romania*. Western-orientated pop/rock and roll took off in Romania as a direct result of the Cliff Richard film *The Young Ones* being screened. Cliff Richard, much maligned by the music critics in twenty-first century Britain, was instrumental in popularising Western pop and rock and roll in the Soviet Union. In the mid-1960s, Ceaușescu operated a more open, liberal policy towards Romania's growing pop/rock music scene but he was heavily influenced by Chairman Mao's "Cultural Revolution" in China, of which Romania was an ally, and by the early 1970s he had enforced a crackdown on all aspects of Romanian life that had a Western influence. Pop music reverted to its more Stalinist interpretation. The republics of the vast Soviet Union did not have the contact and influence from the West that the Balkan countries had through tourism. Although the Soviet government did try to promote Yalta in Crimea as an economic Black Sea holiday resort for Western tourists, it never became quite the magnet that the Balkan resorts evolved into. The Soviet Union absorbed Western-orientated pop/rock music at a much slower rate than its European communist neighbours. As an authoritarian regime, it promoted and preferred its "comrades" to engage more with high culture in music and literature. When it came to a knowledge of Western pop/rock culture, the Soviet regime displayed an obvious ignorance, while its Warsaw Pact neighbours were only too familiar with the likes of the emerging superstar Elvis Presley in the 1950s, even if it was only out of condemnation and contempt. Western-style rock had only a very marginal effect on the Soviet Union. Its isolation from Western Europe and the hangover it suffered as a result of the Stalin years severely impeded the progress of rock music in Azerbaijan and the other Soviet republics. Western-style pop would sometimes crop up on TV and radio emissions. It was also occasionally presented in Soviet films. Local pop singers were far less influenced by Western pop and rock than were their counterparts in Poland, Hungary, and Czechoslovakia. When rock and roll had made its presence felt by the 1970s, a flourishing pop scene existed. The platform for new Bulgarian pop music was the Golden Orpheus Song Festival, held annually at the Black Sea resort of Sunny Beach near Varna. The Soviet Union's Alla Pugacheva took the USSR by storm when she won the Golden Orpheus Song Festival with the song *Arlequino*. She was to be the Soviet Union's greatest pop star. The Bulgarian singer Lilli Ivanova won the Golden Orpheus in 1974 and 1982. She started her career in 1961 singing Bulgarian versions of European and Russian hits. She soon altered her repertoire to sing exclusively Bulgarian songs. All her songs had strong, distinctive, and meaningful lyrics. Her hits of the 1960s and 1970s were often highly orchestrated and dramatic ballads set to complex arrangements. Her popularity stretched far beyond Bulgaria's borders, and it was estimated that she had sold more than ten million records in the USSR. She was often backed by up-and-coming Bulgarian bands that emulated The Beatles. As her original works and style went out of fashion, she adapted to other musical trends and genres including ethno-Bulgarian, rock, and soul. Her compatriot, the Gypsy singer Sofia Marinova, who represented Bulgaria at the 2012 Eurovision Song Contest and narrowly missed out on a place in the Grand Final, is a contemporary pop singer whose vocal range is often compared to that of Lilli Ivanova.

There is a tendency by Western journalists and music critics to blatantly state that no pop music existed in the Soviet Union or other Warsaw Pact nations until the fall of communism in 1989 when what they really mean is that there was no Western pop or rock readily available for public consumption, insinuating that the only music of any value had to come from the West. Until 1989, some Western pop/rock acts were allowed by the authorities to perform at concerts in the USSR but they were only a select few, wiith Cliff Richard, Elton

Eurovision: A Plea for Respect

John, and Paul McCartney among them. ABBA and Boney M were allowed airplay by the regime. Boney M's pan-European *Rasputin* went down well with Kremlin officials because it portrayed the Romanovs in a bad light. Brotherhood of Man's Eurovision winner was also deemed safe enough to be given airplay on Soviet radio. Due to the Soviet Union's isolation from the West, its pop music was more introspective. It has always drawn more from the musical traditions within the many diverse Soviet republics, in particular Russia, and therefore developed differently. Western influence on pop did, however, slowly creep in through various bureaucratic loopholes. The Soviet Union did sometimes win a battle. The Russian Song *Podmoskovnye Vechera*, originally recorded by the young singer Vladimir Troshin, won the 1957 International Song Festival of World Youth and Students in Moscow. It was recorded by Kenny Ball as an instrumental and was a huge hit in the UK and the USA under the title *Midnight in Moscow*. In Azerbaijan's capital, Baku beat music began to appear in dance halls which were normally the residence of waltzers or the tango, popular throughout the USSR in the 1950s. In the early 1960s, Soviet youth caught on to the twist. Many clubs in Azerbaijan banned the new dances from the West. Couples brave enough to go against the ban and take to the floor were usually arrested by the local militia or banned from the club. In Baku, the city made it illegal to dance any of the American-inspired dances that began to infect the country. Party officials in Baku organised alternative youth evenings which included lectures at the root of which was orthodox communist ideology. These were followed by ballroom dancing. The evenings proved to be extremely unpopular and, in the end, guards had to be employed to stop the youth from leaving *en masse*. Many Soviet pop singers slowly introduced Western pop aspects into their repertoire. The twist remained popular in the Soviet Union long after it had become passé in the West. The communist cultural officials bent over backwards to create fresh and innovative Soviet dances for the youth to take an interest in, but it was usually in vain. In late 1968, the annual Sochi Song Festival was held in Georgia SSR. The performing artists' songs and presentations mirrored Western pop and culture. Female vocalists chose to wear mini-skirts and wiggled their bodies, while male singers shook their hips. In the 1990s, Western consumerism and Western rock/pop music arrived in the Soviet Union disguised as democracy but it met with a formidable and established local pop scene. Anglo-American rock/pop was seen by many in the West as an expression of freedom that the people of the USSR had long been deprived of. This, however, gives rise to the question: Was the presence of the omnipotent force of Anglo-American rock/pop a true expression of freedom or just another manifestation of Anglo-American cultural hegemony? The Soviet Union was a regular competitor at the annual Intervision Song Contest in Sopot, the pleasant Polish Baltic resort. The Sopot Song Festival was the idea of the Jewish musician Wladyslaw Szpilman, who survived the Nazi invasion and the Warsaw ghetto. His family was sent to the gas chambers but Szpilman managed to escape and spent the duration of the war in hiding. He had been playing Chopin live on Polish radio as the Nazis invaded. It was to be the last live music broadcast until 1945. Szpilman was the main protagonist in the Polanski film *The Pianist*. Originally, the festival was intended for singers from all around the world who would sing music by Polish composers. In the 1960s and 1970s, when the Eurovision Song Contest was at its height in Western Europe, the Warsaw Pact countries were unable to watch it. They were also not active members of the EBU and so could not take part in the Eurovision Song Contest. They were, however, members of Intervision, the Warsaw equivalent of the EBU Eurovision. By the mid-1970s, the director of Polish TV, influenced by the Western Eurovision Song Contest, was at pains to set up an eastern Intervision counter-contest and, in 1977, the Sopot Song Festival was renamed the Intervision Song Contest. Among the big stars to compete over the years were Poland's Maryla Rodowicz, Russian diva Alla Pugacheva, and Roza Rymbaeva, her compatriot citizen from the USSR (who was born in Kazakhstan). Finland, which was a member of both Eurovision and Intervision, fared much better at the Intervision event than at Eurovision. Marion Rung, who had

Eurovision: A Plea for Respect

sung for Finland at Eurovision in 1962 and again in 1973, coming in sixth, the highest placing ever at that time for Finland, achieved a resounding victory at Sopot in 1980 with her song *What Is Love?* Participating in both the Intervision Song Contest and the Eurovision Song Contest is a manifestation of Finland's position in Europe: physically in the West but psychologically in the East. Castros' communist Cuba was also a regular in the Intervision Song Contest. Soviet singers were often not well received by the Polish audience at the event.

Soviet pop star and heartthrob Lev Leshchenko had been chosen to represent his country in 1971. He had spent months preparing himself, rehearsing for the song contest, but was dropped at the last minute in favour of Maria Kodryanou, a singer from the Moldavia SSR, where Communist Party leader Leonid Brezhnev had been local party chief. Angry Leschenko did get some consolation. Maria came in only fourth at the song contest, an embarrassment for the powerful leader of the communist world in Europe. The following year, Leschenko was dispatched to Sopot and won the contest. The 1977 contest was won by one of Czechoslovakia's greatest stars of the 1970s, Helena Vondračkova, with the song *Malovany Dzbanku*. Along with the actor Václav Nečkar, both of them cultural pop icons, she had starred in the film *The Madly, Sad Princess (Silene Smutna Princesa)*. This witty and highly charming pop art/fairy tale, full of often obvious political jokes, was punctuated with plenty of pop songs, The film centred around a lonely young prince who tries to woo a very difficult and haughty princess. It was made in 1968 but aired on BBC TV on December 15, 1971. It achieved a fair degree of popularity, particularly among younger teenagers in Britain. If recordings of the songs had been released and given airplay in the UK, they most certainly would have become hits. Vondrackva not only achieved success at Sopot but also competed and appeared at the Rio De Janeiro Song Festival, the Tokio Song Festival, the Bratislavska Lyre, the Golden Stag Festival, and the Golden Orpheus Song Festival. In 1978, she sang *Manner Wie Du* in the German Eurovision national final. In 2007, she was approached by Czech TV and asked to sing the Czech debut entry at the Eurovision Song Contest that year, which she agreed to do. Due, however, to several obstacles on the road to the national selection, Helena later declined the invitation. She would appear as a non-participant artist at multiple Sopot Song Festivals, including that of 2020. After the Velvet Revolution of 1989, Vondrackova's popularity waned due to her association with communist rule, as was the case with many other Czech and Slovak stars who had ridden high in the decades of the socialist state. The music-listening public of Czechia absolved her of any guilt through association and she experienced a massive resurgence in popularity and respect, which climaxed in her 2020 album receiving a platinum disc.

One curious aspect of the Intervision Song Contest was its very particular form of voting. The points were awarded according to the load experienced on the electrical network. This was because most of the public in the competing countries were without telephones, so to determine the public vote, viewers would turn the lights on or off depending on if they liked or did not like a song. The year 1980 saw the last of the Intervision Song Contests. Its home in Sopot was not far from Gdansk, where unrest had broken out, and in 1981 Poland declared martial law. Soon, there were ugly scenes as protestors clashed with police and tanks moved in. The song contest behind the Iron Curtain had wanted to prove something to the West: Anything you can sing, we can sing better!

After 1989, the collapse of communism in Europe and the break up of the Soviet Union led to the end of Intervision. That saw former Intervision members applying for membership in the EBU and the majority of them taking part in the Eurovision Song Contest. What happened as a direct result of this was a fusion of the old Western European Eurovision and the former Intervision states of the old Warsaw Pact and the subsequent creation of a new Eurovision. The landlocked former Soviet republic of Byelorussia, now known as Belarus, proudly made its debut at the 2004 Eurovision Song Contest in Istanbul. Its President, Lukashenko, won the country's only

Eurovision: A Plea for Respect

democratic election and has been in power since 1994. He has, throughout his years in office, governed the country in an autocratic manner and has, thus, attracted much international criticism. His government pursued a policy of extremely close ties with the Russian Federation, which led to the creation of the Union State and even greater cooperation. The 2010 presidential election was considered to have been rife with fraud and widespread irregularities and was not recognised by the international community. The song chosen for the following year's song contest was the patriotic sounding *I Love Belarus*, sung by Anastasia Vinnakova, who was considered to be part of the regime's propaganda machine, and the lyrics of the song were construed as pro-Lukashenko. Despite some criticism, the entry went to Stuttgart but did not qualify for the Grand Final. The Norwegian winner from 2009, Alexander Rybak, of Belarussian extraction, had often expressed a desire to write a song for his mother country. In 2014, he composed the song *Accent,* which the singer Miki took to fourth place at the Belarus national final. Belarus competed in every Eurovision from 2004 until 2020. In that period, it reached the Grand Final no fewer than six times, with its highest placing being *Work Your Magic* by Dimitry Koldun, partly written by the Bulgarian-born Russian mega-star Philip Kirkorov, which came in sixth. In 2021, amid rising unrest following yet one more rigged election, as well as general dissatisfaction with the regime both at home and abroad, the Belarus broadcaster, BTRC, sent the song *Ya Nauchu Tebya (I'll Teach You)* by Galazy ZMesta, a blatant political jibe at critics of the regime. It was thus disqualified by the EBU and another somewhat provocative entry was submitted by Belarus in its place. However, that, too, was rejected by the EBU for the same reasons. The EBU voted to suspend Belarus in May 2021 and awaited a response. Belarus did not respond to this and so, in July 2021, Belarus was officially expelled from the EBU. No doubt when the Lukashenko regime collapses, the Eurovision Song Contest will welcome back its Belarus child with open arms. The following are countries outside the "traditional" Europe that could, if they wanted, compete in the Eurovision Song Contest: Algeria, Egypt,

Israel, Jordan, Lebanon, Libya, Morocco, and Tunisia (if we include Eurovision Song Contest participants Armenia, Georgia, and Azerbaijan as Europe). Turkey and Russia are European countries, though most of their land mass is in Asia, which also compete regularly in the contest, as does Israel. Tunisia entered a song for the 1974 song contest but withdrew before the event. Lebanon's state broadcaster, Télé Liban, selected the Francophone song *Quand Tout S'Enfait*, sung by Aline Lahoud in early 2005, to represent Lebanon. It was an orient-occident composition which was due to be presented at the Eurovision semi-final in May. Several problems over the participation and broadcasting of the Israeli song on Lebanese TV resulted in Lebanon breaching the Eurovision Song Contest rules and being forced to withdraw. Lebanon ultimately lost its deposit fee and faced a three-year ban. Morocco entered the contest only once, in 1980. The Vatican is also eligible to participate but never has done so. Kosovo is very enthusiastic about competing in the contest but cannot, as it is still not recognised by the United Nations. Liechtenstein was unable to compete for many years because it had no national broadcaster, although it did once select a song back in 1976. The situation has now changed and the principality can now participate, though lack of finances has stopped them from doing so. Recently, Qatar and Kazakhstan have shown an interest in competing. Egypt, Tunisia, and Turkey were among the EBU's founding members in 1950, so, as senior members, perhaps have more of a right to participate in the Eurovision Song Contest than some of the more recent member countries. It is generally believed that most of the North African and Middle Eastern countries have stayed away mainly because Israel takes part. Morocco debuted in 1980, a year when Israel did not compete. At the same time, the question arises: If they do not compete for this reason, do they boycott the football World Cup because Israel is in that? Or any other major international event? The answer is no, but participation or lack of it may have more to do with religious-cultural reasons. Music, particularly dance music, is frowned upon in Islam and not regarded as proper. On the other hand, local pop music shows flourish on television in many of these countries. Morocco has since established diplomatic relations with the State of Israel, so perhaps a future song from Casablanca may be in the cards.

Azerbaijan, a Muslim country, although many would regard themselves as cultural Muslims, if we are to be 100% accurate, is in both Europe and Asia. The extreme north of the republic is in the Caucasus Mountains, the accepted geographical border between Europe and Asia. The same is true of its neighbour, Georgia, which denotes the accurate geographical description as Eurasian. In the West, due to its Islamic heritage, Azerbaijan was always considered Asian, as opposed to Georgia and Armenia, which, because of their Christian heritage, were considered European. Their Christianity was also behind a more Westward-looking mentality and culture. Armenians are a people scattered over much of Asia Minor and the Middle East. Like the Jewish people, they have often been the victims of horrible genocide, mostly at the hands of the Ottoman Turks. In the Ottoman Empire, they constituted a sizeable minority. From the latter half of the nineteenth century onwards, they were increasingly maltreated, with numerous massacres and expulsions which climaxed in the Armenian Genocide of 1915, when an estimated one million were murdered, many of them on the Death Marches in the Syrian desert. Those events inspired *Face The Shadow*, the Armenian song at the 2015 song contest, sung by Geneology, which commemorated the centenary of the massacre, known as the *Aghet*. It centred around the continual denial of events by the Turkish government, and the chorus resounds with the words *"Don't deny"*. The lyrics of the 2010 Armenian entry, *Apricot Stone*, performed by Eva Rivas, also dealt with the Armenian Genocide. Azerbaijan and Armenia, both emerging democracies, are seeking European Union membership and are members of the Council of Europe. Azerbaijan has a very different history and cultural tradition from many of the Islamic countries in Western Asia. Its constitution contains no reference to a state faith. The current regime is secular nationalist, although most of its population, by tradition, adheres to Shia Islam. Some of the opposition have Shiaism on their agenda. It enjoys a fairly high standard of living due to revenue from oil compared to other former Soviet states and a very low homicide rate as well as an excellent literacy rate. It has both ethnic and historical ties with Turkey. The fact that it is one of the independent Turkic states is manifested in its membership of the Turkic Council and Türksoy. In the earlier part of the twentieth century, Azerbaijan, along with Georgia and Armenia, formed the Transcaucasian Federal Republic. It was to be short-lived and, in 1918, Azerbaijan declared independence. The democracy was the first parliamentary republic in the Muslim world. During this period of independence, Azerbaijan extended suffrage to women, which gave Azerbaijan the honour of being the first Muslim nation to extend, to women, the rights that men already had. A great emphasis was placed on education and the nation prided itself on having the Muslim East's only modern university, which had its roots in a hunger for knowledge and has developed into a healthy respect for education and its institutions. The performing arts were also a major priority in Azerbaijan. It again achieved a first in Islamic Asia by being the first country to build opera houses and theatres. A brave cultural investment is evident in the country's current consumption of high culture as well as a vibrant and diverse music scene that has stemmed from the priority given to the performing arts. The mega-star of pop and opera in the Soviet era, Muslim Magomayev, who enjoyed great popularity throughout the Soviet Union and who greatly influenced the other mega-star of the Soviet era, Alla Pugacheva, was by birth Azerbaijani. Poland Bülbüloğlu, another pop singer, popular throughout the Soviet Union, was credited with composing jazz-based pop songs full of Azeri folk elements. He later went on to be the Azeri Ambassador to Moscow. *Leyli and Majnun*, composed by Azeri Uzeyir Hajibeyov in 1908, was the world's first Muslim opera. Azeri music often features string instruments such as the lute-related Saz and Tar, the garmon (accordion), and the tutek, a small flute. It has been heavily influenced by Persian music; very evident in its Mugam, classical Azeri music which also informs a lot of current pop music, which, in turn, enjoys a dedicated, loyal following and great popularity in this vast, vibrant,

Eurovision: A Plea for Respect

colourful country that embraces the wide range of musical styles within its musical boundaries. Lenin, with a greedy eye on Baku oil, invaded the republic in 1920 and effectively made it part of the Russian-dominated Soviet empire. Its connection with European Russia and other European Soviet republics during its time as part of the vast, multi-ethnic Soviet Union cemented its already Westward-looking, Quasi-European tradition that already existed in the country. This relationship with Soviet Europe saw Azerbaijan's development differ considerably from that of other Islamic countries to its south or east. The events that sprung from Gorbachev's Glasnost led to the dissolution of the old Soviet Union, and Azerbaijan declared independence in 1991. However, in 1993 the democratically elected President Abulfaz Elchibey was ousted in a *coup* which led to the former leader of Soviet Azerbaijan, Heydar Aliyev, being restored to the presidency. Azerbaijan's first few years of independence were marked by conflict with Armenia over the disputed Nagorno-Karabakh territory. Thousands of people died and over a million people were displaced as a result of the war. Thousands of Armenians and Russians, who had formed a large ethnic group before the 1990s, left Azerbaijan. The government of Heydar Aliyev managed to stave off two attempted coups and has brought a degree of prosperity to the country due to the much-improved economy. The regime had, however, received strong international criticism for corruption, abuse of human rights, and election fraud. Azerbaijan's emerging nationhood, as well as its rich musical heritage and the popularity of Azeri pop music, led the country to compete for the first time at the Eurovision Song Contest in 2008. This was also a blatant manifestation of Azerbaijan's European direction and how the country saw itself. Additionally, it was a challenge to the traditional Europe, as Edward Said stated, "not to purge it of all its outer affiliations and connections in order to try to turn it into some pure new thing". Azerbaijan could compete in the song contest because it was a member of the European Broadcasting Union. The country is also a member of the Council of Europe. Its presence at the Eurovision Song Contest and its eagerness to compete veils aspirations to join the European Union. In the case of Azerbaijan, as well as neighbouring Georgia and Armenia, its participation in the song contest was part of a national building process. Nations that are of less international renown or that, for political reasons, might have a less than favourable reputation usually engage in a more sober political approach to the Eurovision Song Contest. It is a chance to stand shoulder to shoulder with nations respected for their democratic institutions, liberal traditions, and sheer power, in a democratic count of equal merit. Former Soviet states that have won Eurovision in the past and then staged the prestigious but costly event, such as Estonia and Ukraine, engaged with the contest as a platform to create some sort of national discourse. Dr Paul Jordan, who holds a doctorate relating to the song contest, tells of how Ukrainian President Viktor Yuschenko wished to make a forty-minute speech at the 2005 Kyiv contest; this was in the immediate aftermath of the so-called Orange Revolution. Ukraine's first participation in the contest in 2005 came about as a suggestion by a PR firm, which, recognising the image-building potential of the event, contacted the government and advised that the country take part. Dr Jordan, in the case of Estonia's participation, stresses the polarisation of the song contest – a direct result of negative publicity in the European press concerning rumours of Estonia's inability to stage the event in 2002 due to financial difficulties. Massive funding and general involvement from the government saw the Eurovision Song Contest effectively evolve into a nation-branding mission. In Azerbaijan, similar image improving tactics, not dissimilar to those employed by General Franco in relation to the song contest in the 1960s, figured very highly in its participation, winning, and, ultimately, staging of the event in 2012, in which there was heavy government involvement. The Eurovision stage, on which it could sing its own chosen narrative, was a chance for Azerbaijan to make its mark on Europe, which its participation has very much done. Since Azerbaijan first participated in the 2008 event, a large number of Western Europeans who were unaware or had little knowledge of the Eastern "Land of Fire" are now very familiar with it, including as a

Eurovision: A Plea for Respect

European nation. To some, it is even a possible future holiday destination, such a powerful impact did its postcard films have on the millions who watched the 2012 Baku Contest live on TV. With the president's wife acting as hostess, Baku welcomed contestants from forty-two countries to the Eurovision Song Contest. The country spent a reported billion on hosting the event, which included the controversial building of a new stadium, the Crystal Hall, especially for it. The Moscow 2009 Eurovision Song Contest was called the Peking Olympics of Eurovision due to the expense and lavish production involved. The Baku hosting of Eurovision 2012 far surpassed the Moscow event and was by far the most expensive and ostentatious contest in Eurovision history. Azerbaijans' debut at the song contest in May 2008 was a song called *Day After Day*, sung in English by the duo Elnur and Smir. It was an interesting, powerful, and very different sort of song which combined opera and the Azeri folk genre *Mugham* with rock elements. There was a degree of opera and theatre in the performance as well – a clear manifestation of the importance and popularity of opera in Azerbaijan. The duo performed the song dressed as the devil and an angel. Elnur and Samir explained the meaning behind the lyrics:

"The main global topic at this very moment is war. Everything evil comes from the devil – who is, in fact, just a fallen angel who was expelled from paradise. He symbolises everything evil which happens on earth but he can turn into an angel again."

Elnur Húseynov, an ethnic Azeri, born in Turkmenistan, worked at the Academic Opera and Ballet Theatre and sang in the choir of the State Philharmonic Society. He moved to Ukraine in 2010 in an attempt to further his singing career. Samir Javadzadeh, as a music student, specialised in the study of *Mugham* before joining the pop group Sheron. The song came in eighth and was a rather memorable debut song for Azerbaijan. Their 2009 entry, sung by Aysel (Tegmurzadeh) and Swedish/Iranian singer Arash Labaf, born in

Tehran, was a less dramatic affair but did considerably better and was placed third. It was an ethno-pop song with a very Eastern feel that charted all over Europe, reaching number three in Sweden and number two in Greece. Arash already had an impressive array of hit songs across Europe. Everyone wants to win the Eurovision Song Contest but Azerbaijan was, more than most, expressly out to win it. As part of their plan, they employed top songwriters and producers, often from abroad. They put all their energy into this aspiration, every avenue that might lead towards victory explored. Since its debut at the contest, Azerbaijan has proved to be one of the most consistently successful contestants. It has qualified from the semi-finals and has placed in the top five on numerous occasions. The Azerbaijan win in Düsseldorf in 2011 was a crowning moment in the emerging nation's post-Soviet history. Elli & Nikki, aka Eldar Gasimov and Nigar Jamal, were the victors. The name changes occurred when it was thought that viewers outside Azerbaijan might interpret Jamal's name as offensive. Their entry, *Running Scared*, was sung in English. During its time as a republic of the Soviet Union, most pop singers from Azerbaijan sang in Russian. Despite the reintroduction of Azeri into official public life after independence, some top artists, such as Aygun Kazimova, still chose to perform in Russian. As of the time of this writing, none of the Azeri Eurovision songs have been sung in Azeri, which is really rather a pity, as it has deprived the rest of Europe of the pleasure of hearing a language very rarely heard on European TV. The English language songs, however, have always served them well. Nikki was based in London at the time of the win. The song, a contemporary international pop song, was devoid of any "mugham" or other ethnic Azeri folk sounds which had featured in the 2008 and 2009 entries. *Running Scared* was written by Sweden's Sandra Bjurman and Stefan Orn and Britain's Ian Farguhanson. In the early twenty-first century, Swedes were giving British songwriters a run for their money both at the Eurovision Song Contest and in the music world in general. The duo was also backed on stage, to add to Swedish involvement, by the group Shirley's Angels, which had competed in the Swedish national final of that year. The

Eurovision: A Plea for Respect

song was a minor hit in several European countries and did not really do at all well commercially, unlike Lena's 2010 German winner, although it did enter the Belgian top ten. The song was a rather saccharin pop ballad. The Azeri entries of 2010 (*Drip Drop*) by Safura and 2012 (*When the Music Dies*) by Sabina Babayeva were far better songs and far more deserving of a winner's trophy than *Running Scared*. *Drip Drop* was written by the same Swedish songwriters who were to pen the winning entry the following year, in 2011, and was featured on Safura's album, *It's My War*, which had some international success. *Drip Drop*, although a favourite to win before the song contest, disappointingly came in fifth. The video of *Drip Drop* was directed by the well-known film director Rupert Wainwright. The choreography for the song at the contest in Oslo was designed by Jaquel Knight, who had also worked for Beyonce and Britney Spears. The 2012 contest in Azerbaijan saw the homegrown girl, Sabina Babayeva, climb up the scoreboard to come in fourth. Loreen's *Euphoria* was a convincing winner but Babayeva's *When the Music Dies* was the most outstanding song out of the rest of the top five. Sabina Babayeva gave a strong performance of this power ballad that relied heavily on elements of Azeri folk. It was written by a Swedish songwriting team, some of whom had already composed the previous Azerbaijani entries. There has been a very strong Swedish connection with the Azeri entries since it started to compete. Arash (2009) was an Iranian by birth but was a resident of Sweden. *When the Music Dies*, a blend of European pop and traditional *Mugham*, featured traditional Azeri folk instruments such as the Kamancha and Balaban. Bayeva later recorded an Azeri version of the song *Gal*. At the song contest in Baku, Babayeva's hometown, the song included heavy "mugham" backing vocals by renowned Azeri singer Alim Qasimov, whom The New York Times described as "One of the greatest singers alive". Qasimov had been awarded the UNESCO music prize in 1999, one of music's highest accolades. His appearance at the 2012 Eurovision Song Contest started as he sang a *Mugham* in the opening sequence of the Grand Final, when the cameras first focused on the stage. Qasimov has toured the world and given concerts in both the USA and Iran. Despite the cosmopolitan nature of his life, he never forgets the peasant roots that formed his musical personality and the harsh life of poverty he came from, nor has he adapted to foreign musical influences, which he dismisses. He dedicates most of his work to pure, traditional Azeri *Mugham*. He does, however, also perform Ashiq, a rural bardic music which developed in Turkey, Azerbaijan, and Iran. Additionally, it is well known that he was highly impressed by the Gazal singer Nusrat Fateh Ali Khan, by whom he has been especially influenced.

The 2013 contest ended in triumph for Denmark but it was very nearly a second win for Azerbaijan, who was the runner-up. Farid Mohammadov's *Hold Me* pushed Ukraine's Zlata Ognevich and her song *Gravity* into third place but was unable to beat off the favourite, Emmelie de Forrest. After the 2013 contest was over, there was anger in Moscow that Azerbaijan had not given any votes to Russia. Foreign Minister Sergei Lavrov lashed out his disapproval. Even Azeri President Ilham Aliyev was in disbelief as he watched the final live in Baku and he subsequently ordered a recount of the votes. The results showed that the joint tele/jury vote had, in fact, given the Russian song, the anthemic ballad, *What If*, ten points but they had mysteriously gone missing somewhere between transferring the thousands of votes into points and the live television delivery of the votes at the song contest. Former Soviet singing star Poland Bulbuloglu, Azerbaijan's ambassador to Russia at the time of the 2013 song contest, made a statement: "I myself, a musician and a man of culture, was surprised. It cannot be that Azerbaijan gives no vote for Russia. Moreover, Russia had an excellent performance and as a result gained fifth place". The problem seemed to lurk somewhere in the complicated vote-counting process, which took place in Germany, and not in the votes' country of origin. The EBU launched an enquiry into the suspicious disappearance of votes. On Azerbaijan's 2011 winning song by Elli & Nikki, both Germany, which had hosted the song contest, and Azerbaijan issued postage stamps in commemoration of the victory.

Eurovision: A Plea for Respect

The staging of the 2012 Eurovision Song Contest in the Azerbaijani capital, Baku, gave Azerbaijan an unprecedented quota of positive public relations with which to promote the country internationally. Criticism of the regime by human rights groups heightened in the weeks leading up to the event. There had been criticism by such groups of the Azerbaijani government's tactics in organising the event. This concerned mainly the forced evictions of Baku residents and the demolition of flats to make way for the construction of a giant stadium built to house the song contest, the Baku Crystal Hall. In the lead-up to the event, the Azerbaijani government, in an effort to clean up its image, released a major dissident. Immediately before the contest, a group of demonstrators, knowing that the eyes of the world were upon them, protested in the streets of Baku. They were, however, halted and arrests were made with the quick arrival of the police. Throughout, the EBU had worked hard to ensure the apolitical nature of the song contest, despite criticism from several organisations in Western liberal democracies, which accused it of not doing enough to highlight the abuses. Some even criticised it for allowing Azerbaijan to host the contest. Most of Europe before the 2012 event was unaware of the issues affecting Azerbaijan. The fact that Azerbaijan won the contest and then hosted the event, however, did much to highlight the country's political issues and even gave a platform to dissident views via the international media. Some members of the EBU proposed that Azerbaijan be banned from the song contest altogether. The leading dissident, Emin Milli, disagreed. He believed that the song contest was a great opportunity to:

Voice public criticism of the government's violations and to demand the release of political prisoners.

On the Monday of the Eurovision week in 2012, the BBC broadcast an edition of *Panorama* which linked the song contest and the EBU to the Azerbaijani regime and questioned if Britain should be taking part at all in Baku. But there had been little mention of abuses inside Azerbaijan in media current affairs before and scant attention paid to any political prisoners. The documentary was very critical of the contest, yet the Eurovision Song Contest was the reason the violations had come to light and were being talked about in the international media in the first place. Emin Milli, strongly against any boycott of the event, stated, "*I was strongly against a move. The event should be used as a platform to show the world what is happening in Azerbaijan.*"

Throughout the BBC documentary, there was much maligning of the Eurovision Song Contest, with the usual diet of British entries plus mediocre and silly songs shown as representative of the contest. The documentary made rather toxic remarks about the event, which it tried to link in some way to the authoritarian regime, almost wanting to make the contest itself seem complicit in human rights abuses. Its apolitical nature was portrayed as giving backing to the regime. An underlying dislike of the song contest lurked throughout rather than a critical report on the political situation and human rights in Azerbaijan. It culminated in reporter Paul Kenyon accosting the British entrant Engelbert Humperdinck as he was leaving an interview with DJ Chris Evans inside the BBC studios. The stunned seventy-six-year-old singer was asked, without giving him time to frame his thoughts, why he was taking part in the event. The Eurovision Song Contest is broadcast by the BBC. Engelbert had been asked to sing for the UK by the BBC, he accepted the invitation, and, as a result, he was ambushed by a team of reporters from *Panorama*, another BBC programme, in the BBC studios. It was somewhat ill-advised and rather sensationalist and would not have happened to an athlete en route to the Beijing Olympics, which also raised questions about human rights abuses. There, of course, had been some programmes highlighting the lack of democracy in China in the months before its Olympics staging but they had focused on that issue in light of the Olympics being held in China. Olympic Games hosts have had bad human rights records, as have some of the regimes that hosted international football matches in recent years. One of the greatest and most enduring aspects of the

Eurovision: A Plea for Respect

Olympic Games is that all countries compete regardless of their regimes. According to the dissident Emin Milli, a former prisoner of conscience who spent sixteen months in jail for speaking up against injustice, it is important to engage with authoritarian regimes like Azerbaijan in order to expose them to processes and factors of change which is far more difficult. As a consequence of the still ongoing conflict between Azerbaijan and Armenia over Nagorno-Karabakh, which has been under Armenian control since 1993, the relationship between the two countries at the Eurovision Song Contest has been, to say the least, somewhat acrimonious. Several controversies between the two have surfaced during the build-up and final of the Eurovision Song Contest since Azerbaijan and Armenia both started to compete. Armenia made its debut in 2006. Despite Nagorno-Karabakh being a "de jure" part of Azerbaijan, it was the birthplace of the Armenian representative Andre, who was born in 1979, when it was Soviet Azerbaijan. The official Eurovision website showed the Armenian singer's place of birth as Nagorno-Karabakh, which gave the impression that it was part of Armenia. Azerbaijan made an official complaint and the place of birth was removed from the webpage. The Republic of Nagorno-Karabakh was, once again, the cause of some tension at the 2009 contest. This time, the object of controversy was the postcard film, the introduction to the Armenian song; it showed in part some of Stepanakert, the capital of the disputed region. Azeris saw it as an insinuation that Nagorno-Karabakh was part of Armenia. A complaint was made and the film was edited but it did not end there. In retaliation, while the Armenian singer Sirusho, who had sung for Armenia in 2008, read out the Armenian votes, she displayed a picture of a monument in Stepanakert, the precise part of the film that had been edited out. Live in the background, in the centre of Yerevan, a large screen displayed the same monument in Nagorno-Karabakh's capital, making the statement that it was Armenian. The postcard that led up to the Azerbaijani entry was also a subject of some controversy. Their film depicted well-known symbols of the Iranian cities of Tabriz and Urumieh. Part of the Azerbaijan region inside Iran has obvious ethnic and cultural ties with the former Soviet republic. Armenia complained that Eurovision was biased in favour of Azerbaijan, as it had forbidden the film of Nagorno-Karabakh but permitted Azerbaijan's postcard of Azeri monuments in Iran. In the true Eurovision spirit of the promotion of peace, despite the acrimony, Armenia awarded Azerbaijan one point; 1,065 Armenians had voted for Azerbaijan. Accusations, however, concerning relations between the Caucasian republics continued to plague the 2009 event. There were allegations that Azeri TV had failed to show the Armenian entry, with another country's shown instead, and no phone number had been shown for Armenia during the televote. After an EBU investigation, Azerbaijan was prosecuted and fined for distorting the TV signal while Armenia was performing and for interfering with the televoting number, making it difficult to read. The worst allegation was that the forty-three Azerbaijan citizens who did vote for Armenia had had their calls traced by the police and were later rounded up and interrogated. The EBU reportedly threatened Azerbaijan with a three-year ban for violation of the contest's rules through isolation, boycotts, or sanctions. The potential for coverage and reporting of the true reality in Azerbaijan put the government in a situation it had never experienced before. However, the problem did not stop thereeee; it persisted in relation to Eurovision. In 2016, the Armenian singer Iveta Mukachayen caused outrage in Azerbaijan by holding the flag of Nagorno-Karabakh while performing on stage. As a result, Armenia was sanctioned by the EBU. The winner of the 2016 song contest, Ukraine's Jamala, was on her maternal side descended from Armenians from Nagorno-Karabakh.

The Eurovision Song Contest itself encountered more direct political issues face on. Once again, relations with Armenia would come to a head at the contest. Despite cool relations, Armenia initially said it would compete in Baku, but following the shooting death of a young Armenian soldier by Azeri forces on the border, several Armenian pop singers launched a campaign to boycott the Baku contest. The Armenian Defence Ministry afterwards stated that the

Eurovision: A Plea for Respect

young soldier had, in fact, been killed by a fellow Armenian soldier and not an Azeri, as was first believed. Due to fears over the Armenian delegation's security in Baku, Armenia finally backed out of the song contest. Armenia was fined for a late withdrawal and told it still had to broadcast the contest live. It was reported, however, that the Armenian commentators spoke over part of the Azerbaijani entry.

Not only did Azerbaijan face Armenian withdrawal but its southern neighbour, Iran, voiced much displeasure over the 2012 song contest. Ayatollah Shabestari and Ayatollah Sobhani were furious at Azerbaijan's hosting of the event, which they branded as "Anti-Islamic behaviour". In retaliation, Azeri protestors in Baku paraded slogans outside the Iranian embassy, ridiculing the Iranian leadership. The administration of Baku advised Iran not to interfere in Azerbaijan's internal affairs. A livid Tehran, in response, recalled the Iranian Ambassador from Baku. The situation escalated and Azerbaijan demanded a formal apology for statements Iran made in relation to the 2012 Baku contest. This led to Azerbaijan recalling its ambassador from Iran. After the 2012 Eurovision Song Contest was over, the Ministry of Security in Baku issued a statement in which it revealed that a plot for a series of terror attacks had been thwarted. The final of the Eurovision Song Contest itself, at the Crystal Hall in Baku, was to be the main target. The Marriot and Hilton Hotels, where most of the delegations were accommodated, were also targets for the plot. It was widely reported that Iran's most powerful leader, Ali Khamenei himself, gave the orders to carry out terror attacks at the Eurovision Song Contest against Azerbaijan and its Western allies. It is, however, interesting to note that although Iran was aggressive against the staging of Eurovision in Baku, it did participate in the 2013 Asian Pacific Song Contest, organised by the ABU in Vietnam.

Baku, with its pleasant sub-tropical climate, was once a Mecca for holidaymakers. Its long, sunny days and dry climate were popular during Soviet times, with workers from other areas of the Soviet Union arriving in the hundreds to relax on its beaches or enjoy the benefits of the many spas that overlooked the blue Caspian Sea – all of which now have the sad, dilapidated air of a bygone era hanging over them. The warm southern winds, the Gilavar, and their cold northern counterpart, the Khazri, blow constantly throughout the season. A legacy from its Soviet days as a great industrial centre, Baku is one of the most polluted cities in the world. It is a separate planet from the wild, inhospitable waves that surround the island of Rockall. Azerbaijan's curious relationship with Europe (or should I say the rest of Europe?) is no more evident than looking over the Caspian Sea towards Iran and the heavily Eastern mosaics that adorn so many traditional buildings and artefacts in the country. Azerbaijan is part of Eurovision. Its geographical position is not quite Europe but not quite Asia, yet it is part of another Europe. The traditional geographical notion of Europe has now given way to a wider Europe: Azerbaijan, a European nation that is part of that great European family, the Eurovision Song Contest. Could it be that participating in the Eurovision Song Contest gives a country the right to be truly European, correctly integrating into what is the true Europe? So great has the impact of the Eurovision Song Contest been that it derives from the old geographical Europe, whose boundaries it now challenges and redefines. The Eurovision Song Contest not only dictates what Europe is but effectively defines it.

Eurovision: A Plea for Respect

CHAPTER 8
"Colonels, Culture, Wine, and Sea"

As a teenager in 1972, I vividly remember Vicky Leandros winning the Eurovision Song Contest with *Apres Toi*, the Luxembourg entry. Vicky had sung for the Grand Duchy before in 1967 with *L'amour Est Bleu*, which went on to be a universal hit. She was a huge international star in the true sense of the word: popular all over the continent and, in particular, in her adopted homeland, Germany. She had moved as a child from her native Corfu with her father, Leo Leandros, AKA Mario Panas, also a singer, to Hamburg to further his singing career. In 1967, Vicky, alias Vasiliki Papathanasiou, had placed fourth. Her second attempt, in 1972, saw a more determined, accomplished, and polished professional Vicky who knew how to act out a powerful ballad. She beat off all opposition from Mary Roos, The New Seekers, Tereza, Jaime Morey, and all the rest. When the presenter, Moira Shearer, former ballerina and star of the acclaimed and somewhat sinister film *The Red Shoes*, proclaimed Luxembourg as the winner from the stage of the Usher Hall in Edinburgh, Vicky became an even bigger international star. Her aptitude for languages saw her record hits in several languages. *Apres Toi* went on to top the charts in several countries and sold six million worldwide. Its English version, *Come What May*, peaked at two on the British charts Her recording of the Mikis Theodorakis Greek song *O kaymos* or *Ich Hab Die Liebe Gesehn* (its German title) in 1973 was also a great international hit and, although never a top ten hit in the UK, the English version, *I Saw The Love In Your Eyes*, entered and re-entered the top fifty several times and lingered around the bottom end of the chart for several months. The very Greek-sounding *Die Bouzouki Klang* was also a huge hit in Germany and throughout Europe, while its English version, *When Bouzoukis Played*, was yet another top fifty hit in the UK. The recording of *Theo Wir Fahr Nach Lodz*, based on a

Polish folk song, also brought her much success across Europe and was number one in the German hit parade. In the 1960s and 1970s, she enjoyed considerable success in Japan with various Japanese language recordings. It is estimated that over the years, she has sold in excess of 150 million discs and earned several gold and platinum discs as well as numerous awards. Vicky's romantic life blossomed and she later married into European nobility. She came in second in the German pre-selection for the 2006 Eurovision Song Contest, beaten by Texas Lightning, and shortly after went into politics in her native Greece. She served as deputy mayor for the port of Pireaus as a member of the PASOK party and in 2003 was decorated by the Greek Orthodox Church for her charity work with children in Africa. This was the highest accolade that the Greek church can bestow on anyone and she is the only woman to have ever received it. Vicky Leandros was the first Greek to win the Eurovision Song Contest. At that time, Greece had never entered the Eurovision Song Contest and was, like Portugal, under a dictatorship. Vicky Leandros first sang on a Eurovision stage in 1967; in the same month, the military staged a successful *coup* in Greece, which was to lead to seven years of government under the Colonels. The world is familiar with the many cases of repression and torture that took place in Greece between 1967 and 1974, with the military being overthrown in 1974. This was thoroughly documented in a four-volume report of the European Commission on Human Rights. Greece's modern history reads like a steady list of coups and dictatorships. This period, however, was significant in that it was the longest period of dictatorship in its post-independent history. The Colonels adorned Greek towns and villages from Macedonia to Crete with revolutionary slogans that identified them with the country's toiling masses. In Britain, hostility to the regime was manifested in a resolution passed by the Labour Party at its 1967 conference, which called for Greece to be expelled from NATO. It also voiced solidarity with the Greek working class in overthrowing the Colonels. The Conservative Party made several moves towards cordial relations with the regime. Cabinet members attended functions at the Greek

Eurovision: A Plea for Respect

Embassy. Most notable was an invitation sent to the commander in chief of the Greek armed forces, who was cordially received during a visit to the UK in 1971.

One of the regime's most distinctive features, apart from the construction of its power base in the army, was the fact that it managed to subjugate the civil service in its entirety. Most high offices in the ministries were replaced by army officials or family members of the Colonels. Nepotism was a chief feature of the dictatorship despite its intention to put an end to corruption in Greek society. The regime also made a great effort to break up the power of many social groups that might have become a force of opposition. In an attempt to do just that, the regime parcelled local government councils and trade unions. In the Greek Orthodox Church, the military replaced the Archbishop of Athens and all of Greece with Jeronimos Kotsonis. This occurred within days of the military takeover. Kotsonis was, at one point prior to the *coup*, a royal chaplain. Other aspects of the church were forced to change to comply with the new regime. Metropolitans who showed signs of non-co-operation were replaced. In 1968, the regime abolished the tenure of judges, which enabled some of the most respected judges to be sacked. Thirty were dismissed and were even denied the right to practice as lawyers. The Greek *coup* of 1967 came at a time of radical changes in Western Europe. The Colonels of Greece, like Salazar's Portugal, contrasted sharply with the quasi-revolutions in Prague and Paris. In Northern Europe, the post-war boom in consumerism had led to an embourgeoisement of the working classes. Divorce became easier to obtain, abortion was legalised, and a new sexual freedom influenced music and the arts. In the UK, the ban against *Lady Chatterley's Lover* was lifted. The UK's Sandie Shaw had brought the Eurovision Song Contest into the Swinging Sixties by appearing on stage in Vienna in a mini dress and bare feet. She sang an up-tempo pop song in a contest dominated by ballads, starchy white shirts, and black bow ties. Students rampaged through the streets of Paris a year later. In Greece, the *coup* was considered by its supporters as a revolution, very different from those in Prague or Paris but one worth equal interest. Through the multicultural plains of Thrace and Macedonia, the mountain villages of the Peloponnese, and the blue shores of the Cycladic Islands, the revolution and its purpose were clearly displayed. Slogans covered the country in blue and white (the national colours). They clearly weaved together several themes that the Greek soul would find appealing:

"GREECE IS RISEN"
"GREECE OF THE GREEK CHRISTIANS"
"COMMUNISTS ARE TRAITORS TO THE NATION"

Anyone who has a Greek background or who has lived in Greece will know the importance of Easter in the Greek Calendar. Back in 1967, Christmas was celebrated to a much lesser extent in Greece than in Western Europe. Easter is a time when families have reunions. Everyone returns to their roots – to their village or island. During Holy Week, many Greeks will fast, which means eating special foods devoid of oil, flour, meat, and other products forbidden during Lent. The festival rises to a crescendo with Christ's Crucifixion in Greek Churches on Thursday evening via the *Epitaphios*, a representation of Christ's body on a bier carried through the streets on Friday night. The celebration climaxes with the *Anastasi* on Saturday at midnight, when crowds gather outside churches, carrying an unlit candle on the stroke of twelve. The priest proclaims, "*Christos Anesti*" (*Christ is Risen*), and the sombre mood of the previous week is shattered by the pealing of bells and fireworks. People kiss and greet each other with the seasonal greeting proclaimed by the priest, "*Christos Anesti*". In 1998, Greek television (ERT) pulled out of the song contest as it coincided with the all-important *Anastasi*.

The Greek word for revolution is *Epantasis*, which if translated literally, is *uprising* and is related to the word *anastasi* – the resurrection. The Colonels were very clever; they knew that the slogan *Greece Has Risen* appealed to not only the patriotic sentiments of the people but also to their inner sentiments of childhood and family associations with the Easter greeting and great Easter hymn, *Christ Has Risen*. This religious phraseology was certainly no coincidence but a special verbal skill employed by the propagandists, many of them ex-communists who had become some of the Colonels' most dedicated supporters. This verbal skill had, in fact, been a feature of the KKE, the Greek Communist Party. Their resistance army, founded in 1942, bore the name/initials ELAS. This acronym is pronounced the same as the Greek word for Greece – Ellas.

Greece had fought a bitter and bloody civil war in the years following her liberation from Nazism. Government forces and communist factions plunged the country into conflict as friends and families in city and village went to war with each other. Nationalist resistance to the communists was strengthened considerably by British support. One of the army's reasons for the 1967 *coup d'etat* was a fear of communist takeover. The chief sources of disagreement and political division in Greece's recent history have been republicanism, monarchism, communism, and nationalism. The main point of conflict, and one which persists even today, in Greek society is the question of whether Greece should be a republic or a monarchy. Since independence, there have been many military interventions in favour of, or in opposition to, one of these viewpoints. One reason why this particular problem refuses to go away is that the great powers of independence felt that a monarchy in Greece would bring some degree of unity to the country; thus, a monarchy was transplanted onto Greek politics, also for the reason that the great powers wanted Greece to remain on good terms with them. The monarchy, due to this, has not developed deep roots in modern Greece. The army often played a big part in the disputes, nearly always on the Republican side. It overthrew King Otto, the Bavarian-born first King of Greece, in 1862 and organised a Republican Revolution in 1909, which left a great imprint on Greece and altered the political role of the army. The crisis that led to the *coup* of 1967 had its roots in 1965. It proved fatal not only to Greece's most cherished national heritage, democracy, but also to the monarchy. King Constantine II succeeded to the throne after the death of his father, King Paul, in 1964. There were several problems at that time, one of which was Cyprus, but the young king had both a loyal government and a loyal opposition. For the first time since the war, a two-party system was beginning to take root and work. In the summer of 1965, a quarrel occurred between the king and his prime minister, George Papandreou. This climaxed in the fall of his Centre Union government. Several conspiracies of both left and right-wing army officers also came to light. Papandreou's son was associated with the left-wing element in all this. These events led to a succession of unstable governments that ended in 1967. It had long been suggested by its opponents that the monarchy had interfered in Greek politics. Under the regime, Greece remained a kingdom with King Constantine as head of state but the royal family, including Queen Frederika, The Queen Mother, went into exile. I quote from her autobiography: "...a time came when ill-disposed voices grew noisy in our lives. I was being accused of wanting political power and of dominating the family. I was accused of using the welfare organisation for our own glory and to influence Greek political life" (*A Measure of Understanding*, p. 239). She would be permitted to return to Greece only after her death, for her controversial internment in Athens in 1980. In a plebiscite, the Greeks voted for a republic after democracy was restored in 1974. King Constantine continued to live in Hampstead, London for many years but after a change in the law was allowed to return to his homeland..

Culture in Western society around 1967 was beginning to show side products. Men and women wore their hair long: an act of defiance against the conventions of society, a sign of refusing to conform. One of the first acts executed by the Colonels' regime was to prohibit long-haired and bearded foreigners from entering the country. This also applied to women wearing mini-skirts. General Pangalos, dictator of Greece in the 1920s, had set a precedent by decreeing that women's skirts were not to rise more than fourteen inches off the ground. Many dictatorial regimes were obsessed with points of sartorial detail. Several years ago, in Bulgaria, trousers that could not be removed while the wearer still had his shoes on were forbidden. Franco's Spain also had strict codes of what to wear. Men had to wear t-shirts when sunbathing on the beaches. Franco's regime, however, did not ban the mini-skirt. The new Western Europe of liberal thinking, freedom of speech, and

Eurovision: A Plea for Respect

dress personified in Sandie Shaw's long, straight-flowing hair, minidress, and bare feet, which triumphed along with her youthful movements and up-tempo pop song, was a victory for the values of the new Europe. The 1967 contest, won by an arch icon of the 1960s, was surely a part of the social, sexual, and political revolution that was sweeping over the Western world. Sandie Shaw's performance was a far cry from the ideals of the Greek Colonels. The regime, however, rescinded the decree on dress code when they realised it would ruin the vital tourist trade. Other decrees followed. It soon became obligatory by law for schoolchildren and teachers to attend church every Sunday. The ancient custom of *Spasimo*, smashing plates in tavernas, was banned. The authorities showed much concern about protecting the youth of Greece from Western culture and values. The regime called it "Western Neo-anarchism". Generals made spine-chilling references to a satanic pit in the USA and Europe. Their institutions with youth associations were overflowing with "free love, intemperate sexuality and hippyism" (*Thesis Kai Ideai*) (1970). This fear of the west was not some phenomenon born out of the *coup* of 1967 or the permissive society that arrived in the West around that time. There was often a dislike of Western values and traditions by Greeks in bygone ages. Suspicion and hatred of the supposed decadence and irreligion of the West have deep roots in the history of modern Greece. The much-respected mystic Athanasios Panos warned parents not to send their children to be educated in the West, as this could lead to atheism or even popery. That was back in the early nineteenth century. Western popular music and other aspects of Western culture in the 1960s and 1970s were often the points of much criticism by Colonel Ionnis Ladas. Ladas made a big name for himself by expressing often unsophisticated socio-cultural views. He personally beat up the author of an article in *Eikones* magazine because he strongly disagreed with its content. He had a particularly strong dislike of long hair; in fact, it would seem it was his pet hate. "Those with long hair represent the degenerate phenomenon of hippyism," he was quoted as saying. His hate of long hair stemmed from the fact that he needed to make a big issue of it in order to contrast it with the short-cropped hair of the military, which, of course, signified upright morality and *Ellas Ellenon Christianon* – Greece of the Christian Greeks. An ambiance like that was hardly the best for the seeds of culture to flourish. The Greek *Rethbetika* – the Greek Blues as it is often known – could be compared to the *Fado* of Portugal. It had been nurtured in the slums of Athens among the immigrants from Asia Minor, where a large percentage of modern Athens has its roots. Among its chief exponents of the last century was Sotiria Bellou, whose earthy, deep voice can still be heard blaring out of traditional Ouzeries and coffee bars. In contrast, the 1960s and 1970s saw the film star/pop singer Aliki Vougouklaki reach her zenith. Rubbished by the followers of sophisticated music, Aliki's popularity earned her the title of *National Star*. Her popularity was partly due to the era in which she first came to prominence. Greece had just recovered from Nazi occupation and its brutal civil war. Its people, many of whom lived in abject poverty, were weary and tired of the hardships of everyday life. Along came Aliki onto the screens of Greek cinemas. In the films, she, like them, was poor and had very little, but she was happy and made the most of things. The Greek public could relate to her. Films such as *Appointment in Corfu* and *S'agapo* became part of the regular diet for Greek TV viewers. When Aliki died of cancer in 1996, she lay in state and was given a state funeral. Her films could be compared to the *Carry on* or *Norman Wisdom* films that were so popular in the UK in the 1950s and 1960s. When King Constantine ascended the throne of Greece upon the passing of his father in 1964, he was seen as a non-political, rather glamourous and charming young monarch. As a crown prince, he had been very popular with the Greek public – a popularity strengthened by his love affair with Aliki Vougouklaki. They made a highly attractive couple but were to part, a decision that was often erroneously blamed on the snobbishness of his mother, Queen Frederika. It is rumoured that if Constantine had married Aliki, her immense popularity across all sections of Greek society would have led to the monarchy's re-establishment after the 1974 referendum. The films of Aliki contrasted sharply with the

Eurovision: A Plea for Respect

content of *Rethbetika* lyrics, which often touched on themes such as jail, social injustice, death, and poverty. *Rethbetika*, named after the *Rebetes*, members of an urban subculture who lived on the edges of society, grew in popularity after 1922, when a huge influx of Greeks fleeing persecution boosted the population of Athens considerably. Greek pop music heard on jukeboxes or in tavernas back in the 1960s and 1970s was a descendant of *Rethbetika*. Much of today's popular Greek music still features the bouzouki, the *baglamás*, and the accordion, which were the main instruments in the *Rethbetik* music of the 1920s. Such was the scene of popular culture's musical angle back in the Colonels' Greece. Popular singers at the time, all of whose music had been influenced by *Rebetika*, included Marinella, Voskopoulos, and Kazandzidis. In a free society, music, film, and other aspects of culture follow their own often changing trends and fashions. Often ephemeral in nature, they are usually remote from politics. Culture in a free society has a lofty and aloof independence in its consumption and creativity. In a totalitarian society, all cultural aspects are under the hand of its rulers. The government decides if you can listen to certain music. There are other societies where the iron grip should be called authoritarian. In political terms, there is little difference between two regimes that ignore the human rights of their citizens. Within a totalitarian regime, all aspects of culture are controlled. This is all part of the realisation of its idea. The party's goal is forced down the public's throat. In an authoritarian regime, the chief aim is that a class of rulers should remain defiantly in charge. Its control over music, theatre, etc. is limited to the suppression of what might endanger its preservation. Regimes like this don't want to mobilise people but to immobilise them. A totalitarian regime tries to mobilize its people via culture. Its regimes are more inclined to encourage good music, for example, an interest in opera, ballet, and classical music, and a solid and free education. This could be seen in the old Soviet Union and Warsaw Pact countries, where such cultural pursuits were within the grasp of working people. The Soviet Union's pop singers had to be graduates of the university of pop music. Authoritarian systems are more likely to promote cheap culture that will keep the masses in a contented trance. This is usually done by providing them with everything from football to pornography, light music to horror movies. An example of this is the Philippines of President Marcos, where beauty contests and bubblegum pop provided the populace with cultural nourishment. What often starts as liberation by totalitarian means may slowly take on the mantle of an established elite interested only in maintaining its power. An authoritarian group may hide under the mask of some ideology which it will use to mobilize people. The group's interest, then, may be associated with, for example, an established church. The Greek dictatorship was indeed a strange blend. It was, of course, an extreme right-wing dictatorship and authoritarian but the Colonels' peasant background and outlook gave it a very much populist flavour. The regime, however, wants to appear progressive, which flavours it with a radical overtone. Their attempt to establish permanent rule over Greece seems to have led them up the path to the creation of an ideology and propaganda that was totalitarian. Pressure from foreign politicians forced them to pay lip service to democracy. All these things gave the regime a strange melange of totalitarianism and much publicised tolerance. Culture was unpredictable. The regime's policy was baffling due to that fact. It would seem, however, that the regime's policy towards culture was mostly authoritarian, which, in many ways, did not upset many ordinary Greeks, especially those who were avid *Panathiakos* or *Olympiacos* supporters or connoisseurs of the more inspid bouzouki music or sentimental hits of the 1930s, if the trend in these or the popularity of the *Aliki* films on TV or radio during the 1967-74 era is anything to judge by. The regime did attempt to suppress more meaningful and deeper forms of culture, intellectuals, and artists who set new standards. In Franco's Spain at the height of the Swing Sixties, several films were made starring a rather pretty young singer called Marisol. The films often featured several new pop songs. None of them were very deep. Most of them wereee catchy or romantic. They were incredibly popular and Marisol was never out of the hit parade. Among her greatest hits were *Mi Rancho* and *Aquel*

Eurovision: A Plea for Respect

Verano. She was in many ways a Spanish equivalent of The Monkees or Aliki Vougouklaki. At the same time, any Gypsy song that went into too much depth on the lot of the Gypsies under Franco would have been frowned upon. There is, indeed, a similarity between Spain and Greece during this period. There are obvious parallels, too, with the clerical/military/semi-fascism of post-Civil War Spain and the Colonels' Greece. Was Marisol a product of authoritarianism? Marisol would probably have gone down very well with Premier Papadopoulos and Colonel Ladas, provided, of course, she had been Greek. Franco's promotion of Marisol for the purposes of his regime was, however, to backfire. Marisol, as she matured, developed an interest in leftist politics. At her wedding, held in communist Cuba, Fidel Castro was the best man.

In 1968, Massiel of Spain won the Eurovision Song Contest in London with the song *La, La, La*. Spain's Salome came in first again in 1969 with *Vivo Contando*. On both occasions, these achievements led to government-sponsored national celebrations in the form of street parties, etc. This was similar to the European Cup football tournament of 1971 in which Greece reached the final; in Athens, the government organised particularly lavish celebrations while the same regime did its best to humiliate Greece's only Nobel Prize winner, the poet George Seferis. Initially, the Colonels showed no interest in sending a Greek entry to the Eurovision Song Contest despite the victory and PR triumph of Franco's Spain. Greek TV did, however, broadcast the 1970 contest. Television did not become widely available in Greece until the 1970s. It is somewhat strange that the Colonels did not see the contest as an event that Greece should take part in. After all, it had been an ancient Greek song festival between the city-states that had given birth to the idea of a Eurovision Song Contest. The Colonels tried very hard to encourage any cultural form that was Greek in essence, yet their censorship led to a loss of many new and interesting developments. The ancient tragedies were very much approved of and even encouraged. They were looked upon as familiar true Greek ground that did not pose any threat. This image of *Greekness* had little to do with the modern post-independent Greece. Here, the Colonels were seduced into the same trap as many before them. Their romantic Greece of Socrates, Aristotle, Athina, Aphrodite, and the tragedies bore no resemblance to the realities of 1960s Greece. It was an artificial representation of what had gone before. The regime's lack of interest in participating in the song contest may have been due to its dislike of Western values and the deep-rooted despisement of all the West's decadence in modern Greek thought. After all, the song contest, in the 1960s and 1970s, had been a very *West* European affair. The regime, however, did not take kindly to anything that could be interpreted as anti-American – even if it came from the States itself. The film *Soldier Blue* was altered so that maltreatment of Native Americans by white Americans was cut out. It may be too much imagination to say that the song contest, which had been set up to stop the Americanisation of Europe, was seen by the Colonels as remotely anti-American. Perhaps it thought that Eurovision would take the limelight from Greece's own Thessaloniki Song Festival. The regime tried to root out anything that was remotely communist. This often meant anything Slavic. *Teach Yourself Russian* and *Greek-Bulgarian* dictionaries were banned. A poet reading one of his works had to remove the word *red* – which in the poem was the colour of his lover's lips. Songs written by communists were prohibited. All music composed by Mikis Theodorakis was prohibited from public performance. This included even his lyrical love songs, one of which, recorded in 1973 by Vicky Leandros, became a huge hit all over the continent. The sale of his records was forbidden and existing stocks of his records were destroyed. The official programme of the Athens Festival was altered at the last moment because it included incidental music composed by Theodorakis. Theodorakis himself was initially imprisoned but released in 1970. Some singers, such as the folk singer Mariza Koch, went into exile, as did many of the country's best actors and directors, among them Cocoyannis and Apassia Papathanassiou. Melina Mercouri had her Greek citizenship taken away from her and recorded a song in which she protested: *Je Suis Greque*. The country's last tragic actress, Anna

Eurovision: A Plea for Respect

Synodinou, refused to give any public performances. George Dalaras, the singer-songwriter from Piraeus whose ballads evoked the plight of the working class, touched the chords of the young and old in docks and village fields alike. Dalaras considered himself the child of the elderly and a brother to Greek youth. Some of his best material was written by Theodorakis. During the dictatorship, Dalaras made a conscious decision to abandon the clubs and play large concert halls and stadiums. Dalaras is often credited with the revival of *Rembetika*, which was considered so revolutionary by the authorities that it was censored between the 1930s and 1950s. According to the actress and politician Melina Mercouri, he was the man who kept alive the spirit of democracy under the dictators. During the dictatorship, many of the country's performers opted for the cabaret circuit. Good money could be earned playing for foreign tourists. Greek artists were doing well all over Europe in the 1960s and 1970s. Jimmy Makoulis, born in Greece, became a big star on the German music scene and represented Austria at the Eurovision Song Contest of 1961. Nana Mouskouri gave an exquisite performance in French at the 1963 contest when she sang *A Force de Prier* for Luxembourg. She did not win and placed only eighth but Nana went on to become one of the world's top-selling artists. In later years, she stood as a New Democracy candidate in the European elections and became a Greek MEP in Strasbourg. One of her tasks was to work for the protection of minority languages. Demis Roussos, a vocalist with the psychedelic Aphrodite's Child, which had hits in the 1960s, became a huge star first in Germany and then throughout the whole of Europe. He topped the charts with *Forever And Ever* and *Goodbye My Love, Goodbye* in several countries. Georges Moustakis, from Egypt but with an Italian-Greek Jewish background, whose parents' roots were in Corfu, became a popular French language artist in the 1970s. A former lover of Edith Piaf, he was especially successful with his song *Le Meteque*, in which he also made a bold social statement.

As a response to growing dissatisfaction throughout the country with the junta, as well as to an abortive *coup* by the Navy, the Colonels embarked on a slow process of liberalisation. The monarchy was abolished in 1973 after a referendum, which was not recognised by the leading political groups. Papadopoulos was declared President. As a result of this radical process of democratisation, students' demonstrations became easier to organise. There had long been a tradition of student protests in Greece but the student sit-in at the Athens Polytechnic in the autumn of 1973 was to be a key event in modern Greek history. There had been several protests at the Athens law school prior to the strike and sit-in at the Polytechnic. The students, influenced by academic opposition to Papadopoulos' attempt to concentrate as much power in his hands as possible on the road to a presidential republic, commenced on November 14. They locked themselves behind the gates of the grand, classical building in central Athens. Parents handed food parcels to their sons and daughters through the bars of the iron gates. Day by day, the protest grew bigger until it got so out of hand that the Papadopoulos government began to panic and responded by sending in the Army to quell the protest. In the wee hours of November 17, 1973, an Army tank drove through the gates of the Polytechnic. This led to the deaths of several of the young protestors. The four-day sit-in was brought to a violent and bloody end. The Polytechnic became a traditional centre of protest around the anniversary of the event and has since then often witnessed violent confrontations between the police and students; the building itself is scarred by its post as a witness to the Greek anti-establishment. Its façade is like the scarred face of an old woman who has lost her children. The heavy-handed fiasco of November 17 led to the end of liberalisation. There was an outcry by many Western European governments at the tragic loss of life and the Colonels needed to win back their approval, or at least some of it. It was thought that a clever idea to appease the criticism and show Greece as a true member of the European family was to enter a Greek song at the Eurovision Song Contest in the spring of the following year.

  Five songs were chosen for the national selection, which was held behind closed doors. Among the top pop/rock groups during the

Eurovision: A Plea for Respect

Colonels' years were Socrates and Nostradamus. Most Western records were available in Greek record shops and there was no restriction on rock concerts by Greek rock bands. The aforementioned. Nostradamus was chosen to sing for Greece in Brighton with the two female vocalists as lead singers. Their song was to be sung in both Greek and English. The junta, however, had other ideas and the members of Nostradamus had their passports confiscated. One of them, Hippocrates Exharhopoulos, was arrested on a trumped-up charge of raping an underage fan – a charge which had been fabricated by the junta. Following this, Marinella was nominated to sing the Greek entry and the experienced songwriters George Katsaros and Pythagoras were invited to compose a song that had to be representative of Greek music as well as exhibit the customs, social aspects, and scenery of the Hellenes and Greece. The song that was the end product of all this was *Krassi, Thalassa Ke T'agori Mou* (*Wine, Sea and My Boyfriend*). This was a long time before the film *Shirley Valentine* was ever thought of. Marinella took to the song, recorded it, and made a short film to be shown all over Europe as a preview. It was shot at Castella in Piraeus, the port of Athens, whose inhabitants object strongly to being called Athenians and whose ferry boats embark to such destinations as the Cycladic islands and Crete. It was often pointed out to me, when passing Castella, that that was where Marinella was filmed for her preview. Marinella herself is considered one of the Great Ladies of Greek Music. Her singing career started in 1957 with her first hit, *Little Helen*. Born in 1938 in the city of Thessaloniki in Greek Macedonia, Kyriaki Papadopoulou – as Marinella – went on to become one of Greek music's most successful exponents. During her long career, she performed a broad repertoire of *Laika* (heavily influenced by *Rebetika*), blues, and pop. The early years of her career were very much caught up with the popular singer Stelios Kazandzidis, with whom she often worked. Marinella recorded her Eurovision entry in early 1974 at a studio in Athens. It was there that she first came into contact with the famous actor and singer Tolis Voskopoulos. Marinella left Athens airport for Brighton accompanied to the departure gate by Tolis. They were later to have a passionate affair which ended in marriage. This led to another musical collaboration and they made several popular recordings as a couple. Marinella's popularity went on to dizzying heights during the 1970s and 1980s. Her albums sold like hotcakes and her shows were box office sellouts. Her shows were to take the Greek music business to new shores with magnificent lighting and presentation techniques which would influence other Greek superstars such as Anna Vissi. Marinella walked onto the stage of the Brighton Dome wearing a striped shirt and trousers. She carried a tambourine in her hand and was accompanied by backing singers and Stelios Zafirou on bouzouki. It was the first time a Eurovision stage had echoed with such a Greek sound. It was the first Greek entry and a very Greek song. Marinella was warmly applauded by the audience but managed to come in only eleventh. The Dutch jury awarded Greece four points, Sweden gave two points, and Portugal one point. There had been a bomb scare shortly before the contest started, which had given the contest a somewhat tense air, even though a search proved that all was well. Marinella's many recordings often turned gold and in 2002 she appeared in concerts with George Dallaras in Greece and abroad. Also, she sang at the closing ceremony of the 2004 Olympic Games in Athens. Marinella's enduring popularity and respect were still very apparent in her seventieth year, with her having performed at the Athenion Arena with Andonis Remos in a highly successful 2006-7 season while two new compilations, *Sti Skim* and *Ta Logia Ine Perita*, were released in 2006. Marinella's tenacity and professionalism have helped her popularity endure for far more years than the authoritarianism of the Colonels or even some of the politicians who succeeded them.

Shortly after Marinella's appearance and the Greek debut at the Eurovision Song Contest, the military was toppled. Taxiarkhos Ionnides, a hard-line member of the junta, overthrew Papadopoulos shortly after the Polytechnic events, in a counter-*coup*. The sheer disfunctionalism and shattered ideology of the Colonels exposed its lack of credibility and re-established itself as a movement with the

Eurovision: A Plea for Respect

interest of the Greek people at heart, through iron-fisted crackdowns and aggressive foreign policies. A junta-backed *coup*, led by the EOKA-B, seized power on the island of Cyprus in the summer of 1974. The President of Cyprus, Archbishop Makarios, was overthrown and went into exile. This event triggered a Turkish invasion of Cyprus which led to much bloodshed and an all-out war between the Turkish Army and the Cypriot forces helped by the Greek Force for Cyprus (ELDYK). The Greek and Turkish nations were an inch away from full war. This situation saw the chief military officials distance themselves from Ionnides. This led to the appointment of a government of national unity, and former prime minister Constantine Karamanlis returned to Athens after years of exile in Paris to continue his premiership. A general election was held in November 1974 and was won by New Democracy, whose leader was Karamanlis. Parliamentary democracy was again restored to the Hellenes.

Cyprus, cradle of the Greek language and Greek religion, had always been of great importance in the changing fortunes of the Aegean Sea. The mega-tourist magnet and rave centre of the twenty-first century, famous for its attractive beaches and bright blue sea, home of Aphrodite, goddess of love, had seen many races and cultures come to its shores long before the present century's youth were drawn towards its sea, sand, and night clubs. Its multiple conquerors have included Romans, Ottomans, Venetians, and British. Such diverse influences can be seen in the everyday customs and culture of the island. Turkey, under its mantle of the Ottoman Empire, put down roots well into Central Europe and Africa. It grabbed the great centre of Greek Christianity and Byzantine civilisation, Constantinople in 1453. An event vivid in the mind of contemporary Greece, its domination of the Balkans would have a profound and lasting effect on the music of that region. Turkish rule first came to the island in 1571. A process of Islamisitation did not come into effect until much later and the island's Greek inhabitants were allowed to practice Christianity without any restrictions. The British first appeared in Cyprus in 1878 when the island was initially leased to the U.K. government. It became a British

Crown Colony during World War I when the Ottoman Empire sided with the kaiser and entered the war as an ally of Germany. After the war, the Turks were given the choice of leaving the island and returning to Turkey or remaining in Cyprus and being British nationals. For years, Greek and Turkish Cypriots lived side by side. Both communities, however, looked upon themselves as different from their mainland cousins. Cypriots considered themselves more liberal, more cultured, more highly educated, and more orderly than their compatriots in Greece and Turkey. One of the factors that contributed to a very separate national identity between the two races was education. This was exploited by the British, whose colonial policies were based on a divide-and-rule ideal which prevented a united Cypriot opposition to colonial rule and further promoted ethnic polarization. There was a fear within NATO that if the British left Cyprus, it would become part of the Eastern bloc with the establishment of a communist regime. Greek resistance exploded on the island in 1950. An attempt by the British authorities to stamp it out was the creation of an all-Turkish police force which the colonial authorities hoped would lead to more ethnic division and polarization. Cyprus gained independence from Britain in 1960 but ethnic violence continued after this. Turks were massacred by Greeks in 1963, which led to the expulsion of thousands of Greeks based in Istanbul, who themselves had suffered considerably due to Cypriot policies in the 1950s. Thousands had already fled in 1955. President Makarios became aware of an EOKB terrorist plot against his government which was backed by the Greek junta. When he expressed his displeasure, the junta ordered the overthrow of the Makarios government and so the Cypriot National Guard staged a *coup*. Makarios escaped and five days later Turkish government troops landed on the shores of Cyprus. Thus, the Turkish invasion of Cyprus began. These events were closely followed by a young Greek folk singer, Mariza Koch. Mariza had been born to a German father and a Greek mother from the island of Santorini in the war-torn Athens of 1944 and was raised on her mother's scenic Cycladic island. A year after the harrowing events in Cyprus, she was

Eurovision: A Plea for Respect

to receive an unexpected phone call from the legendary Greek composer Manos Hadjidakis. He asked her to turn up at a specified radio station and instructed her to bring a mournful song with her that morning. Hadjidakis, a director at the radio station, wanted Mariza to be the composer and singer of the next Greek entry in Eurovision, to be held in the spring of 1976 in The Hague. Hadjidakis felt that the Greek entry should reflect the nation's feelings about the events that had taken place in Cyprus. Mariza Koch and Nikos Xylouris had performed for the many young Greeks who had left their homes to fight in Cyprus. They sang to give them hope and encouragement as they boarded ships bound for Cyprus at the port of Piraeus. Hadjidakis had, therefore, thought it fitting that Mariza should represent Greece. A tower of dialogues over the phone with lyricist Michalis Fotiadis commenced on the very evening that Hadjidakis made that initial call. They came up with the idea of taking the verse from a traditional funeral lamentation of Epirus, the North West area of Greece, to use as the chorus of their new composition, *Panaghia Mou, Panaghia Mou (Madonna Mine, Bring Consolation to My Heart)*. It is often heard at funerals in that region. An apprehensive Mariza set off to meet Hadjidakis the following morning. After she had performed the song in front of him for the first time, he was moved by emotion and demonstrated his approval by embracing Mariza and telling her how much she had fulfilled his wish. The song was performed on the Eurovision stage amid a sea of controversy and fears for the Greek singer's security. Accompanied by Demitris Zoumboulis on the lute and the Metropole Orchestra, she wore a long, plain black dress and a large piece of Byzantine jewellery. Hadjidakis had originally requested that she be accompanied only by a lute but this went against song contest rules. He had also requested that she try not to move about too much during the presentation of the song, which she adhered to. The days leading up to the contest were tense. During rehearsals, thousands of Turks demonstrated outside. They demanded that the song be withdrawn from the contest and claimed it was an insult to the Turkish nation. On the day of the actual contest, Turkish shops and businesses in The Hague closed in an act of solidarity and opposition to the song. Mariza had feared that she would be asked to withdraw from the contest. The Dutch authorities told her that they could not guarantee her safety if she continued to defy the protests and appear on stage at the contest. Although security was high and the venue had been thoroughly searched for bombs, there was the added problem that most of the cleaners at the venue were Turkish and not at all sympathetic to Mariza's mission. Fears began to pile up concerning her safety and climaxed only a few minutes before the start of the 1976 song contest when security arrested an attempted sniper intent on taking Mariza Koch's life. Mariza gave what was to be undoubtedly the best performance of her life. Full of emotion and belief in her great Hellenic mission, her voice, from the heart, echoed around the venue. Shortly after the contest had ended, Mariza was rushed to the airport, where an Athens-bound plane awaited her. It is perhaps at this point, more than any other in the book, that I want to highlight the mocking, derisive, devoid-of-any-worth attitude that the contest acquired in Britain during its Wogan years, when it was almost turned into a standing joke; yet it is precisely moments such as Mariza's interpretation of this controversial song that completely contradict the anti-Eurovision stance in the British media, which at its kindest displays the *"We love it because it's so camp and awful"* attitude. Moments such as those in the contest's history are exactly why it deserves respect and recognition. The brave performance of Mariza Koch, who risked her life to do something she passionately believed in, was equally matched by the first Bosnia-Herzegovina representatives to perform on a Eurovision stage after the breakup of the old Yugoslavia and the turmoil that ensued. Fazla, the nation's representative, avoided gunshots as they crossed the tarmac to board the plane to Ireland in 1993. During the voting, the crackly sound of the Sarajevo juror broadcast the votes from a bunker and was given tumultuous applause from the Irish audience. Mariza's attempt at Eurovision is also of note because of the song's acclaim by Greece's musical aristocracy. At a reception in Athens the day following the contest, Manos Hatjidakis,

Eurovision: A Plea for Respect

Nikos Gatsos, Mikis Theodorakis, Nikos Koundouros, and Nana Mouskouri, all well-respected legends of twentieth-century Greek song, came to congratulate her. *Panaghia Mou, Panaghia Mou* was, by a large section of the British media, considered rubbish. Its lyrics had gone over the heads of a public who rarely knew another European language and who, most likely, knew less than a handful of European singers. While the whole of Europe watched Mariza singing about her country's grievances over Cyprus, Turkey, which did not compete in 1976 but broadcast the contest live, blacked out the Greek entry and showed a clip of ethnic Turkish music instead. At the end of the evening, when all the votes had been cast, *Panaghia Mou, Panaghia Mou* was placed sixteenth out of eighteen entries but Mariza had succeeded in her mission. The Cyprus problem has remained unresolved right into the twenty-first century. Cyprus is part of the European Union and has itself entered a song at Eurovision since 1981, yet the island remains divided.

In the northern Turkish area of the island there now exists the Republic of Northern Cyprus, recognised only by Turkey itself and not by the United Nations. Around the time of Turkey's long-overdue win at the song contest in 2003, a new atmosphere of reproachment between the Greeks and Turks had taken root and was flourishing. Many hoped it would herald a new era. For the first time ever in the song contest, Turkey voted for the Cypriot entry, albeit only one point. Cyprus gave twelve points to Greece but also voted for Turkey, an unprecedented act. The Greek Cypriot TV presenter who read out the votes gave a peace sign before doing so. He then said, "Europe, peace to Cyprus, Turkey eight points". This was a major act of reconciliation. Not only did Cyprus give Turkey a substantial number of votes but their vote was crucial to Turkey's win in that Turkey won the Grand Prix by only two votes ahead of the Belgian song by Urbin Trad. The Turkish breakaway republic of North Cyprus is not recognised and so the Cypriot votes came only from the Greek area of the divided island. There were those in Cyprus who vehemently disagreed with the high vote given to Turkey, looked upon as an occupying force. Some even accused the state-run broadcaster of rigging the vote. In a poll carried out by the newspaper *Semerini Filelethferos*, a percentage thought that it was a mistake for Cyprus to have given Turkey any votes at all. On May 27, the Turkish newspaper, *Hurriyet*, featured this as the front page headline: "*GREEKS REGRET 8 POINTS!*" From whatever side of the divide or viewpoint, there was, however, no getting away from the fact that Turkey's Sertab had won the Eurovision Song Contest on Greek Cypriot votes. It is of note that a month after the song contest, the Turkish Cypriot authorities eased travel restrictions between the north and south of the island. The international media celebrated this by reporting on the reunion of old friends, Greeks and Turks who had once lived side by side. Was it only coincidence, then, that the Turkish song at the following year's contest in Istanbul was sung by a group called Athena?

Mariza Koch continued to take an active interest in the political developments of her country in addition to being one of Greece's foremost folk singers. Among her diverse undertakings was an appearance at the Christmas Bazaar at the German church in Athens. Her wedding dress is on display at the folkloric museum in Santorini. The grave economic crisis that descended on Greece saw the *tragoudistria* play a leading part in the anti-austerity protests which took place in Syntagma Square in Athens.

Mariza still includes her Eurovision song in her repertoire at concerts and has added a verse to the song each year since 1976, something she will continue to do until the Cyprus issue is resolved.

CHAPTER 9
Ein Lied Fur Madrid

The dour, dark, ravaged streets of early-1950s Berlin give way to the subdued grandeur of the Brandenburger Tor. "*It's best we don't speak English here,*" says Hildegard Neff. The film is *The Man Between* and it also stars James Mason and Claire Bloom. It was a classic of the Cold War films and one that illustrated the divided city most accurately at that difficult but mysterious time in the history of Germany's once imperial capital. Like all films based on the Cold War, it dealt with the tense relationship between the West and Stalinist East Europe. It is a film in which there is a constant struggle between good and evil and, as always, in the end, good triumphs in the form of the capitalist West.

* * *

As West Germany rose from the ashes of the Third Reich, it emerged as a phoenix to be a shining example of the prosperity of a democratic free market Western nation. East Germany – the DDR – was the bastion of communism right on the West's door step, and firmly under the grip of the Soviet Union. It grimly emerged from the war in a drab, spartan concrete showcase of Stalinist socialism separated from its Western brethren by an infamous wall that kept the two opposing ideologies firmly apart. Berlin and the DDR, manifested, catacombed in mystique in Western cultural expressions.

* * *

As Germany's Bundes Republik competed year after year in the West's Eurovision Song Contest, the DDR competed at Sopot's Intervision Song Contest. Germany was one of the original seven countries that participated in the first contest in Lugano, and the two German entries were sung by Freddy Quinn and Walter Andreas

Schwarz, respectively. The following year, 1957, the second song contest was held on March 3 in the city of Frankfurt am Main. This was before the tradition of the winning country hosting the next event was established. Rumours abounded that Germany was invited to host the event due to its *supposed* runner-up status the previous year but they were unfounded. Germany competed in, or at least attempted to enter, every song contest since 1956. In 1996, its entry *Planet* of *Blue* by Leon was rejected for the final at the untelevised audio pre-qualifying heat. This was Germany's only absence from a Eurovision Song Contest final which it has won twice: in 1982 and then again eighteen years later in 2010.

* * *

The Soviet dictator, Stalin, died of a cerebral haemorrhage in 1953, shortly before the coronation of the young Queen Elizabeth of the United Kingdom. He left a Europe lodged in thick ice and an atmosphere of unease and potential conflict. The Soviet military remained in the DDR following Stalin's death. There had been a popular uprising against the unpopular regime, but this had been brutally suppressed by the East German police and their allies the Soviet army. From 1948, when the Soviet Union handed over authority to the German Socialist Party until 1989, East Germany was a hardline communist state and one of the most oppressive regimes in the Eastern block.

Culture in East Germany during this period was heavily influenced by the communist regime and its ideology. As the Cold War continued, East Germany felt increasingly insular and was starting to feel more and more threatened. Thus, the government built the Berlin Wall in 1961. This curbed the emigration of often young, highly educated individuals seeking sanctuary in the West.

Berlin was divided between the Soviet sector and the French, British, and American sectors. This meant that West Berlin was surrounded by the Soviet-occupied territory of the DDR, an island of Western capitalism amid a communist sea. The Brandenburger Gate,

Eurovision: A Plea for Respect

the border crossing from West to East, was so near yet also another world away.

****

Spain had won the 1968 Eurovision Song Contest so, by tradition, the 1969 song contest was to be held in Madrid. Germany announced that their national final would present new songs sung by Peggy March, Rex Gildo, Siw Malmkvist, and Alexandra.

Peggy March was a young American who, a few years earlier, in 1963, had enjoyed a big hit in America with *I Will Follow Him*, which sold a million. The song was an English translation of Petula Clark's French language *Chariot*, which had been a hit in the UK the year before. The song was featured many years later in the film *Sister Act*. Peggy March's increasing popularity in Asia and Continental Europe resulted in her leaving the States and setting up home in West Germany, where she recorded mostly in German. Rex Gildo, on the other hand, was a Bavarian *Schlager* singer who proved to be one of the main exponents of the traditional German pop genre in the 1960s and 1970s. He famously duetted on several hit records with the Danish singer Gitte Haenning, another chief *Schlager* exponent of that era.

The *Bundes Republik* – West Germany – enjoyed the prosperity and excesses that came with 1960s life; 1969 was led in the shadow of the legendary student protest movements that had swept over Europe the previous year. West Germany enjoyed a thriving local music scene, from progressive rock to *Schlager*. There was, however, a sinister undercurrent in West Germany that lurked below the visible surface of everyday life. Across the *grenze* in the DDR, life in 1969 was far removed from that of the West. The everyday lives of ordinary people were carefully scrutinised by the state; there was no room to swing in the 1960s, lest you run foul of the authorities. Although East Germany had its own homegrown pop scene, the freedom of thought and expression that went hand in hand with 1960s pop music in the West was absent in the East. East Germany kept its population in line through the operations of the Ministry of State Security, otherwise known by the STASI, a cancer that took root in every aspect of East German society. It had come into being as, more or less, a branch of the Soviet KGB and, although it had been an autonomous body since 1957, the KGB continued to exercise a high degree of authority. The Stasi employed a massive number of staff members as informers, who spied on their fellow citizens. Most were ordinary people whose jobs brought them into contact with the public. They ranged from caretakers to doctors. It is estimated that as many as two million were Stasi spies.

In an attempt to find enemies of the state, spies were placed in apartment blocks, where they informed on every guest who stayed overnight at a tenant's home. They made holes in the walls of flats through which they installed video cameras. One unnerving aspect of the Stasi was that they entered the homes of suspected dissenters and moved around their furniture, tore up letters, tampered with alarm clocks, and emptied or replaced food supplies, such as replacing coffee with sugar. Most of the victims were completely unaware of what was happening and thought they were losing their minds. This led to many cases of breakdowns or even suicide. It was also common for Stasi spies to inform on their spouses. Thus was the reality of life beneath the bright red banners glorifying communism that hung from the drab buildings of East Berlin back in 1969.

While West Germany enjoyed the temptations of consumerism, it was not completely free of the Stasi. There were also Stasi informers in the *Bundes Republik* and it infiltrated West German society at its highest levels. It was also a practice of the Stasi to prey on and marry lonely women to further their purposes or to inform on their activities. It is estimated that a large number of Stasi intentionally targeted West German women living in towns near the border for this practice.

Siw Malmkvist, a native of Landskrona in Sweden, was a popular *Schlager* singer in West Germany in the 1960s. Like other Scandinavian singers, such as Gitte and Wenche Myre, she found the transition to German, from singing in her native tongue, to be an easy task. During her long career, Siw had forty hits in Sweden and twenty in Germany

Eurovision: A Plea for Respect

and is credited with being one of the most successful Swedish artists of the twentieth century. She is also notable, as a Swedish artist, as she was the first to ever have a hit on the American Billboard Hot 100. The Italian song *Sole, Sole, Sole*, on which she duetted with Umberto Marcato, entered the US top twenty in the summer of 1964. The same year, Siw had also hit the number-one slot in the German top ten with *Liebeskummus Lohnt Sich Nicht*. Born on the last day of 1936, Siw Malmkvist embarked on her musical career in Sweden at the age of twenty-one in 1957. Since her early days, Siw had appeared regularly at the Swedish Melodifestivalen and harboured an earnest desire to win the Eurovision Song Contest. In 1960, she won the Melodifestivalen and was sent to London, where she represented her country at the Eurovision Song Contest with *Alla Andra Får Varann*. She placed tenth. She returned to the Swedish national finals again the following year with the song *April, April*. Siw gave a memorable performance, if for all the wrong reasons, as during her rendition, she forgot the words! Unperturbed by this, the juries gave *April, April* a high enough score to qualify it as Sweden's Eurovision entry for 1961. Shortly after the festival, Swedish radio made a statement saying that although *April, April* would go through to the Eurovision Song Contest, Siw Malmkvist would not. Lill Babs would replace her and sing the Swedish entry.

* * *

The Stasi enjoyed a much-privileged position in the Soviet Union, so much so that it was invited by the KGB to set up a branch in Moscow and Leningrad: two cities that were popular with East German tourists. The Stasi could then keep them under surveillance during their sojourn in the Soviet Union. While East Germans were frequent tourists in the Soviet Union, there had been a huge German minority living in various parts of the Soviet Union for hundreds of years. An estimated two and a half million Germans were living in the Russian Empire before the Bolshevik revolution. Germans often formed the upper classes. They were to be found in the royal family, gentry, aristocracy, and leading positions in the army. During World War II, Stalin deported most ethnic Germans living in Soviet Europe to Siberia or Central Asia. In 1948, he declared this temporary banishment irrevocable and forbade them to return to the European Soviet Republic. After the dissolution of the old Soviet Union, many ethnic Germans took advantage of both their newly acquired freedom of movement and liberal immigration laws in Germany and returned to their ancestral land. The German minorities in the Central Asian republics of Kazakhstan and Kyrgyzstan have now all but disappeared. Germans had been a part of life on the Baltic coast since the Middle Ages. Parts of Baltic Russia and the Baltic states were once part of the Prussian Empire, with its original capital at Kalingrad, before it moved to the more central Berlin. Parts of what is now Lithuania were territory of the German Empire and also the Weimar Republic. It enjoyed a brief period of independence in the 1920s and 1930s before being invaded by the Nazis and then being incorporated into the Soviet Union. In the wake of the Soviet takeover and the Nazi retreat, many ethnic Germans fled the area. Many of those who remained were later repatriated. Among the hoards of Germans fleeing at the time with her family to Germany was a child, a certain Doris Treitz, who had been born during the war in Memelland, now in Lithuania. The family settled in the city of Kiel in West Germany. While at school, Doris excelled at languages and exhibited an artistic flair that led to her becoming a fashion designer. Like many newly arrived immigrant families from Eastern Europe, Doris and her family lived in humble surroundings and were employed in various jobs to make ends meet. To make some much-needed money, Doris's family took in a lodger, a Russian immigrant named Nikolai Nefedov, whose intention was to settle in the USA. At nineteen, Doris competed in the Miss Germany beauty contest. This brought her into the limelight and made her a local celebrity. It proved to be a highly enjoyable experience for her. Although she did not win the event and never competed at any more beauty pageants, it was a significant turning point in her life. Meanwhile, the young Doris had fallen in love with her family's

Eurovision: A Plea for Respect

Russian lodger. Although he was twenty years older than her, they married and had a son, whom they named Alexander. The age gap might have been a hurdle in their relationship once the initial feelings of love had gone, for the marriage was short-lived; Nefedov embarked on his ambition to go to America but without Doris or their son.

Doris, who had always enjoyed singing and had a great love of music, began to cultivate this side of her nature and attempted to become a professional singer. She changed her name from Doris Nefedov to Alexandra, which she took from her young son. Early in her career, she appeared as a supporting act for the Belgian singer Adamo. This was an early test of Alexandra's talent. Other female vocalists who had appeared before the main act were badly received by a somewhat hostile crowd. Alexandra wooed them with her unique voice and style of singing. It was a success that would pave the way to a higher degree of acclaim.

Alexandra sang what was known as *Liedermacher*, which differed from the mainstream German *Schlager*. *Liedermacher* is a mixture of French *Chanson* and Central European folk. Alexandra's songs often had a deep, melancholic, even dark side to them. Her record company, however, wanted to promote her as a *Schlager* singer and she was marketed as such. Under the direction of Hans R Beirlein, she toured with the Hazy Oserwald orchestra. The tour included major German cities and also venues in Russia. The tour was a resounding success; Alexandra's singing had not only impressed the Germans but had also captivated the Russian audiences. Berlein had hopes of Alexandra cracking the Russian market. In West Germany in the 1960s, a mystique concerning the countries that lay east of the Odor grew; pop songs with a hint of the balalaika and the flow of the Volga in them often sold well. Alexandra's warm voice lent itself well to her self-penned *Zigeunerjunge*, about an unrequited love for a Gypsy boy, and *Schwarze Balalaika*, her first hits, both of which had a very East European feel about them. The latter, a very Russian-orientated composition, was evocative of the time. German record companies were keen to record anything that was vaguely mysterious with a

Russian rhythm to it, a mystery that only deepened as the Cold War continued. Hungarian-born singer Maria Marky was launched onto the German pop scene in 1966 but was later renamed Edina Pop. German recording bosses saw her as the perfect artist to promote the East European mode that was so in at the time. Records like *Petruschka* and *Zwischen Wolga und Don* reeked of the Samovar. The latter was a German version of Spain's winning 1969 Eurovision winner, *Vivo Cantando*, Russianised, with a good few notes of the balalaika thrown in and the title changed to *Between the Volga and the Don*. None of these recordings did particularly well. An attempt to raise her status in the popularity stakes climaxed when, in 1970, she was asked to take part in the German national selection of the Eurovision Song Contest. In the end, due to illness, she could not take part; her song was sung by a replacement, Mary Roos, who came in second to Katja Epstein. Edina Pop's biggest break came in 1979 when she was asked to be part of the group Djengis Khan, put together for the German Eurovision final to sing the Ralph Siegel song of the same name. Success finally came her way and, with the group, she had several more hits. It is a German Eurovision song that is usually paraded around by the BBC at that particular time of year, a rather crass number despised by those aware of what good songs the Eurovision Song Contest is capable of producing. In the early days of her career, Edina Pop was, however, seen as very much another potential Alexandra who fitted into the East European niche of German pop. Alexandra's manager, Hans R Beirlein, was also the manager of the Austrian superstar and 1966 Eurovision winner Udo Jűrgens. Jűrgens was born on his family estate, Ottmanach Palace in Carinthia, when Austria was a mere federal state in Hitler's Third Reich. His father's family had been Prussians from Russia. His grandfather was the head of the German Junker-Bank in Moscow and had fled to Austria with his family after the 1917 Revolution. As a small boy, he was eager to join the Nazi Jung Volk, which was a junior version of the Hitler Youth. His membership was, however, brief. After a Youth leader, furious that Jűrgens' uniform had not been properly put on, battered him over the head, he quit. This

Eurovision: A Plea for Respect

incident would leave Jűrgens deaf in his left ear. He is credited with introducing piano artistry and clever, introspective lyrics into *Schlager* music. During his career, which lasted five decades, he sold 100 million records. He is credited with being the biggest-selling German language artist ever and holds the world record for being the artist with the longest presence on the charts, fifty-seven years from his first entry in 1957 until 2015. His talents and musical ventures covered a wide spectrum and he would also make forays into classical circles. He composed material for the Berlin Philarmonic Orchestra and also conducted the Vienna Philarmonic Orchestra at the Vienna Festival. Artistic talent ran in the family, as Jűrgens' uncle was the much-acclaimed Dadaist artist Jean Arp.

Hans R Beirlein, like Jűrgens, was a well-known personality of the 1960s and was a major force in guiding Alexandra as an artist; they became friends. Their professional and personal relationship became very bound together and soon they fell in love. She recorded a song entitled *Sehnsucht*; this was the first single released that Alexandra had not written herself. She was said to have disliked it and refused to perform it but this was, however, to be her greatest hit. She also recorded an interesting and memorable version of Serge Gainsbourg's *Accordeon*. Her most dramatic recording must surely be *Illusionen*, which she co-wrote with Austria's Eurovision winner, Udo Jűrgens. It is a tempestuous Central European ballad to end all ballads and one of her most accomplished works. English versions were recorded by Shirley Bassey, Sammy Davis Jr, and Matt Monroe, for whom it became an anthem of his later years. The cabaret singer Dorothy Squires, in her *"scarred by life and experience"* vocals, made one of the best interpretations of the English version: *If I Never Sing Another Song*. The song told of a faded star looking back on a golden heyday.

* * *

The 1969 German prelims for the Eurovision Song Contest took place on February 22 and were presented by Marie Louise Steibauer. The national final was entitled *A Song For Madrid (Ein Lied* Fűr

Madrid) and featured nine new songs written by Germany's most popular composers. It was broadcast from the Hessischer Rundfunk studios in Frankfurt. All of the songs were *Schlager* in genre. Peggy March sang *Karussell Meiner Liebe*, *Aber Die Bleibt bestehn*, and *Hey*. The King of *Schlager*, Rex Gildo, presented three more songs: *Lady Julia*, *Die Beste Idée Meines Lebens*, and *Festival Der Jungen Liebe*. The Swedish Siw Malmkvist offered three more: *Melodie*, *Dein Comeback Zu Mir*, and the very Russian-influenced *Prima Ballerina*, another example of the popularity of this theme in German pop. This was exhibited in the number of votes it received. It was no surprise when it won and was chosen to represent the *Bundes Republik* in Spain's capital.

Where was Alexandra? Had she not been one of the vocalists supposed to compete? Fate had seen fit to change the course of the winds. Alexandra's much-anticipated participation did not materialise. Illness had prevented her from taking part. There were also rumours that the 1966 Eurovision winner Udo Jűrgens had advised against her competing. The Eurovision Song Contest then missed out on one of Germany's best artists of the decade.

Schlager, which had always dominated the German national final and had always met with a reasonable measure of success at the Eurovision Song Contest, was the major musical genre in Germany in the 1950s, 1960s, and 1970s. *Schlager* literally means *"hit"* or *"bang"* and is most often characterised by highly sentimental ballads, often with catchy, uncomplicated melodies and lyrics that dwell on aspects of love. *Schlager* was once a popular genre in both the UK and America. It was frequently heard on the airwaves in the 1950s and was still often found on the charts even in the 1960s but its popularity was surpassed by that of other musical genres that developed after the 1960s. It was also the premier genre of popular music in other German-speaking nations such as Austria, Liechtenstein, and Switzerland. Scandinavia was also an important centre of *Schlager*, in particular, Sweden. The Swedish national Eurovision selection, the Melodifestivalen, often featured *Schlager* songs and was popularly referred to as the *Schlager Festivalen*. The current Swedish national selection enjoys incredible

Eurovision: A Plea for Respect

popularity and is watched by nearly 50% of the country. In recent years, *Schlager* has, however, featured less and less in the festival. The Eurovision Song Contest itself has also seen a dramatic decline in the number of *Schlager* entries. In both cases, this is very much due to the influence of other types of contemporary genres such as R&B, soft rock, ethnic pop, etc. Swedish *Schlager* at the Melodifestivalen in the early years was often heavily influenced by operetta and jazz, as seen in the 1966 entry *Absent Friend* or in Monica Zetland's 1963 entry, *Gang I Stockholm*. That operetta influenced *Schlager* at Eurovision is also evident in the 1961 Austrian entry, *Nur Im der Wienner Luft*. Siw Malmkvist's recordings are very typical of German and Swedish *Schlager* of the 1960s. The face of Swedish *Schlager* changed gradually as the years passed and Sweden became more and more influenced by music from outside its borders.

In the 1970s, Swedish *Schlager* incorporated elements of pop, as did its German equivalent, which even incorporated elements of disco, e.g., Marianne Rosenberg's *Ich Bin Wie du*. Techno, R&B, and rock all informed later decades of Swedish *Schlager*. The Swedish heavy metal group Black Ingvars even made an album in 1998 entitled *Schlager Metal*, so present-day *Schlager* has a very different sound from its past generation.

In neighbouring Finland, *Schlager* always differed slightly from German *Schlager* in that it incorporated elements of local folk as well as traditional Russian folk. Finnish lyrics in *Schlager* songs often border on the melancholic or poetically express feelings of loss and solitude. Many Finnish *Schlager* singers have had consistent success with Finnish adaptations of foreign hits, and a great many Finnish singers have recorded local versions of popular Eurovision songs. Artists like Lea Laven, Seija Semola, and Marion Rung all met with unmitigated success with their homegrown versions of Eurovision songs. Finnish *Schlager* shares a similarity at some levels with the *Schlager* that grew up, quite distinctly, in Serbia, Croatia, and Bosnia in that the two often share elegiac qualities. The Yugoslav Eurovision entry of 1971, *Tvoj Dječak Je Tužan*, sung by Kruno Slabinac, was a classic example of the more mournful sentiments expressed in the *Schlager* of the southern Slavs. The title *The Boy Is Sad* gives some prediction of the melancholia to come.

*"I still think of the one who
doesn't sleep beside me anymore
The night is so long and so dark
just like the pain
The bed is empty and so the table is bare"*

Schlager had taken root in Yugoslavia in the post-war 1940s. Among the great *Schlager* singers whose star glittered in the decades that followed was Lola Novakovič, who sang for Yugoslavia at the 1962 Eurovision and whose recording of Darko Kraljič's *Čamac Natisi* was one of the eternal classics of Balkan *Schlager*. In the Netherlands, where the local *Schlager*, *Levenslied*, enjoyed a high level of popularity. The genre was often accompanied by traditional Dutch instruments such as the barrel organ. Typical *Levenslied* contained highly catchy, simple melodies. Good examples of this are *Het Is Voorbeij* by Corrie and the Rekels, a hit in 1976, and *Mario* from 1972 by Hanny and the same band. *Levenslied*, unlike Serbian or Finnish *Schlager*, was often devoid of melancholy or poetic expressions of solitude and was typicallly feel-good in nature. A topic particular to Dutch *Schlager* was the many songs whose lyrics told of holiday romances in exotic Mediterranean resorts. Imca Marina, in particular, capitalised on the popularity of this sub-genre. Her greatest hits included *Vino*, *Bella Italia*, and *Y Viva España*, which even received airplay on BBC Radio 2. Although true pure Dutch *Levenslied* rarely made it into the Eurovision Song Contest, as it was regarded as too Dutch and not international enough, after a run of disastrous attempts at the Eurovision Song Contest, which had seen a variety of genres from *Nederland*, the Dutch selected their most Dutch-sounding *Levenslied* ever for the Eurovision Song Contest. In 2010, the song *Ik ben Verliefd* by Sieneke was sent to Oslo as the Dutch entry, complete with accordion and barrel organ. An amazingly catchy

Eurovision: A Plea for Respect

number, which did not qualify for the final, it had a lot of people, who thought of themselves as being connoisseurs of more sophisticated pop music, tapping their feet and humming along to it.

In Central and Northern Europe, the term *Schlager* often came to apply to any commercial pop song as opposed to progressive rock or heavy metal. *Deutsche Schlager* hits of 1967 exhibited archetypal high *Schlager* qualities of uncomplicated lyrics and catchy, pleasant melodies. Take, for example, Peggy March's 1967 hit, *Memories of Heidelberg*.

"*Memories of Heidelberg sind memories of you
Und You dieser schönen zeit da träum ich immerzu
Memories of Heidelberg sind memories von glück
Doch die zeit von Heidelberg, Die Kommt Nie mehr zurück*"

Peggy, who had made West Germany her home, was a native of Philadelphia. Ireen Sheer, an English singer from Essex who had once been part of the band Family Dogg along with Albert Hammond (of *It Never Rains in Southern California* fame), moved to Germany in the 1960s and became a popular *Schlager* singer there, recording in German with hits like her 1973 *Goodbye Mama*, establishing her as a German market artist. She was chosen to sing for Luxembourg in 1974 at the Eurovision Song Contest, where she sang a ditty in French with an English title. She represented Germany again at the Eurovision Song Contest of 1978 with *Feuer*. She returned to the Eurovision Song Contest as part of a group of well-known vocalists singing for Luxembourg in 1985 and also reappeared at the German national selection in 2002. Severine, the 1971 Eurovision winner, following some lean years in France, established herself as a German language *Schlager* artist with considerable success. Nana Mouskouri sang a higher form of German *Schlager*, often to Greek music, such as *Weisse Rosen Aus Athene* and *Am Strand Von Corsika*, while Vicky Leandros also recorded *Schlager* with a Greek slant, as is evident in her 1970s hits *Die Bouzouki Klang* and *Kalinichta*.

In the case of Rex Gildo, the melodic, sweet words of his many *Schlager* hits disguised a complicated personal life of trauma, dilemma, and mental health problems. *Schlager* was deeply rooted in German musical culture. The Berlin *Schlager* of the 1920s Weimar Republic produced such popular songs as *Veronika, der Lenz ist da*. By the 1980s, *Schlager* had lost ground to British and American pop and rock. Its popularity was much diminished and seen as passé by a younger generation who looked upon it with disdain. New German singers courted other genres and sang in English.

* * *

Siw Malmkvist was the second Scandinavian to sing for Germany at the Eurovision Song Contest. One year earlier, the Norwegian entry, *Ein Hoch der liebe*, had been sung by Wenche Myhre and placed sixth. She had attempted to sing for Norway in 1961 and again in 1966 but had failed to win the national final although she had just missed out in 1961, coming in second. In 1983, she entered both the German and the Norwegian finals, in which she duetted with Jahn Teigen. She backed out of singing the Norwegian song after it was chosen for Munich, unable to cope with representing two countries. However, this never actually happened, as her German song was not selected as the German entry. Wenche Myhre made one more attempt at Eurovision and competed at the 2009 Norwegian national selection but once again lost out on that occasion to Alexasander Ryback. Siw's *Prima Ballerina* entered the German charts and peaked at thirteen, while the runner-up, *Hey*, by Peggy March, entered the top thirty.

At the 1967 Eurovision Song Contest, the singer-actress Inge Brück sang the pleasing *Anoushka* for Germany, a Hans Blum composition. Blum also composed *Prima Ballerina*. *Anoushka* came in eighth. Alexandra took a liking to the song and recorded it. Her recording reflects a much more *Liedermacher* quality, while the original by Inge Brück was a far more veritable 1960s pop interpretation. By

Eurovision: A Plea for Respect

1969, Alexandra's relationship with her manager was over. She was in the limelight as never before, her musical career taking off to new heights. Her records received excellent reviews from the critics. Some in the music business called her a singer-songwriter beyond her time. She appeared in a seemingly unending catalogue of television appearances. She performed in Rio de Janeiro and worked with the great Brazilian musician Jobin. Alexandra was named best newcomer and was awarded the "Golden Europa". Around this time, she became involved with an individual, a certain Pierre Lafaire, whom she had met in Rio and whose motives, to those surrounding Alexandra, were less than genuine. Despite warnings and advice from her family, Alexandra fell deeper in love with him until smitten; there was no going back. The two decided to marry. Her family strongly opposed the union. By 1969, her health began to suffer as a result of her heavy work schedule and she was advised to take time off and rest. This had coincided with her participation at the 1969 German National Eurovision final, from which she had been forced to withdraw. She released her single, *Erstes Morgen rot*, in the summer of 1969. Her planned wedding to Pierre Lafaire was cancelled due to her family's continued objection. Alexandra's melancholic songs were often an expression of her own life, which had been one of poverty, being misunderstood, and a lack of fulfilment and happiness.

* * *

The guards stood vigilant at Checkpoint Charlie in Berlin as the people of the DDR went about their business.

* * *

In the early 1960s, the East German government created a national version of the twist. State-sanctioned and approved by the Communist Party, it was called the Lipsi. The authorities employed Helga Brauer to launch the new dance with a record entitled, in the English translation, *Tonight, All Young Folks Are Dancing the Lipsi*. The Lipsi was not the success the state had expected and the whole Lipsi plan was soon shelved. Helga Brauer, however, was an example and proof of a lively East German pop scene that was growing up completely separate from that of West Germany.

All East German singers recorded for Amiga, the state-controlled, and only, record company. There was no chart based on sales and it had a very different method of record distribution. Forty-fives often featured a different artist on each side of what were double "A-sided" singles. Albums of only one singer were very rare in the DDR. Beat music was officially banned by the state in 1965, following a riot at a Rolling Stones gig in West Berlin. Anglicisms disappeared completely from East German pop after the ban. In 1964, Ruth Brandin issued the album *Teenager Party Mit Ruth*; following the ban, the words "*teenager*" and "*party*" both disappeared from the lyrics of pop songs. Ruth Brandin remained extremely popular until the end of the decade. Bored with being suspended in a sixteen-year-old bubble and refusing to cooperate with the Stasi, which asked her to spy on fellow artists, Ruth left the music business.

Scandinavian singers in West Germany were outselling their German counterparts. Wenche Myhre, Siw Malmkwist, and Gitte (Haening) were even singing for Germany at the Eurovision Song Contest in the latter part of the 1960s. Gitte would sing *Junger tag* for the *Bundes Republic* in 1973. The East German authorities soon caught on to this and tried to promote their own "East German" Scandinavian star. It proved a difficult mission. It was hard to entice Scandinavian artists to the drab shop windows of East Berlin and the plumbing-related problems in the housing, even though it promised equality to all. In 1968, they found what they were looking for and Norwegian singer Kirstie Sparboe recorded a 45 for the state record company, Amiga. The song *Zwei Augen*, a rather haunting track more in the tradition of *Liedmacher*, was released in the DDR. Kirstie had sung for Norway at the 1965 Eurovision Song Contest when she was only nineteen. Her song *Karusell* came in thirteenth. She participated again in the 1966 Norwegian final and came in second. She again entered the 1967 Norwegian national heats and won her passage to the

Eurovision: A Plea for Respect

Eurovision finals in Vienna with *Dukkeman*; she came in fourteenth. She participated for the fourth time at the Norwegian pre-selection of 1968 with the song *Jag Har Aldri Sa Glad* and was selected to represent Norway in London. Controversy, however, plagued the song since its win at the national heats. It sounded very similar to Cliff Richard's *Summer Holiday*. Cliff Richard was all set to sing for Britain at the same contest and it was thought that this could prove highly embarrassing. The song was eventually disqualified. It was replaced at Eurovision by the runner-up at the National heats, *Stress*, by Odd Børre, a repetitive, tuneless song that was one of the worst-ever songs at the contest. Kirstie, however, had a hit with a Norwegian version of the 1968 winner, *La, La, La*. With her head proudly held high, Kirstie again sang for Norway at the 1969 song contest with the bright, sing-a-long *Oj, Oj, Oj, Så Gled Jeg Skal Bli*, which came in last. Her three Eurovision entries had all together accumulated a total of only four points. Norwegian pop was still a step behind the mainstream European pop of the time and sounded dated to viewers in other countries. Norwegian songwriters had not yet learned international appeal, something they would soon become masters at. Norway boycotted the 1970 Eurovision Song Contest, but Kirstie continued her annual tradition and competed, unsuccessfully, in the German national final. In 1971, she had a hit with her Norwegian interpretation of *Un banc, Un Arbre, Une rue*, that year's winner. *Zwei Augen*, about a girl being watched, was chillingly all too familiar in Stasi-observed DDR. It was to be the only disc cut in the DDR. Kirstie moved on to cultivate a more lucrative career in West Germany, where her recordings enjoyed moderate success.

By 1969, there had grown up in West German society a somewhat condescending attitude towards aspects of East German culture. This was particularly evident in the manner in which East German pop music was looked down upon with disdain as an inferior synthetic copy of its own pop music scene. This, among other reasons, made it difficult for singers from the DDR to break into the West German market. Despite the ban on beat music in 1965, the DDR still managed to produce some very 1960s-sounding quality pop. It is wrong to assume that a country that had produced Johann Sebastien Bach was in some way musically handicapped simply because it did not connect to the Anglo-American cultural empire or its NATO ally, West Germany. The DDR's top male vocalist, Frank Schöbel, was the first East German pop singer to be invited to perform in West Germany. He had started as a teenage star in 1962 and released such sought-after 45s as *Blonder Stern* and *Party Twist*. In 1969, his popularity was at its height. The dream couple in West German pop had been the attractive pairing of Danish singer Gitte and the King of German *Schlager*, Rex Gildo. Their duet recordings were always extremely popular, turning into best sellers. They were viewed by the West German public as the picture of romance, the ideal couple. A similar coupling by the East German record company Amiga brought Frank Schöbel together with Chris Doerk, an East German female vocalist with mediocre success in the DDR. The coupling on singles they released propelled her to superstar status equal to Schöbel's. A real-life fairy tale romance soon blossomed and the pair were married in 1966. This was to the contrary of Rex Gildo and Gitte's romantic duets, the assumed sentiments of which were all a clever marketing ploy by their record company. Off stage, Rex Gildo had married his cousin and was secretly gay. There was never any intention of the couple falling in love. Frank Schöbel and Chris Doerk's combined acting and singing talents excelled in the 1967 musical pop film *Heissen Sommer*, a *Summer Holiday*-orientated film about carefree teenagers in love, lapping up the summer on the soft white sand at the Baltic Sea. With plenty of pop songs, it was a box office mega-hit in the DDR. Chris Doerk had a huge hit with the theme tune. Its popularity endured the demise of communism and the DDR and it was made into a successful stage musical in 2005. Frank Schöbel performed at the 1974 World Cup along with the DDR football team.

Ina Martells, a former laboratory assistant turned singer, dominated the East German pop scene in 1969. She had been launched onto the pop market in 1965, with a German version of Petula Clark's

Eurovision: A Plea for Respect

Downtown. Her recording, *Heute, Bist Du Bei Mir*, was one of her much-acclaimed 1969 successes. Her sad, seductive, sultry voice was one of the most distinctive to be heard on the East German airwaves in the last year of that very special decade.

The DDR was not a member of the European Broadcasting Union and, as such, was never to compete in a Eurovision Song Contest, nor was it broadcast in East Germany. East Germans were, however, more than aware of the event and its prestige. For decades, ordinary citizens had listened secretly to the Eurovision Song Contest on radio from neighbouring West German transmissions of the event. The DDR's relationship with the Eurovision Song Contest, however, was only passive. There was no interaction between the two. It competed for the first time as part of the newly re-united Germany in 1990.

At the height of the communist regime, East Germany hosted its own annual *Deutsche Schlager-Wettbewerb*, a yearly *Schlager* song contest held to find the best *Schlager* song, which also existed in a West German format. The *Bundes Republic*'s 1968 song festival was, again, a reflection of Scandinavian singers' high profile on the West German music scene. Siw Malmkwist won, while in second place was Dorthe, a Danish songstress. French songstress France Gall followed them into third place. At the *Schlager-Wettbewerb*, there was often a cross-over of genres. Artists who were otherwise well-known exponents of other genres, such as France Gall (*Ye, Ye*) and Severine (*Chanson*), performed *Schlager* numbers in German at the festival. In the early 1960s, the song festival occupied a similar position to that of the San Remo Song Festival in Italy. Winning also meant direct entry as the German candidate in the Eurovision Song Contest. By far, the most popular song to win was the 1962 Eurovision entry *Zwei Kleiner Italiener* by Connie Froboess, which sold over a million and went into the history books as a German classic.

The DDR's pop scene in the late 1960s was at least thriving, more East than West, more Intervision than Eurovision.

One of the East German communist party's greatest public relations victories concerned the world of pop music. In 1961, Dean Reed, a young American, charted in the states with *Our Summer Romance*. Reed was also keen to work in films and was signed by Warner Brothers. He embarked on a tour of Latin America, where he found himself exceedingly popular amongst audiences of enthusiastic teenage fans. A popularity poll later proved him to be rated above Elvis, Paul Anka, and Neil Sedaka. He was dubbed the *"Magnificent Gringo"* in the South American press as the adulation he received continued. Due to his overwhelming popularity, it was decided that Reed should relocate to Argentina. During his time there, he became acutely aware of the country's social deprivation and class inequalities. This had a profound effect on him, so much so that he became an avowed Marxist. Political upheaval in Argentina, however, soon forced Reed to relocate again, this time to Italy, where he appeared in several Spaghetti Westerns. While in Leipzig, which then was in East Germany, he met and married a German teacher. In 1973, he set up residence in the DDR, where he made his strong communist sympathies well known.

Throughout the decade, Reed gave concerts in Eastern Europe and the Soviet Union. He toured constantly, usually performing a mixture of country-pop songs, rock n' roll, and protest songs. His gigs often included self-penned revolutionary numbers on the fight of the proletariat. The adulation he received equalled and even surpassed that which he had enjoyed in South America. In Moscow in the spring of 1966, crowds of screaming fans mobbed Reed. Amid scenes of mass hysteria, fans tore wildly at his clothes, leaving him in tatters and a state of panic. He managed to escape into a waiting car but the crowd encompassed the vehicle, rendering it stationary – all this in the Soviet Union while in America he was almost unknown. Reed, on seeking residence in the DDR, had not surrendered his United States passport and was free to return or travel to the West whenever he wanted. In his native country, he became to be regarded as something of a curio.

By the 1980s, Reed's popularity began to flag. During his years in the DDR, he had starred in eighteen films and released fourteen LPs.

Eurovision: A Plea for Respect

At forty-eight, he was a fading star and decided to return to the USA, where he came from, with the idea of repeating his Soviet bloc success there.

A massive wave of publicity in the USA surrounded the news of Reed's return to the States. The singer had hoped to tour American university campuses in the following year, 1987. Western media embarked on a curiosity venture concerning this American maverick artist.

With some confusion as to the whereabouts of Dean Reed, in Berlin the week commencing June 16, 1986, a stunned media reported on July 17 that he had been found dead in a lake outside Berlin. His family and the US government immediately suspected foul play. The official report stated that Reed had taken a sleeping tablet and then proceeded to drive his car. He became drowsy and crashed into a tree. Leaving the site of the accident, he walked over to the lake in an attempt to wash his face, then lost his equilibrium and fell into the lake, drowning. A friend, however, had told the press that Dean had called her the week before and had insisted that something was wrong. She then stated that Reed had phoned her again, insisting that his life was in danger. Despite rumours of Stasi involvement, an official US report into Dean Reed's death found no improprieties and the file was closed. The *New Germany* party newspaper gave him a glorious obituary, praising his dedicated service to Marxism. After the fall of the DDR in 1989, the revelation of Stasi files proved that Dean Reed had not been murdered but had taken his own life after separating from his third wife. Reed had even written a letter to President Honecker, apologising for the suspiciousness his death might cast on the state.

After a cultural thaw in the 1970s, the party revoked the ban on beat music. This led to a host of more Western-inspired bands such as Karat and The Puhdays, who had originally formed as a Beatlesque group in 1965.

While West German airwaves in the 1960s were slowly incorporating British and American rock into their broadcasts, the East German airwaves remained solidly German. As the 1970s turned into the 1980s, German language pop music in the *Bundes Republik* was becoming increasingly elbowed out by British and American music and similar styles. The DDR, on the other hand, continued to heavily promote local German pop acts with a distinct East German sound. By the time of reunification and the decades that followed, German language pop/*Schlager* was almost obliterated from the airwaves in favour of the dominant Anglo-American rock and pop and its English language European equivalents. The question is: Was this really an expression of freedom? What was the worst evil, the biggest threat to German culture? The German language pop enforced by the state, or cultural domination by a foreign culture? Was *Schlager* really only schmaltzy, sentimental trash with pleasing melodies or a mild form of placid protest in the hour of a thunderstorm of foreign cultural hegemony?; A resistance to a musical fascism? After German re-unification, the pop that had dominated the East German scene in the 1960s, 1970s, and 1980s was condemned to a dustbin. Citizens of the new united Germany looked upon it with embarrassment and as an example of their musical naivety and primitive technology.

* * *

Siw Malmkwist had the unusual distinction in Eurovision history of being the first female German representative to perform in a trouser suit. Siw, in keeping with 1960s mode, was almost certain of a huge batch of votes coming her way from her native Sweden. The big shock of the night came when Sweden failed to award her any votes at all: a sign, perhaps, that the Swedes were displeased that one of their top vocalists had chosen to sing for Germany, a foreign rival. An irony was that she came in ninth, a position she shared with her compatriot Tommy Kőrberg and his very 1960s pop song, *Judy Min Vaen*. Siw would make yet one more attempt at the Eurovision trophy in 2004 when she sang *C'est La Vie* at the Swedish Melodifestivalen.

It was good PR for the *Bundes Republik* to be represented by foreign singers. Slamming the door of Nazism firmly closed, West Germany was desperate to be seen as international and reaching out to

Eurovision: A Plea for Respect

a wider Europe. There was a great desire for a Jewish singer to sing for Germany, which would have been a confirmation by West Germany and ratification by Europe that any suspected bacteria of Nazism left in the country was most definitely dead. Ralph Segal penned a German entry of the 1990s, intended for Esther Ofarim, in order to manifest this. The Israeli-born and Jewish Ofarim, already a big star who had topped the UK chart in 1968 and sung for Switzerland in the 1963 contest, however, requested an exorbitant fee. Segal was reluctant to pay up, so the idea was abandoned.

* * *

In the weeks and months that followed Siw Malmkwist's performance of *Prima Ballerina* in Madrid, Alexandra's career went from strength to strength. Her personal life, however, was once again clouded with problems, a fair degree of anxiety, and fear. Much to her family's relief, she had split up from her fiancé, Pierre Lafaire, a somewhat ambiguous character. One early summer evening, she received an anonymous phone call urging her to write out her will in favour of her mother and son. Petrified, Alexandra would not let her son out of her sight for fear of abduction. The lyrics of Kirstie Sparboe's last German hit single *Zwei Augen* were never more horrifyingly apt.

On the last day of July 1969, Alexandra, who had recently moved to Munich, had to travel to the North German city of Hamburg to discuss conditions with her record company. After negotiating at the company office, she embarked on a much-desired summer holiday and drove her car in the direction of Sylt, a North Frisian island connected to the mainland by a causeway and a popular holiday resort of the German jet set. Travelling with her in her Mercedes-Benz were both her son, Alexander, and her mother. On route to the island, she had her car tested at a garage. Shortly afterwards, the car's brakes failed and she collided with another vehicle at a crossing close to the town of Tellingstedt. Alexander, her son, miraculously survived with minor injuries. Tragically, her mother died from her injuries in hospital. As for Alexandra, she had been killed instantly. She was only twenty-seven. There were many allegations that the car had been tampered with; without any proof to the contrary, the case was left open as a death under suspicious circumstances. Alexandra's funeral in Munich was one of the biggest gatherings seen there in years. An estimated 3,000 attended and she was buried at the Westfriedhof cemetery. A stone with only her name, "*Alexandra*", marks where she is buried. A recording of her composition Mein Freund Der Baum Ist Tõt was released shortly after her death.

Thirty years later, the film director Marc Boettcher started to write a biography on Alexandra's life, during which he conducted extensive research into the events surrounding her untimely death. While investigating the circumstances, Boettcher was plagued by numerous anonymous death threats. Curiously, this was in 1999, post-Cold War and post-DDR. His detailed research was published in the autobiography. It revealed that her somewhat insincere, shady fiancé Pierre Lafaire was, in fact, an American spy. Stasi documents cited him as being involved in espionage in Denmark and all evidence pointed to what had been long suspected by many: that Alexandra had been murdered by the East German secret police, the Stasi, in retaliation for her relationship with an American agent.

Illness had prevented Alexandra from competing at the 1969 German Eurovision prelim. However, her particular style of singing influenced many future German and Central European singers who were to yet compete in the Eurovision Song Contest. Her haunting, melancholic style of singing had influenced a whole generation of artists. Alexandra had come from the East at a time when it was shrouded in mystique and intrigue. A melancholic, mystical aura had exuded from her songs and that, in the end, was to tragically become a chilling part of the last months of her own life.

Eurovision: A Plea for Respect

CHAPTER 10
British Attitudes

The Graham Norton Show, May 2013. "Have you heard of the Eurovision Song Contest?" Graham asks some American guests. "No," they reply, "It's very good," he says. Immediately, there is a roar of laughter from the crowd.

* * *

Graham Norton took over as BBC commentator for the event in 2009 when the veteran broadcaster Sir Terry Wogan retired after commentating at every contest since 1980. He had also commentated on the event for radio and television in the 1970s.

* * *

In 2007, Jeremy Vine, the BBC radio chat show presenter, whose father commentated on the 1974 contest and ABBA's win, announced in a strong, confident voice on the Sunday afternoon *Points of View* programme, "The Eurovision Song Contest, which to most people is a joke". The voice of the trusted BBC, the voice of authority and truth, had spoken.

* * *

The United Kingdom arranged a very grand and lavish Festival of British Popular Songs in 1956 that was meant to choose the British entry for the very first Eurovision Song Contest. It ran for months in the form of six heats and then a final. The singers who competed were popular radio stars of the period and included Ronnie Carrol, Petula Clark, Max Jaffa, Anne Shelton, Kathie Kay (who, as a four-year-old, toured with Sir Harry Lauder), Alma Cogan, Ronnie Hilton, Marion Ryan, and the much-loved singer-comedienne Dora Brian. After six months, the final was held and juries voted from all corners of the

United Kingdom. The winner was a song entitled *Everybody Falls in Love with Someone*, sung by Dennis Lotus. The national final had, however, taken so long that by the time the UK had selected its song, the deadline for submissions had passed and it was too late for the UK to take part in the first Eurovision Song Contest. The following year, however, in Frankfurt, the United Kingdom of Great Britain and Northern Ireland made its debut with a song called *All*, sung by the actress Patricia Bredin, who appeared in several British films of the 1950s. Among them were the comedy *Left, Right and Centre*, in which she starred with Ian Carmichael, and *Desert Mice*, for which she worked alongside Syd James. Her talent noticed, she went on to be voted Most Promising Newcomer together with Peter Sellers and Hayley Mills. Patricia Bredin also took over from Julie Andrews in the role of Guinevere on the Broadway production of *Camelot*. The song did not do well and placed seventh out of the ten countries that took part. At the British heats that year, there had been a song entitled *Sycamore Tree*, written by Hubert Gregg, who many years before had composed *Maybe It's Because I'm a Londoner*. Patricia Bredin, at that time unaware of how popular of an event the contest would become, does, however, recall feeling sick at the responsibility of the task ahead while she had lunch with her conductor, Eric Robinson, on the day of the contest. In 1964, she married the Welsh musical theatre singer and actor Ivor Emmanuel, the same year he appeared as Private Owen and led the singing of *Men of Harlech* alongside Michael Caine in the film *Zulu*. They divorced after a few years and Patricia remarried, to the businessman Charles McCulloch. It was a marriage that was to be brief and tragic, as he died on their honeymoon. She is, as of the time of this writing, a best-selling author and a resident in Canada, where she has lived for a number of years. Although *All* had not lived up to the British public's expectations and had been a disappointing debut, it does hold a special significance in the history of the song contest. Whereas songs sung in English are now plentiful, *All* was the very first song sung in English on a Eurovision stage. There was no British entry in 1958 due to the low placing of *All*; the BBC had initially agreed to take part but

Eurovision: A Plea for Respect

withdrew. The Eurovision Song Contest in 1958 was, however, broadcast by the BBC, and the UK returned to compete again in the 1959 song contest in Cannes. The Festival of British Popular Songs was replaced by the Eurovision Song Contest British Final in the search for the UK entry. Among the hopefuls were Rosemary Squires, John Hanson, Lita Roza, and Marion Keens, who had stepped in at the last minute to replace Alma Cogan, who, due to illness, was forced to withdraw. Unlike the national finals in 1956 and 1957, the 1959 affair had only two heats which took place within a week. This was a major change, as the 1956 heats had spanned nearly half a year. At the earlier national finals, more than one singer performed each song but in 1959, it was decided that only one act should perform each song, which brought it into line with the Eurovision Song Contest itself. (This rule had been adopted in 1957.) The format of the Eurovision Song Contest as we know it today was beginning to take shape. Pearl Carr had previously had her own show on the BBC and been voted Best Female Vocalist of 1951 in the Melody Maker poll. Pearl teamed up with her singer husband Teddy Johnson. They sang their way to victory with *Sing Little Birdie* and a passport to the 1959 Eurovision Song Contest. Teddy Johnson, as a teenager, had formed his own band and had worked as a DJ on Radio Luxembourg. He also made several solo recordings for Columbia Records in the 1950s. He is credited with working on the classic children's television programme *Crackerjack*. The pair came in a close second to Teddy Scholten of the Netherlands' *Een Beetje*. Their near victory aroused a lot of media and public interest and *Sing Little Birdie* became a hit. It is perhaps at this point that the UK and, in particular, the British public's long relationship with the Eurovision Song Contest really began. The couple entered two songs in the 1960 national final but were beaten by Teddy's singer-actor brother, Bryan Johnson, with his song *Looking High, High, High*; like his brother's 1959 song, it came in second. These runner-up placings were to start a precedent in British entries. From 1959 to 1970, the UK would come in second on no fewer than seven occasions: 1959, 1960, 1961, 1964, 1965, 1968, and 1970. The parade of British runners-up after 1960 include The Allisons, Matt Munroe, Kathy Kirby, Cliff Richard, and Mary Hopkin. Ronnie Carol came in fourth in 1962 and 1963, as did Clodagh Rogers in 1971. These were considered, at the time, very bad results for the UK. Kenneth McKellar's *Man Without Love*, ninth place in 1966, was, for a good number of years, considered the greatest disaster in British Eurovision history. The New Seekers came in a further second in 1972, while Cliff Richard's second attempt with *Power to All Our Friends* in 1973 came in third. Olivia Newton-John's fourth place, which she shared with Monaco and Luxembourg, in 1974 was another result that the UK considered a serious blow. Ironically, the composition, *Sugar Baby Love*, which would go on to be a UK chart-topper and international success for the Rubettes, an archetypal 1970s pop group that was, in fact, submitted for the UK national heats but subsequently rejected for participation as one of the final six contenders. Another two runner-up places followed, in 1975 and 1977, and again in 1980. Co-Co's 1978 *Bad Old Days* could manage only eleventh and took over from Kenneth McKellar as the worst result ever for the UK entry. Black Lace's 1979 effort was little better, failing to make the top five. The period from 1959 to 1977, in retrospect, was an "Age d'or" for the British entries at the Eurovision Song Contest. The novelty of coming innn second year after year, however, soon wore off and was replaced by an *"always the bridesmaid, never the bride"* discontent. There was also the *"stepping in"* factor in Britain's somewhat tempestuous relationship with the event; that is to say, when a country had struck on victory but its broadcaster could not turn the glory into enough money to stage the following year's contest, or had no facilities to do so, the BBC stepped in and staged the event. This happened on four occasions, in 1960, 1963, 1972, and 1974. Monaco, after having won the 1971 event, was unable to offer a suitable venue for the 1972 song contest. The Monegasque broadcaster had originally put forward the idea that the contest should take place in an open-air venue in Monte Carlo in June but this conflicted with the interests of several broadcasters. Prince Rainier II suggested that the BBC might be able to host it and so the

Eurovision: A Plea for Respect

UK stepped in at the last minute and took on the production of the 1972 contest. They had not anticipated this but, in all fairness, they had won in 1969, along with three other countries, and the Dutch winners had hosted the 1970 contest in Amsterdam, thus removing the responsibility from France, Spain, and the UK. The BBC chose Edinburgh for the venue, which seemed fitting as the winning singer in 1969, Lulu, was Scottish. The BBC's financial contribution to the contest would nevertheless be rewarded, as, many years in the future, after the introduction of semi-finals in 2004, the countries that were the EBU's biggest financial contributors – France, Spain, Italy, Germany, and the UK – would gain automatic entry into the Grand Final without having to qualify. A dissatisfaction with always missing out on the winning slot, by one place, and the BBC having to step in and finance the event when other broadcasters pulled out, unable to shoulder the financial burden, are factors that might have started a rapid maligning in the British media of the contest and a seeming dislike of it at the BBC.

A strike-plagued Britain in the late 1970s, when electric, transport, and garbage strikes were common, led to the winter of discontent. This crisis reached a climax. Politicians were unable to reach agreements with the powerful unions, and relatives, due to striking council workers, had to bury their dead in freezing conditions. The national finals of the Eurovision Song Contest in Britain, A Song for Europe, was blacked out without warning in 1977 and 1979 due to a technicians' strike. Both were to be broadcast live from the New London Theatre and the Royal Albert Hall. Strikes led to the temporary cancellation of the 1977 Eurovision Song Contest at Wembley. The 1980 A Song for Europe was a much more low-key affair and was not transmitted from the Royal Albert Hall, or any other large venue, for that matter. Instead, it was broadcast from the BBC studios in London. This was a notable changing point in Britain's relationship with the song contest and also in the direction the UK and the contest were to go.

Steve Kerr

* * *

In the spring of 2004, on a Radio 2 talk show, a rather bolshy editor of a BBC publication gives her view on the contest in a very abrasive manner:
"*Before, everyone took it seriously; now no one takes it seriously.*"

* * *

A Friday evening tea time edition of the evergreen children's programme Blue Peter *in the late 1990s; the presenters are dressed up as Eurovision Song Contest presenters and make the traditional introduction in French. They go on to exhibit silly lyrics of songs, show some singers singing out of context (making them seem comical), and show footage of some more eccentric dance routines with the message,* "*This is Eurovision!*"

* * *

After Kenneth McKeller had been nominated to represent the United Kingdom in 1966, there were calls that he was not the right singer for such an event. He was a tenor whose repertoire was mainly folk and opera but then, on the other hand, Sweden had been represented the year before by a tenor who had sung a ballad in English: *Absent Friend*. Others found his engaging, pleasant personality endearing and thought him an excellent singer. Ken sang five songs on the A Song for Europe slot in his show, *A Song for Everyone*, with one being performed each week. The songwriters included those who had composed the *Doctor Who* theme and *Scotland the Brave*. In the end, the ballad *A Man Without Love* was chosen to go to the 1966 contest in Luxembourg. It was veering towards the latter half of the 1960s and the big ballad was beginning to make a comeback and have a big impact on the pop scene. In the latter half of the 1960s, Engelbert, Tom Jones, Malcolm Roberts (Eurovision, Luxembourg, 1985), Donald Peers, and Elvis would all enjoy big hits with great, heart-wrenching, creamy ballads. Engelbert was also to have a hit with an Italian song entitled *A*

Eurovision: A Plea for Respect

Man Without Love, but this was a completely different song from Kenneth McKellar's Eurovision entry. On the big night, Ken, who was the last to perform, wore a kilt and a happy smile. The audience was pleased enough with his flawless performance but when the votes were counted, he ended up in ninth place, with the majority of the votes received coming from Ireland. The BBC was forced to go back home and rethink its strategy for the next contest. I have some very special memories of the 1966 contest, as it was the first one I ever watched. My late father, who was a great fan of Kenneth McKellar, forced me to stay up late and watch the contest. As a child, I found the event far too long and tedious. It was on quite late and I was also far too tired to appreciate it. My father was always convinced that the UK should have won that year.

While artists such as Matt Munro, Kenneth McKellar, and Kathy Kirby were very popular, they did not really come from the high altar of cutting-edge pop, for which London was the cathedral. Matt Munro, a former London bus driver, born in Shoreditch, was a vocalist of considerable merit and is now considered one of the finest British ballad singers of his era. He enjoyed a distinguished career with international acclaim and tasted a high level of success with *To Russia With Love* from the James Bond film of the same name. He also relished attainment with the song *Softly as I Leave You*, an English version of the Italian song *Piano*, originally a big hit for Mina. Kathy Kirby, born in Ilford to Irish parents, was voted Top Female Vocalist of 1965 and had a string of hits in the early half of the 1960s, of which the most notable was her version of *Secret Love*. It reached number four in the British charts, where it remained for no less than five months. Her career had waned considerably by 1970, after which a somewhat tempestuous personal life often found her in the tabloids. She was rarely seen on television in the years that followed and her life became shrouded in a Garboesque mystique. In her later years, she was, in fact, diagnosed with schizophrenia. Kathy's niece married Mark Thatcher, son of the British Prime Minister Margaret Thatcher.

The BBC decided, for the 1967 contest in Vienna, that it would do what it did best and send a piece of the Swinging Sixties to spice up the contest. The pop singer Sandie Shaw, who had a string of top ten hits to her credit, was chosen to represent the United Kingdom. The BBC was out to win it! All stops out! Full speed ahead! The five songs for Europe were performed over a period of weeks, in early 1967, on the *Rolf Harris Show*. Songs were submitted by the cream of 1960s British pop composers including Chris Andrews, who had already composed some of Sandie's hits (i.e., *Long Live Love*, *Message Understood*) and had charted himself with *Yesterday Man*. The five songs were *Puppet on a String*, *Had a Dream Last Night*, *Tell the Boys*, *I'll Cry Myself to Sleep*, and *Ask Any Woman*. The best song was *Tell the Boys*, an up-tempo, very 1960s pop number but it was the popular *Puppet on a String* which won the postal vote. The recording of the runner-up, *Tell the Boys*, did, however, go on to become a hit.

Erika Waal introduced the 1967 Eurovision Song Contest with the utmost politeness from the glittering theatre of the Hofburg Palace in Vienna, once home to the Hapsburgs. The stage set was made up of several raised mirrors which reflected the artists as they sang. It was a contest dominated by ballads and, except for Monaco and Germany, bow ties and evening gowns. Then, suddenly, Sandie Shaw exploded onto the scene in a pink glittering mini dress and sang a very up-tempo 1960s pop song in her bare feet, with all movement and mannerisms associated with the genre. The 1967 song contest was a rather formal, stuffy affair which exuded an air of tension – and then, suddenly, a British icon of the 1960s smashed the stage mirrors! Pop had arrived at the Eurovision Song Contest. *Puppet* was a ground-breaking first at the song contest and part of the social revolution that had begun to shake Europe. The song's many critics deride it and it is much maligned by "*the musical elite*" but due to *Puppet*, the starchy, conservative side of the song contest had been turned on its head and the difference between the 1967 contest and the 1968 contest was like day and night.

In the second decade of the twenty-first century, much criticism and prejudice surround *Puppet on a String*. It is clouded with crassness

Eurovision: A Plea for Respect

and exhibited as a classic case of naffness. It is, thus, hard to remember exactly how genuinely popular this song was and how much Sandie Shaw was adored at the time. The song went to number one in the UK and various versions recorded by Sandie in several languages repeated the success all over Europe. *Puppet* had swept the scoreboard with a huge lead over the runner-up, Sean Dunphy of Ireland with the big ballad *If I Could Choose*.

During the voting, Sandie Shaw was filmed backstage with the other artists. The 1966 Austrian winner, Udo Jűrgens, sat next to her and chatted throughout as the votes slowly came in from around the continent. Sandie seemed too engrossed in the numerals being delivered to pay much attention. The Austrian entry that year was the ballad *Warum es hunderttausend Sterne gibt*. Its young singer, Peter Horton, sat on the other side of Jűrgens as he chatted to Sandie. As the voting continued, it became obvious that Austria's victory would not be repeated and was destined for a very low place; Peter Horton began to look more and more depressed. His head drooped lower and lower as each country delivered its votes. His song ended up with only two points and came in fourteenth out of seventeen songs.

Puppet on a String was a 1960s pop song and, in reality, no worse or better than a lot of other contemporary pop at the time, although changing values and tastes dictate to us that it is the worst song ever written. There is as much fascism in music as there is in politics and it is always a good idea to stand up to musical fascists. Much has been written in previous chapters about fascist regimes that once controlled chunks of Europe; this is not confined to politics but exists in a very particular form around music. Taste would seem to be achieved through some oppression or form of suffering which affords a credibility but shallow music critics or self-appointed musical experts are a foul lot propelled by what is trendy and ephemeral, often with little true depth; of an ilk who probably wouldn't talk to you if you weren't wearing the right sort of shoes. Such are the fascists of music who have greatly wounded the Eurovision Song Contest in the UK.

Steve Kerr

In a *Radio Times* interview in 1974, Sandie Shaw's former manager said that Sandie was exploring a new type of music: "Frankly, darling, Sandie is sick of *Puppet* on *a String*". For many years, Sandie disassociated herself from the song. In fact, on more than one occasion she stated that she vehemently hated it. She had become so synonymous with it that she felt her other outstanding works of 1960s pop had been cast into the shadows and that her career as a quality artist in general suffered. After some time out of the limelight, Sandie returned in the mid-1970s with the single *One More Night*, the B-side of which was *Still So Young*, in which she sings, "*I'd rather be dead than have middle-aged spread*". In the mid-1980s, she was back on the charts with the song *Hand in Glove*, which she recorded with The Smiths, Morrisey being a great fan of Sandie. She made several interesting recordings in the late 1980s such as *Please, Help the Cause Against Loneliness* and *Are You Ready to be Heartbroken?*, which reached the top seventy. She even sang a revamped version of *Puppet on a String* on a tour with Jools Holland in 2011 and commented on how good the song was after really listening to its lyrics with a radically different arrangement. *Puppet*, written by Phil Coulter and Bill Martin, was about the ups and downs of a relationship with one person calling the shots, illustrated through fairground metaphors. The song, however unintentionally, reflects the psychodynamics and the sociodynamics of the British attitude towards the song contest as well as the UK's often complex relationship with Eurovision. Through decades of constant ridicule, criticism, and derision in the press, radio, and television, the British public has become conditioned to deride the contest to such an extent that even if the best song ever written entered the contest, accompanied by flawless vocals, it would still be looked upon by a large section of the British public as risible, not because in reality it was but because, due to various factors, they were conditioned to do so. There is also a myth that out of the thousand-plus singers and songs that have competed in the Eurovision Song Contest, the only singers of any merit were ABBA and the only song of any aesthetic value was *Waterloo*. The unconscious conditioning of the public has led them to react "*like a*

Eurovision: A Plea for Respect

puppet on a string" to the contest. They laugh, uninformed, fed by myths from the media. The lyrics of *Puppet* would seem, in recent years, to reflect the United Kingdom's situation in relation to the Eurovision Song Contest. Britain feels rejected by Eurovision's seemingly ambivalent attitude towards its entries and tries hard to understand: "*I wonder if one day that you'll say that you care*". How Britain would deeply love, despite what it might say, for its entries to be approved by the Eurovision family: "*If you say you love me madly I'd gladly be there*".

Britain's 1960s and 1970s golden age in the song contest, followed by its dark age in the first decades of the twenty-first century: "*I may win on the roundabout then I'll lose on the swings*". The UK's theory that there is an Eastern Bloc, ex-Yugoslavia conspiracy in the voting: "*In or out there is never a doubt just who's pulling the strings*". Despite the negativity and derision towards the event, which abounds on its shores, the UK always sends a song. It has been present at all contests except 1956 and 1958, and it does *not* like a low score. It gets frustrated when its entries fail to impress: "*I'm all tied up in you but where's it leading me to?*"

Critics of the song, who have included Sandie herself, accuse it of being nothing but cringeworthy oompah. Sandie, who had her first hit in 1964 with *Always Something There to Remind Me*, was never a sugary/saccharine "*girly*" singer. She was defined by her grit and very distinctive vocals. Let's not forget that she was the first singer to record material by Led Zepplin other than the group itself. *Puppet* sung by any other run-of-the-mill singer might have come under the oompah/*Schlager* tag but Sandie's outstanding and distinctive vocals gave it another dimension. Sandie also became a record breaker with *Puppet* and became the first-ever female vocalist to have three number-one hits since the creation of a British singles chart. Sandie has often vented her dislike of the Eurovision Song Contest but her view seemed to have mellowed when she presented a credible BBC Radio documentary on the event and appeared in the *Making Your Mind Up* song for the Europe British final on which she sang *Puppet*. Emma

Bunting, former Spice Girl, sang a contemporary version of the song at the 2004 British national selection. In a TV interview in 2012, Sandie did go on to criticize the British attitude towards the contest. It would seem, overall, that Sandie had, through the years post-1967, grown to resent both her winning song and the Eurovision Song Contest, as she felt they had hindered her career as a credible singer. It was, however, not the actual song or the contest itself that had dented her career but the attitude of the media in the UK which directed much prejudice towards the song contest with which Sandie was very much associated. It is this attitude that has prevented Sandie from receiving the adulation afforded to Dusty Springfield. Sandie Shaw is very much revered as an arch-icon of the 1960s, of which she is venerated as an important integral component, yet it is the negative perception of the contest that prevents her from elevating to the level that Dusty enjoys in the British music industry. Katrina, the UK's 1997 Eurovision winner, when speaking about her own victory, said that in the eyes of the British music industry she had committed the worst crime by winning the Eurovision Song Contest, and winning the contest has been cited as a major factor in the band's eventual breakup. Dusty, of course, never sang at the Eurovision Song Contest... or did she? The answer is no, but one of Dusty's greatest successes was *You Don't Have To Say You Love Me*, which was originally *Io Che Non Vivo Senza Te* from the 1965 Festival of San Remo, which was in those days also the Italian Eurovision selection. Dusty had, in fact, competed at San Remo that very same year when she was the international act who sang *Di Fronte all'amore* paired with the Italian Gianni Mascolo. This highlights the double standards and ignorance that exist in the UK when it comes to the Eurovision Song Contest. If there had been a degree of respect in the British media's approach to Eurovision, Sandie Shaw would have been afforded the respect she is due and would sit at the right hand of Dusty Springfield, where she so rightly belongs. Sandie, who since the 1990s has practiced as a psychotherapist for stage and sports personalities, announced her retirement from the music scene in 2013 but vowed to continue campaigning for artists' rights.

Eurovision: A Plea for Respect

* * *

No chapter on British attitudes towards Eurovision would be complete without the contribution made by Sir Terry Wogan towards the song contest in the United Kingdom. As a long time radio DJ, television presenter, and Eurovision commentator, Dublin-born Terry Wogan was both a familiar and popular voice and face for over forty years. His early years as a BBC Eurovision commentator saw him take a professional and more formal approach to his commentary, as was manifested at the 1973 contest, which, due to Israel making its debut, was endowed with heavy security; this gave the whole event a rather tense ambience, evident during the live broadcast.

Sir Terry's breakfast radio show was punctuated by light-hearted puns and the DJ had a tradition of playing rather obscure records regularly, thus pathing the way for them to become hits. An example of this was Conrad Veit's *There's a Lighthouse Across the Bay*. In the 1970s, Wogan also regularly played English versions of Eurovision Song Contest runners-up, i.e, Mocedade's *Touch the Wind* (*Eres Tu*). As the years went on, his sporadic commentary turned into a regular annual coverage which, in turn, grew into a tradition. Circa 1980, Wogan began to introduce some humour into the commentary, which at first was welcomed. There was a point in the early days of his commentary when Wogan had struck a happy balance between humour and respect for the contest, which only added to the entertainment. He, however, went on to make more and more fun of some of the more eccentric or low-standard songs.

As the 1990s dawned, Wogan's comments had became more sardonic and he set off on a course of ridiculing the whole event, including not only the songs, which he interrupted, but also the presenters, the jury chairpersons, and the interval acts. At the 2001 Eurovision Song Contest in Copenhagen, he ridiculed the Danish presenters so much that the Danish government demanded an apology. This trashing of all the songs at the contest, regardless of quality,and the extreme devaluation of the event itself began to be known as

Woganism. As the contest continued over the years, it became very much associated, by a large sector of the British public, although not all, with Wogan. In turn, it more or less became the Terry Wogan Show. His ridicule and sardonic approach escalated in the early 1990s, a time when his TV chat show and popularity had begun to flag. The boost of derision and irony in his commentary might have been an attempt to reclaim his popularity. It also coincided with the UK's flagging fortunes in the event.

Wogan always maintained that he loved the song contest but a degree of antagonism born between Wogan and fans of the contest began to grow. He formed a dislike of the fan club, and there was no love lost on the fan club's side, either. On his morning radio show, it became customary to giggle and laugh when Eurovision was mentioned, something which was then repeated throughout BBC programmes in general.

In the eyes of many, Woganism had destroyed any credithe contest had in the UK.Christer Bjorckman ,the Swedish producer of the 2016 event, slammed Wogan and accused him of having created a generation of Britons who viewed the Contest as a kitch ,irrelevancy.He accused Wogan himself of being behind the negative perception that it was mere frivolity. "He did this for 28 years and his commentary always forced the mockery side and there is a grown-up generation in Britain that doesn't know anything better. He raised a generation of viewers believing this was a fun, kitsch show that had no relevance whatsoever"

The participation of more and more Eastern Bloc countries grew throughout the 2000s and their songs became increasingly successful. Simultaneously, the UK's entries became more unsuccessful. Wogan's commentaries grew more cynical on the topic of East European victory and voting patterns. Karen Fricker wrote in The Guardian in 2013, "*Wogan dismissed performance strategies that he could not fit within his understood Eurovision norms*, and *spun increasingly paranoid tales of political voting conspiracies*". As he moaned and groaned about Serbia's and Russia's victories, his commentary had not only derided and

Eurovision: A Plea for Respect

helped turn it into the ridiculous but it had also taken away all expectation, excitement, and joy from the contest for a large section of British viewers. What was worse, he had taken away its whole atmosphere as an event. The late DJ John Peel, an aficionado of the song contest, had stated in 1998 that one reason he liked it was that it had, above other events, the air of being a real occasion. As General de Gaulle so rightly believed, television is a very powerful media that can greatly influence public thought and opinion. Wogan's rants on ex-Soviet and Balkan bloc voting were taken on by a large section of the public and became accepted as fact, even though no official enquiry ever proved it true.

While it was true that many East European and ex-Yugoslav countries qualified from the semi-finals to the finals, many, on the other hand, did not. In 2013, most of the ex-Yugoslav states were in the same semi-final, thus allowing them to also vote in that semi-final. Yet, despite this, none of them qualified for the Grand Final. Of the remaining Balkan countries in the next final, only one went on to the Grand Final. During the 2007 contest, the BBC Radio 2 commentator Ken Bruce abandoned his coverage in the middle of the voting as votes poured in for Serbia, Ukraine, and Russia; he played unrelated music instead. In the latter years of Wogan's commentary, he was given *carte blanche* by the BBC to do what he liked. It all added to the annual tidal wave of negativity that greeted the song contest before a note had even been sung. At the 2008 song contest, Wogan was so convinced of the former Soviet voting conspiracy that he announced, after the Russian win, that he would not be returning to commentate at the 2009 event in Moscow. He was deeply critical of, and strongly disagreed with, the Russian win. The fact that the UK came in last seemed to make it worse. What he failed to mention and seemingly did not take into account was the fact that Dima Bilan, the winner, was the biggest star from the shores of the Pacific to the German border, nor that his winning song, *Believe*, sung in English at the contest, was produced by the American R&B star Justin Timberlake. After the 2008 event, Wogan retired from the commentator's box at the Eurovision Song

Contest and, soon after that, also announced his retirement from his long-running breakfast show, to be replaced by Chris Evans. Wogan's sneers at Johnny Foreigner and xenophobic jibes had begun to wear thin, but had given him enough credit in the eyes of the establishment for him to be knighted. Ironically, in a build-up to a Eurovision interview sometime later, in which Wogan was asked to choose his all-time favourite Eurovision entries, he cited Dima Bilan's Russian winner as one of them. The TV presenter/comedian Graham Norton replaced him at the Eurovision Song Contest as the commentator for BBC Television, joining the lineup of commentators who had presided over the event regularly for the BBC, including broadcasting veterans Pete Murray and David Jacobs.

The year 1980 was a major turning point in the history between the United Kingdom and the song contest, just as significant as 1959 had been in that relationship. The venue, the BBC Theatre, might have been selected in part because of the cameraman strike of 1977 and 1979, both scheduled for the then-major music venues in the United Kingdom. Singer Lyndsey de Paul, who, with Mike Moran, sang *Rock Bottom* in the 1977 national heats, recalled how, on the evening the transmission was due, the artists, who were putting the final touches on their costumes in preparation for the transmission to start at 7.30, were told at 7.20 that the cameramen had walked out on strike and that A Song for Europe would be broadcast only on Radio 2. Lyndsey recalled the great disappointment backstage. Mike Moran had previously been part of Blue Mink and Stone The Crows, two very big music acts of the 1970s. The 1979 A Song for Europe was not even broadcast on the radio after the sound engineers decided to walk out. It would seem that due to the expense put out on the 1977 and 1979 A Song for Europe programmes, and the subsequent waste of money on outside broadcasts that did not happen, the BBC feared another strike and so, to keep costs down, held the contest within the BBC Centre itself. Strikes were nothing new to A Song for Europe. In 1971, the public postal vote was abandoned due to a postal strike. In 1972,

Eurovision: A Plea for Respect

electricity strikes led to blackouts in various areas of the UK at the A Song for Europe final, which resulted in a low turnout of postal votes.

Was there, perhaps, another reason for the downgrading of A Song for Europe? The 1980s were to see the Eurovision Song Contest descend to an all-time low in the UK. It was a decade that did not treat the song contest kindly. At its silver anniversary edition in The Hague in the Netherlands, the Belgian entry, a very contemporary example of synth-pop by the group Telex, was dedicated to the contest itself on its special anniversary and was entitled, of course, *Eurovision*. It ended with the notes of Charpentier's *Te Deum* a la synth-pop. The song came in last and was an uncanny prediction of how the contest would fare in the new decade.

In the United Kingdom, the winter of discontent was followed by the fall of the Labour government led by James Callahan and a general election in 1979. History was made when not only was the Conservative Party voted into power but Britain elected its first-ever woman prime minister, Margaret Thatcher. While the previous Conservative prime minister, Ted Heath, headed a very pro-European administration, the new government, and in particular the Prime Minister, had a much more sceptical view of Europe, which at times bordered on Anti-European in rhetoric. The Prime Minister often made off-the-cuff remarks concerning the French, frequently witty and sometimes little more than humorous jibes when she did not get her own way, but there was a deeper, much more complex set of dynamics behind this. In a response to German unity, its powerful voice in the European community and its past, Thatcher instructed psychologists to undertake an in-depth study of the German psyche. Had she forgotten that the Anglos and Saxons who invaded these islands in the fourth century AD were Germanic tribes that came from what is now Germany? The study of European languages in schools and colleges became a very low priority during the Conservative regime's time in office from 1979-97. Margaret Thatcher's sceptical views on Europe and her stubbornness on the issue were among the factors which led to her downfall, which in reality was little more than a behind-the-doors *coup d'etat*.

The prime minister's scepticism and sheer suspicion of Europe went back a long way and had their roots in her adolescent years. As a bright conscientious student, she had a great determination to succeed in life – something she did not allow even the blitz to interfere with. As German bombs rained down on Britain in the worst of the raids, the young Margaret studied under the kitchen table, unperturbed by the Nazi terror above. This wartime experience did not leave her totally unscarred; it gave her a deep sense of suspicion towards Germany and Europe in general.

During the Thatcher years (1979-90), British public and media interest in the Eurovision Song Contest slowly sank to its lowest ebb. This can be seen as setting in almost as soon as the new administration began to cut its teeth. The downgrading of the national final which started with the 1980 A Song for Europe slowly began to take effect. There was a tie for first place at the 1980 A Song for Europe show. Prima Donna's *Love Enough for Two* and New Faces talent show winner Maggie Moone's *Happy Everything* received the same number of votes but a quick, rather haphazard whip round the regional juries for an outright favourite came out in favour of Prima Donna, though Maggie Moone's ballad was a far superior song. Prima Donna was a group that had been put together specially for Eurovision. The members were Danny Finn, a former New Seeker; Lance Aston, a brother of Jay Aston (Bucks Fizz); Sally Anne Triplett, who later became half of Bardo, runners-up at the 1982 Eurovision with *One Step Further*; Alan Coates; Jane Robbins; and her sister Kate Robbins, who were both cousins of Paul McCartney. Kate later had a major hit with *More Than in Love* and appeared in the long-running soap *Crossroads*. She consistently provided voices for the satirical *Spitting Image*. The comedy shows *Phoenix Nights* and *Last of the Summer Wine* were only a few examples of her diverse roles in music and television. Prima Donna's Eurovision entry was only the fourth British entry to not reach the top forty since 1961 and they disbanded some months after the contest. The creation of groups especially for the A Song for Europe final became the norm, and most of them broke up after the event. Bucks Fizz was created for A Song for Europe to perform the song *Making Your Mind Up*. They did not expect the song to be selected for Eurovision and thought it likely that they would disband shortly afterwards. History, however, has again told a very different tale; not only did they win the green card to Dublin for the 1981 Eurovision Song Contest but they beat Germany and France, which come in second and third, to become the fourth British act to win the contest. One of the group's male vocalists, Mike Nolan, later stated that the dance routine in which the male vocalists pulled off the outer skirts of the two females to reveal mini-skirts and lots of leg helped it rock up the scoreboard. The origins of this stage presentation were, however, of a more practical than sensual nature. The female vocalists, Cheryl and Jay, disagreed on what style of skirt to wear. One favoured a midi, the other a mini; this gimmick was really a compromise. The song was a high-kitsch, retro, 1950s, up-tempo pop number with a dance routine complete with gimmicks. It topped the British charts and established Bucks Fizz as a leading 1980s pop act. They would prove to be the only 1980s British Eurovision act to achieve consistent chart success. This included three UK number-one singles. Cheryl Baker, one of the group's members, had dreamt of winning the Eurovision Song Contest since watching Sandie Shaw win in 1967. Cheryl was runner-up as part of Co-Co at the 1976 A Song for Europe with the song *Wake Up*. Co-Co won in 1978 and went on to sing at the Eurovision Song Contest in Paris. The UK triumph in 1981 fulfilled Cheryl's childhood dream. At the 1981 "Song For Europe" UK selection, she had, in fact, competed against a group called Beyond, which had included two men: one who was later to be Cheryl's fiancè and another who was destined to become her husband. Shortly after the 1981 contest, France complained that the contest had become silly and so did not compete in the 1982 event. Bucks Fizz's output was, more often than not, poppy, kitschy, feel-good music but it would be wrong to say that they were irrelevant and had no influence on the

British music scene. Bucks Fizz's format, choreography, and musical formula very much influenced the British pop scene in the years around the turn of the century, with chart acts such as S Club 7, Steps, Scooch, and others of the era emulating the Bucks Fizz, kitschy, poppy, Eurodance style.

The song *For Only A Day*, performed by Destiny at the 1981 "Song For Europe", ended up coming in last, a disaster for its composers; it did, however, go on to be a top ten hit for Leo Sayer in 1982. Destiny had included Sue Glover, who, along with her sister, formed the well-known session duo, Sue and Sunny. They had backed Lulu in Madrid and had such hits as *United We Stand* when they were both part of the original Brotherhood of Man, which then had a totally different lineup and sound from the 1976 group. Sunny's recording of *Doctor's Orders* charted in 1974, while Sue Glover had teamed up with former Blue Mink vocalist, Madaleine Bell, as the backing singers for Germany's 1975 entry, the soulful *Eine LIEd Kann Eine Brucke Sein* by Joy Fleming. The Scottish folk singer Fiona Kennedy, a Gaelic language star and daughter of the popular Gaelic singer Callum Kennedy, performed the song *So Do I* at the 1985 Song for Europe. As an actress, she worked in various TV dramas and had also appeared in the 1973 film *The Wickerman*. The A Song for Europe finals that spanned the 1980s featured few well-known chart acts. Sinitta, Hazel Dean, Liquid Gold, Scott Fitzgerald, and 1970s *Glam* star Alvin Stardust were among the few established acts who tried but, except for Fitzgerald, none qualified. The artists who took part grew more and more amateurish, obscure, and estranged from the mainstream of British popular music. Many of the British entries during this period were little more than token entries. The songs were not hits and were mostly forgotten by an indifferent public as soon as the Eurovision Song Contest was over.

Due to the obscure nature of many of the artists, a large section of the public lost interest in the contest. They could no longer relate to the mediocre entries sung by singers they had never heard of and would most likely never hear from again. This led to a decline in interest in

Eurovision: A Plea for Respect

the Eurovision Song Contest itself. Gone were the heady days of popular chart-topping acts of the 1960s and 1970s. Even the *Radio Times* stopped featuring the contest on its front page. The BBC did approach Bonnie Tyler and invited her to sing for the UK in 1985 but on that occasion she refused. Lena Zavaroni was also considered but that request was blocked by the music publisher association. The UK plodded on in what had become, in many ways, a wilderness in terms of Britain and the song contest. Was it only a coincidence that the contest had lost so much prestige and had been demoted to such a position under a government that had such a lukewarm agenda towards Europe? One theory is that government policy towards the European Community influenced the BBC in regards to the song contest or that the BBC was given instructions from a higher office. Terry Wogan's sarcastic Euro-scepticism and ridicule of foreigners seems to also have gone hand in hand with how low Europe was on the Conservative Party's list of priorities – a need to belittle or at least devalue Europe's importance in British society. If it is only coincidence that Sir Terry's commentary reflected the values of the government of the day, it was surely a useful tool for an administration that itself was as Eurosceptical as Sir Terry's comments. A knighthood was later bestowed on Wogan, the official seal of approval. After Thatcher's resignation in 1990, she was replaced as Prime Minsister by John Major and a Conservative government continued to govern the United Kingdom, while every spring Woganism became more audacious. Sir Terry had, in his career, interviewed Margaret Thatcher, and on her death was invited to her funeral, which he subsequently attended.

From the mid-1990s, we begin to see a change in direction at the national selection. A music guru was drafted by the BBC and produced several hits. Conservatism was finally defeated at the polls by the promises of Tony Blair's New Labour in 1997 and "Cool Britannia" became the motto in a changing Britain. Tony Blair took up office the day before Katrina and the Waves won the 1997 Eurovision Song Contest with *Love Shine a light*, chalking up a fifth win for the UK. The argument over Europe continued both inside and outside the

Conservative Party. The public was often very sceptical at its mention. The endless rows between Tories on the topic of Europe only heightened after the election of David Cameron as Conservative Prime Minister. The party faithful were split over what seemed to be an obsession with Europe. On the weekend of the 2013 Eurovision Song Contest, a leading Conservative politician declared, "A deep anti Europeanism has infected the party". In the weeks leading up to the 2013 event, the Anti-European Union party UKIP had polled a significant number of votes in local elections and received no less than 25% of the English vote.

* * *

May 2010, at a railway station café somewhere in Yorkshire, the rather surly café owner has a conversation with a younger railway employee reading The Sun newspaper

"My wife's going to a party tonight," says the owner.

"Oh, a Eurovision party, is it?" asks the Sun reader

"Yes, it is. I keep well away from that," says the owner with a great deal of pride.

"Well, it's rubbish, isn't it? I never watch it," says the Sun reader.

"Yeah, that's what it is, rubbish, nothing but rubbish."

"Yeah, rubbish" repeats the Sun reader.

* * *

A quick, superficial look at the history of these islands may begin to explain why attitudes like these towards Eurovision have been formed.

England was conquered by the Normans on an autumn day, as all good pupils know, in 1066. The Normans ascended to the role of aristocracy while the Anglo-Saxons became a subordinate people. England became little more than a French possession. The Normans, however, did not conquer Scotland but married into the nobility and, thus, ended up ruling the country.

Eurovision: A Plea for Respect

In the late twelfth century, the English Crown passed into the Plantagenet family and England became part of the so-called Angevin Empire, a group of territories of varying degrees of autonomy that belonged to the Plantagenets. Among them were the Kingdom of England, Principality of Wales, Duchy of Normandy, Duchy of Brittany, County of Anjou, County of Maine, County of Nantes, County of Poitou, and Duchy of Aquataine. The family also held the title of King of the Romans and were Lords of Cyprus. The Pope, in turn, gifted them the Lordship of Ireland. By 1206, due to wars and several family disputes, the family was left with only England, Wales, and Ireland and their sole French territory, Gascony. During their rule, the Plantagenets changed the face of England from what was little more than a colony, occupied by the French and governed from France, into one of Europe's most powerful kingdoms.

From 1206, the kings of England were resident in England and only made visits to Gascony; the tables were turned. Under the Plantagenets, we see the beginnings of an English identity and a sense of nationalism starting to take root in coastal areas. The Hundred Years' War was a series of battles over the French Crown that took place between France and England from 1337 to 1453. The war was rooted in quarrels over rights of succession. Its most famous conflict, the Battle of Agincourt, was the direct result of the French king refusing to recognise the English claim on the Duchy of Aquitaine and a request of marriage for the King of France's daughter, complete with a dowry. The dowry demanded was full of requests for several French possessions which included Normandy. The French refused and offered the King of England what he considered a far inferior proposal. The English King, Henry V, then waged war on the French and won the battle. It was a crowning moment in English history and the forging of an English identity.

During the Hundred Years' War, the Scots had formed a strong alliance with the French. It was agreed that if France were attacked by the English, then Scotland would use the alliance to also attack England; in turn, France would also attack if the English committed an act of aggression against Scotland. For a great many years, the Scots Guards were the personal army of the French kings and held the same position as the Swiss Guards in relation to the Pope. By the sixteenth century, when the alliance was at it its height, Scotland and France were more or less a united entity with Scots immediately qualifying for French citizenship at birth. Mary, Queen of Scots, was also Queen of France after her childhood marriage to the French Dauphin, Francis, in Notre Dame Cathedral.

As William, Duke of Normandy and the Plantagenets spoke French, the peasants spoke Anglo-Saxon. The fusion created modern English.

In 1525, Henry VIII attempted to divorce his Spanish wife, Catherine of Aragon, with the intention of marrying the vivacious and charming Anne Bolyn. However, the Pope refused. Henry responded by creating the Church of England. His daughter, Elizabeth I, born out of this union, continued the new faith as the state religion and began to bring it more in line with the reformed churches. Another link to Continental Europe was gone. While it had been united as one principality and Gwynedd Fawr at different times, notably under Owain Llywelyn, the loose collection of self-governing lordships and independent kingdoms, which had seen a host of influxes from Continental Europe, in the Celtic lands to the west of England, became the country of Wales and were united with England under Hendry VIII in the Act of Union of 1536. Ireland and Scotland maintained strong ties to the continent through religion, while the Stuarts exiled themselves to France and Italy. Bonnie Prince Charlie, the Stuart pretender, had the backing of the King of France and the Pope; long after he had sailed from Skye, was buried in the heart of the Vatican. After the execution of Mary, Queen of Scots, Philip of Spain sent the Spanish Armada to avenge Mary's death, but the full force of the English fleet saw it off into wild northern waters. During Elizabeth's reign, England, as a sea-faring nation, began to make advances and establish itself as a world power, creating colonies in the New World. This gave it unprecedented wealth. Its frequent conflicts with Spain

Eurovision: A Plea for Respect

and other European rivals in the New World consolidated its isolation and difference from the continent of Europe. Calais had always been considered an integral part of England, and even after the loss of all French territory, Calais remained English. It was a jewel in the English Crown and its treasury was, in particular, of great value to the production and export of wool to a wider market. The fact that it remained English long after Normandy, Anjou, and all the rest, including Gascony, ceded to France in 1453 at the end of the Hundred Years' War, were long gone was caught up with the rival claims of Burgundy and France over the town; both enemies preferred to see it in English hands rather than in the hands of the other. One of the mayors of Calais was the legendary Mayor of London, Dick Whittington. In 1558, during the reign of Mary Tudor, Calais eventually fell into French hands.

Sir Francis Drake and Sir Walter Raleigh were names that kept England free from invasion by Continental Europe and expanded its growing influence in the Americas. Elizabeth I died childless in 1603. Her reign was the most successful, and she was the most popular monarch in English history. She had been instrumental in creating a strong English identity that was distinct from its European neighbours'. The son of Mary, Queen of Scots, James VI, succeeded her on the throne of England and united the monarchy of the two kingdoms. The political union of the two countries to create the Kingdom of Great Britain would occur in 1707. From the sixteenth to eighteenth centuries, due to religious and political conflict in their native land, a huge number of Scots left Scotland and settled in Baltic countries. Thousands set up homes in Poland and there was even a Scottish ghetto in Warsaw. Russia was also a magnet for Scots. Cameron, a Scottish architect domiciled in Russia, was the architect of the new Western-influenced St Petersburg. According to statistics, 100,000 contemporary Russians are of Scottish descent. Others settled in Germany and Scandinavia. The Norwegian composer Greig was a Baltic Scot. At the 1977 Eurovision Song Contest, Sweden was represented by a group called Forbes, whose lead singer was descended.

from Baltic Scots. Exiled Scots also provided mercenaries for a diverse range of European conflicts. Additionally, they were employed by the Bourbons to help keep them in power in Naples. There exists a town, Gurro in Piedmonte, Italy, whose population is descended from Scottish mercenaries and local girls. It is now a popular destination for Scots. Its population cemented its long relationship with "La Scozia" when it built a Scottish Museum. The Montenegran representative at the 2007 Eurovision Song Contest, Stevan Faddy, also claimed to be descended from a Scottish mercenary.

By the late eighteenth century, the union between England and Scotland was beginning to take effect. Bonnie Prince Charlie and the Jacobites had, by now, long been defeated and were slowly ebbing away into the pages of history and the lyrics of romantic songs, while a large section of the Highland population had emigrated to the Americas. Napoleon had crowned himself Emperor of France and was soon making military advances all over Europe. Some of his soldiers landed in Wales. France, England's old enemy, was making trouble once more. The Scottish government, which since the Reformation had maintained cool relations with France (the "Auld Alliance" long over) was now part and parcel of the British government in London. Soon, war existed between Britain and France but the might of the new British superstate soon put an end to Napoleon's skullduggery, and the French army was defeated at the Battle of Waterloo. The Duke of Wellington became a national hero. There had also been another display of naval might when, in 1805, the Royal Navy won a massive defeat over the combined navy of France and Spain at the Battle of Trafalgar under the command of Horatio Nelson, to whom a column was erected in London in a huge central square dedicated to Trafalgar. Between 1652-54, 1665-67, and 1672-74, England and the Netherlands fought naval wars over supremacy of the sea and trade routes. After the Act of Union in 1707, the British government again went to war with the Netherlands, this time from 1781-84, over world trade, which the Netherlands had dominated until 1713. By the late eighteenth century, inroads were being made into India and Canada at

Eurovision: A Plea for Respect

the expense of the French. All these historical events helped to build an anti-Europeanism. *Remember, remember, bonfire night?* November 5 and all that. Was not the Guy on the bonfire the terrorist Guy Fawkes, a sympathiser with Popery? That is to say pro-French, pro-Spanish, pro-Italian, a foreign spy?

England had been in a personal, though not political, union with Ireland since 1542 when Henry VIII was proclaimed king. In 1798, a rebellion tried to end English influence completely and establish an Ireland totally free of Great Britain. This involved a French invasion of the country but the rebellion was put down and two years later, in 1800, the Union of Ireland Act led to the creation of the United Kingdom of Great Britain and Ireland.

The popular song *Rule Britannia*, written in 1740, was put to music by Thomas Arne and had its origins in a *Masque*, a form of courtly entertainment which involved singing and dancing. It was about Alfred the Great and was first performed at the birthday party of Princess Augusta. The lyrics were a poem written by the Scottish poet James Thomson, who lived in England and wanted to create a new British identity that would both include and be influenced by, but at the same time transcend, the traditional identities of the English, Irish, Welsh, and Scots – an identity that would demonstrate the spirit of the new age of unity and cooperation. The song went on to become *"the"* most popular and most sung of all British patriotic songs. It epitomised the confident, robust Britain that would take shape over the next century. Just over 100 years later, most of the world map was coloured red. Britannia now did rule the waves and Queen Victoria was also the Empress of India. At the 1840 Eurovision Song Contest*, the fat lady belted out *Rule Britannia* and won hands down, way ahead of the Netherlands, France, and Spain. Does this, however, mean another Anglo-Dutch naval war? *"Rule Britannia, Britannia rules* the *waves"* was even more popular in the nineteenth century when Britain had a feisty collective national identity. Britain strode on through the nineteenth century and, like the fat lady, belted out her song, leading the world in medical research, new technology, and industry. It felt invincible and maybe even thought it was, taking its faith to all corners of the globe, along with its political traditions, language, and people.

In 1914, once again it's war in Europe, a new kind of war never seen before. The war to end all wars. But it doesn't. Britain is victorious but its confidence is cracked, its spirit broken. It picks itself up and, as resilient as before, bounces back. Again, the world is its oyster. Its territory is an empire on which the sun is never meant to set.

In 1939, the United Kingdom, from which by now most of Ireland had chosen to leave, again found itself at war with Germany as the Nazis marched into one European country after another, installing Quisling governments and occupying once-proud nations.

The Nazis landed on the Channel Islands. The United Kingdom stood alone against the Nazi terror. Night after night, its cities were blitzed from Coventry to Belfast. Lives were lost and streets flattened overnight.

Britain and its allies danced the Hokey Cokey in the streets on VE day and politicians in Westminster formed a consensus to give the British public a better quality of life in appreciation of their war effort. They liberated occupied Europe and the Nazi concentration camps. Slums were demolished. New homes were built with modern amenities and the national health service was set up to give free health care. *"You've never had it so good,"* declared Conservative Prime Minister Harold MacMillan. India and Pakistan gained independence in 1947 and a whole host of British colonies from Cyprus to the coast of South America followed in the 1950s, 1960s, and 1970s. The queen was still the head of the Commonwealth, but the Empire was no more.

In 1963, The Beatles burst onto the scene and the American Rock 'n' Roll that had dominated the music scene around the world gave way to British pop. Its loss of political and geographical empire gave way to something else that compensated the national psyche and gave consolidation to the UK's collective national identity, that is, replacing the old, traditional political empire with the British musical and cultural empire that blared out of transistor radios in Hong Kong and Reykjavik, conquering far more territories than there were ever red

Eurovision: A Plea for Respect

areas on the global map. This was an overcompensation, perhaps, for feelings of unexpressed loss and even anger at the setting of the sun on the political empire. The compensation from the cultural empire, a cream gateau for comfort eating. The social theorist Paul Gilroy attributes Britain's post-colonial melancholy to a failure to properly process and accept the end of the country's status as a world leader.

Margaret Thatcher, a great British patriot who came along at a time when Britain needed a saviour carrying a messianic torch, or UKIP, which arrived at a moment of ultra-Euro-skepticism, both nourished an emaciated British identity that was in crisis.

A stranger prepares to enter a room expecting smiles and welcomes from four people. However, as he enters the room, he is faced with an auditorium packed full of indifferent faces, a cold atmosphere, and a blinding spotlight that highlights his person. Unsure of how he appears, extremely nervous and not sure what to do, he laughs. The Eurovision Song Contest contradicts what the UK still clasps tightly onto: its musical/cultural empire. It feels insecure and no longer recognises itself. It assuages its sorrow, its insecurity, by ridiculing the event and dismisses the music as tasteless and insipid. The fall of the Empire and the post-colonial period that followed saw not only a post-colonial guilt but also a collective national identity dilemma ongoing since the early 1950s. An abhorrence and a tendency to resist European unity or integration has manifested itself as a major factor in this crisis.

The Eurovision Song Contest is where both national pride and Pan-European unity sing a duet, if not in one accord. The Eurovision Song Contest is a part of both our collective national identity and our collective European identity. It is the Arch-European of European organisations and soirees. The European Union fades into second place in the song contest's Arch-European shadows. If only the union could generate as much enthusiasm throughout Europe, even for one night. If only voters in the European elections were as eager to put an X on the ballot paper as they were to pick up the phone and dial a vote for the winning song. It is not an insignificant, silly competition full of instantly forgettable songs. The Eurovision Song Contest is rich in its own traditions and ritual. It is maligned much more by the British media than by any other country's media who stigmatise its artists and the contest itself as an entity. It does not do this to other major international cultural or sporting events. This degrading and derision of the event is often accepted as normal. Can you imagine commentators at the Olympic Games sniggering and making xenophobic jokes about athletes as they warm up on the track? Or the commentator shouting, "She shouldn't be in the lead, she's East European and must be on steroids!" Or for the British Olympic authorities to enter athletes who they know will perform poorly or for the English football association to allow a team from the bottom of the league to go forward to the European Cup? No, the logic behind it is not at all normal and openly states what the Eurovision Song Contest has slowly become in the UK over the last twenty-five years: a scapegoat for unexpressed frustration and anger over a loss of a national identity and the resulting ridicule, as well as a form of revenge and an apparent vagueness at exactly how to react in daring to question who is in charge of the ruling musical and cultural empire.

Lecturer in Dramatic Arts at Brock University, Karen Fricker, states, "*Eurovision became a popular institution where this identity crisis and resultant Euro scepticism played out, and Wogan's commentary was its key conduit. From his earliest days as commentator Wogan positioned the British as bemused participants in a distinctly foreign, European spectacle, one in which the UK none the less enjoyed success and status. He constructed an ideal listener who was like him, male (or masculine defined), British (or British identified) and white*". It is at this crossroads that the lowbrow and the crass of The Sun newspaper meets the highbrow and the refined of The Times being anti-European at the most European of events, the Eurovision Song Contest. This unites them and gives them some sense of a shared national identity. Meanwhile, good marketing ploys and positive publicity have benefited *The X Factor*, at the height of its popularity in the latter part of the 2000s, as well as *Britain's Got Talent*, *The Voice*, and the acts who have appeared on them. The Eurovision Song Contest in Britain is looked upon as being

Eurovision: A Plea for Respect

one step, or even two, lesser in value. The acts on the aforementioned shows are afforded the status of credible or potentially credible acts of merit, yet the acts at the Eurovision Song Contest are often highly polished, established, popular acts who are international stars with a chain of platinum and gold discs to their credit. They are nevertheless written off as drivel compared to the *"real"* talent on offer on the three popular British talent shows. The Eurovision songs and singers are looked upon as subordinate to the artists who appear on them.

Let us imagine that, at the start of each episode of *The X Factor*, Charpentier's *Te Deum* was played as the show's intro music and the Eurovision logo was shown. If, in large letters, the words "Eurovision Song Contest" were displayed on stage for all viewers to see but the programme still showcased the same acts as had appeared on *The X Factor*, a change in attitude would occur and a sudden over-scrutinisation of the acts would begin. The live audience and viewers at home, conditioned by the negativity fed to them about the song contest, would distance themselves from the show. What would they do? Switch off? Say, "It's complete rubbish"? Start laughing at the singers? Laugh at their costumes? It would be interesting to see exactly what would happen. The same change in attitude and perception would occur if we then re-named the Eurovision Song Contest "The X Factor" with the same Eurovision artists and songs showcased at a previous song contest because, you see, then the public would stop ridiculing the acts. They would say how good the singers were, or at least give credit where it was due. More serious commentators would be asked to preside over it. The Eurovision presenters would be afforded more respect and would even be looked up to because, after all, they aren't Eurovision presenters any longer; they have become, after swapping titles, *X Factor* presenters. The winner would be assured of plenty of positive publicity and marketing. Their song would be given plenty of airplay on the radio and would be guaranteed to become a huge hit, as would their follow-up recording. But you see, this is still, in reality, the Eurovision Song Contest even though we have dressed it up in *The X Factor*'s very best evening clothes. There is also the popular misconception in the UK that if a song is a hit in Belgium, Britain, and the USA, it is an *"international"* hit, whereas if it is a hit in Belgium, Norway, Sweden, Spain, Italy, and Greece and not a hit in the UK or USA, it does not qualify as an international hit. Likewise, a European singer who is well known in one European country as well as in the UK and the USA is considered an *international* artist but a European singer who has topped the charts in ten European countries but has not charted in either the UK or the USA is not considered an international artist.

Eurosceptisim, Anti-Europeanism, unexpressed sentiments at the loss of empire, the need to maintain the role as head of a musical, cultural empire, and an identity crisis are at the core of Britain's attitude towards Eurovision. Not only does the scapegoat expect us to compete at an equal level but it also has the sheer bare-faced audacity to suggest that it can play us at our own game and even imply that we should revere or look up to it. Tantamount to an open insult, it implies that it can do what we are so good at doing but even better! To the British, this is completely unacceptable. Ghosts of the Empire and current political paradigms in British society fabricate a prevalent view of the Eurovision Song Contest in the United Kingdom. There are also tinges of Edward Said's notion of otherness at work in the forces that make up British attitudes towards Eurovision. Why do countries of the former Soviet Union and Balkan states take it so seriously? In 2013, Azerbaijan failed to give any votes to their brotherly neighbour, Russia. There were rumours of votes going missing, of vote rigging. Russia supposedly was allocated ten points but the points went missing. The Russian government gave a sober reaction to this and voiced their displeasure in no uncertain terms. The British media were quick to pick up on this and there were several reports on TV news and in newspapers. In more than a few, the degree of irony rose in a giggle from the pages. There was a definite undertone of *"Imagine being so silly as to even think of taking the contest seriously"*. Why do they take it so seriously? Well, of course, the answer is simple: "Because they are so stupid and backward that you can't expect anything else". That is the

Eurovision: A Plea for Respect

perceived mode of thought. While the Empire sent out manpower, politicians, and language to the colonies, it also sent out missionaries by the hundreds to convert the natives. Because of the structure of the United Kingdom, it was often said that England had a political empire but Scotland and Ireland had a spiritual empire, referring to the number of missionaries who came from those places. Now the musical/cultural empire exports music, films, and television programmes, often with an aggressive secularism embedded in them: the reflection of a changed Britain that is more in tune with atheism and agnosticism. An anti-establishment with roots in the 1960s that still feigns anti-establishment status when it has long taken over and has become *the* establishment itself. While sub-Saharan Africa and the Hebrides remain devoutly religious and still regularly express their Christian heritage, there is a tendency, at various levels and quarters of British society, to look upon their spirituality and religious observance as a result of backwardness and ignorance, although this patronising take is rarely articulated. Often, it comes from people who otherwise would consider themselves very far removed from being racist or prejudiced and might even see themselves as somewhat PC. There is even a paradigm that rebukes the Hebrides for their religious observance because they are British and an integral part of the UK; thus, they should know better than to continue to manifest such "*dated nonsense*", while the same paradigm, although not condoning it, regards religious practice in Africa as acceptable and normal because, after all, "*they are in Africa, the dark continent, and can be excused for such backward practices*". Such thinking manifests a still persistent colonialist and racist mentality. From the same pattern of thought comes the idea that other Europeans who have a serious approach to the Eurovision Song Contest do so because they are stupid, have no musical taste, and are most probably tone-deaf. That the Eurovision Song Contest has changed dramatically since 1956 would be an understatement. Its change has mirrored the ever-changing nature of European society in general in sociological, geographical, political, and cultural terms. In 2005, a BBC news reporter stated that it was "*a far cry from the days of *Pearl Carr* and *Teddy Johnson*". The song contest, over the last two decades, has become a far more diverse place reflecting, again, the various aspects of diversity of contemporary Europe. In 2012, the British tabloids featured articles and interviews in which individuals cited racism as a barrier to a British win and the cause of low marks at the Eurovision Song Contest. They seemed to be unaware of the background of that year's Swedish winner: North African Berber. The Swedish entry the year before was sung by Eric Saade, of mixed Swedish/Palestinian heritage. Ukraine, which had some well publicised major racist incidents that year, was also cited as a country that would never enter black singers when, in fact, Ukraine had entered a black singer that very year. East European selection was portrayed as racist when, in reality, the first black singer to win the contest had been an Estonian entry. It also stated that Jade Ewan (UK, 2009), of mixed Sicilian, Scottish, and Jamaican heritage, had lost the contest due to colour when Jade had, in fact, come in fifth, the UK's best result in years and the only entry to reach the top ten since 2002. It all seemed another desperate attempt by the British media to discredit Eurovision and use racism as a cover for the true reason for UK failure: lack of outstanding songs. This seemed a classic example of the bully venting his angst on the scapegoat and accusing it of racism, a practice that he was not completely innocent of and could not come to terms with. There certainly was a lot of prejudice openly displayed in regard to European musicians the UK's media, a bizarre form of musical apartheid. Far-right extremism and ethnic tension exists and is undoubtedly a big problem in several parts of Europe. This still has to be addressed. It has been highlighted in recent documentaries on football hooliganism in Eastern Europe, which heavily featured racist behaviour there. Yet, it is interesting that Ukraine chose a black singer to represent them at Eurovision that year. This would seem to suggest that the Eurovision Song Contest reflects a far broader, inclusive, and more diverse Europe – the *real* Europe, a more *tolerant* Europe, or at least the Europe that it aspires to be. It is a world away from the faceless bureaucracy of Strasbourg and Brussels that the European general

Eurovision: A Plea for Respect

public often have difficulty relating to. In recent years, there has been a very obvious mingling of nationalities on stage at the song contest – a direct reflection of many current European cultures and their ever-changing demographics. Eleni Paparizou, who won for Greece in 2005, was, in fact, a Swede of Greek heritage. Tooji, the Norwegian representative in 2012, was an Iranian asylum seeker. Aleksander Ryback, the 2009 Norwegian winner, was of Belarussian stock.

The 2006 Israeli song was sung by a black singer and, again, in 2020 and 2021, an Israeli of Ethiopian background was selected to sing for them. The 2012 Swedish winner, Loreen, of Moroccan background, danced on stage with a black man. Then there was the 2011 singer for San Marino: Senhit, an ethnic Eritrean who returned to the contest in 2021 with her highly publicised entry, in which she duetted with the 41-year-old African American million-seller, rap star Flo Rida, who had, among his successes, the internationally acclaimed *I Cry*, *Wild Ones*, and *Right Round*. Although the song did not do as well as predicted on the big night and came in twenty-second, it had brought an unprecedented amount of publicity to the tiny state, which boasted of being Europe's oldest republic. The question had to be asked: If a tiny country like San Marino could get a mega-star to sing for them, why was it so hard for the UK to follow suit? While Roumania's 2015 and Albania's 2019 entries centred around mass emigration from their respective countries, Sweden's 2021 entrant, Tusse, a refugee from the war in Congo, was only one of several black artists to be selected for 2020 (the Eurovision that had no competition) and to appear at the 2021 event. The Swedish entrants in 2020 were The Mamas, a black, all-female, American-Swedish group who performed soul and gospel. Meanwhile, the Maltese singer of 2021, Destiny, was the daughter of a Nigerian football player. The Swiss entrant in Rotterdam was of Albanian parentage; his father a Kosovan refugee. The Russian entrant at the 2021 contest, Manizha, a refugee from Tajikistan, was also the acting Goodwill Ambassador for the U.N. Refugee Agency. The Eurovision Song Contest was commended for its refugee inclusivity by the UNHCR, which also hailed the success of those from refugee backgrounds in the contest. The Albanian entry of 2019, *Ktheju Tokes*, which, like many Albanian entries, was a classy rock ballad fused with plenty of elements of Albanian folk and traditional instruments, was selected for Eurovision, as had been the custom since 2004, at the Festival I Kenges, a long-standing song contest that had its inauguration back in the Albania of 1962, when it was still a closed nation shrouded in mystique and visited only by a select few Westerners. The legendary Albanian vocalist with the voice of an angel, Vace Zela, won the festival no fewer than eleven times while the nation languished under the iron authoritarianism of its communist dictator, Enver Hoxa. At its inception and the decade that followed, its songs were mostly about love but that all changed in 1972 when Hoxa accused the participants and organisers of conspiring against the state: endangering the country with immorality and corrupting the youth. The festival then started to reflect government policies and dwelt in an ambiance of severe scrutiny and censorship. Song lyrics were monitored and their themes very limited. This continued until Hoxa's death in 1984, when censorship was relaxed and the following year's festival saw a dramatic change in song topics, which soon became more liberal and progressive each year. Many of the new songs were infused with rock elements which had been absent before. It all manifested a support for liberalisation within the Communist Party itself that the general public was totally unaware of, and the entries began to raise both high expectations and great outrage. The Party was, however, already resigned to the imminent demise of the regime behind closed doors. Albania was the last country in Europe to abandon communism in 1992 but the new orders anthem had already been heralded as early as 1988 through the songs at the Festivali i Këngës. Groups such as Tingulli I Zjarte and vocalists like Parashqevi Simaku and Mrena Reka motivated Albania's new generation with politically progressive lyrics in their entries. In the 1990s, the clothes and the performances often reached new heights of extravagance and singers such as Adelina Ismajli and Ledina Celo became the fashion icons and trendsetters of the transitional democratic era. The latter won

Eurovision: A Plea for Respect

the festival in 2005 and qualified for the Eurovision in Kyiv. After the fall of communism, due to the economic and political instability that ensued, as well as the Kosovar War, thousands of Albanians fled as refugees and continue to do so as economic migrants to other European countries, most notably to Italy and Greece. In the song, its singer, Jonita Maliqi, reflects on the mass immigration from Albania as well as from war-torn Kosovo to Europe and touches on mass immigration throughout the world in general. Communism ended much earlier in Albania's fellow Balkan state of Roumania with the overthrow and execution of its dictator of twenty-four years, Nicolai Ceausescu, in 1989; in its wake, thousands of Roumanians, like Albanians, left their country to find more lucrative employment in the job markets of Western Europe and America. In 2022, an estimated three million Roumanians were employed outside of Roumania. Their 2015 Eurovision song was titled *De La Capat*, sung by Voltaj, and centred around the plight of the many children left behind to be brought up by relatives while their parents were working abroad, which has caused other social problems.

Other aspects of diversity have surfaced at the song contest which have manifested themselves in the number of paraplegic artists competing and providing the interval acts. Notable examples are the finnish punk band PKN, made up of singers and musicians with Down syndrome, who sang for Finland in 2015. Their song deals with the wishes of many disabled people to live a self-determined life and not have decisions that affect them made by others. The Shalva, a group made up of singers with special needs who had qualified from the Israeli final but, due to concerns about performing on the Sabbath, decided to withdraw, did, however, appear at the 2019 song contest as one of the interval acts. Unlike the Olympic Games, which have two separate Olympiads so that paraplegic athletes do not compete against able-bodied athletes, paraplegic musicians at the Eurovision Song Contest compete on an equal parring with the able-bodied at the one event.

The United Kingdom has a history of tolerance towards different cultures and faiths that have come from afar and settled within its borders, much more than in many other European states. This tolerance is partly the result of a deep-rooted democratic tradition as well as contact with different races, faiths, and cultures from far-flung parts of the Empire during colonial times. The same could be said of the Netherlands, also a former colonial power, which has a long-standing tradition of tolerance and liberal attitudes. The British tradition of acceptance contradicts the British attitude towards Eurovision and foreign language music. Although, if we look closer at British society, we see a polarisation of ethnicity, a drawing away from a mainstream culture that minorities might not have initially related to or felt part of, manifested in Polish churches, Irish centres, and Latvian clubs. In the late 1980s, a host of Eurovision Song Contest fan clubs sprang up in the UK. A group of people like all fan clubs with a shared interest? Yes, but Eurovision fan clubs in the UK were more than that. They were a centre point where people whose interests had been marginalised by the mainstream culture were able to express their views on equal terms.

The Netherlands has, since the 1950s, felt comfortable and open to popular music from most of Western Europe. Dutch radio frequently plays songs in Dutch, English, French, Spanish, Italian, and German. Milly Scott, the first black singer at the Eurovision Song Contest, represented the Netherlands in 1966 and sang in Dutch. Milly was a jazz singer from Suriname, then a Dutch colony in South America. Her song was unusual for the time, a Latino number about two Chileans living in the vicinity of Santiago, *Fernando en Fillipo*. She was the first singer to use choreography on stage and entered the stage via steps already into the song's intro. Her backing singers wore sombreros and ponchos. Although the first black singer, she was not the first singer of colour to appear at the event. The Dutch representative at the 1964 contest, Anneke Grönloh, of East Asian extraction, had made that debut hers. Suriname, although in Latin America, is often considered more Caribbean in culture than Latin American. It has a large population of Javanese and Chinese decent as well as inhabitants of African, Native American, and Arab extraction.

Eurovision: A Plea for Respect

It has a small population that is one of the world's most ethnically diverse. In the lead-up to independence in 1975, one-third of its population, afraid that an independent Suriname would fare worse, left the country and settled in the Netherlands. Today, an estimated quarter of a million people of Suriname extraction live in the Netherlands. Of the Netherlands' Eurovision entries from 1970-2010, seven were sung by singers of Surinamese backgrounds, which included Edsilia Rombley's 1998 *Hemel en aarde*, which was the last time a song would be sung in Dutch until 2010. In the song, there are references to the coolness of the weather in relation to the Dutch temperament. Lutgard Mutsaeos, in her essay *Fernando* en *Fillipe*, notes that this created a degree of critical distance to "Nederland", the only time a Dutch Eurovision song mentioned the nation itself, making Edsilia an observer rather than an insider.

Milly Scott, whose homeland of Suriname, like its population, spawned music that was highly influenced by a myriad of cultures, challenged the Eurovision Song Contest of the day with her inescapable blackness, her jazz style of improvising, and her sensual shaking of her hips. Her Eurovision performance also stood up to the accepted mainstream pop music of the time and broke all its rules.

The Estonian winner of 2001, Dave Benton, who had sung his way to victory with Tanel Padar, was from the island of Aruba, part of the Netherlands Antilles; Benton had been a resident of Rotterdam before moving to Estonia. Aruba claimed his victory as a home win, although the Caribbean nation is not a member of the EBU.

Colonialism in Indonesia also led to a substantial number of Indonesian immigrants moving to the Netherlands. After the Dutch East Indies gained independence from the Netherlands and became the republic of Indonesia, it saw an exodus of European and Pro-Dutch inhabitants, most of whom set up home in the Netherlands. The first singer to appear at the Eurovision Song Contest who was not of European heritage was Anneke Grönloh, an Indonesian from North Celebes. Born in 1942, she had spent her formative years in a Japanese concentration camp. At the end of the war and the liberation of the camps' internees, she left Indonesia with her parents for a new life in the Netherlands. While at college in Eindhoven, she made the acquaintance and worked with the popular 1960s Dutch rock group Peter and the Rockets. Grönloh's popularity swelled after she won a talent contest; she became the first Dutch teen idol with a series of record-breaking singles in the early 1960s, such as *Paradiso* and *Brandend Zand*. Although she recorded almost entirely in Dutch, her first record, *Asmara* (1960), was sung in Indonesian. It, however, made no impact on the Dutch music scene. Her Indonesian roots surfaced in her performance of *Kroncong* music, a style that utilised the Indonesian string instrument of the same name. Her 1964 Eurovision song was a ballade, *Jij bent mijn leven*, about a woman who is very much aware of her loved one's unfaithfulness. It was a typical song of the early 1960s in rhythm and arrangement and would not have looked out of place sung in English by Helen Shapiro in the British top twenty of the era. It came in tenth. In the same year, Grönloh married Wim-Jaap van der Laan, a DJ on the pirate station Radio Veronica. Anneke Grönloh was the beginning of a long relationship that the Eurovision Song Contest was to have with Dutch Indonesians. In total, from 1964-2005, no fewer than eight acts of Indonesian background represented the Netherlands at the contest. This was no surprise, as singers of Surinamese and Indonesian blood had often dominated the Dutch music scene, especially in the early years of Dutch pop, the late 1950s and early 1960s. The Netherlands experienced a period of racial discord in the late 1970s due to mass immigration from Suriname and terrorist attacks against the Netherlands by militant Moluccans from Ambon (part of the Dutch East Indies). Predominantly Christian in Muslim Indonesia, their fathers had been recruited as military to subdue discontent in other parts of Indonesia. At Indonesian independence, many Moluccans fought for a separate republic but the insurrection was put down. Former Ambonese, Dutch Army soldiers, abandoned their homeland *en masse* and settled in the Netherlands. The younger European-born Moluccans, angered by what they saw as discrimination and bad living standards, waged attacks on several

Eurovision: A Plea for Respect

targets inside the Netherlands. One of the worst atrocities of the campaign was the hijacking of a train, where they killed several people. In a cry for mutual tolerance and understanding, the Dutch Eurovision entry of 1978 was *It's OK*, sung by a multi-racial group, Harmony, which comprised members of Surinamese, European, and Indonesian blood. In 1989, Justine Pelmelay, a Moluccan singer, sang *Blijf zo als je bent* for the Netherlands in Lausanne. She was the first Moluccan ever to represent them. One of the country's best-known Dutch-Indonesian artists of the twentieth century was Sandra Reemer, who, with her partner Dries Holten, sang for the Low Countries as Sandra and Andres in 1972. They came in fourth with *Als Het om de Liefte gaat*, which was recorded in English as *What Do I Do?* The song was a particular favourite among teenagers in the UK. Sandra and Andres exploded onto the stage in Edinburgh's Usher Hall. Their colourful costumes and, in particular, Sandra's dazzling nail varnish, rose more than a few eyebrows across the continent. In Britain, many fourteen to eighteen-year-olds thought it very contemporary and in keeping with the pop trends of 1972. They considered Dries Holten's composition more contemporary than their own entry, *Beg, Steal or Borrow* by the New Seekers, who, at the time of the Edinburgh Song Contest, were at the zenith of their popularity and could do no wrong. Sandra and Andres split up and Sandra continued to sing solo. She represented her country again in The Hague in 1976; this time, she sang in English. The song was the haunting *Party Is Over*, about a bad dream from which she awakes. It had a very Gypsy feel to it with mandolins in the intro. She delivered her song in a classy evening dress complete with a pseudo-Elizabethan ruff around her neck. The B-side of *What Do I Do?*, on which she also duetted with Andres, was also a very Gypsy-influenced track, even titled *Gypsy Man*. Sandra Reemer returned to sing at the 1979 Eurovision Song Contest, this time as Xandra. The song was sung in Dutch and named after the American state of Colorado. This was the weakest of all her Eurovision entries.

As the 1980s stepped up pace, the Eurovision Song Contest began to be looked upon in the Netherlands as passé, with as much dust and cobwebs as Miss Havisham's wedding cake. It was no longer a great honour to represent the country at the Eurovision Song Contest.

In 1993, the same year the Dutch Surinamese Ruth Jacott sang *Verde* in Millstreet, the gay TV presenter Paul de Leeuw almost single-handedly radically changed the approach to the contest, rejuvenating it and opening its dusty closed windows to the fresh air. The Eurovision Song Contest in the Netherlands was given a makeover and became a place where diversity opened its doors. A new interest in the song contest sprang up.

In 2000, Anneke Grönloh was named *Singer of the Century* in the Netherlands. Back in dear old blighty in 1992, the BBC returned to the selection process that was in operation in the 1960s and 1970s, that of nominating an established artist. The West End musical star and popular personality who had also enjoyed chart success, Michael Ball, agreed to sing for the UK. *One Step at a Time* was the song chosen by public vote. It received 153,792 votes, which gave *As Dreams Go By*, which polled 94,844, the runner-up position. Ball came in second that year in Sweden to Ireland's *Why Me?* Both were rather unremarkable songs that were not deserving of their high placings. The Italian song *Rapsodia* by the late Mia Martini at her best and the song from Greece, *I agape Kai I elpida* by Cleopatra, were among the quality songs that year. They both made the top ten. Malta's Mary Spiteri came in third with *Little Child*, a big but rather dated ballad. The UK continued with the same selection process in 1993. The chart-topping pop singer Sonja accepted the BBC's invitation to sing for the United Kingdom in Millstreet, a village in Cork. One of Britain's leading Eurovision fan clubs had held a poll for who their members thought should represent the UK. Sonja won the poll. As in the previous year, the songs were performed on Terry Wogan's show and then showcased in a final, which the public voted on. It was to be another second for Sonja's *Better the Devil You Know*, another rather mediocre, repetitive pop song. It had polled 156,955 of the total vote, although it was known that Sonja's personal favourite of the eight songs presented was *Our*

Eurovision: A Plea for Respect

World, which came in a distant second at 77,695. Although the Liverpudlian was much associated with up-tempo pop, *Our World* was one of several ballads in the national heats that year. The winner, Ireland's *In Your Eyes* by Niamh Kavanagh, was also another unremarkable ballad, even though it was the best-selling disc of the year in Ireland and turned platinum. Niamh would return to the contest final in 2010 with the very Celtic ballad *It's for You*, which only managed to bubble around the nether places of the scoreboard, but was a far superior song which Niamh performed perfectly.

In 1994, the BBC kept the same format of selecting a singer and then giving them a slot to perform the short-listed potential British entries on Wogan's TV show with a final and public vote. The award-winning singer Frances Ruffelle was chosen to represent the UK but although Frances was Britain's top female "musical" vocalist, with a string of successes in the West End, particularly associated with *Les Misérables*, she was hardly a household name. The song (*We Will Be Free*) *Lonely Symphony* was chosen to go forward to the Eurovision in Dublin. It received 99,946 votes – over 30,000 more than *Sink or Swim*, which 63,417 people had voted for. Frances' true professionalism showed through in her performance on the evening but her song sank to tenth place. This was a rather pleasant contemporary ballad and one of the best British entries. It was disappointing that it did not come in the top five. There was a noticeable difference in the number of public votes cast in A Song for Europe in 1994 compared to the previous two years. Viewing figures for the Eurovision Song Contest itself were down. The BBC considered pulling out of the event. It would seem that the low percentage of televotes cast and the low ratings were due to the singer being relatively unknown to the wider British public, whereas Michael Ball and Sonja had received a much higher number of televotes and BBC viewing figures had been higher.

In 1974, following ABBA's resounding victory at the Dome in Brighton, the Disc music paper proclaimed, "*The Old Eurovision Song Contest is dead, long live the new Eurovision Song Contest!*" It went on to applaud the winning Swedish entry and asked *"But couldn't our Quatro, Sweet or Slade have done it better?"* The question on the music world's lips was: Which top chart act would sing for Britain in 1975? It was a time when the BBC invited an established act to sing six songs on the Saturday night A Song for Europe slot. It was the mid-1970s and most of the top bands on *Top of the Pops* would have leapt at the chance to do the honours. When it was revealed that The Shadows had been chosen to represent the UK, there were gasps! The Shadows, of course, were an important component in the music business, and as pioneers of British pop music they held a special status in the hall of fame of British pop/rock and roll music, but it had been several years since their last hit and, again, they were not at all representative of 1970s pop music. The Shadows left London Airport for Stockholm disgusted and disappointed at the lack of support given to them by the British press. *"Shadows resurrected from the dead for Eurovision"* reported one popular daily. Their song *Let Me Be the One* was a very run-of-the-mill pop song with no special merits but was voted into second place. Runner-up was what, in the 1960s and 1970s, British acts expected to be; anything more was a most welcome bonus, and anything below third was a disaster. The contest had not done The Shadows any harm and their credibility as a legendary British act remained intact. They continued with a career. The postal vote numbers were much down on the previous year, most likely because there was much less public interest in The Shadows than there had been for the 1974 representative, Olivia Newton-John. There was also much objection from sections of the British music industry. Many songwriters were dissatisfied with the BBC selection process and wanted a wider, less restrictive procedure which did not require one act to perform all the songs. An air of change was beginning to creep into the British public's notion of the song contest but would it be a positive one?

    The selection process which had been in operation since 1964 had proved highly successful. The established acts who took part in the Eurovision Song Contest had, during the years 1967-75, all ended up in the top four and none were ever more than seven points behind the

Eurovision: A Plea for Respect

winning entry. The BBC introduced a revamped method of selecting the British entries in 1976. The artists were chosen by the songwriters themselves and the number of songs in the national final, showcased at a very grand A Song for Europe, increased from six to twelve. The first of these national finals was highly successful and produced Brotherhood of Man's *Save Your Kisses For Me*, a popular entry which gave the UK its third win at the Eurovision Song Contest. In its first few years, the new system saw a plethora of chart acts and well-known faces. They included Labi Siffre, Tony Christie, Tammy Jones, Lyn Paul, Guys 'n' Dolls, Polly Brown, Sweet Sensation, and Hazell Dean as well as big names from the 1960s such as Frank Ifield and the Foundations. Brotherhood of Man, Lyndsey De Paul, and Mike Moran, a former member of the prog rock group Stone The Crows, were already well established with several hit records to their credit when they won the national final. By 1978, things had started to go wrong. The UK entries of 1978 and 1979 finished well out of the top five. In 1994, as had happened in 1975, the BBC was forced to look for another selection process. Something a little more contemporary was deemed appropriate and some degree of effort went into the 1995 national final, which was re-named the Great British Song Contest. A major attempt was made to modernise and raise the profile of the event. The contest saw some relatively famous artists such as London Beat, Sox (featuring Samantha Fox), and the Eurodance chart act Deuce. The song chosen for Eurovision was *Love City Groove*, performed by a group of the same name. A rap song, it was an immediate hit in the UK and peaked at seven on the charts but managed only tenth at the Eurovision Song Contest. Despite three follow-up singles, they failed to make much of an impression on the charts and split up. The British/American R&B group Londonbeat, whose career had Dutch roots and who had hit the number-one slot in the Billboard Hot 100, delivered a funky number entitled *I'm Just Your Puppet on a ... String*, which became a minor hit in the British charts, as had other finalists, *Go for the Heart* by SOX and *Dear John* by One Gift of Love. Deuce, whose *I Need You* came in second, also had a top ten hit with their recording. The 1996 Great British Song Contest finalists were previewed by Nicky Campbell on a *Top of the Pops* special and then presented at the actual national contest. As was the case in the previous year, the musical showcase was a very up-to-date production in keeping with the normal *Top of the Pops* format. The final with the short-listed songs was presented by Terry Wogan.

The Australian singer Gina G was the runaway winner with her song *Ooh, Ah, Just a Little Bit*. It went straight to number one on the UK chart and took Europe by storm. It was one of the few Eurovision songs to be a big hit in the USA and was nominated for a Grammy Award for best dance song. One solution the EBU had to the increasing number of countries wanting to take part in the Eurovision Song Contest was to have an audio non-televised semi-final. This was later replaced by a relegation system. Gina G came in third in the semi-final but on the night of the Eurovision Song Contest could manage only eighth. The worldwide hit that it became more than compensated for its disappointing placing. The song was also the first UK Eurovision non-winner to top the charts since Cliff Richard in 1968. The 1997 national final followed the same format and presentation of songs. Although much faith had been invested in Gina G's entry, it had not delivered the goods and failed to bring the trophy back to Britain. The year 1997 was to be a far different affair. The song *Love Shine a Light*, sung by Katrina and the Waves, was selected as a British entry for the Eurovision Song Contest in Dublin. The song had originally been composed by Kimberley Rew as an anthem for the Samaritans to commemorate their thirtieth anniversary. Kimberley Rew, a member of the band, was very against the idea of going in for Eurovision. Although he appeared in the group's lineup at the national final, he was not on stage for the actual Eurovision performance in Dublin. Katrina and the Waves were mostly known for their biggest hit, *Walking on Sunshine* (1985), which became a popular summer anthem played by DJs every year on warm sunny seasonal days. *Love Shine a Light* was to give the United Kingdom a runaway win and receive one of the highest percentages of any winner; it received 78.82% of votes

Eurovision: A Plea for Respect

available on the night. The song also received twelve points (maximum) from ten of the competing nations, the third highest ever, coming behind Italy's 1964 *Non ho l'eta* and Germany's 1982 *Ein Bischen Frieden*. It had beaten Ireland, Turkey, and Italy into second, third, and fourth places. *Love Shine a Light* charted in several European countries and reached number three in the UK. It was not, however, the runaway world hit that Gina G's 1996 entry had been. The group's involvement with the song contest boosted the European pop/cabaret aspect of their career but damaged their image as a credible rock band – a factor that dogged the band members ever since their Eurovision win and an argument that eventually led to the group's breakup in 1999.

There had been a definite effort to raise the profile of the British selection in the 1990s as well as to make it appealing to a younger audience. This had, up to a point, been a success, with one victory and several big hits emerging from the Great British Song Contest. The year 1998 saw the Eurovision Song Contest hosted in Birmingham, the first time since 1982 that it had been held in the UK. There was even a fair amount of positivity coming from the media. On BBC prime time TV in the week of the 1998 event, presenter Carol Smiley announced that the Eurovision Song Contest had *become trendy again*. There was a revived public interest in Europe's most-watched programme throughout the UK. The Israeli singer Dana International, a transsexual, won the 1998 contest with the disco number *Diva*, which received positive reviews in the popular press and became a pan-European hit. Its English version also charted in the UK. Dana International's victory was the third Israeli win. It differed from their 1978 and 1979 wins in that this was not only a victory for Dana International but a triumph for the secular/liberal Israel that she stood for, as opposed to the Orthodox Israel of the ultra-conservative religious who hated Dana International and the Israel she represented. In a CNN interview, she stated, "I do not represent the Jewish state. I represent the Arabs, the Jews, the Christians, I represent all the citizens of Israel, I represent all who want to be represented by me". The ultimate country to perform at the previous year's contest, 1997, held in Dublin, had been Iceland, whose singer, Paula Oscar, as the first openly LGBT participant, had attracted much attention and even ripples of shock in some quarters due to his suggestive performance. Dana International had been born Yarron Cohen of Yemeni/Roumanian stock and had steadily risen to the ranks of Israeli pop to become the most popular singer in the mid-1990s with her blend of Euro dance-trans-pop which more than sometimes had a helping of Hebraic-Yemeni folk in its makeup. At her victory walk on the Eurovision stage, Dana International shone as a member of a minority group, a disenfranchised person who connected with other disenfranchised Europeans such as the East European migrants, the Turkish minorities in Germany and Netherlands, the Portuguese immigrants in France, and others such as the emerging economies and infant democracies of the old Warsaw Pact and former Yugoslavia, the persecuted ethnic, sexual, and religious minorities, those suffering from the economic crisis. All disenfranchised peoples who feel a fuller sense of European citizenship through participating as artists and musicians, televoting at home, viewing the contest on TV, interpreting it as a truly European organisation/event that they are part of in a much wider, more diverse sense as European citizens. The Eurovision Song Contest gives them a veritable sense of inclusion.

After the 1997 event, the Eurovision Song Contest in the UK was almost back in the fast lane – almost but not quite. It still needed a boost to bring it right back to the top. A top established act for 1999? Something to restore its musical credibility in the UK? A new approach to the Eurovision Song Contest in general and not only the British selection at the BBC? The time was ripe. Never had there been a better time to restore the contest's prestige and credibility. Its future relationship with the UK in this respect looked bright. The 1999 entry, *Say It Again*, like 1998's *Where Are You* by Immani, was a very contemporary R&B-pop number that proved to be very popular and was, again, a top ten hit for its singers, the girl group Precious, who declared themselves funkier than the Spice Girls and who would go on

Eurovision: A Plea for Respect

to have relative chart success with three more singles in the top fifty. One of its vocalists, Jenny Frost, continued her musical career with the girl group Atomic Kitten, with which she had three number-one singles.

As Eurovision approached in the late spring of 1999, the last year of the century, the BBC produced a trailer to publicise the event. They had chosen to ridicule it, airing the chorus of Massiel's *La, La, La* to exhibit how banal the songs were, with the theme that it was not awful but fabulous, a time to have a party because it was of no aesthetic value or interest to British viewers as a musical event. If the BBC had seriously wanted to upgrade the Eurovision Song Contest and give it positive marketing and publicity, changing the public perception from that of negativity to that of positivity, it would have acted at that decisive point in time, 1999. The BBC chose not to and only further pushed Eurovision into a corner where there seemed no way of escaping from its dunce's cap and the ridicule from the rest of the class at its status as a joke, the class scapegoat.

There had been some hope and expectation in the period 1992-99 and, in particular, the latter years of 1995-99 but to bring the Eurovision Song Contest into line as a credible, mainstream music event that is of value, the society and cultural structures of the time had to change their perceived notions of the event. The notion of a credible music event was there but that was not enough. Social and cultural forces also had to inhabit it. There was no point for the BBC in arranging a national selection process that veered towards current music trends while the Eurovision Song Contest itself was still ridiculed and deemed tasteless by the organisation generally in its commentary, on its chart shows, and in the regular banter between DJs on their daily output, just as it was equally pointless broadcasting such a programme while the press were still stubbornly embedded in some bubblegum pop songs that had been featured in the contest some forty years before or were obsessed with some of its more kitsch aspects.

Back in 1961, the British national final, A Song for Europe, saw a move away from the light music of the BBC Home Service to a more youth-orientated programme that reflected the rising pop trends of the day. There were, of course, some contenders who, although popular, still slotted into the Home Service mould, such as Bryan Johnson and Anne Shelton. Both were still big names back in 1961. The BBC had asked three leading record companies to present some of their artists for the national selection at A Song for Europe. Among the acts was the former choir boy Mark Wynter who, in 1962, reached number two in the UK with *Venus in Blue Jeans*. Craig Douglas also competed. He'd had hits in 1959 with *Teenager in Love* and *Only Sixteen*. He was to have several hits after the contest as well, the most successful of which was *When My Little Girl Is Smiling*. Another classic teen idol of the era to participate was Ricky Valance, who topped the charts in 1959 with *Tell Laura I Love Her*. One of the entries, *Too Late for Tears*, was composed by Clive Westlake, who later would write the Dusty Springfield hit *I Close My Eyes and Count to Ten*, while *Why Can't We?* was a composition of Eric Boswell, credited with writing *Little Donkey*, a big Christmas hit for Nina and Frederick in 1960 and that is now classed as a Christmas carol sung in churches across the country on Christmas Eve. As the emission neared its end and Katie Boyle read out the remaining votes from the juries, it became clear that *Are You Sure?*, sung by The Allisons, was bound for Cannes as the United Kingdom entry at the 1961 Eurovision Song Contest. Contrary to popular belief, The Allisons were not brothers but consisted of two teenagers, Bob and John Allison (AKA Colin Day and John Alford), whose singing style and presentation were similar to those of the American Everly Brothers. Their career had started in 1960 when they won a talent contest as part of the Boys and Girls Exhibition at Earls Court. Their song *Are You Sure?* was very typical pop of that era. It proved to be the fastest-selling British Eurovision song so far. A week after A Song for Europe, it entered the top twenty and very soon reached number two on the chart. The boys' musical director at A Song for Europe and the Eurovision Song Contest was Harry Robinson, who, as Lord Rockingham, hit the top ten in the late 1950s with *Hoots Man*. The British press of 1961 varied in its critiques of A Song for Europe and the Eurovision Song

Eurovision: A Plea for Respect

Contest. Back then, there was still an ability in the media to criticise songs where criticism was felt appropriate and at the same time give credit where credit was due. There was little of the wall-to-wall preconceived criticism of the contest as a whole which has come into existence today and is a built-in part of the cultural programming as to how the contest is seen in the UK.

Thirty-nine years later, in 2000, A Song for Europe had been demoted to the Sunday afternoon off-peak viewing time of 4.00 p.m. The 2003 national final featured low-quality songs and a farcical attempt at some jury/panel that was meant to represent different areas of the UK. The young duo Jemini's song *Cry Baby*, which was ultimately chosen as the UK entry, received zero votes at the Eurovision final in Riga. Much was made of the politics supposedly behind it but it received '*nil*' points because it was simply a bad song that two inexperienced performing arts students tried their best to perform against an opposition of some formidable international stars such as Sertab and Russia's TaTu. Jemini returned to the UK to a wave of publicity, albeit for all the wrong reasons, and a press blazing with anger at a Europe that had insulted them. Sunday morning current affairs programmes on BBC Radio 4 devoted a considerable amount of discussion time to the Eurovision Song Contest. Jemini quickly and rather embarrassingly crept out of the limelight and, after they split, returned to a fairly mundane life in Liverpool, where, with a new image, they soon failed to be recognised by the general public. But! They had miscalculated the effects of receiving '0' at a Eurovision Song Contest. Record companies and others in the music industry should have seized the opportunity to capitalise on this and use the nil points to catapult them to stardom. As is now generally considered, if you must come in last at Eurovision, it is best to come in last in true style! After the 2003 disaster, a section of the media called for the UK to withdraw from the annual event. One newspaper, of the right, reported that the UK had not received any votes and laughed it off saying, rather angrily, that they just didn't care, when the complete opposite was true.

Jemini, in more recent years, has appeared on Eurovision nostalgia TV shots in the build up to the song contest.

In 2004, the national selection process was again revamped and its profile upgraded. It was moved from the Sunday afternoon graveyard slot to Saturday evening primetime. From 2004-08, the programme was re-titled "Making Your Mind Up" and featured an array of, on the whole, better-known artists, among them some *Pop Idol/X Factor* participants: Andy Abraham, James Fox, and Javine Hylton. Several soloists who were part of highly accomplished groups also featured strongly in the show, such as Anthony Costa (Blue), Kym Marsh (Hearsay), Liz McClarnon (Atomic Kitten), Justin Hawkins (The Darkness), Brian Harvey (East 17), and Daz Sampson, who had charted as part of various outfits. All had enjoyed major hits. Others who attempted to represent the UK during this period were the MOBO award-winning R&B/hip-hop group Big Brovaz, Katie Price (the model Jordan), and Cyndi Almouzni (AKA Cherie), who had already charted big in the US. At the 2007 national final, where the two finalists, Cyndi and Scooch, had completed the final sing-off and the public eagerly awaited the results of the final vote, the presenters, Fearne Cotton and Terry Wogan, simultaneously announced the name of the winner. Wogan proclaimed Cyndi the name of the winner while Cotton proclaimed Scooch as winner. Wogan's voice was the loudest, so most people believed it was Cyndi who had won, when, in fact, it was Scooch. Lots of rumours surfaced around Morrissey, who was reported to have been approached by the BBC to write a song in 2007. In the years that followed, Jarvis Cocker, Primal Scream, and boy band JLS all stated an interest in singing at the Eurovision Song Contest. The period 2004-08 produced several big hits but none of the British entries at the Eurovision Song Contest managed to impress their continental neighbours. None of them placed in the top ten, while Britain came in last in 2008 with Andy Abraham's *Even If.* The UK came in second last in 2007 with Scooch's ultra-camp *Flying the Flag.* Javine's *Touch My Fire* (2005) could reach only twenty-second, while Daz Sampson's catchy *Teenage Life,* which soared up the UK charts,

Eurovision: A Plea for Respect

did not fare much better, ending up nineteenth in Athens in 2006. In this period, there was a surge of interest in the contest and a chain of hits for UK entries despite the UK doing badly at the event. Since the 1990s, fan clubs and websites of the Eurovision Song Contest have increasingly flourished in number and popularity. A solid die-hard core of fans now followed the contest and the national finals leading up to it. The growing number of countries that wanted to take part had now increased so much that two semi-finals had to be created to accommodate them all.

In 2009, Lord Lloyd Webber was drafted to write a song and preside over an *X Factor*-style final that would eliminate so many people each week. Then the finalist acts would sing Lloyd Webber's Eurovision composition and a televote would choose the performer best suited to singing it. This was a completely new format for the British selection and one that was very much in keeping with other talent show formats of its time. Jade Ewen broke down and wept as she sang a reprise of the song *It's My Time*, which she had been chosen to sing at that year's Eurovision Song Contest in Moscow. The song was a big ballad which suited Jade's dress and staging at the national final. However, Jade's dress and the stage presentation on the big night did not complement or blend in well with the song. She slowly walked down steps lined with violinists, which she accidentally banged into, which distracted viewers from the song. Nevertheless, she gave a great performance, which prompted Lloyd Webber to say, "Tonight a star is born". He had, incidentally, played piano on stage while Jade sang. It was to be the best night in seven years for a United Kingdom song and came in fifth, only the second occasion since 1998 that a British entry had placed in the top five. After the accusations of bloc voting from the UK and some other Western "old" European countries, and subsequent complaints to the EBU, it bowed to pressure and reformed the voting process, which had, in the years previous, relied completely on televoting for the results. The new rules installed a jury from each country which would provide 50% of the vote, while the other 50% would come from the public. This would operate until 2013, when another system which still relied on the split jury/public vote but differed slightly from the previous format, came into being. The new split jury/public vote that was implemented in 2009, to appease the UK and others, had given Jade third place in the jury vote but fifth in the combined result. This would, however, backfire in 2011 when Blue came in eleventh with *I Can* in the joint jury/televote result when they had, in fact, come in fifth in the televote, which many British viewers had thought of as biased and favouring Eastern Europe. Jade's recording of *It's My Time* did not reflect its success in Moscow on the British charts, where the song only managed to trouble the top thirty and peaked at twenty-seven. After the Lloyd Webber invitation to write the UK entry, a selection process that had achieved high ratings and a long-awaited, respectable result at Eurovision, much was expected of the 2010 national final. The press had heaped a fair amount of praise on Jade Ewen's 2009 attempt as well as generally welcoming Norway's win, *Fairytale*, which had gone on to be a top ten hit in the UK for Alexandyr Ryback. The BBC announced early in 2010 that Pete Waterman, Mike Stock, and Steve Crosby would write the British entry. Waterman and Stock were consistent hitmakers in the late 1980s. A talent show which had the duration of one night, unlike the previous years, which lasted several weeks, was the designated format in finding an artist to sing the song which, revealed on the same night, was *That Sounds Good to Me*. This selection process was to prove a shambles – the most disastrous national final in years, if not ever. Very amateurish acts, some who could hardly sing, battled it out in a final that the BBC seemed bent on using as a stop-gap between popular reality shows that had obviously been given much more thought and positive marketing, deemed far more important than only the national preliminary of the Eurovision Song Contest. Not only were the acts who performed lacking in talent but the production of the whole show was worth *nil* points. The whole organisation of the show from beginning to end had sunk to an all-time low. The nineteen-year-old Josh Dubovie was by a long chalk the best artist of the bunch and had the most engaging and professional approach. He easily won the ticket

Eurovision: A Plea for Respect

to Oslo and the right to sing *That Sounds Good to Me*, a very poor song, which he sang well. It provoked a backlash of questions as to why top hit writers of the 1980s had been asked to compose the British entry in 2010. As betting odds in the days before the Eurovision Song Contest looked fairly hopeless, Pete Waterman distanced himself from the song, saying it was not a great number. Josh gave a well-performed rendition a the Grand Final but came in last. Its commercial success fared just as badly, not even entering the top hundred. Its sales allowed it only a number 175 on the charts. Josh smiled and took the result like a good sport, then went away and reinvented himself as Josh James with a change of image.

The stigma in the UK of singing at the Eurovision Song Contest, plus the fact that no matter how good follow-up material might be, it is still likely to be snubbed, has often led to a stalling in artists' careers. Jade Ewen, who rode on the crest of a wave when she was elected to sing for the UK, went on to join the Sugababes shortly after her Eurovision participation, seemingly abandoning a solo career. This, despite the fact that Jade was one of the most promising new female vocalists to emerge in 2009. If she had won *The X Factor*, her solo stardom plus a run of hit singles to accompany it would have been assured, simply because the then-current cultural value systems dictated that *The X Factor* was superior to the Eurovision Song Contest. *The X Factor* is not stigmatised, which I tried to illustrate earlier in this chapter. Artists such as Jessica Garlick, who came in third to Will Young in the *Pop Idol* quest and then went on to compete in the Eurovision Song Contest. James Fox, who came in fifth on Fame Academy in 2003 and the following year sang for the UK in Istanbul. Javine, whose R&B vocals earned her a part in *Popstars the Rivals*, where she narrowly missed out on a place in Girls Aloud. Andy Abraham, the former dustman from London who was runner-up in the 2005 *X Factor* and who represented the UK in Belgrade. All of their careers, sadly, floundered due to negative attitudes that abound in Britain towards Eurovision. Jessica Garlick left the music business a short time after. Several had done reasonably well before having competed at the song contest. James Fox later branched out to become one of the top "musical" artists post-Eurovision and has toured Europe and the USA, singing in successful musicals. Javine had several top twenty hits in the years leading up to her Eurovision adventure. Andy Abraham had a platinum album to his credit, and all of them were serious singers. A few came from the soul-R&B genre as opposed to bubblegum pop. The song *Flying Your Flag* by Scooch, unashamedly in the tradition of *Boom-Bang-Bang* and *Jack in the Box*, was, at the time, a bit of fun. The exception was James Fox, who made his way successfully into the likes of *Chess* and *Jesus Christ Superstar* and charted again with a football anthem for Cardiff FC. The others either failed to make an impact or made a limited one on the British charts after Eurovision. Some were, however, still very much around. It became customary for them to be interviewed prior to an imminent Eurovision Song Contest. Their personal lives were also followed with interest by the tabloid press, which focused on Javine in particular. Even James Fox has, interestingly, channelled his talent into stage musicals rather than mainstream pop. Michael Ball's career flourished after his participation in Eurovision but this was, again, due to his overwhelming success in West End musicals. Perhaps musical theatre was more tolerant of Eurovision artists. Daz Sampson, whose *Teenage Life* was one of the biggest hits from the UK entries that decade and whose populist "*voice of the working man*" approach endeared him to many, soon faded from the pop scene. However, he did submit a song, *Goodbye*, for the 2007 British pre-selection, on which he duetted with former T-Pau vocalist Carol Deckar. The BBC rejected it on the grounds that it was too soon for him to represent the UK again. Daz Sampson had nine hit singles in the top forty which included *Ai No Corrida* as part of Bust Stop, Uniting Nations, and others. His performance at the 2006 Eurovision Song Contest was watched by half the country, marking the highest UK viewing figures for a Eurovision Song Contest in its history. Daz attempted Eurovision again in 2021 when he teamed up with Katya Ocean and competed in the Belarus national final but did not qualify for the song contest. The artists found themselves treading through

Eurovision: A Plea for Respect

thick mud to continue with successful careers – a reflection of the British attitudes towards Eurovision. They were victims of the bully and the scapegoat scenario and caught in a cultural mouse trap.

The 2007 Eurovision winner, *Molitva*, from Serbia was probably the most outstanding winner of the decade. It was a powerful song with rousing music and fitting lyrics dramatically performed by Maria Serafovič. The French newspaper *Le Monde*, not normally known for its pro-Eurovision views, stated that at last a classy song had won the Eurovision Song Contest. Yet, despite being one of many songs from the 2007 Eurovision Song Contest to enter the British top hundred, it was widely snubbed by radio stations. Morning TV shows were full of derogatory criticism of the Serbian entry. In terms of taste, anything "*tarnished*" by Eurovision was bad – even if it was good. While the British top fifty over the last forty years has supposedly been full of great musical masterpieces with deep, meaningful lyrics, if that is so, then there have been more than a few exceptions! The song *You're Gorgeous* by British band Babybird, a hit in 1996 and for many years given plenty of airplay, is an example of a much overrated piece of work that is deemed good music. The song contains elementary repetitive lyrics and no melody. Yes! The song's lyrics dwell on the exploitation of models by photographers, but many Eurovision entries have also focused on exploitation. A prime example is the 1977 Austrian entry *Boom Boom Boomerang*, about the exploitation of artists and musicians by the music industry. This is usually paraded on TV documentaries as an example of how awful Eurovision is when the whole meaning behind the song and its performers, *Schmetelinge* (a Brechtian theatre group), are ignorantly overlooked. One tool in underrating and undermining music featured in the Eurovision Song Contest is to play only small parts of songs completely out of context on documentaries and current affairs programmes related to Eurovision. Researchers will have searched for as many *La, La, La's, Li, Li, Li's,* and *Na, Na, Na's* without seemingly having carried out much in-depth research into the contest. Simultaneously, the 2013 chart topper *La, La, La* by Naughty Boy and the 1991 *Gypsy*, a house hit from Crystal Waters which featured a rather tuneless chorus of *"Da, Da, Di, Da, Di, Da"* might also qualify for derision in the stocks. There is a strong emphasis on over-emotion by showing a clip, a split second long, of extreme soprano or sentiment. Several radio/TV documentaries on the Eurovision Song Contest have surfaced over the years. In 1996, an excellent documentary was made for the BBC by John Peel. He made another in 1998. Around the same time, a behind-the-scenes documentary was made about the 1998 song contest in Birmingham. *Naked Eurovision* portrayed the event in a favourable light. Ten years later, Sandie Shaw narrated another radio documentary which explored the theory that *X Factor/Pop Idol* shows satisfied the need for mini contests in the absence of the Eurovision Song Contest for most of the year. In the first decade of the twenty-first century, most documentaries had been aimed solely at ripping the contest apart, conveying the idea, once more, that it was a joke of only comical merit. They are generally presented by trendy young things who, in a culture that puts too much emphasis on irony and wit, express it to the full, fabricating a nonsensical sixty minutes and compounding negative attitudes. Given wrong and biased presentation through television and the press, it would be possible to turn anything into the ridiculous and valueless. What has been done to the Eurovision Song Contest could easily, given the right circumstances, happen to any other music or television production. There is a capability to turn anything that is respected and classy into the complete opposite. The popular turn-of-the-century music hall artist Marie Lloyd popularised songs that are still well known today and that are now embedded in London folklore. Social reformers of the time were very against the music hall for what they saw as its decadence and potential corruption of youth. They were particularly against Marie Lloyd, whose popular, often bawdy songs epitomised the behaviour found at the music hall. Accused of indecency, Marie Lloyd was summoned to sing in front of the Licensing Council committee. Among the songs she performed was her classic *A Little of What You Fancy Does You Good*, which she sang softly and sweetly. The council could not find anything in the lyrical content that was indecent or

Eurovision: A Plea for Respect

offensive. Feeling insulted by having to appear in front of the Council Committee and angered at what she saw as their unnecessary interference, Marie Lloyd then insisted on performing Alfred Tennyson's sentimental parlour ballad *Come Into the Garden Maud*, complete with her stage trademark of winks and nudges at each suggestive innuendo. The bawdy manner in which she performed the song both shocked and affronted the council but Lloyd maintained that the rudeness was in their minds alone and had nothing to do with the lyrics of the song or her singing style.

In 2011, the BBC opted for another selection process. This time, both the artist and the song were selected internally. A large section of the public agreed with this. They felt the public had not been able to choose a song that Europe would appreciate and that the public tended to vote for kitsch, poppy numbers. This move by the BBC was seen as a veering away from this and a chance to enter a better quality of song. Blue's contemporary song *I Can* was sent to Dusseldorf. Blue had been a chart-topping boy band only a few years before. Viewing ratings again shot up. *"Britain Falls Back in Love with Eurovision"*, a major tabloid's headline read the Monday following the event. The DJ Chris Evans asked, with a degree of irony, *"Was it ever in love with it?"* Karen Flicker, in her 2013 Guardian article, praised the British entries of more recent years: *"There is another way of looking at the UK's recent relationship to Eurovision, that offer of a more optimistic vision of the country's ability to tolerate difference, particularly the internal differences and characterise the post-imperial, multi cultural, multi national British state. The UK's Eurovision acts since 1997 have been stylistically varied, and the country has been represented by a notably diverse group of artists. The last time the UK won the Eurovision was with Katrina and the Waves; an immigrant (the American Katrina Leskanich) led the act"*. The Eurovision Song Contest certainly has reflected the rich diversity in the vibrant new United Kingdom as it makes an effort to adapt to the premier pan-European cultural extravaganza, which itself is adapting to new conventions at an astonishing rate or is what in reality the continent itself has become: a multicultural conglomerate. That Britain was once joined geographically to the European continent appears to be an irrelevant ancient story seen through the mists that trail back to a time inconceivable to post-modern inhabitants. The current indigenous peoples of these islands are the intermarried descendants of the many different settlers who arrived from the continent long after Great Britain had become an island, if we also remember that Cornwall was previously joined to Africa and that there was also Doggerland, a land mass in the North Sea which, like Atlantis, was submerged under the waves. Among the peoples who came here were the Danes. Was King Canute not a Dane? Was not Old King Cole an ancient Briton? Normans, Anglos, Saxons, Jutes, Romans, and Norse, many of them preceded by Iberians and followed by waves of other Europeans from what are now Germany, France, and the low countries. The United Kingdom's past and a large percentage of its population are inextricably linked forever with Europe, like the ethno-pop/dance songs from Greece, Cyprus, and Turkey which fused local musical culture with Anglo-American pop dance. The English language as we now know it is a fusion of Anglo-Saxon and Norman French. The Normans and French but with Norse forefathers. Irish, Italians, and a host of Commonwealth citizens have, since the nineteenth century, settled in the United Kingdom, whose culture is at a crossroads in its history. In 1998, one year after Katrina's win, the first black singer to represent Britain, Immani, came in second at the Eurovision Song Contest. Since then, seven acts of mixed race or colour and groups comprising singers from a plethora of ethnic backgrounds have represented the United Kingdom. The genres have ranged from R&B to pop, from soul to rap and urban dance.

In 2012, the BBC selected seventy-six-year-old Engelbert Humperdinck to sing *Love Will Set You Free* and in 2013 selected, for the third year internally, Bonnie Tyler. Flickers comments on recent UK entries again in The Guardian on a positive note.

"Performers' ages have ranged from 19 to 76. This year's entry, Bonnie Tyler, is evidence of both intra-national diversity and age acceptance, being a 61 year old woman".

Eurovision: A Plea for Respect

The BBC and the United Kingdom are obviously getting some things right. Perhaps they have found the right track in the stadium. Now, all they have to do is run on it. Ripples of shock, or at least bemusement, went through some of Eurovision's aficionados when it was announced that the veteran singer, the legendary Engelbert Humperdinck, would sing for the UK. He had been absent from the scene in Britain for some years, although active in the States and Europe. Engelbert's song was composed by top contemporary writers Martin Terefe and Sasha Skarbek. If the artist was more associated with the 1960s then the songwriters were certainly right at the summit of the 2012 music scene. Terefe, a Swedish composer based in London, had been brought up in Venezuela, where he developed an interest in music and learned to play the guitar. This Latin influence was very evident in *Love Will Set You Free*. Terefe set up a home in London in 1996 and established a recording studio in the Kensal Town area. Before composing for Eurovision, he had written and produced albums for Jamie Callum, K. T. Tunstall, and Martha Wainwright, among others, and was also a member of the group Apparatijk. Sasha Skarbek trained as a classical musician. This background also showed through in *Love Will Set You Free*. He had written songs for a broad range of artists, notably Adele's *Cold Shoulder* and Samantha Mumba's *Body to Body*. He co-wrote with James Blunt on his international hit *You're Beautiful* and songs on the multi-platinum album *Back to Bedlam*. Skarbek's success led to him being awarded a Grammy and an Ivor Novello Award Although the best years of Engelbert's career in the UK were behind him, his classic hits were still remembered and whistled in the streets or sung at karaokes by a public who remembered him with great affection. Although he had never entered the Eurovision Song Contest before, he had enjoyed chart success with *Another Time Another Place*, one of the six finalists that Clodagh Rogers had sung at the 1971 Song for Europe UK national final. Engelbert had also represented the United Kingdom at the 1966 Knokke Song Contest and the same year hit the Belgian charts with *Dommage, Dommage*. His number-one single, *The Last Waltz*, was also covered in French by

Mireille Mathieu. Her version was a minor hit in the UK. The Anglo-Indian, alias Gerry Dorsey, originated from Madras but had moved with his family to Leicester at the age of ten. His illustrious career had included a host of 1960s hits, many of which were English cover versions of continental, especially Italian, hits. He counted several Grammy Award winners among them. As his chart entries began to dry up, he branched out into an American cabaret market. This led to his name adorning the Holywood Walk of Fame. He entered the top five on the British album charts in 2000, and another album released five years later was also extremly succesful. Known as the *King of Romance*, he was certainly one of the biggest names to represent the UK in a long time. There was, however, a degree of criticism from some quarters about the BBC's choice of Engelbert for 2012. They felt that the ageing vocalist was too old and too disconnected from the current British music scene and that he was generally regarded as a retro 1960s artist. A river of his 1960s and 1970s hits returned to the BBC airwaves and Engelbert found himself yet again at the centre of the British limelight in the lead-up to Eurovision. A spokesperson for the BBC said that he had been chosen for Eurovision on the grounds of his popularity in Eastern Europe, a major force when it came to voting. Critics, however, argued that if the BBC had wanted to put in a 1960s legend who was also still well-known in Europe, why hadnt they asked Tom Jones to sing for the UK? After all, he still had the occasional hit, appeared regularly on TV, and was a mentor in *The Voice* talent show. Nevertheless, after selection, there was a fair degree of revived public interest in Engelbert. The song *Love Will Set Us Free* was a dusky ballad and Engelbert is an artist who knows how to sing that particular genre extremely well. He is one of the very few British male vocalists, along with the late Matt Munro, who can perform it *par excellance*. This was a good song and an unexpected and unusual change in style for a UK entry. Of course, the song did face criticism from celebrities in and out of the music business who, for some reason, felt they had to make derogatory remarks about it to maintain their status as *trendy* and *hip*. It was nevertheless generally well received by DJs on BBC Radio.

Eurovision: A Plea for Respect

Engelbert performed *Love Will Set You Free* as the first entry of the evening on the stage of the Grande Final in Baku, complete with a medallion given to him by Elvis for good luck. His performance was that befitting a big star of nearly fifty years of experience and an energetic feat for a seventy-six-year-old man. He came in twenty-fifth, second from last. It seemed that a combination of factors led to such a low placing. His age? Others have said a big *no* to that, stating that second place was taken by a group of singing grannies from Russia. The public, however, adores cuddly, sweet grannie figures more than they do ageing sex symbols. The Russian entry also arrived at the song contest on a wave of publicity and media interest which had accompanied them since their win over mega-star Dima Bilan at the Russian final some months before, which had given them an added advantage. Other theories for Engelbert's failure to impress the juries were block voting – but then some of his few votes came from Eastern Europe, while most of the Western countries, with a few exceptions, had failed to vote for him at all. Another theory was that he performed first. Maybe it could have been partly to blame but then again, in 1976, Brotherhood of Man had performed first in The Hague and had won.

The fact is, despite the Russian runners-up placing, the contest mirrored the tastes of a younger generation, which was clearly shown in the notes of the Swedish winning song *Euphoria*, a good contemporary pop song that, prior to winning, was already in the top ten of several European countries. Like recent winners in previous latter-day contests, i.e., *Believe* (2008), *Fairytale* (2009), and *Satellite* (2010), it had received plenty of airplay and was popular all over Europe. Their airplay and success Europe-wide were not because of politics; it was because they were very popular. In the official UK charts, Engelbert reached only sixty. The recording did, however, fare much better on the Amazon chart.

The contest was over and the BBC was in a huff. Loreen's *Euphoria*, which entered the charts at number three, was snubbed. Engelbert's CDs were put back on the shelf. *Love Will Set Us Free* withdrew self-consciously from the airwaves and the Eurovision Song Contest ceased to be mentioned.

Engelbert already had a long and distinguished career ranging from European chart success to packed cabaret shows in Las Vegas. Eurovision was unlikely to affect his career. BBC commentator Graham Norton mentioned on his radio show that he felt Engelbert should have received an honour in the New Year's list. Perhaps yes, but given that he had represented the BBC in Eurovision, couldn't they have given him a TV series to endorse his comeback in Britain? The same applies to some of the young, up-and-coming acts that have, in recent years, represented the UK. After all, they have literally laid their careers on the line to sing at the Eurovision Song Contest and should be given some support in maintaining or even furthering their careers. Couldn't Josh Dubovie have been given a *Ready, Steady, Go/Top of the Pops*-style programme to introduce while at the same time using it to showcase some of his own repertoire. In the case of Andy Abraham, The Beeb could have given him a soul/R&B-oriented TV show to host, complete with special guests from that genre and singing some of his own material.

A year after Engelbert's attempt, the BBC again selected both song and singer internally. Bonnie Tyler had sent the BBC a copy of her new album, *Rocks and Honey*. They chose the track *Believe in Me*, written by top American songwriter Desmond Childs, and invited her to sing it as the British entry in the 2013 Eurovision Song Contest in Malmo. Bonnie had declined a BBC invitation to sing at the event in 1983 but on this occasion she accepted. There were, again, critics in some quarters who criticised the selection of an artist whose major hits had been in the 1970s and 1980s; yet Bonnie Tyler, although absent from the British chart for some years, had always been around on TV shows and radio interviews. Her classic songs *It's a Heartache, Total Eclipse of the Heart*, and *Holding Out for a Hero* still received frequent airplay. Again, the BBC defended its selection and insisted it was due to Bonnie's popularity on the continent, where she still played to packed houses. She had won the 1979 Tokyo Song Festival with *Sitting*

Eurovision: A Plea for Respect

on the edge of the Ocean and in 2004 topped the French charts with *Si Domain...*, a duet with Kareen Antonn, a French vocalist. The song was an arrangement of her 1980s hit *Total Eclipse of the Heart*. The follow-up, *Si, tout s'arrete*, also a duet with Antonn and the French vesion of *It's a Heartache*, was another top twenty hit. Bonnie also recorded a song in German and English with Matthias Reim, *Vergiss Es*. She again teamed up with Reim in 2010 on the single *Die Wilden Tranen*. Bilingual recordings with continental singers were nothing new. In 1992, she recorded *Pethano Stin Erimia* with Greek singer Sofia Arvaniti from the album *Parafora*, composed by Michalis Rakintzis. A year later her third album, *Silhouette in Red*, was a great success in Norway, Switzerland, Austria, Germany, and Sweden. As a result of Bonnie's success on the continent, she was awarded the Golden Europa award and won Best Female Vocalist at the RSH gold awards.

On the *"Bonnie Tyler was part of the eighties"* criticism, Bonnie responded by saying, "I'm not part of the eighties. I'm part of now!" The *"She's popular in Europe"* explanation was, however, greeted by *"We've heard that one before!"* Due to relatively positive publicity in the media, her congenial personality, and her popularity during the song contest week in Malmo itself, the singer gathered a lot of support. Viewing figures in Britain were, once again, up. Despite "giving it welly", she came in only nineteenth, which was, at least, a slight improvement on some of the recent placings. Her song, *Believe in Me*, was a rather turgid ballad performed halfway through the contest. It was not that it was a partcularly bad song but that it was not a particularly good one. Of the other tracks from the album *Rocks and Honey*, the ESC Insight felt that a far better entry for Eurovision would have been *This Is Gonna Hurt*, a meaty rock number which exhibits Bonnie's distincive vocals to the full. Something echoed retrospectavly post-contest all over Europe.

In 2006, there had been a firm belief in certain sections of online sites that Anthony Costas' *Beautiful Thing* should have been sent to

Athens. Instead, he had to compete and then came in second place in the British national final.

It reflected the national final songs such as Finland's 2013 entry, *Marry Me*, which qualified for the Grand Final but did little else on the big night, lingering in the lowest areas of the scoreboard. It was a very up-tempo, contemporary number which a lot of Finns believed should have stayed at home and been replaced by the classy, orchestrated ballad from the national selection, Mikael Saari's *We Should Be Through*.

The Eurovision Song Contest's image problem in the United Kingdom makes the search for a current chart artist who is both an established and international star, as well as being held in high esteem across the board, no easy mission. On a BBC morning chat show some years ago, which followed yet another British disappointment at Eurovision, the singer Aimy Macdonald was asked by the presenters if she would be prepared to represent the UK, to which she replied no. She was then asked, if she didn't want to perform, would she write a song for it? She again declined. When asked why, she explained that its image was "just too naff". Around the same period, the singer Sophie Ellis Bextor was also asked to sing for the UK. She declined. In 2009, Boyzone star Ronan Keating wrote the Danish entry. In 2012, jazz singer-songwriter Jamie Callum composed the German entry and, in 2013, Black Sabbath's Tony Iommi wrote the Armenian entry. In 2013, viewing figures both in Britain and across Europe soared, only consolidating further the song contest's status continent-wide. The "image" problem, which is again a result of the "scapegoat" situation in Britain, is a major stumbling block to the UK's success and credible reputation at the song contest. It is totally unacceptable to simply wipe off this arch event of pan-Europeanism as a tasteless joke at a time in its history when academics are beginning to discover the importance of the event to the continent of Europe at various levels and its great contribution to European history – a significant force in the socio-cultural and political development of the continent. In the lead-up to, and in the wake of, the 2013 event, there was a growing trend among

Eurovision: A Plea for Respect

those in the know in the media to take the stance that the contest was no longer about music, no longer a song contest, but all about stage presentation and glitzy production, which only increased an already embedded negative perception in the British psyche. To rebuild British faith in the song contest in the hangover of decades of derision, criticism, Woganism, and scepticism towards Europe in general would take a mammoth marketing and publicity drive. However, it is a task that is by no means impossible. A contemporary established chart act, a consistent hitmaker, must be selected. The difficulty in finding the correct one may be compounded by record companies that might, as the situation stands, not be so keen on their recording artists competing in such an event. The selection of Engelbert and Bonnie Tyler, established, popular, and household names, has in many ways been a positive move. Although in terms of the singles chart of late neither could really be seen as the cutting edge of British pop, their selection nevertheless has paved the way for other established and newer, more contemporary acts to compete, who just might be able to inspire an interest in the contest among a younger generation in Britain, just as it has in several continental countries. A vocalist who merits a special mention in this chapter is the British entrant of 2018, SuRie, who sang *Storm in Lisbon*. She had already sung backing vocals and acted as MD in previous contests for the Belgian entrants Blache and Loic Nottet but had been chosen from six other acts in the *Eurovision: You Decide* national selection show

While the up-tempo version of the pop/dance song was dispatched to Lisbon, SuRie also recorded an acoustic ballad version, on which she accompanied herself on piano. This manifested what a quality song this was and might have fared better if it, rather than the up-tempo, had been performed on the big night. SuRie came from a notable family; her grandfather was Sir Hans Kornberg, the well-known German-born biochemist and a Master of Christ's College Cambridge. Sir Hans, a Jew, fled Nazi Germany in 1939 but, heartbreakingly, had to leave his parents behind to face the horrors of the Holocaust. His parents were later murdered under the Nazi regime.

SuRie's brother is the singer-songwriter Benedict Cork. While SuRie was performing her song on the Eurovision stage in Lisbon, the protestor Dr ACactivism jumped onto the stage, aggressively grabbed the microphone from her hand, and yelled, "Modern Nazis of the UK media. We demand freedom. War is not peace". This came at a time when there had been multiple terrorist attacks worldwide, several within Europe by supporters of ISIS. To many in the arena and watching on television, it initially appeared to be the start of some terrorist-related disruption. SuRie stood on stage bruised and in a state of shock as the perpetrator was dragged off by security men. In the tradition of "the show must go on", SuRie charged into what was a powerful and stupendous performance. She declined a request to sing the song again and was justly satisfied with the performance she had given. The voters did not warm to the song and no sympathy vote seemed to surface. The song came in a poor twenty-fourth with forty-eight pints. After the scenario in Lisbon, SuRie had proven what a credible solo artist she was and should have been given more support from the media and music industry in furthering her career to the more dizzy heights that she merited.

"Trendy" is a word I have referred to before in this book, often in derogatory terms. "Trendy" often adheres to very narrow confines of how to be and is often aggressively intolerant of that which it does not include in its boundaries. Yet, "trendy" is what, in order to redefine itself in the British psyche, the Eurovision Song Contest must become. If the United Kingdom still regards itself as the Nirvana of pop, it must prove this by consistently entering excellent songs and star acts – even if they come in last. It has at least credited itself with knowing that it has made an effort with only the best. It is imperative in the revamping of the event in Britain that this happens. That's if the powers that be really want this. In 2012, a well-known British quality newspaper featured an article on the bad fortunes of the UK after Engelbert's disappointing second-to-last placing. The journalist responsible for the article proposed a national final in the tradition of Children in Need and Red Nose Day that would raise money for charity, with all

Eurovision: A Plea for Respect

proceeds of the show going to charitable causes. This would attract top chart acts who, keen to do their bit for charity, would have an incentive to take part. All proceeds from the winning song that qualified for Eurovision would also go to charity. The journalist was baffled as to why this type of selection method had never been considered. Others in the music business advocated an *X Factor*-style talent contest to select the Eurovision entry, but would the type of act to come out of such a selection process be strong enough to bring a surge of prestige to Eurovision, or would it just descend into a mediocre talent show? This is literally what happened to A Song for Europe in the 1980s. While the BBC chose to select internally in 2011, 2012, and 2013, it presented the news of the song and the act to the public via YouTube. It would have been a far better arrangement if they had presented the act and the song in a musical showcase with a Eurovision theme. In 2007, the Greek prelims centred around three songs, prospective Eurovision entries within a musical extravaganza with special guests, some of that year's opposition from the rest of Europe, songs put together for the show that were parodies of an ultra-ethnic entry which featured musical rhythms and a plethora of instruments from every area of Greece, all of which created a very well-produced and interesting show. To re-establish its prestige and value in the United Kingdom, it is important that not only the British entry be seen as a work of quality but also that the Eurovision Song Contest be seen as an event capable of producing quality, "*trendy*", contemporary music. In 2013, nine Eurovision entries entered the British charts in the week after the contest itself. The winner, *Only Teardrops*, was played regularly on radio and climbed to fifteen on the official charts. The Maltese song entered the top fifty and also entered the indie top ten even though it was given no airplay. While the 2012 winner Loreen re-entered the chart with *Euphoria*, the Swedish, Norwegian, Finnish, German, Dutch, and Russian entries also became minor hits, but again they suffered setbacks in their chart progress due to a total lack of airplay. If these mostly contemporary pop recordings, which already had a lot in common with so much material in the top ten, had been frequently played on the airwaves, it would have raised the contest's profile and boosted its prestige, bringing it to the average pop station listener's attention as well as manifesting to established acts the impact it was making on the British music scene. It is also most likely that if the charting Eurovision songs of 2013 had been allocated more airplay, they would have also been much bigger hits. It was particularly unfair to the Norwegian entry, *Feed Me Love*, that, as a song very much reflecting current trends, it had lots of potential to be a big hit in Britain. The Eurovision Song Contest is, let us not forget, a contest. Therefore, it will always attract the excellent, the awful, the mediocre, the good, the bad, and, if I may be forgiven for quoting a cliché, the ugly. Note that among the preceding words were also "the excellent" and "the good". There is the camp and the kitsch but look under that and there is also the quality. It is also a fact of popular music that good songs do not always win song contests. In 1984, bubblegum pop soared up the scoreboard at the expense of quality. Sweden's Herreys won with *Diggi-Loo Diggi-Ley*, which triumphed over the best songs in that year's event. It was a disaster PR-wise for the contest. Its win was a victory for critics of the event, who could, once more, use it as an example that Eurovision was only a contest of nonsense. That same year, Italy's *I treni di Tozeur* by Franco Battiato and Alice, *Avanti La Vie* by Belgium's Jaques Zegers, and *Silêncio e tanta gente* by Portugal's Maria Guinot were examples of high-quality songs that did not make nearly as much impact on the scoreboard of 1984 as they should have. In 2013, in Malmö, the Israeli entry, *Rak Bishvilo* by Moran Mazor, a well-sung, emotive ballad, made little impact on the jury or the European public and failed to qualify from the semi-finals. The consistent chart climber Cascada's *Glorious* for Germany, although in a similar vein to *Euphoria* and a possible contender for the Grand Prix, managed only a very disappointing twenty-first place, while at the same contest the British media focused heavily on the Romanian entry by Cesar, a falsetto pop opera/dance number, who at the Romanian final wore an evening suit in keeping with the opera theme. At the actual final, contest producers had changed the image completely and dressed

Eurovision: A Plea for Respect

him as Dracula, standing on a smouldering volcano, which only served to make the song seem ridiculous and easy fodder for the hypercritical media. There was no interest by the British media in the Israeli song or any of the other quality entries, Anouk's *Birds* for the Netherlands and *L'Essenziale* by Italy's Marco Mengoni, which had also won San Remo, both of which were placed in the top ten. France's excellent *L'enver et moi* by Amandine Bourgeois, although favoured by the juries of 2013, received a very low televote score and, thus, was rather unfairly placed at the bottom end of the scoreboard. The British media seized on the chance to show the Romanian entry at every opportunity when they really should have engaged more with some of the classier entries. In documentaries about the song contest made in Britain, there is no expression of the quality, no outlet for the more tasteful aspects of the music on display. And it is that which is a hindrance to efforts to positively present the contest to the Great British public and influence how they want them to perceive it. We are constantly blitzed each year with *"The Worst Songs of Eurovision"*, *"The Funniest Moments of Eurovision"*, *"The Worst Hairstyles of Eurovision"*, and *"The Stupidest Lyrics of Eurovision"*, both on TV and online, but in Britain there are never any documentaries made on *"The Best Ballads of Eurovision"* or *"The Most Outstanding Artists of Eurovision"*. For the last decade, Eurovision artists have dominated the MTV Awards, with the likes of Eleni Paparizou, Sergey Lazarov, and the Turkish singer Kenan Doğulu all being nominated. The Russian Eurovision winner Dima Bilan won Best Russian Act for several consecutive years. In 2010, the Italian singer Marco Mengoni was awarded the MTV Best European Act, the first Italian artist to win it. Israeli singer Shiri Maimon and Bosnia's Laka, both former Eurovision contestants, are among several others to win MTV Awards.

During the choosing of the British Eurovision entry, there has been the added obstacle in recent years of sinking sand between the selection and contemporary superstars who shun the contest, not only for reasons associated with its image problem but also for simply not wanting to be in competition at their level of stardom. Seen in comparison, Italy's San Remo Song Festival is approached with a very different attitude, like the Swedish national prelim, the Melodi Festivalen, which reads like an annual "who's who" of Swedish pop. Likewise, San Remo is an annual showcase of the best established and new acts on the Italian music scene combined with the cream of contemporary songwriters. Mega-stars compete year in and year out, often to be heavily defeated but they resiliently pick themselves up and return again and again, aiming for their goal: first place. Italian superstars Adriano Celentano and Mina have competed with other music stars, who did not necessarily enjoy the same status in Italian music as they did but nevertheless allowed themselves to be put in competition with them. The old guards of San Remo were Domenico Modugno and Claudio Villa, vintage competitors on its ever-revolving stage, their careers bound by the orchestral strings of San Remo. Then there is Peppino Di Capri, who won the contest with *L'ultimo Romantico* and who competed no fewer than twenty-two times. As the twenty-teens continued, it seemed inconceivable, unfortunately, that British music stars of that decade, i.e., One Direction, Adele, and Olly Mars, would ever participate in a musical competition against each other let alone in a British Eurovision prelim.

Song festivals have never really, except for Eurovision, been an integral part of British culture. On the continent, song festivals have come and gone and sprouted branches: the Thessaloniki Song Festival, the Sopot Song Festival, the Liublijana Song Festival, the Bratislavska Lyra, the Benidorm Song Festival, the Castle Bar Song Festival, and the Venice Song Festival, to name only a few. Nor are they confined solely to Europe. The Rio De Janeiro Song Festival and the Viña del Mar Song Festival in Chile, along with the Tokyo (Yamaha) Festival in Japan, all have kept the nation in suspense while sometimes supplying international hits such as The Three Degrees' *When Will I See You Again?*, which won the Tokyo event in the early 1970s. The French singer Romuald, who sang for both Monaco and Luxembourg at the Eurovision Song Contest in the 1960s and 1970s, sang for Andorra at the Rio Festivals of 1968 and 1969 and came in fifth on both occasions.

Eurovision: A Plea for Respect

A few years on and he changed nations, singing for France at the 1973 Viña del Mar festival in Chile. His song *Lasse moi le temps* came in second. Some months later, it was recorded in English as *Let Me Try Again* by Frank Sinatra, which became the theme song for the latter years of The Voice's singing career.

In recent years, there has been an attempt to establish an Asian-Pacific Song Contest, much inspired by the Eurovision Song Contest and aimed at countries in the Asia-Pacific region that avidly viewed the Eurovision Song Contest but were not eligible to compete, i.e., Australia.*

In 1961, ITV arranged a rival song contest to the Eurovision British prelims which displayed the most popular of home acts around. Among those who competed were Matt Munroe, who had to cancel his participation in the British Eurovision heats that year on account of his appearing at the ITV Song Contest, which was scheduled around the same time. Others who took part included Mike Preston and Frank Ifield. It was won by Mike Preston with *Marry Me*. Matt Munroe came in second with *My Kinda Girl*.

ITV also broadcast the British Song Festival, a one-off extravaganza which was initially intended to be an annual event but, due to a big financial loss, rowdy teenage fans in the audience, and a small scandal picked up on by the press when the wrong winner was announced, it was decided to not hold another festival. It had featured some of the UK's top pop acts of the Swinging Sixties. Lulu, The Moody Blues, The Ivy League, Manfred Mann, Elkie Brooks, Helen Shapiro, and Billy J Kramer all competed.

There was yet another attempt at a song contest organised by ITV in the early 1970s. The Opportunity Knocks Song Contest was held for new non-professional songwriters whose songs were performed by singers who had appeared in the talent show in previous years. It was won by Stuart Gillies with the song *Amanda*, which became a hit. An American song contest which would see all of the fifty states compete against each other is planned for 2022, and a similar concept for a Canadian song contest is expected to premiere in 2023

A bit of bravery in the competitive field by some of the UK's superstars might not go amiss in an attempt to redefine the Eurovision Song Contest for a new generation. Discrimination again raises its ugly head. A former member of the boy band Brother Beyond, who had attended a previous contest in the role of manager, was asked on a breakfast news programme on the Sunday morning following the 2013 contest why the UK did not send outstanding songs to Eurovision. He put it in a nutshell when he said that British songwriters refrained from sending high-quality material because it would not be played on the radio. Beforehand, the presenter who had conducted the interview had giggled at a clip of the Azerbaijan entry from the previous evening, already providing an answer to the question they were about to ask.

Back in the 1960s and 1970s, common complaints about the Eurovision Song Contest by British viewers in magazines such as the *Radio Times* were often of a linguistic nature: "Why can't all the songs be sung in English?" wrote one viewer. This was in the days when only the British and Irish entries were in English. All the other participants had sung in their national languages. Other viewers found fault with a contest full of songs whose lyrics they could not understand, which made it, so they said, *boring*. English in the 1960s did not have quite the hold on pop that it does now. English was not the international language that it is today. In countries like Italy and Spain, few of the record-buying public spoke or understood the lyrics of British and American pop and rock songs. The Beatles recorded some of their earlier material for the German market in Germany. In Italy and Spain, Sandie Shaw recorded most of her material for release in their respective languages. She was regarded as an Italian recording artist in Italy. Her British hits *Always Something There to Remind Me* and *Long Live Love*, familiar tunes to those fans of 1960s music, rode on the Italian airwaves in *"Italiano"*. At San Remo, British and American artists performed their songs in Italian. In Spain in 1964, Sandie Shaw recorded *Always Something There to Remind Me*, which she also released in Italian, French, and German; however, she failed to record it in Spanish. This void was filled by Karina, who had a hit with a Spanish

Eurovision: A Plea for Respect

cover, *Siempre Hay Algo Que Me Recuerda A Ti*. Sandie continued to release original English versions of her hits in Spain but, due to the high demand for her material, she started to record all her output for the Spanish market in Spanish, with *Non Lo Comprendi* (*Message Understood*) and *Viva El Amor* (*Long Live Love*) becoming particularly popular. At the 1967 Eurovision Song Contest, which Sandie won, Spain, surprisingly, did not give her any votes. This has often been put down to suspected tactical voting by the Spanish, who had entered their biggest international star of the 1960s, Raphael. Sandie's winning song was almost immediately recorded in Spanish as *Marionetas En La Cuerda*, which brought her her greatest-ever Spanish success. Her recording was followed by a plethora of diverse Spanish versions by Spanish singers, so great was its popularity. Cilla Black's career owes probably more to Italian music than does that of any other British singer outside opera. The dramatic emotive ballads so popular in Italy in the 1960s were perfect for her style and vocal abilities. Her hit *Don't Answer Me* was an adaptation of Donatella Moretti's *Ti Vedo Uscere*, while *A Fool Am I* came from Fabrizio Ferretti's *Dimmelo Parlami*. Cilla also had top twenty hits in the UK with *I Only Live to Love You*, originally Umberto Bindi's *Cosa Sifa Sta Sera Non Se Domain*. Another song by Tony Dallara became *Where Is Tomorrow* when recorded in English by Cilla. It proved to work both ways. Her 1968 recording of *Step Inside Love*, a song written for Cilla by Paul McCartney, which was to become her theme tune, was adapted to Italian. *M'innamoro*, as it became, was Cilla's only Italian language hit.

Some British groups like The Rokes were based in Italy and were primarily "Italian" groups. Another artist to make it big in Spain and Italy but who never managed to break into the British scene was Andee Silver. A Londoner, Andee concentrated on the demands of the Italian and Spanish markets, delivering a more continental pop sound such as that found on her single *L'Amore Dice Ciao*, which was the soundtrack to a 1968 film. Clodagh Rogers, who had been on the periphery of the music scene for some time, had her first UK hit in 1969 with *Come Back and Shake Me*. She released the Italian version, *Il Cuore Nell A*

Rete, shortly afterwards but it did not crack the Italian hit parade, as hoped. Two years later, Clodagh released another single in Italian, her 1971 Eurovision entry *Pupazzo*, while some years before, Kathy Kirby had recorded her 1965 Eurovision entry, *I Belong*, for the Italian market. In its adapted form, it was known as *Tu Sei Con Me*. Kathy also released the runner-up at A Song for Europe, *I'll Try Not To Cry*, and launched it in Italy as the Italian language *Non Piangero Per Un altra*. In the 1960s and 1970s, Britain had a much better track record of promoting its Eurovision entries on the Continent than it does today. It was far better at promoting its entries than a lot of other competing countries of that era. British Eurovision songs, as a result, were often well-known and even popular before the night of the song contest. The fact that top contemporary stars represented the UK also helped promote the recordings. Mary Hopkin, who was a Welsh folk singer of the 1960s, was propelled to international stardom with her eerily haunting 1920s retro *Those Were the Days*, produced by Paul McCartney, which topped the charts worldwide. It reached number two on the Billboard Hot 100, kept from the top by *Hey Jude* by The Beatles. The Russian ballad *Dorogoi Dinnoyiou*, written by the Soviet composer Boris Fomin and the poet and lyricist Konstantin Podrevsky, was originally recorded circa 1925 by both the Georgian vocalist Tamara Tsereteli and the Russian Aleksander Vortinsky. It had featured in the 1953 British film *Innocents In Paris*, in which it was sung by the Gypsy singer Ludmila Lopato. Mary Hopkin's version sold no fewer than ten million copies. One year before her appearance at the 1970 Eurovision Song Contest in Amsterdam, Mary competed at the San Remo Song Festival. In those days, foreign artists at the festival sang their entries in Italian. Her entry, *Lontano Dagli Occhi*, was a perfect example of her pop-folk genre, which her clear voice complimented so much. The Italians appreciated this beautiful song and it entered the Italian top twenty shortly after San Remo.

Marianne Faithfull, who participated at San Remo in 1967, was a frequent visitor to the British top ten, having had a hit in 1964 with Mick Jagger and Keith Richards, *As Tears Go By*. Her third top-ten hit,

Eurovision: A Plea for Respect

Little Bird, was cut in Italian and released as *Un Piccolo Cuore*. Her song at San Remo, *Cè Chi Spera*, was not placed and was overlooked by the British media. Her performance of *Cè Chi Spera* was overshadowed, as was the whole of the 1967 festival, by the tragic death of Luigi Tenco. Despite all this passage of tragic events, the only topic that the British press could focus on concerning San Remo was the romance between Marianne Faithfull and Mick Jagger. The Italian balladry complimented Faithfull's voice and fitted into her stage presentation. Other British Eurovision acts of the 1960s to record foreign language material for the European market were Lulu and Mary Hopkin. Lulu did not release her first UK hit, *Shout*, in Germany but in 1966 recorded *Wenn du da bist*, her debut German single. It made little impact on record buyers in the old *Bundes Republic* and Lulu would not make any inroads into Germany's music scene until 1969, when *I'm a Tiger* caught the attention of the public and reached number four. A few months later, in the spring of 1969, shortly after her all-white snow-capped wedding to Bee Gee Maurice Gibb, Lulu won the Eurovision Song Contest. Her song *Boom Bang a Bang* was the only one of the four winners to make much of an impact on the German chart. At the British national selection, it had emerged as the winner over Elton Johns' *I Can't Go On Living Without You*, which came in last. Among the other songwriters in the British final were Ken Howard and Ian Blaikley, who were credited with writing the classic 1960s songs *The Legend Of Xanadu*, *Last Night In Soho*, *Zabadak*, and *Bend It*, which were all hits for Dave Dee, Dozy, Beaky, and Mick and Titch.

Sandie Shaw's *Puppet on a String* had been a hit in its original English version in Germany and also a hit in her German version. Lulu attempted to repeat this but the German version of *Boom Bang a Bang* failed to enter the German charts. She made three more German language singles in 1971: *Warum test du mir Weh* and *Ich brauche deine liebe*, both of which were produced by Giorgio Moroder, although none of them made much of an impact. Lulu sang in particularly good German and Italian, unlike some other British singers, whose pronunciation left a lot to be desired. Many of them had recorded their songs phonetically. Her follow-up single to her Eurovision winner was her debut with her new record company, Atlantic. It was cut in Italian and released in Italy. *Provera Me*, one of her more powerful ballads, arrived on the British market as *Oh Me, Oh My*.

Jackie Trent, who, with her husband Tony Hatch, formed one of the 1960s' most prolific songwriting teams, had topped the British charts with *Where Are You Now?* in the mid-1960s but as a solo singer was struggling to maintain a chart career in the UK. One idea thought up to change her luck was to record an Italian song. Both Cilla Black and Dusty Springfield, the UK's top female vocalists of that era, had amazing success with Italian ballads. The solution was their own composition, which they penned with not only the British market, but also the Italian market, in mind. *Il Mondo Degli Altri* was inspired by the leggero music of Italy and reflected the ballads of San Remo that had been so successful worldwide. A heavily orchestrated ballad with just a touch of British beat music, *Il Mondo Degli Alri* built up dramatically to a great chorus.

In January 1964, The Beatles recorded two of their legendary hit songs in a Paris recording studio. What was different about them was that they were to be the only recordings The Beatles ever made in German, the only foreign language that they ever recorded in, except for *Michelle*, which was only partly sung in French. The two songs were *Sie Liebt Dich* (*She Loves You*) and *Komm Gib Mir Deine Hand* (*I Want to Hold Your Hand*). The instrumental tracks were the originals used in their English versions. The Beatles already had strong connections with Germany, in particular Hamburg, where, as up-and-coming musicians, they had lived while performing in the city's vibrant club scene. The German producer of EMI, Otto Demler, desperate for The Beatles to release German language material, flew the Luxembourg-born singer-songwriter Camillo Felgan to the George Hotel, Paris where the Fab Four were staying while in the city for a concert. Felgan, who was also a programme director for Radio Luxembourg, was given the task of writing the German lyrics and coaching the group

Eurovision: A Plea for Respect

phonetically in German in the limited time of twenty-four hours. Paul McCartney, some years later, recorded a rough, incomplete version of *Get Back* in German, a language he was familiar with; *Geh Raus Nach Deinem Haus* was, however, never released.

Camillo Felgan is the link in a major Eurovision connection with The Beatles. Felgan, as was mentioned in a previous chapter, sang for his native Luxembourg in Luxembourgish, not German, in 1960 and again performed a song in French in 1962. During World War II, Felgan worked as a German translator for the Germans in occupied Luxembourg. His musical career blossomed after he joined Radio Luxembourg as a chorus singer. One of his greatest hits was *Sag Warum* in 1959, which was taken from a melody by Phil Spector. Under the pseudonym Lee Montague, Felgan wrote songs for popular British groups of the 1960s, The Honeycombs and The Searchers. He also penned songs for Petula Clark, all of which were recorded in German for release in the *Bundes Republik*. Sadly, the idea of a top British band recording in German nowadays would be looked upon only as laughable by a large section of the public.

* * *

For many years, a market flourished every Sunday on Athenas Street in the heart of Athens. It was a sprawling mass of exotic products that stretched from Omonia Square to Monastiraki. It was a cosmopolitan mix of fruit, vegetables, carpets, music cassettes, sweets, clothes, linen, and hundreds of other artefacts from the Balkans, Central Asia, the Middle East, Russia, and Africa. The origins of the goods were reflected in the faces of the multi-ethnic traders. Gypsy children sold enormous bags of vegetables at ridiculous prices. Russian women in scarves sold embroidery. Egyptians bartered with customers over cushion covers embroidered with gold threaded camels. In the mid-1990s, there was a disagreement between rival Gypsy traders and a shoot-out ensued which ended in the deaths of several people. The great chaos of exotic colours and smells that had been a landmark of Athens weekends was promptly closed by the Athens Council. A replacement emerged in the port of Pireaus shortly after but it was never to be the same, the distinct character of the old Sunday market gone. Like the Athens market, the Eurovision Song Contest is a meeting place, a cross-cultural centre point, a tower of Babel of different languages and lingua francas. Like in the Athens market, there is chaos and confusion as traditional European cultural and ethnic rhythms mix with sounds that do not have their roots in Europe: African rhythms, Central Asian Makahms. In the chaos and exotic mish-mash of the cultural arena, the genres mix and interlope. Italian pop, a descendent of opera, sits side by side with Turkish melodies heavy on Arabesque. It is a meeting place where both kitsch and the "low" popular music culture as well as the serious "high" popular music culture come together. It is where taste is challenged, defended, and questioned. In this market of cross-cultural references that stretch far beyond the continent of Europe, there is a freedom of musical expression that extends beyond the boundaries of set genres and national rhythms. It is here that, in the various stalls of exotic artefacts, rich with the aroma of rich Eastern spices, genres are toyed with, bent, formed, and re-formed. Here in the stall with the vendors of kitsch stands the British attitude, unable to comprehend the vendors in the other stalls, so blinded by the kitsch, which both disgusts and fascinates it, that it is unable to see the vast array of precious goods available in the other parts of the market, so blinded by the glittering array of dazzling kitsch that its take on the market is framed only by this one single stall. It is a necessity and even an accepted norm in Britain that entries such as the 2008 Latvian *Wolves of the Sea*, by Pirates of the Sea, should be presented as the accepted "mediocre", "trashy" norm that is Eurovision or that Ralph Segal's *Djengis Khan* was the archetypical Eurovision song of a past era when, in fact, it was not. There are few, if any, references to British examples of a "high" pop culture domiciled to the song contest or to what, at any rate, are artists and songs normally considered outside of Eurovision as "quality". The press and other expressions in media are never so enthusiastic at conveying to the public aspects of the participation of the likes of Jamie Callum and

Eurovision: A Plea for Respect

Tony Iommi of Black Sabbath, whose song was sung by the Led Zepplin-inspired Dorians. They never crop up on chat shows, discussions, documentaries, or other Eurovision-related items on television. I make reference to them in the background of a sea of quality Eurovision songs and artists who remain in a state of anonymity, kept hidden from the British public like some secret KGB file. The Kosovo-born British-Albanian singer Rita Ora shot to fame after the release of her debut album, which reached the number-one slot in the British album chart. The successful album gave birth to an impressive three singles that topped the UK singles chart, which qualified Rita Ora as the artist with the most number-one singles on the charts in 2012. Before recording her debut album, Ora had appeared as a hopeful in the 2009 British Eurovision heats of the Eurovision Song Contest. Taunted with doubts about her experience and ability, she withdrew from the Lord Lloyd Webber-supported programme. In 2013, Ora made several statements to the press indicating that she was much relieved to have not competed in the Eurovision Song Contest because she felt it would have destroyed any chance of furthering her career. The Eurovision Song Contest, in the past few years, has shown us a portion of a perfectly united Europe keen on diversity, an inspiration at least of how it should be. Different angles of minority culture have shown through in ethnic pop, endangered languages, multi-ethnic artists, kitsch, and camp, all of which are an expression or embody an element of a European minority of one sort or another. The somewhat questionable and even trendy approach to presenting it as a festival of camp and kitsch has been the domain of some of the British media. Even The New York Times has described it as kitsch. While camp and kitsch are supposedly an expression of a minority group, a question must be asked; Does the media accept this as a valid cultural expression of an oppressed group or does it say that minority groups are of any value only if they are comical and can be laughed at? If camp and kitsch are the cultural expressions of an oppressed minority, do they not merit a place under the canopy of quality culture similar to soul or blues or Greek rethbetika, which was only very recently upgraded to being quality culture.

If Rita Ora had, in 2013, in the aftermath of her victorious musical ventures in 2012, sang for Albania in her mother language, would she still be considered a credible serious artist? Would she have been stamped as camp and kitsch? And if she had, would these classifications have come with credibility? Or would she simply have been condemned to the bargain bin of music stores? It would seem that camp and kitsch can be expressions of an oppressed minority when the media feels fit to adapt them to that or it, like a chameleon, can turn just as quickly to rendering an aspect of culture as ridiculous and comical, as seems to be this double-knifed usage of the two terms when referring to the Eurovision Song Contest. It is also a rather bewildering state of affairs that such highly respected media specimens as The New York Times and CNN, which, as American media-based bodies in a country that has been a melting pot for numerous European cultures, whose regard for the content has been held in high esteem, and whose mother countries have contributed some very classy material to the song contest, i.e., Italy and Serbia, should take on the British point of view in all this and project what is little more than Woganism onto the American psyche. In his article *"Why Americans can't afford to miss out on this cultural spectacular"* in The Guardian in May 2013, the American journalist Jason Fargo warns Americans, *"Ignore the British"*. Then he goes on to explain why: *"I am here to tell you, on this occasion at least tune them out. Perpetual losers of the Eurovision Song Contest the British have tried to assuage their pain by dismissing the show as a big joke"*.

* * *

The fascination with the song contest has featured in many other forms of British popular culture. The play *Gogol and the 1969 Eurovision Song Contest* was written by Tim Luscombe. It revolves around an author, Stuart, who has written a book about Gogol. He is, however, trapped in a world of agoraphobia. He gradually descends

Eurovision: A Plea for Respect

into a breakdown and, as problem compounds upon problem, slowly slips into insanity. Lenny Kuhr, the sweet Dutch singer-songwriter, leads him back to the 1969 Eurovision Song Contest, where Laura Valenzuela, the Spanish TV host, is waiting to welcome him. Possessed by the spirits of the Russian writer Gogol and some characters found in his books, Stuart has a bleak-looking future.

Benny Hill also created a very credible parody of the 1969 Eurovision Song Contest, which in his version was won by Britain's Singing Postman with the ditty *Ting-a-ling-a-loo*, a take-off of Lulu's *Boom Bang a Bang*. The song was recorded as the B-side of his 1970 number-one hit *Ernie*. In Benny's parody, caricatures of well-known continental singers, whose names have been only slightly changed, represent their countries. Flora Loles, Mireille Maton, and Omonia Rodriguez are thinly disguised take-offs of Spain's Lola Flores, France's Mirrielle Mathieu, and Portugal's Amalia Rodrigues.

The late, great Dick Emery was a master of guises and one of Britain's funniest comedians of the 1960s and 1970s. His weekly show provided the British public with a number of wonderful characters to both laugh at and take to their hearts, such as Hetty, Mandy, and Lampwick. In one 1973 sketch, Lampwick, a World War I veteran, takes his granddaughter on a visit to the Tower of London to see the Crown Jewels. This was filmed the week of Princess Anne's participation in the British team at the Olympic Games. As she views the priceless objects, Lampwick explains to her what each one is. His granddaughter points to a shimmering, jewel-encrusted artefact in the corner behind the Crown. "*What's that?*" asks the little girl. "*That's what Princess Anne won at the Eurovision Song Contest*", Lampwick answers, a reference to Princess Anne's participation in her own particular international event.

Monty Python's Flying Circus, a zany progressive comedy show of the 1960s and 1970s, featured a Euro-police Song Contest in a 1970 episode. It was won by Inspector Zatapathique of Monaco. *Boom-Bang-a-Bang*, a play by Jonathan Harvey named after Lulu's classic song, deals with a Eurovision party hosted in a flat in a Kentish town and the issues and problematic lives of some of the guests. Through comedy, we see Eurovision songs represented by nonsensical ditties with banal titles. These shows were made around the turn of the decade, 1969/70, at a time when the song contest, except for the British entries, was solidly made up of ballads. Therefore, the whole perception of Eurovision songs was conceived of with knowledge of only the British entries. This was due to an introverted cultivation of national pop music which generally excluded continental genres such as *Chanson*, even though numerous hits by British artists of the 1960s had been adaptations of continental songs. The first days of the stranger's appearance and the bully senses something different about him, something the bully does not understand. He fails to make head or tail of the stranger, who soon becomes the scapegoat. The bully is out to destroy something that is beyond his understanding.

One other comedy show that the Eurovision Song Contest featured in was *The Lily Savage Show* in the 1990s. Lilly (Paul O'Grady), head of a dysfunctional family, asks her next-door neighbour, Katie Boyle, if she can borrow one of her "*old Eurovision frocks*" to wear when the social worker visits her home.

In more recent Eurovision Song Contests, a wealth of quality acts have appeared on British TV screens. Among the most notable have been singers from the Balkans, in particular Serbia's top male vocalist, Želko Joksimović, who was runner-up with his folk-ethnic pop ballad, *Lane Moje*, in 2004. He wrote the Serbian entry again, the haunting ballad, *Oro*, performed by Jelena Tomašević, which placed sixth. The song was supported by the renowned flautist Bora Dugič. The lyrics of the song end with the line "*Wake me on Saint Vitus' Day to look at him again*", which refers to the traditional Serbian holiday of June 28. Joksimović was one of the co-presenters of the 2008 Eurovision Song Contest in Belgrade. He represented his country again in 2012 with yet another Balkan ballad, *Nije Ljubav Stvar*, which once more soared up the Eurovision scoreboard to third place. Before the 2012 Eurovision Song Contest, Serbian television showcased his repertoire in a special programme which aired the song for the first time. The song was chosen internally and not at the Belgrade Song Festival, which had been the norm for previous Serbian entries. *Synonym*, an excellent English version of his Eurovision song, was also presented. Joksimović also composed the Bosnian entry of 2006, again a Balkan ballad so typical of his compositions. *Lejla*, performed by Hari Mata Hari, came in third in Athens. It was to be Bosnia's highest placing at the song contest. While ignoring and employing existing prejudices towards these songs and artists, the media are denying the British public access to a well of talent that is already established and polished. Performing only a few songs before Serbia's 2012 entry was FYRMacedonia's Kaliopi singing the classy Balkan-rock ballad *Crno I Belo*. A much-experienced artist, Kaliopi had started singing and training in classical music, then branched out into popular music. After the breakup of Yugoslavia, she won the 1996 Skopje Festival, which was the FYRMacedonian national final for Eurovision. It was the first year FYRMacedonia had submitted an entry as an independent state. Her song *Samo Ti* was sent to the first Eurovision semi-final, set up to cope with the influx of new participants, which was an audio, non-televised, selection. Her song was rejected. Kaliopi had sung with successful groups in Yugoslavia before that but during the Yugoslav war was absent from the music business. Kaliopi's stupendous performance at the 2012 Eurovision Song Contest proved she was by far the best vocalist at the event. Despite such credentials, it seems a nigh impossible task to persuade the British public and its media of the positive quality of these acts and songs.

In this book, the term "mainstream" has been much used but several Eurovision artists have not been part of the mainstream music scene. Examples include Austria's *Schmettelinge* in the 1970s and, more recently, Turkey's singer from 2012, Can Bonomo, who performed, shortly after Eurovision, at New York's Carnegie Hall. It is of prime importance to re-establish the Eurovision Song Contest as a potent entity in British music and to select as the home entry someone who is an established member of the contemporary pop establishment.

During the Eurovision tide of 2013, the British journalist Fraser Nelson wrote in the Telegraph, *"The Eurovision Song Contest has provided a musical metaphor for Britain's relationship with the European Union. We submit an appalling entry chosen because we assume the continentals like trash. They don't, and tend to give us nil points".* It was one of the most accurate and intelligent articles from the British press in Eurovision week 2013. Nelson even shows a pro-Eurovision outlook on the event. Concerning Wogan's rants taken on by the press at large, he says, *"We then assume we are victims of discrimination that Eurovision is an exercise in Continental intrigues that we wouldn't want to win anyway".* He agrees with the theory that Eurovision is a collision of power, politics, music, and culture and that a Eurovision victory goes to those who best adjust to the fact. Right at the epicentre of the contest exists the creation of the world's most powerful medium of all, a tune. In his article, Nelson firmly believes, *"It is a battle that Britain could win every year — if we stopped to see what's happening."* While Britain has gone for internal selection in the last few years, the Swedes choose their song partly through the votes of a foreign jury, which in 2013 caused a nationwide uproar when the international jury at Melodi Festivalen overturned the Swedish decision. Other national selections are real musical extravaganzas and gala evenings. Other entries are chosen at well-established song festivals such as San Remo, the German *Schlager* Festival, the Belgrade Festival, and the Skopje Festival. With an absence of an equivalent of such an event, the nearest thing to an established music festival is probably Glastonbury. The prospect of choosing a British Eurovision entry at Glastonbury; now that would be an interesting option for the BBC to consider. Ludicrous? Out of the question? If so, then why? It was announced in the spring of 2022 that the Ukrainian group Go_A, which had been chosen to sing Solovey at the 2020 contest and came in third in Rotterdam in 2022 with Shum, both electro-folk numbers heavy with traditional Ukrainian instruments, had been invited to perform at Glastonbury the following summer. While Robin's Swedish entry was watched at the 2013 finals by over half the country, his song was number one on the Swedish

Eurovision: A Plea for Respect

chart. The 2013 United Kingdom entry had not even entered the top ten. Who can blame Continental Europe for showing reluctance to pick up the phone and vote for a UK song? Nelson, in his article, has this to say: *"We're guaranteed a place in the final, to the fury of other nations, who can see how little effort we put into it. Our entry is chosen by an anonymous BBC official, who evidently has a supercilious disregard for the whole show."* The seeming lack of understanding, and even blindness, by the BBC, which seems to be met with a measure of obstinacy, blurs the way to finding a truly good song that needs to cut through barbed wire borders, politics, culture, and language. According to Nelson, Eurovision is an annual search for a common currency ... everyone wants to do a *Lilli Marlene*. Both *Volare* and *Lilli Marlene,* as we have seen, have been the most evocative and covered songs of the last hundred years, while *Waterloo* still sounds as fresh as a newborn spring baby's cry and exudes a certain power. A poll conducted in the United Kingdom showed that a whopping 75% of the public interviewed believed that the UK was discriminated against at Eurovision and firmly put the blame on block voting.

While Terry Wogan retired from the contest after 2008, Woganism still haunts attitudes towards Eurovision in early twenty-first century British society. Its interpretation of the event, as a parade of foreigners there to be laughed at, is still evident, yet very much dated.

While the UK does badly at the contest, top British artists were responsible for four of America's top-selling albums in recent past years. One in six CDs sold in Germany is by a British artist. There were even calls in the British press for the contest to be taken over by another broadcaster in the United Kingdom because, once again, a BBC attempt had failed. It should be noted that out of the big five, Spain, Germany, and Italy managed a place in the top ten at the 2012 Eurovision Song Contest, while in 2021 Italy and France occupied the two top placings

Another of the more "enlightened" articles to appear in the press in the fever of the 2013 event was The Spectator's *"Why Eurovision needs to be saved from the BBC".* The article strongly contradicts the *"joke"* approach prevalent in Britain. *"If Eurovision was a night of inconsequential trash, then Iran would not have recalled its ambassador from Azerbaijan last year in protest at the tolerance Baku authorities were showing to gay fans. Vladimir Putin would not have hailed Dima Bilan's win in 2008 as a triumph for all of Russia, and the Baku police would not have tracked down and questioned people who used their mobile phones to vote for Armenia, which is supposed to be Azerbaijan's mortal enemy".* Indeed, as the twenty-first century progressed, winning Eurovision became one of Putin's major objectives and a vital propaganda tool to show Russia as the major world power both internationally and at home. At the 2016 song contest, Russia battled it out face-on with Ukraine at a time when relations between them were at a very low ebb. The Ukrainian song that year was *1944*, sung by the jazz singer Jamala, whose paternal grandparents had been Tatar Muslims from Crimea who had been forcibly resettled in the Central Asian republic of Kyrgystan by Soviet authorities in the 1940s, where she herself had been born. On her mother's side, she was Armenian. Its melancholic sound and critique of Russian maltreatment of the Tatar minority in the Crimea under Stalin in 1944 raised more than a few eyebrows in Moscow. Russia's broadcaster had sent one of the greatest names in the Russian music industry, Sergey Lazarev, who had originally been a member of the band Smash!!, whose disc sales soared to millions in the Russian Federation and throughout the Commonwealth of Independent States. Lazerev had knocked mega-star Dima Bilan into second place at the awards for Top Male Vocalist of 2006. Russia had employed top songwriters to compose his song, *You Are the Only One*, which became, in the pre-contest betting odds, the favourite to win. In a tense standoff during the final minutes of the televoting, Russia and Ukraine, so symbolic and reflective of current politics and what was to come, anxiously awaited the outcome. Lazerov was given a massive 361 and won the televote but did not have enough points to win the overall contest, which meant that Jamala's *1944* had won the 61st Eurovision Song Contest. A furious Moscow responded with complaints that the song was too political and should have been disqualified. Never before

Eurovision: A Plea for Respect

had the contest witnessed such a political drama manifested on its very stage at the most crucial of moments. Cliff Richard's *Congratulations*, runner-up in 1968, despised by a musical elite as crass and synthetic, has become a standard song at birthdays, weddings, and bar mitzvahs. It was sung at Queen Elizabeth's Diamond Jubilee celebrations and outside the hospital as Prince William and the Duchess of Cambridge left with their newborn son. *Congratulations*, love it or hate it, has replaced other celebratory anthems of a past era such as *For He's a Jolly Good Fellow*. It has surpassed the world of pop and embedded itself in British folk but due to the current attitudes and trends, this would now be impossible for a British Eurovision entry to achieve.

While the BBC may snigger at the mere mention of Eurovision and employ people who loathe it to arrange a national selection, there is a wider world of both original Eurovision nations and former Intervision states that take it very seriously. They are not simple-minded nor are they idiots or tone-deaf. They are intelligent people with musical tastes. As to why Britain takes this snobbish, pompous attitude toward the Eurovision Song Contest, sometimes reminiscent of the pomposity accorded to American popular culture, like some invited guest at a party who refuses to mix with the other guests in the living room because he/she is too aloof and so hangs around in the kitchen all evening, The Spectator's answer to all this is because it is run by a bureaucracy as puzzled by popular culture as the old Intervision once was. Neither Sky nor ITV suffers such problems. It is time for the BBC to let someone else find a song for Europe. Indeed, the BBC's snubbing of Eurovision songs is tantamount to banning them from the airwaves. Its efforts to discredit the song contest speak of similarities seen in the Soviet Bloc and the authorities' approach to the new pop/beat culture that swept Europe in the early 1960s. Attitudes toward Eurovision bear some comparison to the introverted nationalist-centred cultural policy of Mussolini's Italy. This has evolved into an ignorance of the musical attributes of Eurovision and European pop in general – an ignorance that there is even a substantial degree of pride in. This bewilderment at European pop music and inability to fathom the Eurovision Song Contest is related to Euroscepticism, which is then lashed out onto the scapegoat – a theory which Karen Fricker so aptly applies to the British attitude. Earlier in this chapter, I wrote about how the lyrics to *Puppet on a String* had uncannily become a metaphorical narrative of Britain's relationship with the Eurovision Song Contest. Perhaps, then, the lyrics of Mary Hopkin's 1970 entry ring true on the night as the song contest itself commences and the UK wallows in self-pity as it sees itself: the Eurovision Pariah:

"Tears of rain run down my window pane
I'm on my own again good evening sorrow
Sit and dream of how things might have been ..."

But with a bit of effort, things could be so very different.

- The 1840 Eurovision Song Contest never existed and no one belted out *Rule Britannia*. This is a metaphor for Britain's world position at that time.

Eurovision: A Plea for Respect

CHAPTER 11
Through the Star Spangled Sky

Through some mystical illusion, Britain's first-ever Eurovision contestants, Pat Bredin, Pearl Carr, and Teddy Johnson, travel on a mystical magic carpet ride from the 1950s through the starry skies of time into future contests and a myriad of cities. They travel on through the dark skies and view past contests below them, on their way. They hover over the song contests of the early 1960s and glimpse Grethe and Jorgen Ingman's *Dansevise* triumph over Esther Ofarim's *Te'n va pas* after she had already been declared the 1963 winner – a year to remember

The picture of a youthful Sandie Shaw emerges, singing barefoot in Vienna in 1967, so memorable to so many Britons of that era, and the contest's character is beginning to change through the cultural and social revolution of the Swinging Sixties. They travel on yet further through the many invisible barriers of time. The stars of the distant planets, light years away, glisten in the night sky of a thousand Eurovision Song Contests, those past and those still yet to come. The magic carpet hovers for a time over the Madrid of Franco's Spain and the confusing drama of the 1969 contest and its four winning songs. Farther over the English Chanel and along the seaside towns of the English Riviera they speed until they reach the Brighton of April 1974 and hear Katie Boyle announce ABBA's historic win as the event takes on yet another guise. They travel on and further still into time through even more change and advances in technology, into a new Europe and the glittering productions of the song contests of the twenty-first century. They arrive at 2013 and the magic carpet hovers over Malmo. Then they journey on in the velocity of time to the atmospheric 2019 contest in Tel Aviv, the last before the great Covid pandemic would consume the world and the 2020 Eurovision Song Contest: The Song Contest That Never Was.

Steve Kerr

The Eurovision Song Contest they once knew and competed in has changed beyond belief. It is, in every respect, unrecognisable. As for the hundreds of singers and songwriters who once competed in its Brigadoon kingdom, well, they too have changed, aged on its sojourn. For many, their youth, like the contests of old, is now long gone. For the many who have followed the journey of the Eurovision Song Contest through the years, there will also be personal memories and associations with each event. Those who have grown up with the early song contests are now also growing old.

Pat Bredin, The UK's first-ever Eurovision singer, is now well into her eighties. Teddy Johnson died in 2019 and was soon followed by his wife of many years, Pearl Carr, in 2020, both well into their nineties, gone to eternal Eurovision glory and immortality.

In its years of existence, the Eurovision Song Contest has seen singers perform at the event who have risked their lives to do so. It has seen a role call of Europe's most accomplished and respected singers, musicians, and songwriters. It has seen artists whose entries have changed the direction of music in their respective countries, such as Italy's Domenico Modugno. There have been groundbreaking achievements, revolutionary and beyond their time, as witnessed in Serge Gainsbourg's participation. Its music has seen genres descended from opera and medieval Provencal troubadours; it, therefore, has a historic and genetical connection to the continent's past. It challenges boundaries and what the definition of Europe really is. In recent years, it has become a reflection of a fast-changing Europe and an arena for the merging of old and the creation of new genres.

In Britain, where it enjoys a somewhat curious, even tainted, relationship with the public, it is still, nevertheless, recognised as an institution; part of the credo of annual British culture, competing: a tradition. Its situation in Britain has seen the contest take on a baton of anti-establishment and bravely stand up to a musical fascism which it does not pander to, nor does it dance to its commands. While taste is dictated to the public, the Eurovision Song Contest in Britain does its own particular, unique, credible thing. Taste is, after all, merely a

Eurovision: A Plea for Respect

dictatorship; no one in all honesty has the right to dictate taste, which is ultimately only a confident point of view. As has been proven time and time again, confidence does not always equal credibility.

The Eurovision Song Contest celebrated its diamond anniversary in 2015 and we found Europe once again looking towards the beautiful old imperial city of Vienna, forty-eight years after its transmission from the Hofburg Palace, as it took on the mammoth task of staging the 2015 event. Conchita Wurst, alter ego of Tom Neuwirth, whose 2014 victory *Rise Like a Phoenix*, an autobiographical composition about Neuwirth's often painful past in a small town in Upper Austria, gave Austria its first win since Udo Jűrgens' heart rendering *Merci Cherie*, has proved to be the contest's most controversial winner. The singer not only challenged traditional concepts of gender but stretched tolerance to its limits. Held up as a beacon of tolerance and liberalism in the West as opposed to the reactionaryism of the former communist East, she was vehemently denounced in the East and in particular Russia as a blatant example of Western decadence and immorality. Her victory catapulted her into being a potent political symbol.

Despite condemnation by political leaders in Russia and other former Soviet republics, Conchita's powerful Bondesque ballad, which brought the house down at the contest, went on to enjoy great popularity in those very countries, many of which had also voted for her song. It *was* a sad irony that the ever-young, suave Udo Jűrgens, the biggest-selling German language singer in history, died very suddenly at the end of 2014, the year in which Austria gained its second, long-awaited victory.

Netta Brazilai's 2018 winning entry for Israel, *Toy*, was inspired by the Me Too movement, which no doubt Turkish singer and women's rights activist Aijda Pekkan would have very much given *douze points*. The song would do well in the USA, and Netta would return as a guest at the 2020 Europe Shine a Light event to sing the thought-provoking *Coockoo*, which dealt with mental illness. The song contest has had, throughout its history, an excellent track record on women's equality and a historical reputation for promoting female vocalists. From its fledgling days in 1956 right up to 2022, its jurors/televoters have voted overwhelmingly for female acts. In a male-dominated music business, no fewer than forty-three winners were either solo female vocalists or lead vocalists in a group, while the contest was introduced every year from 1957 to 1977, and sporadically after that, by a sole woman presenter.

A more concentrated effort went into the UK entry of 2020, which I prefer to call *the Eurovision Song Contest that never was*; the Eurovision Song contest, without a competition if we take into account the fact that forty-one Eurovision songs were submitted and that some sort of official substitute event did take place that unforgettable May. The BBC had asked James Newton, brother of singer John Newman, who had topped the UK charts with *Love Me Again*. James, although not well known as a singer, was, however, one of Britain's most prolific contemporary songwriters and had penned a number of successes for the likes of Little Mix and Kaiser Chiefs, as well as Olly Murs, among others, and had a Brit Award to his credit. His 2020 Eurovision song, *My Last Breath*, was a rather strong number which charted in the UK but we will never know how it would have fared at Eurovision. He was invited back to sing for the UK the following year with the song *Embers*, which, despite receiving much airplay on Radio 2, became only a minor hit while it received both nil points in the jury vote and the televote. He much endeared himself to viewers continent-wide for his good-natured acceptance of his last placing at Eurovision.

The aftermath of the 2021 event saw the success of the Italian winners, Maneskin, and their meteoric rise in popularity throughout Europe, which included success in the UK, where they had three hits in the British top twenty at the same time and were given the backing of Mick Jagger. The group also gathered a following in the USA, where they also charted. One year later, their vocals were heard on the soundtrack of the newly released Elvis movie.

No fewer than eight Eurovision entries entered the UK top ten in the week following the Grand Final, with three in the top fifty. They

Eurovision: A Plea for Respect

included four non-English language songs. Once again, however, none of the songs, apart from the UK entry, received any airplay and, as a result, like in previous years, they did not climb any higher on the charts despite having the potential to do so. This is in stark comparison to the Swedish charts, where Maneskin hit the number-one slot and eleven other entries made the top fifty, or the Dutch chart, where twelve entered the top fifty and twenty reached the top ten. In Finland, twenty entries got into the top twenty alone – the same number as entered the Norwegian hit parade. Despite deeply rooted negative attitudes in the British media, it would nevertheless seem that within the last year, the Eurovision Song Contest has been embarking on a change in direction and is moving into unchartered territory and maybe even, in Britain, a respect for its uniqueness in European history and for its musical quality. The recent Netflix film *Eurovision: The Story of Fire Saga*, a parody or, as some have seen it, a tribute to the contest, has also reestablished a wider interest in the event in the pandemic years across Europe and beyond. A turn in musical credibility and calibre would surely be a massive step toward improving the contest's standing in Britain and a turn in the country's relationship with the event in terms of how it is valued. However, even that will not change the general lack of respect the public has for it if popular DJs on prime-time radio insist on continuously referring to its songs or the overall event as "bonkers" or ridiculing those who avidly follow it for its music as some peculiar curiosity without doing any valid research. In the early years of the 2020s, the BBC DJs Scott Mills and Rylan Clark, both avid fans of the contest, have made some effort to raise its profile in the UK. Other BBC presenters and DJs, i.e., Rev. Kate Bottley, Michelle Visage, Anna Matronic, OJ Borg, Sophie Ellis Bexter, and Ritchie Anderson, all aficionados, have promoted the contest via their broadcasts. Times change and there is no place in our current culture for a tense, overly serious festival that the contest might have once been in some parts of Europe. It is a positive development that the song contest has acquired associations with Euro-parties, the celebration of pan-Europeanism, and general fun but this should not distract or subtract from taking it seriously aesthetically or respecting it. In the build-up to the event, the airwaves are often full of people who mark the event with a camp, OTT fancy dress party while seeing it as of no artistic worth or musical merit, often backing the most banal and frivolous of songs. They attach little value to European music, viewing it with a degree of contempt, and see the whole thing as little more than a massive comic spectacle. It is important that a fine balance be struck between enjoying it as a great celebration, a great show, and a time to party, and respecting it, holding it in high regard for its musical quality, inclusive values, and pan-European camaraderie. Nor, on the other hand, must all credible Eurovision songs be emotional, melancholic ballads. There is plenty of room for up-tempo, joyful compositions, which also have their place in life and in the contest, never more true than in the Covid pandemic and Ukrainian war period. The fact that songs are of a feel-good nature does not necessarily mean that they are trash, although Eurovision has suffered from the image of being only bouncy compositions with meaningless lyrics, which this book has proved to be false. In the lead-up to the 2022 contest in Turin, the Kansan Katrina Leskanich; much inspired by the West Coast sound whose 1997 winner had been the uplifting *Love Shine A Light*, featured in the EBU's *Europe Shine A Light* (The Eurovision Song Contest That Never Was), during lockdown, stated in a BBC interview that on that evening, the song brought out an emotional aspect that she herself had, until then, not noticed and that every time she has sung it since then, she has felt moved by it because it is really quite a moving song. She went on to say, in praise of the song, "*And it's testimony: My god, if only we could all live like that in this crazy old world*".

    The Eurovision Song Contest has proved to have a unique ability to truly unite Europe, even when it is falling apart. It can express a healthy nationalism coupled with an international belonging and identity. In 2022, a BBC Radio 2, listener commenting on a popular daytime programme, described major British success in the contest as "*All so restorative for us as a country and for human nature in general*".

Eurovision: A Plea for Respect

As we gradually approach the contest's Platinum Jubilee, we look forward in hope as we see its centenary slowly appear on the horizon. In praise and affection of this great event, member states of the EBU that compete should officially name the day that the contest falls on as Eurovision Day.

In the end, it is all about the music, and, yes, some truly good songs have appeared in its vast arena. If only those who move to the rhythm of musical fascism could adjust themselves to a pull of the string at the insecurities of being *"like a puppet on a string"* and dig deeper to discover.

EPILOGUE

I myself have much-cherished memories of the Eurovision Song Contest as a family event and, in particular, the event of 1967. My parents thought it too late for me to stay up and watch, so I went to bed with the radio on my bedside table broadcasting the contest, through some sort of exotic mystery, all the way via Eurovision from the far-off city of Vienna. I listened in the darkness. By the time the voting started, I was asleep. Mum and Dad woke me up to tell me that Sandie Shaw had won and I joyfully sang along to the winning reprise. My interest in music and geography deepened as I entered my teens and I could not wait to travel the world and savour its music on the way. If I had to name the best Eurovision winner then I would to have to say Monaco`s 1971 winning entry *Un Banc,Un Arbre ,Une Rue"* ;sung so passionately by Severine .On Christmas Day 1972, my parents gave me my first tape recorder, with which I recorded the 1973 contest. It is to their memory that I dedicate this final chapter.

As for the Eurovision Song Contest, when history is written, it will be interesting to see who, in the end, has the last laugh.

* * Australia was invited to compete at the contest in 2015 as a one-off, in celebration of the contest's sixtieth anniversary, but has continued to do so in the years following that event. It was the first nation to debut at Eurovision, outside of Eurasia, since Morocco in 1980. Australia has made several valid contributions to the contest and was the runner-up in 2016.

BIBLIOGRAPHY

1. Eurovision(The Greek Entries) Makis Delaportas :Orpheas ,May 2006

2. AFistful Of Gitanes: Sylvie Simmons ,Da Capo, Sept. 2002

3. Post War French Popular Music: Adeline Cordier

4. Pop Music In France From Chanson to Techno: H.Dauncey/P.Le Guem, Ashgate 2003

5. French Chanson: R.Haworth Jan 2008

6. Thesis Kai Idea :Athens 1970

7. Performing The New Europe: Karen Fricker, Palgrave , Macmillan ,May 2013

8. A Song For Europe :Popular Music and Politics In The Eurovision Song Contest.: R.D. Tobin & I Raykoff , Routledge , June 2007

9. Nul Points : Tim Moore , Vintage Press, May 2007

10. The Modern Fairy Tale:Paul Jordan ,University of Tartu Press,Dec.2014

11. Songs For Europe ,The 1970s (volume 2): G.Roxburgh ,Telos, April 2014.

12. Songs For Europe ,The 1980s (Volume3):G.Roxburgh ,Telos Jan.2017.

13. Songs For Europe ,The 1950s and 60s(Volume1):G.Roxburgh,Telos May 2012.

14. Portugal:Revolution Of The Carnations: K.Maxwell ,Oxford University Press, 2009.

15. Portugal: A Twentieth Century Interpretation:T.Gallacher,Manchester University Press.

16: Rock Around The Block:T.W.Ryback,Oxford University Press, Jan 1990.

17. Europop Music :Italy ,France ,Germany Online Magazine and Encyclopedia. .

18. The Italian Canzone and San Remo Festival:Change and Continuity In Italian Mainstream Pop Of The 1960s :R.Agostini, Cambridge University Press ,2007.

19. >Europop music ;Spain Under Franco,Online Magazine.

20. Fascism In Spain 1923-1977: S.G.Payne ,University Of Wisconsin ,1999.

21. The Spanish Civil War :S.G. Payne ,Cambridge University Press, 2012.

22. Carving Out A Bold Destiny For Fado:Larry Rohter ,New York Times , March 2011.

23. Fado And The Place Of Longing : R.Elliot ,Ashgate ,2010.

24. A measure Of Understanding : H.M. Queen Frederika Of The Hellenes.Macmillan ,1971.

25. >State,Identity And Politics Of Music:Eurovision And Nation Building In Azerbaijan.M .Ismayilov,Journal Of Ethnoism And Ethnicity 2012.

26. :Eurovision Song Contest Can Be A Tool For Change In Azerbaijan: British Plolitics And Policy ,L.S.E.

27. The Eurovision Song Contest: A useful Barometer Of Europe As Any,The International, ,2012.

28. When The Music Dies:News Makoob,2013.

29. Little Harmony In Europe Before Eurovision Song Contest, CNN,2013.

30. Eurovision Dominates UK Charts, The Eurovision Times,May 2013.

31. Why Americans Can`t Afford To Miss Out On This Cultural Spectacular: Guardian,2013.

32. It`s Time To Stop Laughing At Foreigners: K. Fricker,The Guardian, 2013.

33. Lilli Marleen : The Song That Became A legend ,BBC Radio,8th January ,1982.

34. Lahle Andersen,Die Lilli Marleen : Magnus_ Andersen , Universitas Verlag,1985.

35. Greek Music –Greek Songs : Online Encyclopedia./Magazine.

37. Diggyloo Thrush : Website of Eurovision Lyrics,www.diggyloo.net

38. A special Thankyou to Eija Hokkanen of Kuopio ,Finland who provided invaluable contributions on Finnish music and the Finish National Finals of the 1970s.

39. Eurovision Song Contest:Brand Impact Report ,EBU ,August 2022.

40. Orientalism: Edward W Said , Penguin ,2003

www.ingramcontent.com/pod-product-compliance
Ingram Content Group UK Ltd.
Pitfield, Milton Keynes, MK11 3LW, UK
UKHW030640040325
4842UKWH00027B/263